Scent from the Garden of Paradise

Islamic History and Civilization

STUDIES AND TEXT

VOLUME 140

The titles published in this series are listed at *brill.com/ihc*

Scent from the Garden of Paradise

Musk and the Medieval Islamic World

By

Anya H. King

BRILL

LEIDEN | BOSTON

Cover illustration: Moschus chrysogaster by H. Milne Edwards, (1868–1874), Recherches pour servir à l'histoire naturelle des mammifères: Atlas (1868–1874), Plate 20.

Library of Congress Cataloging-in-Publication Data

Names: King, Anya H.
Title: Scent from the Garden of Paradise : Musk and the Medieval Islamic World / by Anya H. King.
Description: Leiden ; Boston : Brill, [2017] | Series: Islamic history and civilization ; volume 140 | Includes bibliographical references and index.
Identifiers: LCCN 2016046844 (print) | LCCN 2016048246 (ebook) | ISBN 9789004336247 (hardback : alk. paper) | ISBN 9789004336315 (E-book)
Subjects: LCSH: Musk. | Aromatic compounds. | Perfumes—History. | Perfumes—Social aspects. | Medicine, Medieval. | Islam and culture. | Commerce—History—Medieval, 500–1500.
Classification: LCC QD331 .K54 2017 (print) | LCC QD331 (ebook) | DDC 668/.542—dc23
LC record available at https://lccn.loc.gov/2016046844

Typeface for the Latin, Greek, and Cyrillic scripts: "Brill". See and download: brill.com/brill-typeface.

ISSN 0929-2403
ISBN 978-90-04-33624-7 (hardback)
ISBN 978-90-04-33631-5 (e-book)

This book is printed on acid-free paper and produced in a sustainable manner.

Printed by Printforce, the Netherlands

To my parents

Peregrinos ipsa mire odores et ad exteros petit: tanta mortalibus suarum rerum satias et alienarumque aviditas.

Pliny *HN* XII.38.78

Contents

Preface IX
List of Illustrations XI
List of Abbreviations XII

Introduction 1

1 **About Musk and Its Terminology** 11
Musk and Its Origin 11
Exploitation of Musk 19
Terminology for Musk 23

2 **Commodities of Further Asia and the Islamic World** 37
Introduction 37
Central Eurasia 44
China 50
India and the Indian Ocean 59
Southeast Asia 70
Place of Origin, "Brand", and Rank 75
The Impact of Commodity Knowledge in *Adab* 78
Conclusion 82

3 **History of Musk and the Musk Trade: From Further Asia to the Near East** 85
Introduction 85
Musk in China 86
Musk in India 93
Musk in Tibet 108
Musk in Central Asia 112
Aromatics in the Persian World 121
Musk and Aromatics in Sasanian Persia 126
The Westward Spread of Musk in Late Antiquity 132
Conclusion 145

4 **Islamicate Knowledge of Musk and Musk Producing Lands** 147
Arabic Terminology Relating to Musk 147
Persian Terminology for Musk 157
Sources of Musk: Middle Eastern Knowledge of the Geography of the Musk Producing Lands and the Origins of Musk 157

Types of Musk and their Rankings 187
Toponyms, Ethnonyms, and Sources 192
The Islamicate Understanding of the Production of Musk 207
Conclusion 217

5 **The Merchant World and the Musk Trade** 219
Musk Producers and the Trade 221
Tribute and Royal Gift-Giving 224
Routes and Emporiums 227
Merchants 249
Islamic Merchants 250
Data on the Commerce in Musk 259
Perfumers and Pharmacists 260
Adulteration and Imitation of Musk 262
Conclusion 271

6 **Musk in Daily Life in the Early Medieval Islamic World** 272
Introduction 272
Arabic and Persian Perfumes 272
Incense 277
Unguents 278
Scented Powders 281
Oils and Waters 282
Musk and Men 283
Musk and Women 292
Musk and Medicine: Pharmaceutical Specifications of Musk 303
Pharmaceutical Applications of Musk 309
Musk in Food and Drink 317
Conclusion 323

7 **The Symbolic Importance of Musk in Islamic Culture** 325
The Primacy of Musk among Aromatics in Islamic Culture 325
Symbolic Meanings of Musk in Medieval Islamic Culture 328
Musk and Kingship 336
Musk and Islam 340
Musk and the Garden 352

Conclusion: Worldly and Otherworldly 366

Bibliography 369
Indices 414

Preface

Scent from the Garden of Paradise: Musk in the Medieval Islamic World is an expanded and thoroughly revised version of my 2007 Indiana University Ph.D. dissertation, "The Musk Trade and the Near East in the Early Medieval Period". This book is not intended as an exhaustive history of musk, but rather attempts to explore its role in the early medieval Islamic world. In order to establish the circumstances of the introduction of musk into the Near East, however, this book also includes an examination of the history of musk prior to its westward spread that goes well beyond the boundaries of Islamic studies. I am acutely conscious of the limits of my expertise in these areas and trust that the specialized reader will consider that it was better to attempt to cover this ground than to omit it entirely and thus give an incomplete picture.

I have benefited from the aid of many scholars and friends. I must first acknowledge the generous help of my dissertation advisors, Christopher I. Beckwith, Suzanne P. Stetkevych, Paul Losensky, and Gyorgy Kara. I must also thank those who read all or portions of the manuscript of the present book for their valuable help: Christopher I. Beckwith, Jennifer A. Davis, Valerie Hansen, Cynthia and William King, and the anonymous reader for Brill. David Brakke, Leslie Cortsen, William E. McCulloh, and C. Randall Newton gave needed assistance at different points in the research that led to this book. I have also benefited from discussions with Anna Akasoy, Frederique Darragon, Mauro Maggi, and James McHugh. Of course, all errors remain my own responsibility.

I must also thank the Islam and Tibet project at the Warburg Institute for inviting me to participate in a most stimulating conference in 2006. The College of Liberal Arts at the University of Southern Indiana gave me a Liberal Arts Research Award that facilitated completion of the manuscript. And finally, I must thank the libraries and interlibrary loan staff at Indiana University, Bloomington, and the University of Southern Indiana.

Transliteration and Conventions

I have employed the Library of Congress system for romanization of Arabic with a few modifications. While I have followed the system of D. N. Mackenzie for Pahlavi, I have transcribed Persian following the Arabic script rather than the Persian pronunciation. The main advantage of this procedure for this work is that the employment of identical spellings of terms makes their identity with the Arabic terms obvious to the reader in a way that they would not be if a system hewing to the modern pronunciation were used. The transliteration employed for other languages generally follows those used by specialists in those fields to the best of my ability. I have employed the Pinyin romanization for Chinese. For Tibetan, I have used the system suggested to me by Christopher I. Beckwith.

Translations quoted are my own unless noted. I have given the familiar names of places and dynasties in their usual English forms rather than use diacritics.

Weights and measures[1]

dāniq	= 0.52–0.74 grams
dirham	= 3.125 grams
mithqāl	= 4.46 grams
ūqiyah	= 33.85 grams
raṭl	= 406 grams
mann	= 816.5 grams

Disclaimer

This book is a historical study and in no way condones the exploitation of musk or of any other substance derived from an endangered animal or plant.

1 These are approximate and correspond to the weights used in Iraq. See W. Hinz, *Islamische Masse und Gewichte* (Leiden: Brill, 1955), for a study of weights and measures.

List of Illustrations

1.1a–b Musk deer. From Charles H Piesse, Piesse's Art of Perfumery, 5th ed. (London: Piesse and Lubin, 1891), 258 and 264 14

1.2 The musk apparatus. From Piesse 259 16

1.3a–b Musk pods. From Piesse 269 17

List of Abbreviations

BAI	Bulletin of the Asia Institute.
BSO(A)S	Bulletin of the School of Oriental (and African) Studies.
CAJ	Central Asiatic Journal.
Dietrich DT	A. Dietrich. *Dioscurides Triumphans: Ein anonymer arabischer Kommentar (Ende 12. Jahrh. n. Chr.) zur Materia medica*. 2 vols. Göttingen: Vandenhoeck & Ruprecht, 1988. Cited by item number.
EI	*Encyclopaedia of Islam*. 2nd ed. 12 vols. Leiden: Brill, 1954–2004. 3rd ed. Leiden: Brill, 2007–.
EIr	*Encyclopaedia Iranica*. New York: Bibliotheca Persica, 1982—and online at www.iranicaonline.org.
EQ	*Encyclopaedia of the Quran*. 6 vols. Leiden: Brill, 2001–6.
Garbers	K. Garbers. *Kitāb Kīmiyāʾ al-ʿiṭr wa-t-taṣʿīdāt: Buch über die Chemie des Parfüms und die Destillationen. Ein Beitrag zur Geschichte der arabischen Parfümchemie und Drogenkunde aus dem 9. Jahrh. P.C.* Abhandlungen für die Kunde des Morgenlandes 30. Leipzig, 1948. Ingredients cited by item number, formulas cited by "formula #".
Ibn Juljul	*Die Ergänzung Ibn Ǧulǧul's zur Materia medica des Dioskurides*. Ed. and trans. A. Dietrich, Göttingen: Vandenhoeck & Ruprecht, 1993. Cited by item number.
Ibn Kaysān	Sahlān b. Kaysān. *Mukhtaṣar fī al-ṭīb*. Ed. P. Sbath "Abrégé sur les arômes." *Bulletin de l'Institut d'Égypte* 26 (1943–4): 183–213.
Ibn Mandawayh	Ibn Mandawayh. *Risālah fī uṣūl al-ṭīb wa-l-murakkabāt al-ʿiṭriyyah*. In M. Dānishpazhūh. "Du risālah dar shinākht-i ʿiṭr." *Farhang-i Īrān-Zamīn* 15 (1347/1967): 224–253.
Ibn Māsawayh	Ibn Māsawayh. *Kitāb Jawāhir al-ṭīb al-mufradah*. Ed. P. Sbath, "Traité sur les substances simples aromatiques." *Bulletin de l'Institut d'Égypte* 19 (1936–7): 5–27.
JA	Journal Asiatique.
JAOS	Journal of the American Oriental Society.
JESHO	Journal of the Economic and Social History of the Orient.
JMBRAS	Journal of the Malayan Branch of the Royal Asiatic Society.
JNES	Journal of Near Eastern Studies.
JRAS	Journal of the Royal Asiatic Society.

LA	Ibn Manẓūr. *Lisān al-ʿArab*. 18 vols. Beirut: Dār al-Kutub al-ʿIlmiyyah, 2003.
LN	Dihkhudā, ʿAlī Akbar. *Lughat Nāmah*. 33 vols. in 42 parts. Tehran: Dānishgāh-i- Tihrān, 1946–73.
LSJ	H. G. Liddell, R. Scott, and H. S. Jones. *A Greek-English Lexicon*. Rev. ed. Oxford: Oxford University Press, 1996.
Maimonides	M. Meyerhof. *Moses Maimonides' Glossary of Drug Names*. Trans. F. Rosner. Philadelphia: American Philosophical Society, 1979. Cited by item number.
Monier-Williams	M. Monier-Williams. *A Sanskrit- English Dictionary, etymologically and philologically arranged with special reference to cognate Indo-European languages*. Oxford, 1899; repr. Delhi: Motilal Banarsidass, 1993.
Nuwayrī	Al-Nuwayrī, Shihāb al-Dīn Aḥmad b. ʿAbd al-Wahhāb. *Nihāyat al-arab fī funūn al-adab*. 33 vols. Cairo: Dār al-Kutub al-Miṣriyyah, 1923–97.
MRDTB	Memoirs of the Research Department of the Toyo Bunko.
PSAS	Proceedings of the Seminar for Arabian Studies.
T	*Taishō shinshū Daizōkyō*. 85 vols. Tōkyō: Taishō Shinshū Daizōkyō Kankōkai, 1988.
Ṭabarī	Abū Jaʿfar al-Ṭabarī. *Taʾrīkh al-rusul wa-l-mulūk*. Ed. M. J. De Goeje, et al. 15 vols. Leiden, 1879–1901; repr. Leiden: Brill, 1964–1965.
WKAS	*Wörterbuch der klassischen arabischen Sprache*. Wiesbaden: Harrassowitz, 1957–.
ZDMG	Zeitschrift der Deutschen Morgenländischen Gesellschaft.

Introduction

> The Ipariye Mosque was built by a wealthy merchant trading with China. He had seventy loads of musk of Khotan mixed into the mortar for its walls, which is why it is called the Ipari—that is, musk-scented—mosque. And indeed, if one sniffs, in humid weather, near its walls, one's nose gets perfumed with the fragrance of musk. Its domes are not covered with lead. However, it has a large congregation and is never without attendance in the morning or evening.[1]

So wrote the Ottoman traveler Evliya Çelebi about one of the mosques of Diyarbekir, in eastern Anatolia on the upper Tigris. Even though the building is still extant,[2] it is unclear if the mortar was actually mixed with musk. The idea of a building constructed with mortar of musk is striking, and this image captures an important aspect of Islamic civilization: the association of musk with the sacred. No other aromatic had the importance of musk in Islamic civilization, not even the beloved rose, let alone the incenses of Arabia—frankincense and myrrh—which were valued so highly in the ancient Mediterranean lands but much less so in Arabia herself. Musk was the scent of the Prophet Muḥammad, and it maintains its pride of place down to this day.[3]

---

1 Evliya Çelebi, *Evliya Çelebi in Diyarbekir*, ed. and trans. M. van Bruinessen and M. Boeschoten (Leiden: Brill, 1988): 136–7.

2 It is shown on pl. XIb of the above-cited book.

3 No comprehensive work on musk in the Islamic world has yet been attempted (the present work merely sets out to survey the first few centuries of Islam). A recent article on the interconnections between musk in the Islamic World and Tibet is A. Akasoy and R. Yoeli-Tlalim, of the Warburg Institute Islam and Tibet project: "Along the Musk Routes: Exchanges between Tibet and the Islamic World," *Asian Medicine* 3 (2007): 217–40. One of the symposia held by this project has resulted in Akasoy, et al., eds. *Islam and Tibet: Interactions along the Musk Routes* (Farnham: Ashgate, 2011), which includes the present author's "Tibetan Musk and Medieval Arab Perfumery." This symposium highlights musk as a key symbol of the trade and interaction between Tibet and the Islamic world. Accounts of musk exist in reference works and studies of materia medica. These include A. Dietrich, "Misk," *EI* 2nd ed. s.v.; F. Heyd, *Histoire du commerce du Levant au Moyen-Âge*, 2 vols. (Leipzig: Harrassowitz, 1885–6; repr. Amsterdam: Hakkert, 1959), 2.636–40; Garbers, *Kitāb Kīmiyā' al-ʿiṭr wa-t-taṣʿīdāt: Buch über die Chemie des Parfüms und die Destillationen. Ein Beitrag zur Geschichte der arabischen Parfümchemie und Drogenkunde aus dem 9. Jahrh. P.C.*, ed. and trans. with commentary by K. Garbers, (Leipzig, 1948), #72; S. Anṣārī, *Tārīkh-i ʿIṭr dar Īrān* (Tehran: Wizārat-i Farhang wa Irshād-i Islāmī, 1381/2002–3), 51–5; H. Schönig, *Schminken, Düfte und Räucherwerk der Jemenitinnen: Lexikon der Substanzen, Utensilien und Techniken* (Beirut: Ergon Verlag, 2002),

DOI 10.1163/9789004336315_002

What is particularly striking about the importance of musk is that it had to be imported from Further Asia, that portion of the Eurasian landmass beyond the eastern frontiers of the early world of Islam.[4] No musk deer lived within the boundaries of the medieval Dār al-Islām. Musk was unknown to the ancient Greeks; it appeared in the west only in the latter days of the Roman Empire. The incorporation of musk into the pharmacology and perfumery of medieval Europe is the result of its enthusiastic adoption in the Islamic period. Few imported goods can claim the importance—both practical and symbolic—of musk in Islamic civilization; with the exception of the massive importation of human labor from the Steppelands into the Islamic world, only silk can come close. The story of musk is thus an important part of the story of how Islamic civilization interacted with its neighbors to the east during the medieval period.

To some extent the moralizing judgments on perfume familiar since antiquity have given the impression that the study of aromatics is a subject of little importance. Yet when one attempts to understand social history something as commonly used as perfume becomes very significant, and the differences between perfumes used in various lands and throughout time reveal many important facets of the cultures that used them. Aside from the general neglect of aromatics and the difficulty of applying our modern conceptions of perfume to the pre-modern world, where aromatics took physical forms very different from modern "perfumes" and were more a part of medicine, perfume was of vast economic significance. No one doubts the great importance of trade in Arabian aromatics into the rest of the Near East and Mediterranean in antiquity.[5] Much less, however, has been done on the trade and use of aromatics since antiquity despite the significance of aromatics in medieval and modern Islamic societies.[6] Musk, the most important aromatic in the medieval

184–7; E. Lev and Z. Amar, *Practical Materia Medica of the Medieval Eastern Mediterranean According to the Cairo Genizah* (Leiden: Brill, 2008), 215–7, and M. A. Newid, *Aromata in der iranischen Kultur unter besonderer Berücksichtigung der persischen Dichtung*, (Wiesbaden: Reichert, 2010), 62–103.

4 I have employed the term "Further Asia" to avoid cumbersome expressions such as "Inner Asia, East Asia, South and Southeast Asia beyond the realm of Islam".

5 W. W. Müller, "Weihrauch," in A. F. v. Pauly and G. Wissowa, *Real-encyclopädie der classischen Altertumswissenschaft, Supplement 15* (Munich, 1978), 701–777; N. Groom, *Frankincense and Myrrh: A Study of the Arabian Incense Trade* (London: Longman, 1981), A. Avanzini, ed., *Profumi d'Arabia* (Rome: Bretschneider, 1997), D. Peacock and D. Williams, eds., *Food for the Gods: New Light on the Ancient Incense Trade* (Oxford: Oxbow, 2007).

6 A new survey is Z. Amar and E. Lev, "Trends in the Use of Perfumes and Incense after the Muslim Conquests," *JRAS* (2013): 11–30.

Middle East,[7] came exclusively from Central and Eastern Eurasia. It was thus entirely imported. Given its high price, this must have meant wealth being traded to the east, as well as money to be made for the merchants dealing in musk within the Middle East.

The neglect of aromatics in history is changing with a growing interest in the role of scent and smell. For the Near East, an important study of the role of scent in Rabbinic Judaism has been written by Deborah Green.[8] Anthropological work has made important strides in understanding the role of perfume and scent in the modern Middle East. Aida Sami Kanafani's *Aesthetics and Ritual in the United Arab Emirates* offers a perspective into the practices associated with aromatics in the United Arab Emirates, and Dinah Jung's *An Ethnography of Fragrance: The Perfumery Arts of ʿAdan/Laḥj* provides a detailed study of the uses of aromatics in modern Yemen from a historical perspective.[9] A literary perspective is employed in Mehr Ali Newid's *Aromata in der iranischen Kultur unter besonderer Berücksichtigung der persischen Dichtung*, which contains a survey of Persian aromatics, their uses, and roles in Persian poetry.[10]

In the present work, musk is used as a prism to allow the discernment of many aspects of medieval Islamic civilization. Musk was considered the best of aromatics and used in perfumes by men and women. It was present in the highest echelons of society, at the caliphal court. As a drug, musk appeared in a striking range of medical applications and prescriptions. And since it was so valued, a sizable body of lore grew around it; consumers, and potential consumers, of musk sought information about their cherished aromatic. The mere fact of musk's existence as an import forced medieval writers to elaborate their understandings of Further Asia. The tales of its exotic origin became part of its mythos.

As an aromatic and drug alone, musk would have been of great importance. But musk was the chosen aromatic of Muḥammad, a perfume-lover who preferred musk above all other scents. And Muḥammad's life, of course, became the *sunnah* or model for pious Muslims to emulate, assuring musk a permanent

---

7 The great importance of the musk trade has been stressed by C. I. Beckwith, "Tibet and the Early Medieval Florissance in Eurasia: A Preliminary Note on the Economic History of the Tibetan Empire," *CAJ* 21 (1977): 100–1.

8 D. A. Green, *The Aroma of Righteousness: Scent and Seduction in Rabbinic Life and Literature* (University Park: Pennsylvania State University Press, 2011).

9 A. S. Kanafani, *Aesthetics & Ritual in the United Arab Emirates* (Beirut: American University of Beirut, 1983); D. Jung, *An Ethnography of Fragrance: The Perfumery Arts of ʿAdn/Laḥj* (Leiden: Brill, 2011).

10 M. A. Newid, *Aromata in der iranischen Kultur unter besonderer Berücksichtigung der persischen Dichtung* (Wiesbaden: Reichert, 2010).

place in the Islamic world. Even today, when the natural musk is highly endangered and rare, synthetic musks continue to hold a high place.[11]

The time period covered in this book extends from the first attestations of the use of musk in antiquity and extends roughly to the 12th and 13th centuries, essentially to the time of the Mongols. This cut-off works fairly well for the relevant scientific literature, for by this date the basic texts on musk that would be used over and over by later savants had been written. In addition, fashions in perfume seem to have changed, to judge by the perfume formulas present in later works. Musk may have been getting harder to come by because it was not called for in the profligate quantities of earlier formulas, and in addition civet, another animal-based perfume, was increasingly replacing it.[12]

Sources

The bulk of the source material used in this book is written in Arabic, the major language of science in the early medieval Islamic world. Additional Near Eastern materials are found in languages such as Persian, both Pahlavi and Classical Persian, and in Syriac, but these sources are nowhere near as extensive as the Arabic literature,[13] which must form the foundation for this study. Sources in other languages such as Chinese, Sanskrit, Old Turkic, Tibetan, Greek, and Latin are also of great importance in tracing the history of musk in Further Asia and its westward spread.

While this book speaks of the "Islamic world", I will employ the term "Islamicate" rather than "Islamic" to refer to the technical literature on aromatics and science as a whole, and to the types of perfumes used. There are two reasons: first, a number of the authors of the works we will use were not Muslim even though they often worked for Muslim patrons. Second, the aromatics and types of compound perfumes that we will examine in this book

11 Jung 37.

12 Musk continued to be a very valuable component of trade; for the history of the musk trade through early modern times see Heyd 2.640, R. Ptak, "Moschus, Calambac und Quecksilber im Handel zwischen Macau und Japan und im ostasiatischen Seehandel ingesamt (circa 1555–1640)," in *Portugal und Japan im 16. und 17. Jahrhundert* (Frankfurt: Verlag der Interkulturelle Kommunikation, 1998), 72–95, and P. Borschberg, "Der asiatische Moschushandel vom frühen 15. bis zum 17. Jahrhundert," in J. M. dos Santos Alves, et al., eds., *Mirabilia Asiatica: Productos raros no comércio marítimo*, vol. 1 (Wiesbaden: Harrassowitz, 2003), 65–83.

13 At least in the period in question; the Classical Persian literature on aromatics becomes quite sizable later.

were used in common by the different communities who lived within the early medieval Islamic world.

The Arabic literature on aromatics seems to have been fairly extensive, but most of it is now lost.[14] Of the nine works mentioned by Ibn al-Nadīm (d. 990 or 991) in his *Fihrist* (Index), the only possible survival may be the book attributed to al-Kindī that is discussed below.[15] All physicians and others, especially alchemists, who were expected to produce interesting compounds, had to have knowledge of the preparation of perfumes. It is perhaps not surprising then that our best sources of information on aromatics come from scientifically-minded writers rather than from the *adībs*. Our earliest detailed sources date to the 8th and 9th centuries.

The Christian Abū Zakariyyā' Yūḥannā b. Māsawayh (c. 776–857), who was one of the greatest physicians of the early medieval Middle East, worked for several of the Abbasid caliphs.[16] His treatise on aromatics is the *Kitāb Jawāhir al-ṭīb al-mufradah* (Book of the Properties of Simple Aromatics).[17] This work consists of descriptions of individual aromatics and does not include formulas for compound perfumes. Ibn Māsawayh gives a little information on their medical properties as well. He is also credited by al-Nuwayrī (discussed below) with a formula for a henna fruit oil, so various unattributed formulas may also derive from him.

Perhaps the earliest extant perfume formulas are contained in the *Firdaws al-ḥikmah* (Paradise of Wisdom) of ʿAlī b. Rabban al-Ṭabarī, which dates from 850.[18] Al-Ṭabarī's book is a compendium of medical knowledge and includes numerous formulas for compound drugs as well as perfumes and cosmetics.

Abū Yūsuf Yaʿqūb b. Isḥāq al-Kindī lived c. 795–865; he flourished especially in the reign of al-Muʿtaṣim.[19] Two books by al-Kindī on perfumes, *Kitāb*

---

14 Cf. the survey in M. Ullmann, *Die Medizin im Islam* (Leiden: Brill, 1970), 314–16.

15 Ibn al-Nadīm, *al-Fihrist*, ed. G. Flügel (repr. Beirut: Maktabat al-Khayyāṭ, 1966), 317. Trans. B. Dodge, *The* Fihrist *of al-Nadīm*, 2 vols. (New York: Columbia University Press, 1970), 2.742.

16 F. Sezgin, *Geschichte des arabischen Schrifttums* (Leiden: Brill, 1967–), 3.231–6.

17 Ibn Māsawayh. See also the trans. by M. Levey, "Ibn Māsawaih and His Treatise on Simple Aromatic Substances," *Journal of the History of Medicine* 16 (1961): 394–410.

18 ʿAlī b. Rabban al-Ṭabarī, *Firdaws al-ḥikmah fī al-ṭibb*, ed. M. Z. Siddiqi, (Berlin: Sonne, 1928). See also M. Meyerhof, "ʿAlī aṭ-Ṭabarī's 'Paradise of Wisdom, One of the Oldest Arabic Compendiums of Medicine," *Isis* 16 (1931): 6–54 and Sezgin, 3.236–40.

19 Biographical information in F. W. Zimmermann, "Al-Kindī," in *Cambridge History of Arabic Literature. Religion, Learning, and Science in the ʿAbbasid Period*, ed. M. J. L. Young, et al. (Cambridge: Cambridge University Press, 1990), 364–9.

al-Ṭīb (Book of Aromatics) and *Kitāb Kīmiyāʾ al-ʿiṭr* (Book of the Chemistry of Perfume), are mentioned by Ibn al-Nadīm.[20] Among the extant works attributed to al-Kindī is the *Kitāb Kīmiyāʾ al-ʿiṭr wa-l-taṣʿīdāt* (Book of the Chemistry of Perfume and Distillation) that may be the same as the book referred to by Ibn al-Nadīm.[21] This book contains numerous formulas for compound perfumes and also some methods of making imitations of rare substances; it is perhaps the most comprehensive early work on the subject. One formula in the book (#33) is explicitly credited to al-Kindī. Even if the core of the book was originally by al-Kindī, it was apparently redacted at a later time, perhaps by one of his students. Al-Kindī's *Aqrābādhīn* (Formulary) includes aromatic substances in its formulas, though not as abundantly as other pharmacological works.[22] Finally, his epistle on music includes a short section on perfumes.[23] The work treats the psychological effects of different notes and then goes on to discuss colors and the harmonies of perfumes.

The Baṣran polymath Abū ʿUthmān ʿAmr b. Baḥr al-Jāḥiẓ (c. 776–868 or 9) provides important information on musk in his *Kitāb al-Ḥayawān* (Book of Animals).[24] Another important work, the *Kitāb al-Tabaṣṣur bi-l-tijārah* (Book of Discernment in Commerce), was traditionally ascribed to al-Jāḥiẓ, although this is considered doubtful because the style of the writing is not really like his.[25] Regardless of its origin, the range of subject matter within the book is not beyond the range of al-Jāḥiẓ's wide interests. The book seems to reflect approximately his time period, the height of the Abbasid caliphate. The work contains general opinions on commerce and evaluations of different commodities and their varieties. It includes a section on aromatics that has partly disappeared in a lacuna in the manuscript.

---

20 Ibn al-Nadīm 317; Dodge 2.742.

21 Garbers. Cf. Sezgin, 4.6 and D. M. Dunlop, *Arab civilization to A.D. 1500* (London: Longman, 1971), 229–31.

22 Al-Kindī, *The Medical Formulary or Aqrābādhīn of al-Kindī*, ed. and trans. M. Levey (Madison: University of Wisconsin, 1966).

23 Al-Kindī, *Risālat al-Kindī fī ajzāʾ khabariyyah fī al-mūsīqā* (Cairo: al-Lajnah al-Mūsīqiyyah al-ʿUlyā, 1963) and trans. H. G. Farmer, "Al-Kindī on the 'Ēthos' of Rhythm, Colour, and Perfume," *Transactions of the Glasgow University Oriental Society* 16 (1955–6): 29–38.

24 Al-Jāḥiẓ, *Kitāb al-Ḥayawān*, ed. ʿAbd al-Salām Muḥammad Hārūn, 8 vols. (Cairo: Muṣṭafā al-Bābī al-Ḥalabī, 1966).

25 *Kitāb al-Tabaṣṣur bi-l-tijārah fī waṣf ma yustaẓraf fī al-buldān min al-amtiʿah al-rafīʿah wa-l-aʿlāq al-nafīsah wa-l-jawāhir al-thamīnah*, ed. Ḥasan Ḥusnī ʿAbd al-Wahhāb al-Tūnisī (Cairo: Maktabat al-Khānjī, 1994); trans. Ch. Pellat, "Ǧāhiẓiana, I. Le Kitāb al-Tabaṣṣur bi-l-Tiǧāra attribué à Ǧāḥiẓ," *Arabica* 1 (1954): 153–65.

Another important early writer who dealt with aromatics was Aḥmad b. Abī Yaʿqūb b. Jaʿfar al-Yaʿqūbī (d. 897). Two works of this writer are extant, his *Tārīkh* (History) and his geography, the *Kitāb al-Buldān* (Book of the Lands). It is not clear what work the extensive quotations from al-Yaʿqūbī preserved in the *Niḥayat al-arab* of al-Nuwayrī and the works of others came from, but they show great familiarity with aromatics and their origins.[26]

A wealth of information on aromatics comes from the *Murūj al-dhahab wa maʿādin al-jawhar* (Meadows of Gold and Mines of Jewels) of Abū al-Ḥasan ʿAlī b. al-Ḥusayn al-Masʿūdī (c. 896–956).[27] His accounts of aromatics such as musk and ambergris were enormously influential and widely copied into later books. They are related to the work of his early contemporary Abū Zayd al-Sīrāfī. Al-Sīrāfī prepared an addition to the mid-9th century work now known as the *Akhbār al-Ṣīn wa-l-Hind* (Book of the Reports on China and India) that contains important information on musk and ambergris, among other aromatics.[28] We will compare these two accounts in Chapter 4.

Another pair of interrelated writers is Abū ʿAlī Aḥmad b. ʿAbd al-Raḥmān b. Mandawayh (d. 1019) and the Egyptian Christian Abū al-Ḥasan Sahlān b. Kaysān (d. 990). The works of these two writers, the *Risālah fī uṣūl al-ṭīb wa-l-murakkabāt al-ʿiṭriyyah* (Epistle on the Principal Aromatics and Perfumed Compounds) and the *Mukhtaṣar fī al-ṭīb* (Concise Guide to Aromatics) respectively, are nearly identical and thus pose difficulties that we will consider in Chapter 4.[29] The Ibn Mandawayh-Ibn Kaysān tradition is very significant because its information is often unparalleled elsewhere and seems almost totally independent of the works enumerated above.

Muḥammad b. Aḥmad b. Saʿīd al-Tamīmī (d. 980) wrote a book on aromatics called variously *Kitāb Jayb al-ʿarūs wa-rayḥān al-nufūs fī ṣināʿat al-uṭūr* (The Bosom of the Bride and the Sweet Flower of the Souls of the Making of Perfumes) or *Kitāb Ṭīb al-ʿarūs wa-rayḥān al-nufūs fī ṣināʿat al-uṭūr* (The Perfume of the Bride and the Sweet Flower of the Souls of the Making of Perfumes).[30] Only part of this work, consisting of some chapters of the

26 Cf. Ullmann 315.

27 Al-Masʿūdī, *Murūj al-dhahab wa-maʿādin al-jawhar*, ed. B. de Meynard and P. de Courteille, revised by C. Pellat, 5 vols., (Beirut: Manshūrāt al-Jāmiʿah al-Lubnāniyyah, 1966–74).

28 Al-Sīrāfī, *Riḥlah*, ed. ʿAbdallāh al-Ḥabashī (Abu Dhabi: Manshūrāt al-Majmaʿ al-Thaqāfī, 1999). For the original part of the *Akhbār*, without al-Sīrāfī's supplemental chapter, see Akhbār aṣ-Ṣīn wa l-Hind: *Relation de la Chine et de l'Inde*, ed. and trans. J. Sauvaget (Paris: Les Belles Lettres, 1948).

29 Texts in Ibn Mandawayh and Ibn Kaysān.

30 Ullmann 315.

formulary, is extant; it has recently been edited.[31] Extensive quotations of it are found in the *Nihāyat al-arab* of al-Nuwayrī, and it is quoted by other works as well. The important description of musk appears only in quotation.

The Spanish physician Abū al-Qāsim al-Zahrāwī (d. shortly after 1009), known as Abulcasis in Europe, dealt with aromatics and cosmetics in the nineteenth book of his great medical work, *Kitāb al-Taṣrīf li-man ʿajiza ʿan al-taʾlīf* (A Presentation to Would-Be Authors on Medicine), which exists in manuscripts but has not been edited except for certain sections.[32] Book 19 contains brief discussions of individual aromatics including musk and extensive formulas for aromatic compounds. Other works of medical literature contain accounts of musk and other information about the usage of musk in medicine. The great polymath Abū al-Rayḥān al-Bīrūnī (d. c. 1050), among his numerous scientific works, has a *Kitāb al-Ṣaydanah* (Book of the Pharmacy) that includes an important article on musk.[33]

The encyclopedias of Shihāb al-Dīn Aḥmad b. ʿAbd al-Wahhāb al-Nuwayrī (1279–1332), the *Nihāyat al-arab fī funūn al-adab* (The Fulfilment of Desire in the Arts of Culture),[34] and Shihāb al-Dīn Aḥmad b. ʿAlī al-Qalqashandī (1355–1418), *Ṣubḥ al-aʿshā fī kitābat al-inshāʾ* (The Blind Man's Illumination in the Art of Chancery Communication),[35] contain highly important accounts of aromatics derived from earlier sources, especially al-Yaʿqūbī and al-Tamīmī. The *Rabīʿ al-abrār* (Springtime of the Pious) of Abū al-Qāsim Maḥmūd b. ʿUmar al-Zamakhsharī (1075–1144) contains a chapter on aromatics with

---

31 Al-Tamīmī, *Ṭīb al-ʿarūs wa rayḥān al-nufūs fī ṣināʿat al-uṭūr*, ed. Luṭf Allāh Qārī and Aḥmad Fuʾād Bāshā (Cairo: Maṭbaʿat Dār al-Kutub wa-al-Wathāʾiq al-Qawmiyyah bi-al-Qāhirah, 2014).

32 Abū al-Qāsim al-Zahrāwī, *al-Taṣrīf li-man ʿajiza ʿan al-taʾlif*, 2 vols. (Frankfurt am Main: Institute for the History of Arabic-Islamic Science, 1986) is a facsimile of Süleymaniye Beşirağa collection ms. 502; Chapter 19 appears in vol. 2, 38–62. A summary in S. Hamarneh, "The First Known Independent Treatise on Cosmetology in Spain," *Bulletin of the History of Medicine* 39 (1965): 309–25. The manuscripts are described in S. K. Hamarneh and G. Sonnedecker, *A Pharmaceutical View of Abulcasis al-Zahrāwī in Moorish Spain* (Leiden: Brill, 1963), 137–47.

33 *Kitāb al-Ṣaydanah*, ed. and trans. H. M. Said (Karachi: Hamdard National Foundation, 1973), and ed. ʿAbbās Zaryāb, (Tehran: Markaz-i Nashr-i Dānishgāhī, 1991). See also the excerpts from al-Bīrūnī's works (including the passage on musk from the pharmacopoeia) collected in Z. V. Togan, *Bīrūnī's Picture of the World*, (New Delhi: Archaeological Survey of India, 1937).

34 Al-Nuwayrī, *Nihāyat al-arab fī funūn al-adab*, 33 vols. (Cairo: Dār al-Kutub al- Miṣriyyah, 1923–97), 12.1–141.

35 Shihāb al-Dīn Aḥmad b. ʿAlī al-Qalqashandī, *Ṣubḥ al-aʿshā*, 14 vols. (Cairo: Al-Muʾassasah al-Miṣriyyah al-ʿĀmmah li-l-Taʾlif, 1964), 2.119–31.

quotations from ḥadīths and other literature.[36] There is also a similar section in ʿAbd al-Hamīd b. Hibbat Allāh b. Abī al-Ḥadīd's (1190–1257 or 8) *Sharḥ Nahj al-balāghah* (Commentary on the Way of Eloquence).[37] The *Maṭāliʿ al-budūr fī manāzil al-surūr* (Risings of the Full Moons in the Mansions of Pleasure) of ʿAlāʾ al-Dīn ʿAlī b. ʿAbdallāh al-Bahāʾī al-Dimashqī al-Ghuzūlī (or Ghazūlī) (d. 1413) contains a chapter on aromatics with the usual quotations from ḥadīths as well as some formulas from an unknown source or sources.[38]

Several Islamic period manuscript fragments touching on perfumes have been published. One is an Arabic papyrus from Egypt which contains an account listing quantities of aromatics including cloves, musk, ben oil, roses and rosewater, Arabian jasmine, and others.[39] It has been paleographically dated to the middle or latter half of the 9th century.[40] Another is a pharmacological fragment from the Cairo Geniza, an accumulation of documents covering a wide range of dates during the medieval period, many of which are of great importance for the musk trade.[41] The document itself gives no indication of date, so it cannot be dated except in a general way. It probably dates from the 11th or 12th century. It consists of two folios of Arabic written in Hebrew script and contains formulas for various drugs, including some perfume formulas. An interesting feature of this manuscript is that some of the perfumes have identifications in *laṭīnī*, which here refers to an early form of Spanish. The Geniza manuscripts include many documents on *materia medica* and pharmacy; the

---

36 Al-Zamakhsharī, *Rabīʿ al-abrār wa-fuṣūṣ al-akhbār*, eds. ʿAbd al-Majīd Diyāb and Ramaḍān ʿAbd al-Tawwāb (Cairo: al-Hayʾah al-Miṣriyyah al-ʿĀmmah li-l-Kitāb, 2 vols., 1992–2001), 2.213–25.

37 Ibn Abī al-Ḥadīd, *Sharḥ Nahj al-balāghah*, 20 vols. (Cairo: ʿĪsā al-Bābī al-Ḥalabī, 1960–4), 19.341–51.

38 Al-Ghuzūlī, *Maṭāliʿ al-budūr fī manāzil al-surūr*, 2 vols. bound as one (Cairo: Maktabat al-Thaqāfah al-Dīniyyah, n.d. [2000]), 1.75–9.

39 G. Levi Della Vida, "A Druggist's Account on Papyrus," in *Archaeologia orientalia in memoriam Ernst Herzfeld* (Locust Valley: Augustin, 1952), 150–5. The fragment is at the University of Pennsylvania Museum.

40 Della Vida 151.

41 R. Gottheil, "Fragment on Pharmacy from the Cairo Genizah," *JRAS* (1935), 123–44. Genizah documents are studied in the great work of S. D. Goitein, *A Mediterranean Society: The Jewish Communities of the Arab World as Portrayed in the Documents of the Cairo Geniza*, 6 vols. (Berkeley: University of California Press, 1967–1993; repr. 1999). Of special relevance to the present work are the documents pertaining to trade, specifically the India trade, published in his *Letters of Medieval Jewish Traders, Translated from the Arabic with Introductions and Notes* (Princeton: Princeton University Press, 1973) and in S. D. Goitein and M. A. Friedman, *India Traders of the Middle Ages: Documents from the Cairo Geniza: "India Book"* (Leiden: Brill, 2008).

study of these will undoubtedly reveal many interesting facts concerning the history of aromatics in the coming years.[42]

Because of the ubiquitous nature of aromatics in Middle Eastern Islamicate society they may be met with in almost any text. The works of the geographers are particularly rich in information about aromatics, their trade, and local distribution. The principal writers utilized in this work are Ibn Khurradādhbih (d. c. 911), Ibn al-Faqīh al-Hamadhānī (fl. c. 903), and the anonymous Persian language *Ḥudūd al-ʿālam* of 982–3.[43] Historical texts, *adab* works, and especially poetry give a wide-ranging amount of information on the uses and importance of aromatics.[44] Under these circumstances, much said below must be considered impressionistic. There are undoubtedly many hundreds of incidental sources which have not been consulted, as there is no vade mecum for the study of aromatics in Arabic literature,[45] though the situation is a better for Persian literature.[46]

---

42 H. D. Isaacs and C. F. Baker, *Medical and Para-Medical Manuscripts in the Cambridge Genizah Collections* (Cambridge: Cambridge University Press, 1994), Lev and Amar, and L. N. Chipman and E. Lev, "Syrups from the Apothecary's Shop: A Genizah Fragment Containing One of the Earliest Manuscripts of the Minhāj al-Dukkān," *Journal of Semitic Studies* 50 (2006): 137–68.

43 Ibn Khurradādhbih, *Kitāb al-Masālik wa-l-mamālik*, ed. M. J De Goeje. (Leiden: Brill, 1889, repr. 1967), Ibn al-Faqīh al-Hamadhānī, *Mukhtaṣar kitāb al-buldān*, ed. M. J. De Goeje (Leiden: Brill, 1885; repr. 1967), and Anonymous, *Ḥudūd al-ʿālam min al-mashriq ilā al-maghrib*, ed. M. Sutūdah (Tehran: Dānishgāh-i Tihrān, 1983), and trans. V. Minorsky, *The Regions of the World, A Persian Geography 372 A.H.–982 A.D.* 2nd ed. (London: Gibb Memorial Series, 1970).

44 D. Agius has made a case for utilizing *adab* as a source of documentation of the vocabulary of material culture: *Arabic Literary Works as a Source of Documentation for the Technical Terms of Material Culture* (Berlin: Klaus Schwarz, 1984). On poetry as a source for aromatics see A. King, "The Importance of Imported Aromatics in Arabian Culture: Illustrations from Pre-Islamic and Early Islamic Poetry," *JNES* 67, no. 3, (July 2008): 175–89, C. van Ruymbeke, *Science and Poetry in Medieval Persia: The Botany of Nizami's* Khamsa (Cambridge: Cambridge University Press, 2007), and Newid.

45 Garber's work on pseudo-al-Kindī, mentioned above, contains detailed coverage of the substances used in the formulas. See also H. Schönig, *Schminken, Düfte und Räucherwerk der Jemenitinnen: Lexikon der Substanzen, Utensilien und Techniken* (Beirut: Ergon Verlag, 2002), which is focused on modern perfumery and cosmetics in Yemen, but also contains information on earlier uses. Neither of these works delves into the role of aromatics in literature, and this remains a great desideratum.

46 *Farhangnāmah-i Adab-i Fārsī*, vol. 2 of *Dānishnāmah-i Adab-i Fārsī* (Tehran: Mu'assasah-i Farhangī wa Intishārāt-i Dānishnāmah, 1996), esp. 241–266, S. Anṣārī, *Tārīkh-i ʿIṭr dar Īrān* (Tehran: Wizārat-i Farhang wa Irshād-i Islāmī, 1381/2002–3), which covers the entire history of perfume in Iran from antiquity, and Newid, which is similarly wide-ranging but focused on poetry.

CHAPTER 1

About Musk and Its Terminology

Musk and Its Origin

Strongly scented substances that are described as musky are produced by many different types of plants and animals throughout the world. Among plants the best known is *Hibiscus abelmoschatus*, also called abelmosk, ambrette, and the musk mallow.[1] There are musk roses and musk melons; hints of muskiness can be detected in their scents. The root of plants such as spikenard (*Nardostachys jatamansi* or grandiflora), also called *sumbul*, from Arabic *sunbul*, and angelica (*Angelica archangelica*) have a strongly musky scent.[2] Numerous animals also have musky odors. The musk ox of northern climes is well known. There is even a musk duck, *Anas moscata*, found in Ghana, and there is the American musk tortoise or turtle (*Sternotherus odoratus*, also called the stinkpot).[3] The muskrat *Ondatra zibethicus* is probably the best known of these animals excluding the musk deer, but it is a species that originated in the Americas and was later introduced into Eurasia and so need not enter into our discussion.[4] Thus, pre-Columbian references to "musk rats" in southern Eurasia most likely refer to the Asian musk shrew (*Suncus murinus*)

1 G. Watt, *The Commercial Products of India* (London: John Murray, 1908), 629. Ambrette is used in perfumery for its musky qualities, see S. Arctander, *Perfume and Flavor Materials of Natural Origin* (Elisabeth, N.J., privately published, 1960), 58–60.

2 Arctander 592–3 and 603, on spikenard and sumbul oil; also see Garbers #109, 222–5, W. Schmucker, *Die pflanzliche und mineralische Materia Medica im Firdaus al-Ḥikma des Ṭabarī* (Bonn: Selbstverlag des Orientalischen Seminars, 1969), 248–9, K. Nielsen, *Incense in Ancient Israel* (Leiden: Brill, 1986), 64, Dietrich DT I.6, H. Schönig, *Schminken, Düfte und Räucherwerk der Jemenitinnen: Lexikon der Substanzen, Utensilien und Techniken* (Beirut: Ergon Verlag, 2002), 276–7, and E. Lev and Z. Amar, *Practical Materia Medica of the Medieval Eastern Mediterranean According to the Cairo Genizah* (Leiden: Brill, 2008), 289–93. *Nardostachys jatamansi* grows in the Himalayas but is now cultivated in India, China, and Japan as well. However, "spikenard", like the ancient and medieval terms for it such as Arabic *sunbul*, can denote other species of *Nardostachys*, as well as Valerian. See Arctander 62–6 on angelica.

3 M. Capula, *Simon & Schuster's Guide to Reptiles and Amphibians of the World* (New York: Simon & Schuster, 1989), #92.

4 R. Nowak, ed., *Walker's Mammals of the World*, vol. 1, 5th edition (Baltimore: Johns Hopkins University Press, 1991), 753–4; Grzimek's *Animal Life Encyclopedia* 16, 226. It has been possible since the 1940's to extract the musk from the musk rat, but this does not seem to have

 | DOI 10.1163/9789004336315_003

in the south[5] or, in the north, to the Russian desman (*Desmana moschata*).[6] There are musky rat-kangaroos (*Hypsiprymnodon moschatus*) in Australia.[7] However, the scent of the various species of *Moschus*, the musk deer, is by far the most famous and was the most widely used of the musk scents throughout Eurasia and the only substance from a "musk" animal to be used in perfumery.[8]

The genus *Moschus* is considered by zoologists to be related to the true deer or Cervidae, though they are regarded as more primitive. They are placed into their own family, the *Moschidae*, the taxonomy of which has been much discussed.[9] The eponymous musk deer, *Moschus moschiferus Linnaeus*, is not the source of the most famous Tibetan musk, though it is frequently cited as such. *M. moschiferus*, also called the Siberian musk deer (*M. sibiricus*), is native to eastern Siberia, northern Mongolia, Heilongjiang, Xinjiang and Gansu in China, Korea and also Sakhalin (the Sakhalin musk deer is also considered a subspecies of *M. moschiferus*). The predominant musk deer of the Himalayas stretching into China, which produced the most famous musk, is M. *chrysogaster Hodgson*. The different species of musk deer, their common names, and ranges are set out in the table. Apart from cranial measurements, color of coat and other biometric criteria, these species are also distinguished by the altitude range in which they live.

---

ever been done commercially. Cf. also C. H. Piesse, *Piesse's Art of Perfumery*, 5th ed. (London: Piesse and Lubin, 1891), 255–6.

5 *Walker's Mammals* 1.161–2. There are many shrew species of *Crocidura* commonly designated varieties of musk shrew as well.

6 The desman is especially valued for its fur, but it is said to be used in perfumery sometimes as well, cf. Grzimek's *Animal Life Encyclopedia* vol. 13, 200. If so, this is most uncommon, at least for Europe, for it is unlisted in the usual guides to aromatic substances, such as Arctander and Poucher, as well as the older work of Piesse.

7 Grzimek 13.69–72.

8 This stipulation excludes the two animal aromatics most similar to musk: castoreum and civet; neither is called musk nor comes from an animal described as a "musk" animal. We will consider them below. An entertaining introduction to musk and its sister aromatic ambergris can be found in E. W. Bovill, "Musk and Amber," *Notes and Queries* 198 (1953): 487–9, 508–10; n.s. 1 (1954): 24–5, 69–72, 121–3, 151–4. General accounts of musk from the perspective of the perfumer are in Piesse 256–75, Arctander 422–4, N. Groom, *The New Perfume Handbook*, 2nd ed. (London: Blackie Academic and Professional, 1997), 219.

9 C. P. Groves, Wang Yingxiang and Grubb, P., "Taxonomy of Musk Deer, Genus Moschus (Moschidae, Mammalia)," *Acta Theriologica Sinica* 15:3 (1995): 181–97.

TABLE 1.1 *Species of* Moschus

Species	Common name(s)	Range
Moschus anhuiensis	Anhui musk deer	Anhui, China[a]
Moschus berezovskii	Forest musk deer, dwarf musk deer	China, Vietnam[b]
Moschus chrysogaster	Alpine musk deer, Himalayan musk deer	Bhutan, China, India, Nepal[c]
Moschus cupreus	Kashmir musk deer	Afghanistan, Pakistan, India[d]
Moschus fuscus	Black musk deer, dusky musk deer	Bhutan, China, Myanmar, India, Nepal[e]
Moschus leucogaster	Himalayan musk deer	Bhutan, China, India, Nepal[f]
Moschus moschiferus	Siberian musk deer	China; Kazakhstan; Korea, Democratic People's Republic of; Korea, Republic of; Mongolia; Russian Federation[g]

a Y. Wang and R. B. Harris, "Moschus anhuiensis," *The IUCN Red List of Threatened Species 2015*, http://www.iucnredlist.org/details/136643/0. Accessed March 22, 2016.

b Y. Wang and R. B. Harris, "Moschus berezovskii," *The IUCN Red List of Threatened Species 2015*, http://www.iucnredlist.org/details/13894/0. Accessed March 22, 2016.

c Y. Wang and R. B. Harris, "Moschus chrysogaster," *The IUCN Red List of Threatened Species 2008*, http://www.iucnredlist.org/details/13895/0. Accessed March 22, 2016. An overview of its geographical distribution can be found in M. Green, "The Distribution, Status and Conservation of the Himalayan Musk Deer Moschus chrysogaster," *Biological Conservation* 35 (1986): 351–8.

d R. J. Timmins and J. W. Duckworth, "Moschus cupreus," *The IUCN Red List of Threatened Species 2015*, http://www.iucnredlist.org/details/136750/0. Accessed March 22, 2016.

e Y. Wang and R. B. Harris, "Moschus fuscus," *The IUCN Red List of Threatened Species 2015*, http://www.iucnredlist.org/details/13896/0. Accessed March 22, 2016. Present in Vietnam according to Dao Van Tién, "Sur quelques rares mammifères au nord du Vietnam," *Mitteilungen aus dem Zoologischen Museum in Berlin* 53 (1977): 326–8. See also J. Blower, "Conservation Priorities in Burma," *Oryx* 19 (1985): 80–1.

f R. J. Timmins and J. W. Duckworth, "Moschus leucogaster," *The IUCN Red List of Threatened Species* 2015, http://www.iucnredlist.org/details/13901/0. Accessed March 22, 2016.

g B. Nyambayar, H. Mix, and K. Tsytsulina, "Moschus moschiferus," *The IUCN Red List of Threatened Species 2015*. http://www.iucnredlist.org/details/13897/0. Accessed March 22, 2016; Groves et al. 191. For Korea see also W. R. Carles, "Recent Journeys in Korea," *Proceedings of the Royal Geographical Society* (1886): 307.

ILLUSTRATION 1.1a–b *1a (left) and 1b (right). Musk deer.*

The historic range of these species of musk deer encompasses most of the forested and semi-forested highlands of Eastern Eurasia. They live at high elevations; 1500 meters seems to be as low as they are found.[10] Musk deer are now quite endangered throughout their range because of both habitat destruction and exploitation for musk.

The musk deer is a small animal with a head and body length of about 70 to 100 cm., a height averaging 50 to 61 cm. at the shoulder, and an average weight of 7 to 17 kg.[11]

It has very long muscular hind limbs and moves with a motion resembling a rabbit more than a deer. It has flexible toes ending in sharp, hard hooves. It has a thick, bristly coat which is variable in color, but predominantly brown and mottled. The musk deer is well adapted to life in rugged mountainous terrain; it is not at home anywhere else. Musk deer prefer high altitudes with forest or shrub vegetation and rocky cover; they are not steppe animals nor do they live

10 Green, "Distribution, status and conservation of the Himalayan musk deer".

11 This description is based on the following: Flerov 14–45, R. Nowak, ed., *Walker's Mammals of the World*, vol. 2 (Baltimore: Johns Hopkins, 1991), 1364–5, V. G. Heptner, et al., *Mammals of the Soviet Union Volume 1*, trans. P. M. Rao, Washington, 1988, 101–24, M. J. B. Green, "Some ecological aspects of a Himalayan population of musk deer," in *Biology and management of the Cervidae: a conference held at the Conservation and Research Center*, National Zoological Park, Smithsonian Institution (Washington, 1987), 307–19, idem "Distribution, status and conservation of the Himalayan musk deer", and V. Homes, *On the Scent: Conserving Musk Deer- The Uses of Musk and Europe's Role in Its Trade* (Brussels: TRAFFIC Europe, 1999).

in desert conditions.[12] As a result of its preference for high terrain, its populations tend to be very uneven over a wider area, clustering in areas with higher altitudes. They are shy, elusive and usually solitary, mostly nocturnal animals. Their diet consists of vegetation; more than 130 species of herbaceous plants are known to be eaten by Siberian musk deer, and in the north they eat lichens as well.[13] Apart from its production of musk, another remarkable feature of the musk deer is that the males possess large canine teeth which protrude from their upper jaw; this fact was keenly noted by the Arabic zoographers. In addition, neither sex of the musk deer grows antlers.

Musk is produced only by the male deer and may be used to attract females for mating. It is present in the urine of the males and likely also functions to indicate the territorial range of the male animal. The glands that produce the musk are located in a special pouch of skin in front of the preputial orifice; these glands reach about four centimeters in breadth and four to six centimeters deep in the adults. As we will see, many pre-modern sources regarded this gland as the navel of the animal. Each musk vesicle or "pod", as it is commonly known,[14] can weigh from twenty to fifty-two grams and contain fifteen to thirty grams of musk; twenty-five is average.[15] Musk is mostly produced by the deer in the Himalayas from April to May. A yellowish, milky secretion gathers in the musk vesicle and over the course of a month ripens into a reddish-brown

12 Cf. F. Markham, *Shooting in the Himalayas: A Journal of Sporting Adventures and Travel in Chinese Tartary, Ladac, Thibet, Cashmere, &c.* (London: Richard Bentley, 1854), 89–90, and Green, "Some Ecological Aspects", 312–13 and C. Jest, "Valeurs d'Échange en Himalaya et au Tibet: Le amber et le musc," in *De la voûte céleste au terroir, du jardin au foyer: mosaïque sociographique: textes offerts à Lucien Bernot* (Paris: Éditions de l'École des hautes études en sciences sociales, 1987), 230, who cites a Tamang informant stating that the musk deer prefer birch and rhododendron forests.

13 Heptner 113.

14 In older English, the musk pod is also called a musk "cod". In the 16th century, the Portuguese term papos was used for musk pods, e.g., Tomé Pires, *The Summa Oriental of Tomé Pires and the Book of Francisco Rodrigues*, trans. A. Cortesão (London: Hakluyt Society, 1944) 1.96 and n. 4, R. Ptak, "Moschus, Calambac und Quecksilber im Handel zwischen Macau und Japan und im ostasiatischen Seehandel ingesamt (circa 1555–1640)," in *Portugal und Japan im 16. und 17. Jahrhundert* (Frankfurt: Verlag der Interkulturelle Kommunikation, 1998), 74, and Jan Huygen van Linschoten, *The Voyage to the East Indes*, ed. A. C. Burnell and P. A Tiele, 2 vols. (London: Hakluyt Society, 1885), 2.94. See also P. Borschberg, "Der asiatische Moschushandel vom frühen 15. bis zum 17. Jahrhundert," in J. M. dos Santos Alves, et al., eds., *Mirabilia Asiatica: Productos raros no comércio marítimo*, vol. 1 (Wiesbaden: Harrassowitz, 2003), 67.

15 Green, "Distribution", 363.

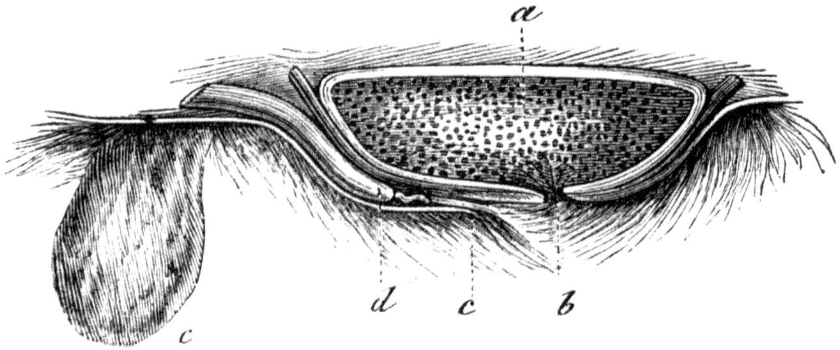

ILLUSTRATION 1.2 *The musk apparatus. a: vertical section of musk pod, b: orifice of the pod, c: not specified, d: gland carried by the filiform prolongation into the urethra.*

color which darkens further when removed from the animal.[16] The term "grain musk" refers to the dry granular musk excreted from the musk pod.[17] It is dark reddish-brown in color, but when removed from the pod it eventually becomes almost black.

The musk fragrance is derived from a substance called muskone/muscone which is present in the musk; it is highly fragrant and the scent is perceptible even when diluted 1/3000.[18] Some people, however, cannot smell musk at all; they are described as musk anosmic. Musk is used for its own scent and as a fixative in perfumery. The mid-20th century scientist of perfume Jellinek states that the scent of true musk has three nuances: one that is pungent—ammonia-like, one that is sweet-nutty, and one that is bitter-animalic.[19] The complex musk smell is very similar to the musky odor produced by human axillae and also found in urine. The odor in humans is due to the presence of certain ste-

16 K. M. Nadkarni, *Indian Materia Medica*. 3rd. ed. by A. K. Nadkarni (Bombay: Popular Book Depot, 1955), 196–205, 197; M. Green, "Musk production from musk deer," in R. J. Hudson, et al., eds., *Wildlife Production Systems: Economic utilisation of wild ungulates* (Cambridge: Cambridge University Press, 1989), 405–6.

17 Cf. Markham 88, who describes the grains as being "from the size of a small bullet to small shot, of irregular shape, but generally round or oblong, together with more or less in coarse powder [sic]."

18 B. Mukerji, *The Indian Pharmaceutical Codex: Volume 1- Indigenous Drugs* (New Delhi: Council of Scientific & Industrial Research, 1953), 149.

19 P. Jellinek, *The Psychological Basis of Perfumery*, ed. and trans. J. Stephan Jellinek (London: Blackie Academic and Professional, 1997), 40. (Originally published 1951).

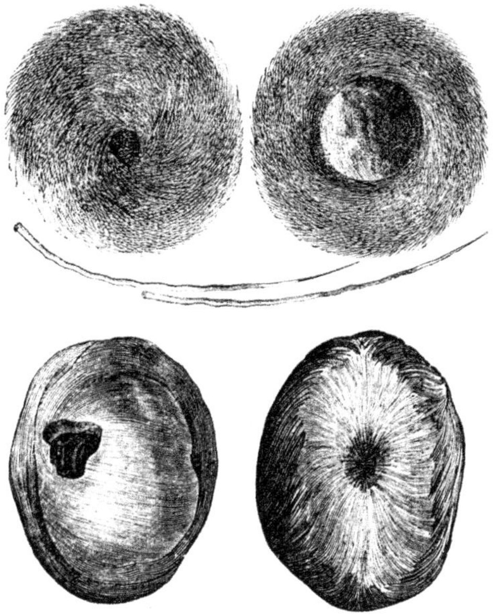

ILLUSTRATION 1.3a–b
3a (top) and 3b (bottom). Musk pods.

roids, notably 3α-androstenol. While the scent of these chemicals is like musk, they are not chemically similar.[20] As Stoddard argues in his *The Scented Ape: The Biology and Culture of Human Odour*, these chemicals play an important role in human sexual attraction. There is no doubt that it is the similarity of musk to these human pheromones that has made it such an important ingredient in perfumery. Jellinek argued that all perfumes for women should have this animal erogenous component underlying them.[21] This dictum is no longer so important, and indeed, natural musk is hardly used in modern perfumery because of its rarity and the protection of the musk deer. Synthetic musk does not have the complex aroma of the naturally derived musk; Jellinek describes its scent as being exclusively on the sweet-nutty continuum.[22]

In perfumery, musk can be used in several forms.[23] A standard tincture of 3% musk is prepared from it, and an absolute can also be made through

20 D. M. Stoddart, *The Scented Ape: The biology and culture of human odour* (Cambridge: Cambridge University Press, 1990; repr. with corrections, 1991), 64–6.

21 J. S. Jellinek in Jellenik, 253–4. However, see Stoddart, 163, who sees the whole of perfume in sexual terms.

22 Jellinek 75–6.

23 Cf. Arctander 44 and 422–3.

alcohol extraction. The absolute is dark brown and viscous; it is the most valuable form of musk because it is so highly concentrated. The tincture can be used directly in perfumes.

Musk has several properties in perfumery. It acts as a base, that is, an element of a perfume that has a slow rate of evaporation and underlies a compounded scent; in this way it also functions as a fixative. Musk need only be used in small quantities because of the strength of its scent. Robert Boyle described in 1675 an alcohol-based essence of musk which he prepared; a single drop or two was enough to scent "a pint, or perhaps a quart" of wine in such a way that "the whole body of the wine would presently acquire a considerably musky scent, and be so richly perfumed both as to tast [sic] and smell, as seemed strange enough to those that knew the vast disproportion of the ingredients."[24]

But musk also, in small quantities, enhances many different perfumes in a way that is difficult to describe. In the words of the perfumer Steffen Arctander, "It gives a distinct 'lift' or 'life' to almost any well balanced perfume base when used at the proper concentration, i.e. just above the level of perception, or at the level where the effect is a perfect 'rounding-off and levelling out' of the perfume. The animal note should, in most cases, not be distinctly perceptible."[25] Indeed, musk was used in a great many aromatic preparations during the 19th century and into the 20th, but most would scarcely be described as musky because of the small quantity of musk used.

As Boyle maintained, the scent of musk, like the scent of civet, was best appreciated from a distance, in a less concentrated state, because of the rank strength of the urinous and putrid elements of its scent. In a dilute form the scent of musk becomes pleasant to the modern nose. Boyle also reported that an acquaintance of his noted that a dunghill could have a musky scent at a certain distance from it; this gave way to the "stink proper to such a heap of Excrements" at a closer distance.[26] Indeed, other aromatics also have these properties. The chemical indol, which is found in human feces, is present in small amounts in the scent of flowers such as jasmine, and this is considered one of the reasons for the erogenous appeal of certain floral scents.[27] In small quantities, mixed with other scents and out of its familiar context, it forms an important component of perfume.

---

24 R. Boyle, "Experiments and Observations about the Mechanical Production of Odours," in *The Works of Robert Boyle*, Vol. 8, ed. M. Hunter and E. B. Davis (London: Pickering and Chatto, 2000), 387.

25 Arctander 423.

26 Boyle, "Experiments," 385.

27 Jellinek 42.

Musk has always been an expensive item; indeed, it is perhaps the most expensive substance derived from an animal. The value of musk can fluctuate with the supply. No more than several hundred kilograms of musk were available per year during the 1960's, 1970's and early 1980's; the average was about 325 kg a year by then.[28] At about 25 g of musk per musk pod, this would represent the killing of 13,000 male musk deer annually, not counting the females and juveniles killed incidentally. In recent times, the value of musk rose from about a quarter of its weight in gold to three times its weight in gold between the 1850's and 1970's.[29] In 1979 the international trade rate for musk was $24,000 per kilogram, with the higher grade granular musk selling for up to $45,000 a kilogram.[30] During the 1990s a kilogram of musk absolute sold for nearly $700,000.[31] Today, the bulk of the world demand for musk comes from East Asia where it is used for traditional medicine.

Exploitation of Musk

Various methods were used in different areas to hunt for musk deer. In the former Soviet Union, musk deer were taken with dogs, by stalking or by driving the animals.[32] Snaring was probably the most common method of taking musk deer,[33] and the medieval Arabic writers mentioned it. The Tamang People of the slopes of Ganesh Himal in Nepal northwest of Kathmandu built barricades of branches to funnel musk deer towards snares placed in ditches.[34] The captured deer were then smothered against the earth; for the Tamang, spilling the blood of the musk deer would cause further hunting to fail. The best

28 Green, "Distribution", 362, 359.

29 Green, "Musk production from musk deer", 402.

30 A chart comparing the price of musk in Japan, the world's largest importer of musk, with the price of gold is in Green, "Distribution", 360.

31 Jellinek 253.

32 V. E. Sokolov and N. L. Lebedeva, "Commercial hunting in the Soviet Union," in R. J. Hudson, et al., eds., *Wildlife Production Systems: Economic utilisation of wild ungulates* (Cambridge: Cambridge University Press, 1989), 180.

33 Jean-Baptiste Tavernier, *Travels in India by Jean-Baptiste Tavernier*, trans. V. Ball and ed. W. Crooke, 2 vols. (London, 1889, repr. New Delhi: Munshiram Manoharlal, 1995), 2.114, and Markham 95.

34 Jest 231. This article contains an account given to Jest by one of the hunters with much interesting information on the taboos and religious aspects of the hunt. The deity of the hunt, Drablha Meme, had to be propitiated as part of the proceedings. In addition, a secret language was used for terminology related to the musk hunt.

time for the hunt in the Himalayas was in April.[35] In eastern Nepal, traditional hunters prepare bamboo spears that are coated in a poison made from plants.[36] Spears are set up in areas where the animals congregate, such as rocky outcroppings where signs of musk deer have been observed. Musk deer were often located through their droppings. Often fires are set to drive the animals into an area for hunting; these fires are often very destructive to the habitat as well.[37] In other areas of Nepal, hunters use barricades, nooses or dogs, or a combination of them, or poison, to catch the musk deer.[38] In China, deer are noosed, caught in nets and trapped.[39] While the snared animals may be pilfered by predatory animals, the musk itself is not eaten because it is unpalatable, and some of it can often be recovered from the site.[40]

There is at least one example of an animal snare recovered from an ancient site. It was found at the Dunhuang limes by M. A. Stein; it is dated to the 1st century BCE.[41] It is made of fiber formed into a ring with sixteen wooden teeth inside. Stein's informant T. A. Joyce explained its use: "the ring is placed over a hole in the ground with a noose over it, the end of which is attached to a tree or stake. The animal on putting its foot into the ring attempts to shake it off

---

35 Jest 230.

36 R. Jackson, "Aboriginal Hunting in West Nepal with Reference to Musk Deer Moschus moschiferus moschiferus and Snow Leopard Panthera uncia," *Biological Conservation* 16 (1979): 65, and Jest 232.

37 Jest 232. On the prohibition of setting fires to drive out animals in a Tibetan law of 1733, see Shih-yü Yü Li, "Tibetan Folk-law," *JRAS* (1950): 139.

38 P. S. Jamwal, "Collection of Deer Musk in Nepal," *Journal of the Bombay Natural History Society* 69:3 (1972): 647–9. Cf. Markham 95–6 and Jackson 65.

39 Zhang Baoliang, "Musk deer: their capture, domestication and care according to Chinese experience and methods," *Unasylva* 35 (1983): 16–24; available online at http://www.fao.org/documents/show_cdr.asp?url_file=//docrep/q1093E/q1093e02.htm (accessed 4/26/05). Conservationists studying the musk deer recommend netting to prevent injuring the musk deer during capture for surveys, see B. Kattel and A. William Alldredge, "Capturing and Handling of the Himalayan Musk Deer," *Wildlife Society Bulletin* 19:4 (Winter, 1991): 397–9. R. B. Harris, "Conservation Prospects for Musk Deer and other Wildlife in Southern Qinghai, China," *Mountain Research and Development* 11:4 (1991): 355, notes that in Qinghai musk deer were snared or shot with firearms.

40 Markham 96–7.

41 Its inventory number is T.xv.a.i.009. Publ. M. A. Stein, *Serindia: Detailed Report of Explorations in Central Asia and Westernmost China*, vol. 2 (London, 1921; repr. Delhi: Banarsidass, 1980), 704, 782, pl. 54. There is a color illustration in S. Whitfield and U. Sims-Williams, *The Silk Road: Trade, Travel, War and Faith* (Chicago: Serindia, 2004), 179.

and so pulls the noose tight."[42] This is identical to the sort of snare used to hunt musk deer that Frederick Markham describes in the Himalayas.[43]

In the 19th century, musk pods were cut from the animal with the skin, and then placed upon a stone heated in a fire or near a fire to dry it.[44] This causes the attached flesh to dry and it shrinks into the compact form of the musk pod. Markham, who observed the process, noted that it was better for the musk to dry naturally.[45]

Musk deer are very difficult to keep in captivity and breed; the male musk deer are solitary and territorial. The Leipzig Zoo appears to be the only zoo outside of Asia that has bred musk deer successfully. There have been extensive projects in China since 1958 to breed musk deer and to raise them commercially.[46] A recent survey revealed a total population of about 1400 animals, with only 450 males capable of producing musk, yielding about 6 kg of musk per year.[47] In India as well there are non-commercial facilities working on the breeding of musk deer in Himachal Pradesh at Kufri and in Uttar Pradesh at Kanchula Kharak and Meroli.[48] These projects have met with mixed results. Musk may be harvested from live animals annually;[49] the yield is not as great compared with the yield of a slaughtered animal. The musk produced commercially in China is considered inferior to the wild product.[50] In addition, an enormous number of captive animals would be required to satisfy world demand for musk. The biologist Green has suggested removing the musk from live caught wild musk deer and releasing them; this would be especially appropriate in areas where musk deer hunting forms a major part of income.[51] The prognosis for the future survival of the wild musk deer in many parts of its range is thus rather doubtful in the face of continuing world demand for musk even though the trade in poached musk is, of course, illegal, and many

42 Stein 782.

43 Markham 95–6.

44 Markham 97–8; Jest 231 describes drying musk pods by the fire for the late 20th century among the Tamang hunters.

45 Markham 97–8.

46 Zhang 16–24.

47 R. Parry-Jones, "TRAFFIC Examines Musk Deer Farming in China," http://www.traffic.org/traffic-dispatches/traffic_pub_dispatches16.pdf (accessed 6/24/16).

48 M. S. Jain, "Observations on the Birth of a Musk Deer Fawn," *Journal of the Bombay Natural History Society* 77 (1980): 497–8.

49 Flerov 33.

50 Green, "Musk Production", 404.

51 Green, "Distribution", 369–70.

restrictions are placed on the musk trade. The musk deer also faces the loss of habitat as the human population extends further.

Trade in musk is regulated by the Convention on International Trade in Endangered Species of Wild Flora and Fauna (CITES). Himalayan musk is prohibited outright under this convention because of the endangered status of the Himalayan musk deer populations, but Chinese and Russian musk is still traded, although with regulation.

Other musky scents derived from animals and used in perfumery include castoreum and especially civet. These are distinguished by the Arabic writers from musk, which was always regarded as superior. Castoreum is produced from the castor sacs of the beavers such as *Castor fiber*, the European beaver, which is widespread in Eurasia. Castoreum has not been used so much in perfumery because musk is considered superior; nevertheless, it appeared in some perfumes although now synthetic musky scents are generally used.[52] Castoreum was used medicinally since the time of the ancient Greeks, as we will see below. The Arabic word for castoreum is *jundbādastar*, which is a loanword from Persian.[53] Within the Middle East beavers were apparently restricted to some of the highlands of Iran and the Caucasus, and even so castoreum was really an imported product for the Islamic world. Al-Nuwayrī said that it occurred only among the Qipchaqs, a Turkic people of Central Eurasia.[54]

The animals that produce civet are members of *Viveridae*, and although they are commonly called "civet cats" they are not members of *Felidae* at all but are more similar to mongooses.[55] The principal species exploited for civet are the African civet, *Civettictis civetta*, and civets of South and Southeast Asia such as the Large Indian civet, *Viverra zibetha*, and the Small Indian civet, *Viverricula indica* (= *V. malaccensis*). The Arabic word for civet is *zabād* (also used in Persian), the ultimate source of the English word. The civet "cats" could be raised in the Middle East, and indeed they were, starting probably in

52 Piesse 249–50, Arctander 136–7, and Groom 57.

53 Dietrich DT II.22, A. Dietrich, *Die Dioskurides-Erklärung des Ibn al-Baiṭār. Ein Beitrag zurarabischen Pflanzensynonymik des Mittelalters* (Göttingen: Vandenhoeck & Ruprecht, 1991), 97, and *An Eleventh-Century Egyptian Guide to the Universe: The Book of Curiosities*, ed. and trans. Y. Rapoport and E. Savage-Smith (Leiden: Brill, 2014), 522, on the equivalence of castoreum and jundbādastar. See also Newid 60–1 for castoreum in perfumery.

54 Nuwayrī 10.318.

55 Arctander 173–6; H. F. Rouk and H. Mengesha, *Ethiopian Civet* (*Civettictis civetta*) (Addis Ababa: Imperial Ethiopian College of Agriculture and Mechanical Arts, 1963), K. H. Dannenfeldt, "Europe Discovers Civet Cats and Civet," *History of Biology* 18:3 (1985): 403–31, Groom 68, and J. McHugh, "The Disputed Civets and the Complexion of the God," *JAOS* 132:2 (2012): 245–73.

Abbasid times. Civet has continued in use in perfumery into the 20th century although, like musk, it is now also replaced almost completely by synthetics.[56]

Like musk, both castoreum and civet must be highly diluted in perfume for their sweetly fragrant qualities to emerge. They smell quite rank in their natural state.

Terminology for Musk

In the traditional range of the musk deer there are various indigenous terms for the animal. Information on the musk deer goes back further in China than anywhere else; the earliest attestations will be discussed in the following chapter. In Chinese the deer is she 麝; Middle Chinese ☆*źia*^{3} (Early Mandarin Chinese *ʑia*^{h})[57]< Old Chinese **liah*[58]~**ljAks*[59]~**ᵇm-lak-s*.[60] Musk is called *shexiang* 麝香, "musk-deer aromatic" and the vesicle *sheqi* 麝臍 "musk navel." Japanese *ja* 麝 is simply the Chinese word. Musk deer are also called *shefu* 射父 and *xiangchang* 香麞 "fragrant water-deer." There was a folk etymology for the word *she* 麝. Li Shizhen 李時珍 (1518–93), author of the most famous Chinese pharmacopoeia, the *Bencao Gangmu* 本草綱目, believed that the Chinese character, which combines the "deer radical" 鹿 with the phonetic element *she* 射, meaning, "to shoot," derived from the fact that the scent of musk projected far. He also suggested that it could mean that the aroma came in a shot or blast.[61]

In Tibetan[62] the musk deer is called *glaba*, the substance musk is *glartsi*. A variety of terms are formed from the root gla besides these: *glamo*, a female musk deer, *glaphrug*, a young musk deer, *glabai lteba*, musk pod, literally "navel of the musk deer;" another term used for the pod is *dritsan-lteba*.[63] This root

---

56 Piesse 250–5, Arctander 175, and Groom 68.

57 E. J. Pulleyblank, *Lexicon of Reconstructed Pronunciation in Early Middle Chinese, Late Middle Chinese, and Early Mandarin* (Vancouver: UBC Press, 1991), 279.

58 S. Starostin, *Rekonstruktsia drevnekitaiskoi fonologiceskoi Sistemy* (Moscow: Nauka, 1989), 563.

59 W. H. Baxter, *Handbook of Old Chinese Phonology* (Berlin: Mouton de Gruyter, 1992), 786.

60 L. Sagart, *Roots of Old Chinese* (Amsterdam: Benjamins, 1999), 73.

61 Li Shizhen 李時珍, *Bencao gangmu* 本草綱目 (hence BCGM), *Guoxue jiben congshu* edition (Taipei, 1968), Ch. 51, 38.

62 I have followed the transcription system suggested by C. I. Beckwith; see his article "The Introduction of Greek Medicine into Tibet in the Seventh and Eighth Centuries," *JAOS* 99 (1979): 308–9 n. 2.

63 Y. N. Roerich, *Tibetan-Russian-English Dictionary with Sanskrit Parallels*, vol. 4 (Moscow: Nauka, 1983–9), 215, H. Jäschke, *A Tibetan-English Dictionary* (London, 1881; repr. Delhi: Motilal Banarsidass, 1995), 262a.

also applies to some plants with a scent resembling musk: *gladara*, which is *Delphinium moschatum*, and *glasgan*, a medicinal root.

Musk is attested in three early medieval Eastern Iranian languages. Abū Rayḥān al-Bīrūnī, who was a Khwārezmian, records a likely corrupt Khwārezmian word for musk in Arabic script in his *Kitāb al-Ṣaydanah*.[64] This word does not seem to have appeared in any of the ancient Khwārezmian manuscripts discovered so far, nor has any word which could be identified with musk.

Two probably related words for musk come from Khotanese and Sogdian. In Late Khotanese Saka musk is *yausa*.[65] This term is used to translate Buddhist Sanskrit *kastūra*, "musk."[66] It appears in both medical and religious texts. The Sogdian word for musk is *γγs yaxs/ γγš yaxš*.[67] It was first identified by W. B. Henning in a Sogdian text published by Émile Benveniste, a Sogdian translation of the Chinese *Amoghapāšamantrahṛdayasūtra*.[68] Benveniste's understanding of the Sogdian was influenced by a translation of the Chinese original provided to him by Demiéville that is unpublished except for brief excerpts, including the passage in question. The text makes the comparison of the aromatics sandalwood and *γγš* and camphor to the dharma: *ZKw čntn 'PZY γγsh* ZY *ZKw kp'wr*.[69] The identification of *γγš* with musk is hypothetical. The line in the Chinese is "take sandalwood or aloeswood (*chen* 沈) and musk, etc. (取栴檀香或沈麝等 *qu chantan xiang huo chen she deng*)."[70] There is no doubt that *čntn* is sandalwood (from Sanskrit *candana*) or that *kp'wr* is camphor; the latter was added by the Sogdian translator. In place of the aloeswood and musk of the Chinese, there is only the single word *γγsh*.[71] Benveniste thus

64 Z. V. Togan, ed., *Bīrūnī's Picture of the World* (New Delhi: Archaeological Society of India, 1938), 136 reads *'kt mnjl* as the Khwārezmian word for musk, H. M. Saʿīd's edition of the *Kitāb al-Ṣaydanah* (Karachi: Hamdard National Foundation, 1973), 345 has *'kt bnjl* and ʿAbbās Zaryāb's edition (Tehran: Markaz-i Nashr-i Dānishgāhī, 1991), 577 has *'kt bnkhl*.

65 H. W. Bailey, *Dictionary of Khotan Saka* (Cambridge: Cambridge University Press, 1979), 343b.

66 *Jīvakapustaka* in H. W. Bailey, ed., *Indo-Scythian Studies being Khotanese Texts*, vol. 1 (Cambridge: Cambridge University Press: 1969), 178–9. f.97v4.

67 B. Gharib, *Sogdian Dictionary* (Tehran: Farhangan Publications, 1995), #10956, #11099.

68 É. Benveniste, *Textes Sogdiens édités, traduits et commentés* (Paris: Geuthner, 1940), #7, 93–104. The *Amoghapāšamantrahṛdayasūtra* is T 1095.

69 Benveniste, #7.108.

70 T. 1095 in vol. 20, 406c.

71 There is a compound perfume *chenshe* 沈麝, see T. Morohashi, *Dai Kan-Wa Jiten*, 13 vols. (Tōkyō: Taishūkan Shoten, 1955–60), 6.969; no information on its composition is immediately available. Demiéville (apud Benveniste, 213) has translated similarly to my

read musk for the *yγsh* in this place, although elsewhere he translated it as aloeswood.[72] In his review of Benveniste's book Henning agreed that *yγsh* meant musk and adduced a passage from the Sogdian Ancient Letter II (line 58) that specifies a shipment of thirty-two *yxsyh* sent to Dunhuang by the merchant Nanai-vandak.[73] The word *yγš* also appears in the third of Benveniste's texts, an interesting document that includes notes on the supernatural properties of stones, various rites, and other such things.[74] It contains a list of aromatic and medical substances: *kp'wr* "camphor," *čntn* "sandalwood," *črpγwδn* "ointment,"[75] *wẓprnh* "safflower?"[76] *'kwšty* "costus," *β'rγwn?*, *nβ'rγtk nwš''tr* "crushed sal ammoniac,"[77] *kwrkwnph* "saffron," and *yγš* "musk." Another Sogdian text found by Henning has: *pr tw' ''γwndytww cχš'm yxs' w'ryṯw cn kpwry šnyštw* "May I anoint you with the eye-salve *čaχšām* (prepared from the seeds of *Cassia absus L.*) may I rain (on you) the perfume *yaχsa*, may I snow camphor on you."[78]

There are two Old Turkic words for musk: *kin* and *yıpar*. The relationship between them is not entirely clear. The word *kin* is most often attested in Old Turkic in the compound *kin yıpar*. *Kin* is likely the original term for the musk pod.[79] It is used in some modern Siberian Turkic languages in the sense of

---

translation: "prenant du parfum candana ou de l'aloès et du musc." Given that the Sogdian translation has three items, it is perhaps easiest to imagine that the translator has preserved the idea of three items in the original. The meaning of Sogdian *yysh* as musk is secure by analogy with Khotanese *yausa*, which is glossed with Sanskrit *kastūra*.

72 See his glossary, Benveniste 277.

73 W. B. Henning, "The Sogdian Texts of Paris," *BSOAS* 11:4 (1946): 727. See now N. Sims-Williams "The Sogdian Ancient Letter II," in *Philologica et linguistica: historia, pluralitas, universitas: Festschrift für Helmut Humbach zum 80. Geburtstag* (Trier: Wissenschaftlicher Verlag, 2001), 272–3 and de la Vaissière, *Sogdian Traders: A History*, trans. J. Ward (Leiden: Brill, 2005), 43–5.

74 Benveniste #3.174; #7.108 .

75 Cf. Benveniste 197 n. 171; Gharib #3258.

76 Henning, "Sogdian Texts", 727.

77 Gharib #5868.

78 I. Gershevitch, *A Grammar of Manichean Sogdian* (Oxford: Blackwell, 1954), §807 quotes Henning's trans. of M137 V.

79 I. Hauenschild proposes that *yargun* in the *Irk Bitig* means musk deer based on the context, see *Die Tierbezeichnungen bei Mahmud al-Kaschgari. Eine Untersuchung aus sprach- und kulturhistorischer Sicht* (Wiesbaden: Harrassowitz, 2003), 107, and cf. G. Clauson, *An Etymological Dictionary of Pre-Thirteenth Century Turkish* (Oxford: Oxford University Press, 1972), 963. A word unattested in Old Turkic but found in later Turkic for musk deer is *tabırya*.

musk and secretions of other animals that are musky.[80] Bazin has read *kin* instead of *kün* in an Old Turkic Yenissei inscription and interpreted the word as "navel" (and hence "clan"); thus *kin yıpar* is literally "navel perfume".[81] That *kin yıpar* meant musk is demonstrated by such texts as the *Suvarṇabhāsottamasūtra* chapter 7,[82] which includes a formula for a magic bath that was to dispel evil influences from the planets, stars, birth and death, etc. The list evolved as the text was translated from Sanskrit into Chinese. The Sanskrit list, as one might expect given the late period in which musk came into use in India, does not include musk, but rather consists of different plant and mineral materials. The Chinese translator added musk and made several other changes. The Uyghur version is a translation of the Chinese, and here we find Chinese *shexiang* translated by *kin yıpar*. The Chinese version also gives the original Sanskrit names for the ingredients; for musk we find *mojiapojia* 莫迦婆伽, *mahābhāga*, "illustrious"[83] which does not apparently mean musk in Sanskrit.[84] However, in the *Bencao gangmu* it is said that *mohepojia* 莫訶婆伽, certainly an alternate transcription of the same word, is the Indian word for musk.[85] The term *kin yıpār* is translated by Maḥmūd al-Kāshgharī with Arabic *nāfijat misk*,

80 W Radloff [=V. Radlov], *Versuch eines Wörterbuches der Türkdialekte*, 4 vols. (St. Petersburg, 1893; repr. The Hague: Mouton, 1960), 2.1344 and Clauson, *Etymological Dictionary*, 725.

81 L. Bazin, "Un nom 'turco-mongol' du 'nombril' et du 'clan'," in M. Erdal and S. Tezcan, eds., *Beläk Bitig. Sprachstudien für Gerhard Doerfer zum 75. Geburtstag* (Wiesbaden: Harrassowitz, 1995), 1–12.

82 *Jin guang ming zui sheng wang jing* 金光明最勝王經, Taibei: Fo jiao chu ban she, Minguo 85 [1996]), 245; *Sutra zolotogo bleska: Tekst uigurskoi redaktsii*, ed. V. V. Radlov and S. E. Malov (Saint Petersburg: Tipografiia Imperatorskoi Akademii Nauk, 1913–19), text 476. Cf. J. Nobel, "Das Zauberbad der Göttin Sarasvatī," in *Beiträge zur indischen Philologie und Altertumskunde. Walther Schubring zum 70. Geburtstag dargebracht von der deutschen Indologie* (Hamburg: Cram, De Gruyter & Co., 1951), 123–39, 130, and D. Maue and O. Sertkaya, "Drogenliste und Dhāraṇī aus dem 'Zauberbad der Sarasvatī' des uigurischen Goldglanzsūtra," *Ural-Altaische Jahrbücher* n. F. 6 (1986): 87.

83 Monier-Williams 798a.

84 J. Nobel, "Das Zauberbad der Göttin Sarasvatī," in *Beiträge zur indischen Philologie und Altertumskunde. Walther Schubring zum 70. Geburtstag dargebracht von der deutschen Indologie* (Hamburg: Cram, De Gruyter & Co., 1951), 130 #8. F. Edgerton, *Buddhist Hybrid Sanskrit Grammar and Dictionary Volume II. Dictionary* (New Haven: Yale University Press, 1953; repr. Delhi: Motilal Banarsidass, 1993), 424, quotes this passage and suggests that it is the name of a medicinal plant.

85 BCGM 25. 51. 38.

which means "musk pod", in his Turkic lexicon, the *Dīwān Lughāt al-Turk* (completed 1077).[86]

By itself, the Old Turkic word *yıpar* is often difficult to interpret. Sometimes it evidently means perfume in general,[87] but other times it means musk: it is considered to be the source of the Mongolian word for musk, *ciğar*.[88] But Mongolian also has the word *küdäri* for the musk deer; probably this is an old name for the musk deer in Central Eurasia.[89] *Yıpar* most certainly had a positive connotation; compounded with the word *yıd* "scent, smell" it means "a pleasant smell."[90] The earliest appearance of the term *yıpar* in Old Turkic is in one of the most famous of the Orkhon inscriptions of the early 8th century. A badly damaged section of the Bilgä Kaghan inscription contains a description of the funeral of the dead kaghan that mentions several actions taken by the Chinese to honor him. Just prior to the passage in question is the statement that Lisun Tai Sengun brought for the memorial *koklık*, which has often been translated as "perfumes."[91] Clauson, however, regarded this *hapax legomenon* as being uncertain in meaning and denies that it could be derived from the verb *kok*—in this sense.[92] The passage which includes the word *yıpar* is: II S 11 *yoγ yıparıγ kelürüp tike berti čindan ıγač kälürip*. This reading seems secure from the printed photograph.[93] The sentence has been variously interpreted. The decipherer of Old Turkic runes, Vilhelm Thomsen, translated: "Ils apportèrent du musc(?) pour les funérailles et le placèrent, et ils apportèrent du bois

86 Al-Kāshgharī, *Maḥmūd al-Kāšγarī: Compendium of the Turkic Dialects* (Dīwān Luγāt at-Turk), trans. R. Dankoff and J. Kelly, 3 vols. (Cambridge, MA: Harvard University Press, 1982–5), 1,171 (= 171).

87 E.g., F. W. K. Müller, "Uigurica," *Abhandlungen der Preußischen Akademie der Wissenschaften* (1908) Nr. 2, I.29 where it translates *xiang* 香.

88 G. Clauson, *Turkish and Mongolian Studies* (London: Royal Asiatic Society, 1962), 232.

89 P. Pelliot, *Notes on Marco Polo*, ed. L. Hambis. 3 vols. (Paris: Imprimerie Nationale, 1959–73), 2.742, #249, connects it with Kirghiz *küdörö*, which he suggests means a civet cat and Kalmyk *küdr̥*, which refers both to the musk deer and the desman. *Küdäri* is attested by Marco Polo with variant spellings in the different manscript traditions of his work, see Pelliot, and H. Yule and H. Cordier, *The Travels of Marco Polo* 3rd ed., 2 vols. (London, 1929; repr. New York: Dover, 1993), 2. 45 and n. 5 on 48–9, in his description of the musk deer of Tibet.

90 Clauson, *Etymological Dictionary*, 883a.

91 E.g., V. Thomsen, *Inscriptions de l'Orkhon déchiffrés* (Helsinki, 1896), 130; T. Tekin, *A Grammar of Orkhon Turkic* (Bloomington: Indiana University, 1968), 279.

92 Clauson, *Etymological Dictionary*, 610.

93 A. Heikel, et al., *Inscriptions de l'Orkhon recueillies par l'expédition finnoise 1890* (Helsinki: Société Finno-Ougrienne, 1892), plate 30, line 47.

de sandal ..."[94] S. E. Malov translated "they brought funerary smoking candles (*pogrebal'nye kuritel'nye svechi*) and set them up."[95] Talat Tekin translated: "They (also) brought scented candles (?) for the funeral and set them up for us. (Moreover) they brought sandalwood and ..."[96] Gerard Clauson notes that this sentence is obscure and suggests that it means: "brought perfumes and had the funeral feast set up." There are two difficulties with the first part of this sentence. The first is the placement of the word *yoγ*; placed before *yıparıγ* it seems likely to be an adjective describing that word, as noted by Clauson, who wonders if it is a Chinese loanword with an accusative ending.[97] The second problem is the object of *tike*. The verb *tik*—is used in the sense of "to set up, to erect". This is why Tekin suggests that they set up candles, which would be a plausible explanation if the traditional Chinese vehicle for communion with the spirits of the departed was the candle and not the incense burner. There is no reason why this passage cannot refer to an incense burner. In the Bilgä Kaghan inscription the word *yıpar* most likely means incense or fragrance in general rather than musk specifically, although Chinese compound incense often included musk.

The terms for musk from Chinese, Tibetan, Khotanese, Sogdian, and Old Turkic are unrelated to the words used for musk in the Middle East or Europe, and India has its own special nomenclature, which we will consider below. The English word "musk", however, is part of a large group of words used in the Middle East and ultimately in Europe which originates in Persian. Some of these words include:

Pahlavi *mušk* (spelled *mwšk*)
Classical and New Persian *mushk*
Arabic *misk*[98]
Ottoman and Modern Turkish *misk* (derived from Arabic)

94 Thomsen 130.

95 S. E. Malov, *Pamiatniki Drevnetiurkskoĭ Pis'mennosti Mongolii i Kirgizii* (Moskva: Izdatel'stvo Akademii Nauk SSSR, 1959), 23.

96 Tekin 279–80.

97 Clauson, *Etymological Dictionary*, 878–9.

98 K. Vollers, "Beiträge zur Kenntniss der lebenden arabischen Sprache in Aegypten. II. Ueber Lehnwörter, Fremdes und Eigenes," ZDMG 50 (1896): 649, 652, A. Jeffrey, *The Foreign Vocabulary of the Qur'ān* (Baroda: Oriental Institute, 1938), 264, W. Eilers, "Iranisches Lehngut im Arabischen," in *Actas IV congresso de estudos árabes e islâmicos: Coimbra-Lisboa ... 1968* (Leiden: E. J. Brill), 609, and A. Asbaghi, *Persische Lehnwörter im Arabischen* (Wiesbaden: Harrassowitz, 1988), 257. W. W. Müller, "Namen von Aromata im antiken Südarabien," in A. Avanzini, ed., *Profumi d'Arabia* (Rome: Bretschneider, 1997), 209–10,

Armenian *mušk*[99]
Syriac *mušāk*[100]
Ethiopic *mesk*[101]
Greek μόσχος (*móskhos*)[102]
Russian *muskus*
Latin *muscus*[103] and later, *moschus*
Spanish and Italian *musco, muschio*[104] (derived from Latin)
French *musc* (derived from Latin)[105]
German *Moschus* (derived from Latin)[106]

The Oxford English Dictionary says the ultimate source of the English word musk is "perhaps" Sanskrit *muška* [*muṣkaḥ*], Hindī *mušk* "scrotum, testicle." In this they follow a widely accepted etymology.[107] The problem is that Sanskrit has several words for the substance musk, and none of them is *muṣkaḥ*. The usual Sanskrit word for musk is *kastūrī* and also the derived *kastūrikā*,[108] which

---

proposes that Arabic *misk* came from Old South Arabian, directly from India with no Iranic intermediation.

99 H. Hübschmann, *Armenische Grammatik. Erster Teil: Armenische Etymologie* (Heidelberg, 1897), 196 #415.

100 R. Payne Smith, *Thesaurus Syriacus*, 2 vols. (Oxford: Oxford University Press, 1879–1901), 2.2055.

101 Jeffrey 264.

102 LSJ 1148a; M. Brust, *Die indischen und iranischen Lehnwörter im Griechischen* (Innsbruck: Instituten für Sprachen und Literaturen, 2005), 467–70.

103 Jerome, *Adversus Jovinianum*, in J. P. Migne, ed., *Patrologia cursus completus, series latina* (Paris, 1883), 23.311.

104 *The Oxford English Dictionary*, "Musk," s.v.

105 A. Dauzat, *Dictionnaire étymologique de la langue française* (Paris: Larousse, 1938), 492.

106 German also has *Bisam* for musk, and hence *Bisamtier* "musk deer"; this word is derived ultimately from the Hebrew besem "pleasant scented." See F. Kluge, *Etymologisches Wörterbuch der deutschen Sprache*. 20th edition, ed. W. Mitzka. (Berlin: De Gruyter, 1967), 79.

107 H. Yule and A. C. Burnell, *Hobson-Jobson* (New Delhi: Munshiram Manoharlal, 2000), 599, and K. Lokotsch, *Etymologisches Wörterbuch der europäischen (germanischen, romanischen und slavischen) Wörter orientalischen Ursprungs* (Heidelberg: Winter, 1927), 122, who cites the more immediate source of the European words for musk as Persian *mušk*, but still retains the derivation from the Sanskrit. It had been doubted already by P. Haupt, "The Etymology of Egypt. ṯsm, greyhound," *JAOS* 45 (1925): 319, who argued it was Semitic.

108 Monier-Williams 266. Cf. B. Mukerji, *Indian Pharmaceutical Codex* (New Delhi: Council of Scientific & Industrial Research, 1953), 149, also for terms for musk in modern Indian

are loanwords from the Greek καστόρειον,[109] "castoreum." Greek also has κάστωρ (*kástōr*) and καστόρειος (*kastóreios*), "beaver". The Sanskrit word *kastūrī* is well attested by the middle of the first millenium CE; the Greek word κάστωρ has a pedigree going back at least to Herodotus.[110] The Byzantines were even aware that the Indians used this term, though they thought it applied to the animal itself.[111] The application of the Greek term καστόρειον to musk suggests a Greek influence on Indian culture dating to perhaps Hellenistic times, probably by way of medicine where we may assume that musk, available in trade from the Himalayas, came to be used in Greek prescriptions calling for castoreum, which was unavailable in India.[112] Another term used for musk in Sanskrit is *darpa*, which is perhaps the indigenous term and seems less common than *kastūrī*. Sanskrit, rich in synonyms, also has numerous derived words which are used for musk; these include such words as *mṛga* "deer," *mṛganābhi* "deer navel," *mada* "honey, intoxicating drink," *mṛgamada* "deer liquor," *nābhigandha* "navel perfume," and others.[113]

As stated above, the term *muṣkaḥ* means testicle in Sanskrit. The application of the term for testicle would not be surprising because of the round, hairy appearance of the musk vesicles and their location on the body of the animal, but did this transfer happen in Sanskrit? Manfred Mayrhofer argues that it is more likely for Persian *mušk* to have originated from an Iranic root **muṣka*

languages- all are from Sanskrit *kasturi* or the compound *mriganabhi*, from Sanskrit, meaning "deer navel."

109 M. Mayrhofer, *Kurzgefaßtes etymologisches Wörterbuch des Altindischen*, 4 vols. (Heidelberg: Winter, 1956–80), I.192; LSJ 882b; H. Frisk, *Griechisches etymologisches Wörterbuch* (Heidelberg, 1960–70) I.799–800.

110 Herodotus, *The Persian Wars*, ed. and trans. A. D. Godley, rev. ed., 4 vols. (Loeb Classical Libary. Cambridge, MA: Harvard University Press, 1926), IV,109.2. καστόρειον appears in the Hippocratic corpus.

111 Cosmas Indicopleustes, *Topographie Chrétienne*, ed. and trans. W. Wolska-Conus, 3 vols. (Paris: Les Éditions du Cerf, 1968–73), XI.6.

112 Various loanwords entered Indic during the Hellenistic period, when the so-called Indo-Greek state existed. Many are military terms, e.g. σῦριγξ "pipe, subterranean seige passage," giving Sanskrit *suruṅgā*, O. Stein, "Σῦριγξ und suruṅgā," *Zeitschrift für Indologie und Iranistik* 3 (1925), 280–318, 345–7; more loanwords are listed in C. Töttössy, "Graeco-Indo-Iranica," *Acta Antiqua Academiae Scientiarum Hungaricae* 25 (1977): 133–5.

113 Some of these and others are listed in J. McHugh's "The Incense Trees from the Land of Emeralds: The Exotic Material Culture of Kāmaśāstra," *Journal of Indian Philosophy* 39 (2011): 91, and see McHugh, *Sandalwood and Carrion: Smell in Indian Religion and Culture* (Oxford: Oxford University Press, 2012), 176.

cognate with the Sanskrit root.[114] In New Persian the term *khāya* or *tukhm*, both originally meaning egg, or the Arabic *bayḍah*, also meaning egg, or the Arabic *khuṣyah* are used for testicles, so if this older root existed, it has disappeared in the sense of "testicle".[115] Since the word *mušk* in the sense of musk first appears in the Iranic area, and there is no comparable usage in India, it is best to abandon the derivation of the musk family of words from Sanskrit.

Pahlavi and Persian *mušk*, like Sanskrit *muṣkaḥ*, is itself ultimately derived from the Indo-European root for mouse.[116] The similarity of the word *mušk* to such words for rodents as *mūš* and *mūšak*, coupled with the similarity of the musk vesicle to a small, hairy, round mouse, undoubtedly underlies the zoological confusion in many early sources on musk, for example, the Arabic *fa'rat al-misk* "musk mouse" (discussed below) and Jerome's statement that musk was the "skin of a foreign rodent" (*peregrini muris pellicula*).[117] Similarly, in an early medieval Syriac source we find musk counted by skins (*meške d-mūšāk*); this probably means pods.[118] It seems less likely that the term arose specifically from musk shrews that are found in many parts of Africa and Asia. The musk shrew does not seem to have ever been used as an aromatic and is generally considered a nuisance.[119] However there was evidently continuous zoological

114 Mayrhofer, 2.657 n. See also Brust, 468–9. Christopher Beckwith argues against the Indo-Iranian theory in *Empires of the Silk Road* (Princeton: Princeton University Press, 2009), 363–9.

115 Brust 469.

116 Mayrhofer gives a parallel: Middle High German *miuselîn* means both *membrum virile* and little mouse. II.657.

117 Jerome in *Adversus Jovinianum*, in J. P. Migne, ed., *Patrologia cursus completus, series latina* (Paris, 1883), 23.311.

118 E. A. W. Budge, ed. and trans., *The History of Alexander the Great, being the Syriac version, edited from five manuscripts, of the Pseudo-Callisthenes* (Cambridge, 1889; repr. Amsterdam: Philo Press, 1976), text 200. Mention of musk in the Syriac Alexander Romance either originates with the Syriac translation, or may come from an intermediary version in Arabic which is not extant. Musk is not found in the best-known Greek text, as translated by R. Stoneman, *The Greek Alexander Romance* (Harmondsworth: Penguin, 1991).

119 Cf. Novak, *Walker's Mammals* 2.161–2, *The Imperial Gazetteer of India: The Indian Empire Vol. 1: Description.* New Edition (Oxford: Oxford University Press, 1909), 225. See also Piesse, 255–6. T. C. Jerdon notes the belief in India that the musk odor of the shrew could penetrate corked bottles and render the contents undrinkably musky, see *The Mammals of India: A Natural History of all the Animals known to inhabit Continental India* (London: John Wheldon, 1874), 53–4. But Linschoten 1.303 says regarding rats in India: "There is another sort of Rattes that are little and reddish of haire: They are called sweet smelling Rattes, for they have a smell as if they were full of Muske," not implying negative effects.

confusion between the musk deer and the musk shrew, largely because the musk deer is not native to the Middle East.

The Pahlavi word *mušk* appears in several texts. Unfortunately, all of these are works that took their final form in Islamic times well after the fall of the Sasanian Empire. Perhaps most likely to contain information about musk from Sasanian times is the *Bundahišn*, a cosmological text which includes elaborate zoological classifications of the different animals known to the Iranians.[120] These do not seem affected by zoological ideas current in Islamic times, although the *Bundahišn* does contain material from the Islamic period in other places. The ninth category of animals in the good creation of Ohrmazd[121] are the *mušk* animals: "*nōhom mušk hašt sardag. ēk ān ī pad muškīh āšnāg ud ēk ān mušk ī nāfag ān-iz-ī pad gand zanišn.*[122] *bīš-mušk kē bīš xwarēd ud mušk ī syā čiyōn hamēstār ī garzag pad drayā(b) kustag wēš bawēd ud abārīg mušk sardag.*"[123] "Ninth: eight species of *mušk* [animal]. One of them is the one known for musk, and one of them is the musk animal having a navel (*nāfag*), in which is the origin of the scent, and the *bīš mušk* which eats *bīš*, and the black *mušk* which is the opponent of the serpent which is poison in the region of the sea, and other species of *mušk*."

The author of the *Bundahišn* had probably never seen a musk deer, and does not seem to have known what sort of animal it was. He certainly associated it with the musk shrews and rodents (zoologically speaking, though, shrews are not rodents), because later he says that the musk animals live in a burrow

---

120 F. Pakzad, ed. *Bundahišn: Zoroastrische Kosmogonie und Kosmologie*, vol. 1: *Kritische Edition* (Tehran: Center for the Great Islamic Encyclopaedia, 2005), *The Bondahesh, Being a Facsimile Edition of the Manuscript TD1* (Tehran: Iranian Culture Foundation, 1970), and *Dastnivīs-i TD2/ MS. TD2: Iranian Bundahišn & Rivāyat-i Ēmēt-i Ašavahištān*, 2 vols. (Shiraz: Asia Institute of Pahlavi University, 1978). There is also a transcription and translation by B. T. Anklesaria, *Zand-Ākāsīh, Iranian or Greater Bundahišn* (Bombay, 1956).

121 In Zoroastrianism, creation is divided into the Good Creation of Ahura Mazda and the Evil Creation of Angra Mainyu. On the subdivisions of this system and parallels with other Indo-European traditions, see H.-P. Schmidt, "Ancient Iranian Animal Classification," *Studien zur Indologie und Iranistik* 5–6 (1980): 209–44.

122 Reading of Pakzad and M. Bahār, *Vāzhihnāmah-i Bundahish* (Tehran: Iranian Cultural Foundation, 1345), 384, Anklesaria 122 reads *dū-zangān* which he translates "having two legs" and interprets as the jerboa and TD1 79 has *dōk zangān* which would mean "spindle-legged"?

123 *Bundahišn* XIII:21 in Pakzad 173–4, and see TD2 pg. 96 lines 13–4 (Facsimile p. 97), TD1 Fol. 39r lines 16–7 (Facsimile p. 79), and Anklesaria 122–3.

(*sūrăg*);[124] this is true of most rodents and also of musk shrews.[125] The list of the *Bundahišn* gives four different forms of the musk animals. The first two apparently refer to animals from which musk was derived. The distinction between the musk animal which is "known for musk" and "the musk animal having a navel" probably reflects two different types of musk: musk which arrived in Iran within its pod and musk which came separate from the vesicle. Ḥamdullāh al-Mustawfī Qazwīnī, in his *Nuzhat al-Qulūb* (written 1339–40), says that the *bīsh mushk* resembles the rat. Qazwīnī writes, "Its flesh antagonizes poison and overpowers it, whether taken before the poison or after it."[126] Qazwīnī does not connect this animal with a *bīsh* plant, but the Egyptian zoographer Kamāl al-Dīn Muḥammad b. Mūsā al-Damīrī (1341–1405) says that it lives among the roots of the *bīsh* plant, which is a poisonous plant identified with aconite.[127] The black *mušk* of the *Bundahišn* may be a musk shrew, if not a mongoose, well known for its antagonism to snakes, as Anklesaria translates.[128]

In any case, while some of the animals of the ninth category of the good creation are rodents or shrews, these two musk animals are not stated to be rodents, and the explicit designation of the "navel" of the animal indicates some familiarity with the true origin of musk. The *Wizīdagīhā-ī Zādspram*, from Islamic times, has an account of animals which is probably derived from the same source as the *Bundahišn*, or from that text itself, and it is similar with regard to the musk animals.[129]

The Persian word for musk, while attested in several places in Pahlavi literature (these will be discussed in Chapter 3), has not been found in any document

124 *Bundahišn* XIII.25: Pakzad 175, TD1 39v l.8, TD2 97 l. 8, Anklesaria 123.

125 Novak, *Walker's Mammals*, 1.161–2.

126 Ḥamdullāh al-Mustawfī al-Qazwīnī, *The Zoological Section of the Nuzhatu-l-Qulūb*, ed. and trans. J. Stephenson (London: Royal Asiatic Society, 1928), text 19, trans. 13. The translation is Stephenson's.

127 A. Siggel, *Arabisch-Deustches Wörterbuch der Stoffe* (Berlin: Akademie-Verlag, 1950), 22. It is Arabic *bīsh*. Aconite is said to have a musky scent, see Dietrich, *Dioscurides Triumphans* 2.581; W. Dymock, *Pharmacographia Indica: A History of the Principal Drugs of Vegetable Origin met with in British India* (London: Kegan Paul, Tench & Trübner, 1890–3; repr. in one volume, Karachi: Hamdard, 1972), 1.4, says it has a scent like hyraceum, the aged and hardened dung of the Rock Hyrax, which has a musky scent. The Persian and Arabic words come from Sanskrit *viṣa* "poison".

128 Anklesaria 123. The 8th category of animals in the Bundahišn includes the sable and weasel, so perhaps we should not look for the mongoose here. See also Schmidt 223.

129 Zādspram, *Wizīdagīhā-ī Zādspram*, ed. and trans. P. Gignoux and A. Tafazzoli, *Anthologie de Zādspram*, Paris: Association pour l'avancement des études iraniennes, 1993), 50–3 (transcription and translation) and 206–9 (text).

recovered from Late Antique Iran or its environs. It has so far been found only in texts from the early Islamic period when the Pahlavi literature assumed the form in which it is extant today. It is doubtful, however, that all the occurrences of the word *mušk* reflect Arab-Islamic influence in this literature, although at least one passage discussed in Chapter 3, from the *Wizīdagīhā-ī Zādspram*, probably does. Musk is also mentioned in accounts of the Sasanian court preserved in Arabic literature. It is likely that this lack of musk in Sasanian sources stems from the types of sources that are still extant; nothing resembling a comprehensive overview of Sasanian culture is possible using the few miscellaneous sources that have survived. While newly discovered Pahlavi documents have been appearing in recent decades, none mentioning musk, or any other aromatic, has yet appeared.

From Persian, *mušk* passed into Syriac.[130] The Syriac term *mušāk* is said to be found in a work of the 4th century writer Ephrem (d. 373); if so, it would be a very early attestation of musk, but the presence of the word cannot be confirmed and must be regarded as doubtful.[131] The 5th century Syriac writer Isaac of Antioch (d. c. 460) mentions it.[132] It also appears in the Syriac translation of the Pahlavi version of the *Pañcatantra*, the *Kalīla wa Dimna*, a work that dates to the 6th century.[133] It is also used in the Syriac translation of the pseudo-Callisthenes Alexander Romance (dating to the early Islamic period).[134] Apart from musk itself, we also find the word *prmbštā* in the 10th century lexicon of Bar Bahlūl, glossed in Arabic as musk scented perfumed powder (*al-dharīrah al-mumassakah*).[135] In addition, the Syriac Book of Medicines has *nrmwšq/*

---

130 See C. Ciancaglini, *Iranian Loanwords in Syriac* (Wiesbaden: Reichert, 2008), 204 and P. Gignoux, *Lexique des termes de la pharmacopée syriaque* (Paris: Association pour l'Avancement des Études Iraniennes, 2011), 56.

131 C. Brockelmann, *Lexicon Syriacum* (Halle: Max Niemeyer, 1928), 408a. I must thank David Brakke [personal communication] for attempting to verify Brockelmann's reference in the Harvard library. Ciancaglini repeats Brockelmann's reference.

132 Isaac of Antioch, *Homilae S. Isaaci Syri Antiocheni*, ed. P. Bedjan (Leipzig: Harrassowitz, 1903), 472 line 10.

133 Ed. G. Bickell, *Kalilag und Damnag, alte syrische Übersetzung des indischen Fürstenspiegels* (Leipzig: Brockhaus, 1876), 37 l.10.

134 Budge, *History of Alexander*, 200, 201; cf. M. Feldbusch, *Der Brief Alexanders an Aristoteles über die Wunder Indiens. Synoptische Edition*, Beiträge zur klassischen Philologie 78 (Meisenheim am Glan: Anton Hain, 1976), 153b. The second passage attests the belief that musk came from the navels of the animals; Syriac *šūrā* is cognate with the Arabic *surrah*.

135 *Lexicon syriacum auctore Hassano bar Bahlule voces syriacas graecasque cum glossis syriacis et arabicis complectens*, 3 vols., ed. R. Duval (Paris: Republicae typograhaeo, 1888–1901), 2.3267.

nrmwšk, which Gignoux identifies as an otherwise unattested Persian loanword **nar -mušk* "male musk".[136]

In Arabic, the vowel of *mušk* was changed to an "i" and spelled with *sīn* instead of *shīn*, a well-attested sound change.[137] The Arabic word *misk* is well attested in pre-Islamic Arabic poetry as well as in the Qur'ān and Ḥadīth.[138] In addition, a business document on wood from pre-Islamic Yemen mentions the shipment of kinds of musk (*'mskn*).[139] *Misk* is ubiquitous in early medieval Arabic literature, as we will see in the following chapters. Many Arabic philologists claim it to be a loanword. Mawhūb b. Aḥmad al-Jawālīqī (1072–1145), who wrote the most important early surviving work on loanwords in Arabic, the *Kitāb al-Mu'arrab min al-kalām al-a'jamī*, gives the following concise entry on the word misk: "*al-misk*, it is an aromatic (*al-ṭīb*), it is Arabized Persian."[140] Al-Tha'ālibī, according to al-Suyūṭī, also claims it is Persian,[141] as does al-Khafājī.[142]

Since the term *misk* appears in the Qur'ān, many Islamic writers stated it to be a true Arabic word under the belief that the Qur'ān contained only true Arabic words.[143] An example of the justification of this can be seen in al-Sarī al-Raffā''s (10th century) statement: "As for musk, it is called musk because it is retained (*masaka*) by the gazelle in its navel."[144] In spite of this folk etymology

136 P. Gignoux, "Les relations interlinguistiques de quelques termes de la pharmacopée antique," in D. Durkin-Meisterernst, et al., eds., *Literarische Stoffe und ihre Gestaltung in mitteliranischer Zeit* (Wiesbaden: Harrassowitz, 2009), 95, Gignoux, *Lexique* 61.

137 Mawhūb b. Aḥmad al-Jawālīqī, *al-Mu'arrab min al-kalām al a'jamī 'alā Ḥurūf al-Mu'jam* (Tehrān, 1966), 7, says, "they may exchange *sīn* for *shīn*, as they say for desert, *dast*, and it is *dašt* in Persian." See also Jeffrey 264 and Eilers 609.

138 E.g., Imru' al-Qays' *Mu'allaqah*, in *Dīwān*, ed. Yāsīn al-Ayyūbī (Beirut: al-Maktab al-Islāmī, 1998), lines 8 and 41 on pp. 46 and 68 and Qur'ān 83:26. This subject is discussed in greater detail below.

139 J. Ryckmans, W. W. Müller, and Y. M. Abdallah, *Textes du Yémen antique inscrits sur bois* (Louvain-la-Neuve: Institut Orientaliste, 1994), 65–6, 102–3. These documents are believed to date, based on paleographical evidence, to the first three centuries CE, making this document, along with Talmudic evidence and Jerome's passage, the earliest attestations of musk in the west.

140 Jawālīqī 325.

141 Al-Suyūṭī, *Muhadhdhab fīmā waqa'a fī al-Qur'ān min al-mu'arrab* (Beirut: Dār al-Kitāb al-'Arabī, 1995), 87.

142 Al-Khafājī, *Shifā' al-ghalīl fīmā fī kalām al-'Arab min al-dakhīl* (Cairo: al-Maktabah al-Azhariyyah li-l-Turāth, [2003]), 271.

143 Cf. Jeffrey 6–9 and W. Fischer, "Mu'arrab," *EI²* s.v.

144 Al-Sarī al-Raffā', *Al-muḥibb wa-l-maḥbūb wa-l-mashmūm wa-l-mashrūb*, 4 vols., ed. Miṣbāḥ Ghalāwinjī, (Damascus: Majma' al-Lughah al-'Arabiyyah, 1986–7), 3.253. Cf. Haupt,

most philologists rightly preferred to regard *misk* as a Persian word; this is a good indication that this idea of its origin was widely accepted.

The Greek term comes into use in the 6th century.[145] Musk is mentioned in medical works by Aetius of Amida (fl.530–60) and Alexander of Tralles (525–605). It is also mentioned by the traveler Cosmas Indicopleustes (fl. first half of the 6th century). Cosmas' mention of musk strongly suggests that musk was generally known in the Byzantine world during his time. Musk is mentioned in the *Adversus Jovinianum* of Jerome (d. 419 CE), and, if genuine, this is the earliest reference to musk in the Graeco-Roman world. All of these works are discussed in Chapter 3.

The fact that the substance called musk travelled throughout the Near East and into Europe bearing that name indicates that musk spread through a single process. That process can be traced to Sasanian Iran, from which the word musk appears in Armenian, Syriac, Arabic, Greek and other languages. The Pahlavi term *mušk*, undoubtedly the closest one to the origin of this family, is an Iranian formation unrelated to the terminology for musk in India. While musk was certainly traded through India, the Near Eastern and Mediterranean merchants purchasing it became familiar with the substance through contact with Sasanian Persia, and not in India.

319, who also derives it from *masaka*, but because "the odor of musk … is the most lasting of perfumes." There is also a theological reason to stress that the blood was retained to make the musk, because flowing blood is *ḥarām* or illicit in Islamic law, as we will discuss below, 211–13.

145 LSJ 1148 notes: "borrowed from Pers. *mušk*".

CHAPTER 2

Commodities of Further Asia and the Islamic World

Introduction

It is well known that the medieval Islamic world played a key role in the trade of Eurasia and Africa. The image of the *sūq* or *bāzār* stocked with exotica has even penetrated the popular conception of the Middle East. Yet not enough attention has been given to trade and commerce in early medieval Islamic culture despite the general acknowledgement of its great importance.[1] The figure of the merchant has been studied,[2] and merchants' importance in the development of Islamic jurisprudence and culture in general is acknowledged.[3] Yet the fact that these merchants dealt often in goods that came from the ends of the earth as known at the time perhaps needs greater emphasis.[4] What knowledge was gathered in the early medieval Middle East about Asia and Africa beyond the boundaries of the *Dār al-Islām* was largely filtered through the lens of this mercantile culture. This commercial orientation is notable throughout the geographical literature, as well as in references to Further Asia in other works. As one of the investigators of Islamic literature on Further Asia writes: "This part of the world was the source of a large quantity of spices and drugs used in Arabic medicine and therefore the Arabic reading public had a certain

---

1 See especially G. Heck, *Charlemagne, Muhammad, and the Arab Roots of Capitalism* (Berlin: De Gruyter, 2006), 41, and M. Shatzmiller's more cautionary "Economic Performance and Economic Growth in the Early Islamic World," *JESHO* 54 (2011): 132–84.

2 E.g., G. Wiet, "Les Marchands d'épices sous les sultans mamlouks," *Cahiers d'histoire egyptienne* 7 (1955): 81–147, S. D. Goitein, "The Rise of the Middle-Eastern Bourgeoisie in Early Islamic Times," *Journal of World History* 3 (1957): 583–604; A. K. Lambton, "The Merchant in Medieval Islam," in *A Locust's Leg* (London, 1962), 121–30 and M. Rodinson, "Le marchand musulman," in D. S. Richards, ed., *Islam and the Trade of Asia* (Philadelphia: University of Pennsylvania, 1970), 21–35. On Sogdian merchants, see É. de la Vaissière, *Sogdian Traders: A History*, trans. J. Ward (Leiden: Brill, 2005).

3 H. J. Cohen, "The Economic Background and the Secular Occupations of Muslim Jurisprudents and Traditionists in the Classical Period of Islam," *JESHO* 13 (1970): 16–61. See also Heck.

4 A bold step in the right direction is André Wink's emphasis on the economic aspects of the early Islamic presence in Asia in his *Al-Hind: The Making of the Indo-Islamic World*, vol. 1 (Leiden: Brill, 1990).

 | DOI 10.1163/9789004336315_004

interest in it."[5] Thus the literature on Further Asia, "geographical" or otherwise, contains accounts, often detailed ones, of the commodities of Further Asia, especially drugs, spices, and perfumes. Another factor in the commercial orientation of the literature is that the accounts of Further Asia used by the geographers and other writers came into the Islamic world through merchants who traveled in Further Asia in order to acquire these commodities. No wonder, then, that the literature on Further Asia has a commercial orientation.

This knowledge was largely a product of the great heyday of long-distance voyaging in the 9th–11th centuries. The data collected during this time formed the basis of subsequent literature and was scarcely added to until Mongol times, but was transmitted, mostly faithfully, even when it was archaic, much later. This corpus of information, as it penetrated into the writings of the geographers and other *adībs*, was concerned primarily with the commercial and secondarily with the religious (and thus legal) situation of the countries involved in trade. While much information can be found on the different varieties of trade goods, or the beliefs and laws of the peoples of Asia and Africa, little is available on the political situation beyond its impact on commerce, and even that information tended to become fossilized during its transmission through the centuries. The commercial information was of such interest that it was widely disseminated in many branches of literature, and certain features of it reached the point that they became stock images.[6] Above all, since this literature was meant for literati, it had the function of entertainment, and thus was filled with interesting information and tales of curiosities involving the lands and their products. From such come several ideas involving the musk deer and the production of musk, which will be discussed in Chapter 4.

Since the Arabic and Persian literary materials on Asia beyond the world of Islam are commodity-centered within the framework of each text, it is worthwhile to discuss these products of Asia and their cultural impact to provide a context for the trade in musk. It will be evident from the following that the materials and finished goods obtained from Further Asia were of great

---

5 G. R. Tibbetts, *A Study of the Arabic Texts containing material on South-East Asia* (London: Royal Asiatic Society, 1979) 3.

6 Christine van Ruymbeke shows just how much overlap was possible between the scientific and literary modes of writing: see *Science and Poetry in Medieval Persia: The Botany of Nizami's* Khamsa (Cambridge: Cambridge University Press, 2007). Her analysis of the works of the 12th century Persian poet Niẓāmī demonstrates his familiarity with scientific literature and use of imagery derived from it in his works. Niẓāmī may be an exceptional case in some respects, but the overlap between scientific literature on musk and poetry and other fields of Arabic literature, as we will see in the following chapters, is not negligible.

diversity and real importance in medieval Islamicate society. What needs especial emphasis is that despite being famed for silk, the trade with Further Asia consisted of much more,[7] and the drugs, spices, perfumes, and dyestuffs were perhaps of even greater significance than the textiles.

The trade between the Near East, Mediterranean world, and Asia goes back, of course, to remote antiquity. It evolved and expanded through time, as new products were discovered, or markets for old ones developed in the Near East and Mediterranean. The development of the Eurasian overland trade networks developed in tandem with steppe empires.[8] The increasing arrival of silks from East Asia in the Near East and Central Asia resulted from the formation of the Xiongnu Empire, which was centered in the Mongolian Plateau and Northern China. The Xiongnu received large quantities of silk from Han China as part of the relations between the two powers.[9] The Xiongnu sought the establishment of border markets for trade through which the Han also acquired goods that they needed.[10] Some of this silk ended up heading west overland through the royal exchange of gifts or through the direct activities of merchants, probably financed (or supplied) to some extent by members of the Xiongnu court, as later occurred during the Mongol Empire.[11] Steppe empires continued to acquire silk from China throughout their history, and this silk fueled the exciting events of the 6th century, when the Turks and Sogdians sought western markets with first the Sasanians and then the Byzantines.[12] The Uyghur court

7 As emphasized by V. Hansen, *The Silk Road: A New History* (Oxford: Oxford University Press, 2012), 5.

8 C. I. Beckwith, *Empires of the Silk Road: A History of Central Eurasia from the Bronze Age to the Present* (Princeton: Princeton University Press, 2009).

9 A. F. P. Hulsewé, "Quelques considérations sur le commerce de la soie au temps de la dynastie des Han," *Mélanges de Sinologie offerts à Monsieur Paul Demiéville* (Paris: De Boccard, 1966–74), 2.117–35, M. G. Raschke, "New Studies in Roman Commerce with the East," in *Aufstieg und Niedergang der römischen Welt. Band II.9.2.* (Berlin: De Gruyter, 1978), 604–1361, esp. 606–22, De la Vaissière 28–32, and N. Di Cosmo, *Ancient China and its Enemies: The Rise of Nomadic Power in East Asian History* (Cambridge: Cambridge University Press, 2002), esp. 247–9. This is not to deny that some Chinese silk reached western Asia before this time; it surely did.

10 For a list of steppe products desired by the Han, see *Yen T'ieh Lun: Discourses on Salt and Iron*, trans. E. M. Gale (Leiden: Brill, 1931; repr. New York: Paragon, 1967), 14–5.

11 See T. T. Allsen, "Mongolian Princes and their Merchant Partners," *Asia Major* 3rd ser. 2 (1989): 83–126.

12 See Menander Protector, *The History of Menander the Guardsman*, ed. and trans. R. C. Blockley (Liverpool: Cairns, 1985), 111–27, K. Hannestad, "Les relations de Byzance avec la Transcaucasie et l'Asie Centrale aux 5e et 6e siècles," *Byzantion* 25–7 (1955–7): 421–56, and de la Vaissière 228–32.

sold much-needed horses to the Tang dynasty for silk; this fact is noted even in an Arabic source of the early 9th century.[13] In effect, the formation of steppe empires drove the Central Eurasian economy and facilitated the development of long-distance trade.

The first great heyday of the maritime trade was in late Hellenistic and early Roman times, spurred on by the discovery of the use of the monsoon winds that made the voyage to India easier.[14] A particularly important document of this trade is the *Periplus Maris Erythraei*, a first century CE text written by a Greek sailor with experience on the two routes of the Indian Ocean: to the eastern coast of Africa and to India.[15] This work is a descriptive guide to the commerce of the Indian Ocean written by a merchant for the use of merchants, and it includes much information on the commodities of the time. These voyages made use of the monsoons, but other, short-haul voyages skirted the coastlines of the Red Sea to Ethiopia and Arabia, where both locally produced and imported commodities were acquired. At this time, the Arabs of the southern Arabian Peninsula, especially from Yemen, were active in trade throughout the

13 V. Minorsky, "Tamīm ibn Baḥr's Journey to the Uyghurs," *BSOAS* 12:2 (1948): 279, 283, 298–9. On the Uyghur trade in horses for Tang silk, see C. I. Beckwith, "The Impact of the Horse and Silk Trade on the Economies of T'ang China and the Uighur Empire: On the Importance of International Commerce in the Early Middle Ages," *JESHO* 34 (1991): 183–98.

14 The date at which the use of the monsoon winds was discovered is controversial, but perhaps it was as early as the late second century BCE, see Raschke 660–2. For Roman commerce with India and Asia, a useful overview is L. Casson, "Rome's Trade with the East: The Sea Voyage to Africa and India," in his *Ancient Trade and Society* (Detroit: Wayne State, 1984), 182–98, see also the monographs of E. H. Warmington, *The Commerce between the Roman Empire and India*, 2nd ed. (Cambridge: Cambridge University Press, 1974; repr. New Delhi: Munshiram Manoharlal, 1995). Raschke, S. Sidebotham, *Roman Economic Policy in the Erythra Thalassa 30 BC–AD 217* (Leiden: Brill, 1986) and his *Berenike and the Ancient Maritime Spice Route* (Berkeley: University of California Press, 2011), G. K. Young, *Rome's Eastern Trade: International Commerce and Imperial Policy 31 BC–AD 305* (New York: Routledge, 2001) and the papers, mostly on archaeological evidence, in V. Begley and R. D. De Puma, eds., *Rome and India: The Ancient Sea Trade* (Madison: University of Wisconsin Press, 1991).

15 L. Casson, ed. and trans., *The Periplus Maris Erythraei* (Princeton: Princeton University Press, 1989).

region and India.[16] In late antiquity the Sasanian Persians came to dominate the seaborne trade of the Indian Ocean, and Arabia seems to have declined.[17]

Most scholars of long-distance ancient trade have focused on the Greek and Roman trade with India and the lands further to the East. In effect, these ancient trading economies were seen against the backdrop of European trading interests in Asia in the Early Modern and Modern Worlds. The Near East itself has seldom been appreciated as having its own significant place in this trade, perhaps because for the Graeco-Roman writers (as for later European traders), it was mostly an obstacle to be overcome in the sense that political and military affairs interfered with the ability to trade freely with the spice lands of India or the silk lands of China, or they saw it as a region to be traversed, literally sitting between them and their destinations. For the purposes of this investigation, however, it is Europe that sits on the periphery. The societies of the Near East, going back to ancient Mesopotamia and Egypt, traded with India and surrounding lands for spices, woods, and precious stones that they used in their own cultures. The Near East was itself thus a significant market, and the great prosperity of the Sasanian Empire meant that the Sasanians had the same appetite for the silks of China or the spices of India as the Romans of the late empire or the Parthians had. This appetite was present in early Islamic society as well, fed by a culture of merchants who were happy to strive to satiate it.[18]

While much of the Middle East is in the Asian side of Eurasia, we wish to make a distinction between the lands of the Middle East and Asia within the *Dār al-Islām* and those beyond the *Dār al-Islām*. This distinction is particularly relevant to the musk trade, for the highlands of eastern Central Eurasia that produced musk remained (and today still remain) almost entirely beyond these boundaries. Musk was always something which had to be brought into the world of Islam. Many other commodities—especially natural products that could only be grown in specific regions—fall into the same situation. For the sake of simplicity, we call this Further Asia as a way of avoiding awkward

16 G. F. Hourani, *Arab Seafaring in the Indian Ocean in Ancient and Early Medieval Times*, rev. ed. J. Carswell (Princeton: Princeton University Press, 1995), 13–50. See also the papers in A. Avanzini, ed., *Profumi d'Arabia: Atti del Convegno* (Rome: Bretschneider, 1997) and D. Peacock and D. Williams, eds., *Food for the Gods: New Light on the Ancient Incense Trade* (Oxford: Oxbow Books, 2007).

17 This is discussed in Chapter 4.

18 On the Sasanian background, see M. Morony, "The Late Sasanian Economic Impact on the Arabian Peninsula," *Nāmah-i Īrān-i Bāstān* 1:2 (2002): 25–37.

expressions such as "China, Eastern Central Eurasia, India, and Southeast Asia" or the potentially confusing "Asia" alone.[19] We will also retrospectively apply this term to antiquity in consideration of the lands beyond the empires of the Iranian Plateau and Mesopotamia: the Achaemenids, Seleucids, Parthians, and Sasanians, which straddled the Near Eastern/Mediterranean worlds and the lands of Central Asia and India.

There is a fashion for calling just about any trade with Asia "Silk Road" trade, in the context of first the overland "Silk Road" (a conception dating only as far back as the 19th century),[20] and later into the maritime trade of the Indian Ocean, sometimes called the Maritime Silk Road or the Silk Road of the Sea. We have eschewed this terminology for two reasons.[21] First, the term "Silk Road" overemphasizes the importance of silk. Many other commodities traveled across the "Silk Road", including the musk we discuss. And while silk was certainly one of the most important of these, especially as far as Europe was concerned, the Islamic world (not to mention the Byzantine Empire) could produce its own silk even if it still cherished the silk of China. Second, it is difficult to distinguish a "Silk Road" from the very complex interconnected network of trade routes, both overland and maritime throughout Eurasia, the Indian Ocean Lands, and Africa. In effect, what has been described as the "Silk Road" is the Central Eurasian economy and its ramified network of intercultural trade in the pre-Columbian Old World. As a symbol of this trade or heuristic device to cope with this complexity, the term "Silk Road" perhaps has its uses, but it appears too limited to deal with the problems associated even with the musk trade between eastern Central Eurasia and the Middle East.

The corpus of commodities traded between Further Asia and the Near East was more limited in antiquity compared with Islamic times. Late antiquity and the early Islamic period witnessed an astounding expansion in the variety of goods of Further Asia imported into the Middle East and, ultimately, into the Mediterranean world. Some measure of this expansion may be seen in a book by the Andalusian physician Ibn Juljul (944–94).[22] This work

---

19 But cf. the fine colloquium ed. by D. S. Richards, *Islam and the Trade of Asia* (Oxford: Bruno Cassirer, 1970).

20 For a recent scholarly survey of the history of the overland routes, see V. Hansen, *The Silk Road: A New History* (Oxford: Oxford University Press, 2012).

21 Note also Hansen's remarks on 5–7.

22 Ibn Juljul. I have discussed the goods mentioned in this text in my paper "The new *materia medica* of the Islamicate tradition: the pre-Islamic context," *JAOS* 135, no. 3 (2015): 499–528. See also Z. Amar, E. Lev, and Y. Serri, "Ibn Rushd on Galen and the New Drugs Spread by the Arabs," *JA* 297 (2009): 83–101 and their "On Ibn Juljul and the meaning and

catalogues the *materia medica* missing from the work of the first century physician Dioscorides, the standard authority of the classical world on pharmacology. These sixty-two items consist of imported herbs and spices, introduced fruits and vegetables, as well as animal products such as musk and ambergris. Some have become very familiar, such as nutmeg, cloves, coconut, orange, lemon, jasmine, camphor, musk, ambergris, sandalwood, bananas, cucumber, eggplant, and spinach. Many others are plant substances used in medieval medicine that are not well known today. As the identification of ancient plants is seldom precise, it is probable that Dioscorides did indeed know some of the plants mentioned by Ibn Juljul, but most of these plants have dubious attestations (if any) in antiquity and reflect the expanding horizons of late antique trade. Most are from India and the lands around the Indian Ocean, but some, like spinach, musk, and rhubarb, come from Central and Western Eurasia. The assimilation of these substances into western medicine, pharmacology, perfumery, and agriculture was usually due to the agency of the physicians and druggists of the Islamic world.[23] Of course much more than these *materia medica* were imported into the Middle East. The commodities imported from Asia were many and varied, and they were put to many different uses in Islamic culture, as we will discuss below.

For the purposes of this overview, the Further Asian trade will be divided into four regions that generally produced distinctive goods, although each also traded in the commodities of the other regions as well. Central Eurasia is the largest and most important of these regions; trade from Central Eurasia was almost entirely conducted overland. For the purposes of this work, we will focus on the lands of Central Eurasia beyond the boundaries of the lands of Islam: Tibet, Inner Asia (Eastern Turkestan and Mongolia), and Siberia. In the east, China and its neighbors Korea and Japan were known, even if vaguely, to early medieval Islamicate scholars. Their commodities were available in the Middle East through both the overland and maritime routes. The other lands were reached by sea and cluster around the Indian Ocean. India was right on

---

importance of the list of medicinal substances not mentioned by Dioscorides," *JRAS* (2014): 529–55; the latter appeared too late to be cited in my paper.

23 The introduction of some of these plants is discussed in detail in A. M. Watson, *Agricultural innovation in the early Islamic world: The diffusion of crops and farming techniques, 700–1100* (Cambridge: Cambridge University Press, 1983). On this work, see M. Decker, "Plants and Progress: Rethinking the Islamic Agricultural Revolution," *Journal of World History* 20 (2009): 187–206, who concludes that Watson overplays the contribution of Islamic civilization and shows that many of these crops have deeper roots in the Near East.

the frontier; the Umayyads had taken Sind, and the Islamic world had fairly direct access to the markets of northeastern India. Beyond India was Southeast Asia, source of spices and aromatics. The Arabs thought of it as an extension of India, as indeed it was culturally to some extent.[24] Direct sailing from the Middle East to Southeast Asia and China brought Muslims into contact with this region. The trade with the east coast of Africa is not to be distinguished from the general trade of the Indian Ocean, of which it was an integral part since antiquity. But here we will not enter into Africa's trade with the Middle East, through which musk flowed to Africa, in order to focus on the Asian trade through which musk arrived.

Central Eurasia

The Arab conquests focused especially on securing the territory of Khurāsān, the old Sasanian realm of the east, and Central Asia proper, which soon became parts of the lands of Islam. In early medieval times under the Arab Empire, "Khurāsān" meant Central Asia including what is now Afghanistan, Tajikistan, Uzbekistan, and Turkmenistan, up to the frontiers of the Turk peoples and Tibet of Central Eurasia, and up to India in the south. The area of Marw was the colonial capital of the Arab Empire's Central Asian territory and was a focal point for Arab settlement. It gave the Arabs immediate access to the Sogdians, based in Transoxiana, who were probably the most important of the Eurasian trading peoples in the early medieval era.[25] Adjoining steppe territory inhabited by nomadic empires, Khurāsān had access to the products of the peripheral parts of Eurasia, as well as the goods of the northern forests and steppelands and mountains within it. Many of the important products of Central Eurasia were natural: raw materials used in manufacturing or medicine. The precious metals produced in Transoxiana (called *Mā warā' al-nahr* in Arabic, lit. "that which lies beyond the river") helped drive the early medieval economy,[26] while rubies and lapis lazuli from the mountains of Afghanistan adorned jewelry in the periphery of Eurasia.

From beyond the boundaries of the *Dār al-Islām* came an even greater range of commodities. Of the animal and plant products, musk is perhaps the most

---

24 Wink 190–2.

25 H. Kennedy in G. Herrmann, ed., *Monuments of Merv: Traditional Buildings of the Karakum* (London: Society of Antiquaries, 1999), 28 and de la Vaissière 270–6.

26 I. Blanchard, *Mining, Metallurgy and Minting in the Middle Ages Vol. 1. Asiatic Supremacy. Mawara'an-nahr and the Semiryech'ye 425–1125* (Stuttgart: Steiner, 2001).

important, but Central Eurasian rhubarb, with a vast canon of medicinal uses, remained in great demand until modern times as well. Central Eurasia also produced manufactured goods that were eagerly sought by Muslim merchants, especially textiles and metalwork.

Aside from musk, there were a great variety of natural products from Central Eurasia. Bezoars, concretions from the liver or digestive system of herbivores believed to have great medical (and magical) power, came from Khurāsān as well as China and India.[27] Undoubtedly many came in branches of commerce from the herding peoples of Central Eurasia. Furs were one of the most important commodities in medieval Eurasia.[28] Most of the furs originated in the forest lands of the north such as Russia and much of the trade with the Siberian forests went through Khwārazm,[29] although there were, of course, other routes. According to the *Kitāb al-Tabaṣṣur bi-l-tijārah*, Khwārazm exported musk, furs, and aromatic reeds: almost all of this (and certainly all of the musk) was transshipped through Khwārazm rather than originating there. The most valuable furs sought in the Middle East were sable (*sammūr*),[30] ermine (*qāqūm*),[31] squirrel (*sinjāb*), and marten (*fanak*). Many of these were also produced within Central Eurasia itself. One of the most famous points of origin was the fur trading state of the Kimäk.[32] The anonymous 10th century

---

27 Ibn al-Bayṭār, *al-Jāmiʿ li-mufradāt al-adwiyah wa-l-aghdhiyah*, 4 vols. (Būlāq, n.d.), 1.81. On bezoars, see G. F. Kunz, *The Magic of Jewels and Charms* (Philadelphia: Lippincott, 1915; repr. New York: Dover, 1997), 201–21.

28 R. K. Kovalev, "The Infrastructure of the Northern Part of the "Fur Road" between the Middle Volga and the East during the Middle Ages," *Archivum Eurasiae Medii Aevi* 11 (2000–01), 25–64 and J. Martin, *Treasure of the land of darkness: the fur trade and its significance for medieval Russia* (Cambridge: Cambridge University Press, 1986).

29 Al-Iṣṭakhrī, *Kitāb masālik al-mamālik*, ed. M J. de Goeje (Leiden: Brill, 1870; repr. 1967), 305 and al-Jāḥiẓ (attr.), *Kitāb al-Tabaṣṣur bi-l-tijārah fī waṣf mā yustaẓraf fī al-buldān min al-amtiʿah al-rafīʿah wa-l-aʿlāq al-nafīsah wa-l-jawāhir al-thamīnah*, ed. Ḥasan Ḥusnī ʿAbd al-Wahhāb al-Tūnisī (Cairo: Maktabat al-Khānjī, 1994), 28. See also T. Noonan, "Khwārazmian Coins of the Eighth Century from Eastern Europe: The Post-Sasanian Interlude in the Relations between Central Asia and European Russia," *Archivum Eurasiae Medii Aevi* 6 (1986–8): 243–58.

30 Al-Damīrī, *Ḥayāt al-ḥayawān al-kubrā*, 2 vols. (Cairo, 1963 repr. of Būlāq ed.), 2.34; cf. also Qazwīnī, *Nuzhat al-Qulūb*, ed. and trans. J. Stephenson, *The Zoological Section of the Nuzhat al-Qulūb of Ḥamdullāh al-Mustaufī al-Qazwīnī* (London: Royal Asiatic Society, 1928), text 25 and trans. 18 where *sammūr* is called the most expensive of furs.

31 Damīrī 2.239.

32 Cf. *Ḥudūd al-ʿālam min al-mashriq ilā al-maghrib*, ed. M. Sutūdah (Tehrān, 1983), 85. On the Kimek, see B. E. Kumekov, *Gosudarstvo Kimakov IXXI vv. po arabskim istochnikam* (Alma-Ata: Nauka Kazakhskoi SSR, 1972).

Persian *Ḥudūd al-ʿālam* mentions that much sable and squirrel were imported from the area around the Issık Köl.[33] Likewise, the marten of Kashgar and the squirrel of the Kirghiz were proverbial for their excellence.[34] These furs were undoubtedly expensive, but they seem to have been widely used to make clothing warmer for use in the winter. The fine northern furs were doubtlessly too expensive for all but the wealthy or powerful to afford.[35] Fur-lined garments are regularly associated in Arabic literature with the court and elite.

There were also a variety of herbs and other plant products imported from Central Eurasia. Rhubarb from eastern Central Eurasia and western China was perhaps the most important of the *materia medica* of plant origin, developing an extensive catalogue of uses over the centuries.[36] It was recommended for use in medicines for insanity, epilepsy and cold ailments as early as al-Kindī.[37] The 10th century Persian pharmacologist Abū Manṣūr Muwaffaq al-Harawī calls it *rīwand ṣīnī* "Chinese rhubarb" and gives a description of its medical properties. He also notes a Khurāsānī variety of rhubarb.[38] There was much adulteration of rhubarb because the best of it was produced solely in Inner Asia and Northwest China.[39] Rhubarb is mentioned by Dioscorides[40] and is

---

33 *Ḥudūd al-ʿālam* 27.

34 Al-Thaʿālibī, *Laṭāʾif al-maʿārif* (Cairo: Dār Iḥyāʾ al-Kutub al-ʿArabiyyah ʿĪsā al-Bābī al-Ḥalabī wa-Shurakah, n.d.), 235–6. Cf. also Ibn al-Faqīh, *Mukhtaṣar kitāb al-buldān*, ed. M. G. De Goeje (Leiden: Brill, 1885), 255 on the marten of the Turks.

35 Cf. Damīrī 2.34.

36 On rhubarb, see Ibn al-Bayṭār 2.129–34, and also C. M. Foust, *Rhubarb: The Wondrous Drug* (Princeton: Princeton University Press, 1992), which is inadequate for the Near Eastern history of rhubarb, but is a very interesting account of its trade in the early modern world. See also Akira Haneda, "On Chinese Rhubarb," in *The Islamic World: Essays in Honor of Bernard Lewis* (Princeton: Darwin Press, 1989), 27–30 and E. Lev and Z. Amar, *Practical* Materia Medica *of the Medieval Eastern Mediterranean According to the Cairo Genizah* (Leiden: Brill, 2008), 259–60.

37 Al-Kindī, *The Medical Formulary or Aqrābādhīn of al-Kindī*, ed. and trans. M. Levey (Madison: University of Wisconsin, 1966), 205. Cf. also Sābūr b. Sahl, *Dispensatorium Parvum* (*al-Aqrābādhīn al-Ṣaghīr*), ed. O. Kahl (Leiden: Brill, 1994), index under *rāwand*.

38 Abu Manṣūr Muwaffaq al-Harawī, *Kitāb al-Abniyah ʿan ḥaqāʾiq al-adwiyah*, ed. A. Bahmanyār (Tehran: Dānishgāh-i Tihrān, 1992), 165–6. Cf. B. Laufer, *Sino-Iranica: Chinese Contributions to the History of Civilization in Ancient Iran, with Special Reference to the History of Cultivated Plants*. Field Museum of Natural History Anthropological Series 15:3 (Chicago: Field Museum, 1919; repr. New York: Kraus, 1967), 547–50 on rhubarb; he considers al-Harawī the earliest Persian source.

39 Cf. ʿAbd al-Raḥmān b. Naṣr al-Shayzarī, *Kitāb Nihāyat al-rutbah fī ṭalab al-ḥisbah*, ed. al-Sayyid al-Bāz al-ʿArīnī, 2nd ed. (Beirut: Dār al-Thaqāfah, 1969), 42–3.

40 Dioscorides, *De Materia Medica*, ed. M. Wellmann. 3 vols. (Berlin: Weidmann, 1958), III.2.

attested in China as early as the Han dynasty and remained especially a product of the northwest.[41]

Saffron was another important plant product in the Islamic world; it was used in medicinal, culinary, and perfumery applications as well as for a dyestuff. It was exported from Ṣaghāniyān and Wāshjird into the Middle East;[42] of course saffron was grown in other regions, especially Iran, as well.[43]

Another famous botanical product of Central Eurasia was a special wood for making bows, arrows and other equipment called *khadhank/khadang*.[44] The word is found in a variety of spellings in Arabic and Persian literature.[45] It is usually identified with the white birch or the white poplar.

Central Asia is to this day famous for its fruit.[46] Al-Ma'mūn (r. 813–33) must have become acquainted with a Khwārazmian melon called *bāranj* while in Central Asia.[47] When he moved the caliphate back to Baghdad, he brought a taste for these melons which persisted at least through the reign of al-Wāthiq (842–7). Al-Thaʿālibī notes that they were packed in ice inside leaden boxes and then taken to Baghdad. He states that a melon that survived the journey was worth seven hundred dirhams. Perhaps it is the same as the melon called *al-ma'mūnī* by the 13th century pharmacologist Ibn al-Bayṭār.[48] The pears (*kummathrā*) of Khurāsān were considered to be very sweet by Muʿammil, an

41 É. Chavannes, *Les documents chinois découverts par Aurel Stein dans les sables du Turkestan Oriental* (Oxford: Oxford University Press, 1913), 115. It is, however, absent from the Mawangdui medical manuscripts. E. Schafer, *The Golden Peaches of Samarkand: A Study of T'ang Exotics* (Berkeley: University of California, 1963), 180.

42 Iṣṭakhrī 44.

43 D. Waines and F. Sangustin, "Zaʿfarān" in *EI* 2nd ed. s.v., Garbers #130 on 381–4, W. Schmucker, *Die pflanzliche und mineralische Materia Medica im Firdaus al-Ḥikma des Ṭabarī* (Bonn: Selbstverlag des Orientalischen Seminars, 1969), 217, H. Schönig, *Schminken, Düfte und Räucherwerk der Jemenitinnen: Lexikon der Substanzen, Utensilien und Techniken* (Beirut: Ergon Verlag, 2002), 319–24, Lev and Amar 270–3.

44 G. Doerfer, *Türkische und mongolische Elemente im Neupersischen Band III* (Wiesbaden: Steiner, 1967), #1164 on 183–4; Z. V. Togan, *Ibn Fadlans Reisebericht* (Leipzig, 1939), 211–15.

45 E.g. *Ḥudūd* 14 on 80, 25:79 on 116.

46 Cf. Ibn Ḥawqal, *Kitab ṣūrat al-ʿarḍ*, ed. J. H. Kramers (Leiden: Brill, 1938), 465. Compare the famous "golden peaches (apricots) of Samarkand" of Tang times: Schafer 117.

47 Thaʿālibī, *Laṭā'if*, 226. On the name see D. A. Agius, A*rabic Literary Works as a source of documentation for technical terms of the material culture* (Berlin: Schwarz, 1984), 199–201. *Bāranj* means coconut in Persian, so C. E. Bosworth, in *The Book of Curious and Entertaining Information* (Edinburgh, 1968), 142 n. 157, suggested that this was a melon that resembled a coconut. Agius suggests *bā-ranj*, Persian "with color."

48 Agius 201.

acquaintance of al-Jāḥiẓ.[49] They were apparently available in Iraq, but perhaps this refers to the type of tree rather than actual fruit imported from Khurāsān.

A variety of minerals also came from Central Eurasia. Sal ammoniac (*nūshādhir*) was a specialty of Sogdiana.[50] Ibn Ḥawqal claimed that he had only seen it there and in Sicily, but that the Sicilian was inferior.[51] Sal ammoniac was produced in other areas, but that of Sughd was indeed quite famous, and al-Masʿūdī preserves an account of its procurement from volcanic vents.[52] It had numerous applications in medicine: al-Kindī includes it in eight different prescriptions for uses as diverse as eye medicine, sore throat treatment, and even dentifrice.[53] Transoxiana also produced gold, silver, and mercury,[54] while Badakhshān produced precious stones including rubies (English "balas" ruby is ultimately derived from the name of Badakhshān)[55] and lapis lazuli.[56] Tibet also had sources of gold.[57] Finally we must mention a stone whose name to this day bears witness to its association with Central Asia: turquoise, deriving from the name of the Turks.[58] Turquoise was produced throughout Iran, Khurāsān, and Tibet; it is likely that most turquoise (called *fīrūzaj* in Arabic, from the Persian) used in the Middle East came from Iran and Khurāsān.[59]

Central Eurasia was distinguished for more than natural resources. A variety of manufactured goods were imported from Central Eurasia. Al-Ṭabarī records al-Maʾmūn's gifts (*al-taʿẓīm wa-l-hadāyā*) from the riches (*ṭuraf*) of Khurāsān to the caliph al-Amīn before their falling out: household goods, vessels, musk, livestock, and weapons (*al-matāʿ wa-l-ānīyah wa-l-misk wa-l-dawābb wa*

---

49 Al-Jāḥiẓ, *Kitāb al-Bukhalāʾ*, ed. Ṭāhā Ḥājirī (Cairo: Dār al-Maʿārif, n.d. [1958]), 98.

50 Schafer 218.

51 Ibn Ḥawqal 465.

52 Al-Masʿūdī, *Murūj al-dhahab wa maʿādin al-jawhar*, ed. B. de Meynard and P. de Courteille, revised by C. Pellat, 5 vols. (Beirut: Manshūrāt al-Jāmiʿah al-Lubnāniyyah, 1966–74), 1.163–4 (§383).

53 Al-Kindi, *Medical Formulary*, 341. See F. Käs, *Die Mineralien in der arabischen Pharmakognosie*, 2 vols. (Wiesbaden: Harrassowitz, 2010), 1.1100–5 for an overview of sal ammoniac in Islamic medicine.

54 Ibn Ḥawqal 464.

55 *Oxford English Dictionary*, "Balas", s.v.

56 Al-Bīrūnī, *Kitāb al-Jamāhir fī maʿrifat al-jawāhir*, ed. F. Krenkow (Haydarābād: Dāʾirat al-Maʿārif, 1355/1936), 81; W. Barthold, A. Bennigsen, and H. Carrère-d'Encausse, "Badakhshān," in *EI* 2nd ed. s.v. and Schafer 231.

57 *Hudūd al-ʿālam* §11 on 73. See Chapter 4 for the stories of gold-digging ants.

58 *Oxford English Dictionary* s.v.

59 Al-Bīrūnī, *Kitāb al-Jamāhir*, 169–71. See on turquoise in Asia in general B. Laufer, *Notes on Turquois* [sic] *in the East* (Chicago: Field Museum, 1913).

al-silāh).[60] Weaponry and armor from Tibet was especially famous in the early medieval Islamic period.[61] Fine Tibetan armor (*kashkhūdah*) saved the life of a khāqān of the Türgish.[62] Frequent mention is made of the bucklers (*turs*/ *tirās*) of Tibet[63] as well as other Tibetan shields.[64]

Central Eurasian textiles played a very important role in the medieval Islamic Middle East. The urban centers of Central Asia produced fine textiles in silk and cotton. The Transoxianan tribute to the early Arab conquerors consisted of silks (*ḥarīr*) and garments (*thiyāb*).[65] Iṣṭakhrī noted that their excellent cotton garments (*thiyāb al-quṭn*) were exported to distant lands (*al-āfāq*).[66] Some of the most famous of all medieval textiles were the Zandanījī fabrics mentioned often in early medieval Arabic and Persian literature. These were produced in a village of Bukhārā and exported to all countries, according to the 10th century writer al-Narshakhī.[67]

60 Ṭabarī III.775.

61 See J. Clarke in D. J. LaRocca, *Warriors of the Himalayas: Rediscovering the Arms and Armor of Tibet* (New York: Metropolitan Museum of Art, 2006), 22. For some examples of decorative horse tack and other ornaments, see A. Heller, "Archaeological Artefacts from the Tibetan Empire in Central Asia," *Orientations* 34:4 (April, 2003): 55–64.

62 Ṭabarī II.1521–2. Cf. C. I. Beckwith, *The Tibetan Empire in Central Asia: A History of the Struggle for Great Power among Tibetans, Turks, Arabs and Chinese during the Early Middle Ages*, 2nd ed. (Princeton: Princeton University Press, 1993), 109 n. 4.

63 Ibn Ḥawqal 472.

64 Ibn al-Faqīh 255; Mas'ūdī 4.279 (§2657).

65 Al-Balādhurī, *Kitāb futūḥ al-buldān*, ed. M. J. De Goeje (Leiden, 1866; repr. Leiden: Brill, 1968), 408.

66 Iṣṭakhrī 288.

67 Al-Narshakhī, *Tārīkh-i Bukhārā*, ed. Mudarris Raḍawī, 2nd ed. (Tehrān: Intishārāt-i Tūs, 1984), 21–2. A fragment of fabric once thought to bear a Sogdian inscription read as Zandanījī cloth has been preserved in the Collegiate Church of Notre Dame at Huy in Belgium: D. G. Shepherd and W. B. Henning, "Zandanījii Identified?," in R. Ettinghausen, ed., *Aus der Welt der islamischen Kunst* (Berlin, 1959), 15–40. The inscription has now been identified as Arabic and does not mention zandanījī, see N. Sims-Williams and G Khan, "Zandanījī Misidentified," *BAI* 22 (2008)[2012]: 207–13 which also has good photographs. The motifs on this silk and the others identified with zandanījī are also incongruous with the art of Sogdiana. For discussion, see B. I. Marshak, "The So-called Zandanījī Silks: Comparisons with the Art of Sogdia," *Central Eurasian Textiles and Their Contexts in the Early Middle Ages* (Riggisberg: Abegg-Stiftung, 2006), 49–60 and R. N. Frye, "Bukhara and Zandanījī," ibid. 75–80.

One of the most historically significant products of Central Asia for the development of the Islamic world was paper.[68] Paper provided a somewhat cheaper alternative to papyrus for writing and undoubtedly facilitated the growth of literacy. While paper was imported all the way from China,[69] it was also produced in Transoxiana. The manufacture of paper was introduced into the Islamic world in the middle of the 8th century by the Chinese.[70] Samarkand was particularly renowned for it, and the *Kitāb al-Tabaṣṣur bi-l-tijārah* attributed to al-Jāḥiẓ mentions it as a product of Samarkand.[71] The earliest example of an Arabic document on paper has been found at Mt. Mug in Tajikistan, along with documents in Chinese and Sogdian.[72] By the 11th century al-Thaʿālibī could write that paper (*kāghad*) had caused the abandonment of papyrus (*qirṭās*) and parchment due to its excellence, suppleness, usefulness, and appropriateness for writing.[73] Even at that time it was still a specialty of Central Asia, although paper making had spread further into the Middle East by then.

China

While much trade went overland through Central Eurasia to China, China and India were two of the main goals of Arab-Islamic maritime trade. During the 8th–10th centuries sailors from the Middle East made the long voyage all the way to China.[74] In later times, this trade came to be increasingly segmented,

68 J. M. Bloom, *Paper before Print: The History and Impact of Paper in the Islamic World* (New Yaven: Yale, 2001).

69 *Kitāb al-Tabaṣṣur* 26.

70 Thaʿālibī 217. On the contacts between Chinese craftsmen and the Islamic world, cf. P. Pelliot, "Des artisans chinois a la capitale abbasside en 751–762," *T'oung Pao* 26 (1929): 110–12. Cf. also W. Barthold, *Turkestan down to the Mongol Invasion*, 2nd ed. (London: Gibb Memorial Series, 1958), 237; Du You, *Tongdian* (Beijing: Zhonghua shuju, 1988), 193: 5279–80; Beckwith, *Tibetan Empire*, 140.

71 *Kitāb al-Tabaṣṣur* 28.

72 I. Iu. Kračkovskiĭ, "Drevneĭshiĭ arabskiĭ dokument iz Sredneĭ Azii," in *Sogdiĭskiĭ sbornik* (Leningrad: Akad. NAUK SSSR, 1934), 52–90. See also J. M. Bloom, "Revolution by the Ream," *ARAMCO World* 50:3 (May/June 1999), 26–39, 29 and his *Paper before Print*, 42–5.

73 Thaʿālibī 218. Cf. also Ibn Ḥawqal 465.

74 The earliest voyages must have antedated 758 CE, because in that year ships of the Arabs (*Dashi*) and Persians (*Bosi*) sacked Canton, see *Jiu Tangshu* 10:253. Evidence of commercial voyaging to China in an Ibadite source comes from around this period, see T. Lewicki, "Les premiers commerçants arabes en Chine," *Rocznik Orientalistyczny* 11 (1935): 173–86.

with most Middle Eastern merchants going only as far as India to exchange cargoes with merchants plying the eastern seas, thus reducing the risk of a lengthy voyage. During the early medieval period, when China was unified under the powerful and cosmopolitan Tang Dynasty (618–907), those in the Middle East were well aware that China, *al-Ṣīn*, was one country whether reached by land or by sea. This knowledge came to be obscured after the increasing segmentation of long-distance trade from the Middle East and the diminution of the long-haul voyages to China. The division of China in the 10th century into the realms of the Song in the center and south (refered to as *al-Ṣīn*) and the Kitan Liao Dynasty (c. 907–1122) in the north, called usually *al-khiṭā'* in Arabic and Persian sources, hence English Cathay, Russian *Kitai*, etc.,[75] also contributed to this process. The Kitan Liao Empire was known primarily through the overland route,[76] and many of the Kitans themselves moved west onto the fringes of Central Asia and established the Karakhitāy state once their dynasty was overthrown by the Jurchen Jin dynasty.[77]

For the medieval Muslims, China was a land famous for its wealth and high level of craftsmanship. Arabic and Persian writers praise the technical skills of the Chinese. The Abbasid littérateur al-Jāḥiẓ considered the Chinese supreme in handicrafts (*ṣināʿāt*) even as the Byzantine Greeks were supreme in wisdom and letters, the Sasanians in kingship and the Turks in warfare.[78] He writes further:

> [The Chinese are] masters of casting and smithing, pouring and smelting, of wonderful dyes, masters of turning, sculpting, depiction, manuscripts and calligraphy and are gentle of hand in everything they are entrusted

75 See A. King, "Some 10th Century Material on Asian Toponymy from Sahlan b. Kaysan," *Archivum Eurasiae Medii Aevi* 16 (2008): 121–6, and "Early Islamic Sources on the Kitan Liao: The Role of Trade," *Journal of Song-Yuan Studies* 43 (2013): 253–71.

76 For Kitan Liao foreign relations, see V. Hansen, "The Kitan People, the Liao Dynasty (916–1125) and their World," *Orientations* 42:1 (January-February 2011): 34–42, M. Biran, "Unearthing the Liao Dynasty's Relations with the Muslim World: Migrations, Diplomacy, Commerce, and Mutual Perceptions," *Journal of Song-Yuan Studies* 43 (2013): 221–51, and V. Hansen, "International Gifting and the Kitan World, 907–1125," *Journal of Song-Yuan Studies* 43 (2013): 273–302.

77 M. Biran, *The Empire of the Qara Khitay in Eurasian History: Between China and the Islamic World* (Cambridge: Cambridge University Press, 2005).

78 Al-Jāḥiẓ, *Rasā'il al-Jāḥiẓ*, ed. ʿAbd al-Salām Muḥammad Hārūn, 2 vols. (Cairo: Maktabat al-Khānjī, 1964), 1.67.

> with and assist, regardless of the difference in its substance, distinction of its manufacture, and the disparity in its value.[79]

Likewise, Abū Zayd al-Sīrāfī praises the skills of the Chinese craftsmen and gives an account of the grueling standards of excellence that were imposed upon them.[80] This reputation must have been based on some knowledge of Chinese products, for the craftsmen of Islam during the height of the Abbasid caliphate were themselves skilled. The Chinese themselves attributed this skill to their influence. Following the so-called Battle of Talas in 751 (fought at Aṭlakh), Chinese prisoners of war entered the Middle East and supposedly taught "weavers of *ling*-silk brocade, silver and goldsmiths, and painters (綾絹機杼金銀匠畫匠)."[81]

Many kinds of manufactured luxury goods were attributed to China. In 134/751–2, an Arab raid on the town of Kish, modern Shahrisabz, to the south of Samarkand, yielded many Chinese commodities:

> Abū Dāwūd took from the Ikhshīd[82] and his companions when he killed them painted and gilded Chinese porcelain vessels the likes of which had never been seen, and also some Chinese saddles, and goods of China consisting of brocade and other things, and many things from the riches of China. And Abū Dāwūd gathered it all for Abū Muslim, who was at Samarkand.[83]

The *Kitāb al-Tabaṣṣur bi-l-tijārah* lists as imports from China brocades (*firind*), silks (*ḥarīr*), vessels (*ghaḍā'ir*), paper (*kāghad*), ink (*midād*), peacocks, horses (*barāḍīn furrah*), saddles, felt (*lubūd*), cinnamon (*dārṣīnī*) and rhubarb (*rāwand*).[84] The text also mentions that the sable of China was the best, as was

---

79 Jāḥiẓ, *Rasā'il*, 1.69.

80 Abū Zayd al-Sīrāfī, *Riḥlah*, ed. ʿAbd Allāh al-Ḥabashī (Abu Dhabi: Manshūrāt al-Majmaʿ al-Thaqāfī, 1999), 60.

81 According to the information of Du Huan 杜環, who returned to China and wrote a now lost book called the *Jingxingji* 經行記, quoted frequently by his relative Du You, *Tongdian* (this passage is at 193: 5280). See P. Pelliot, "Des artisans chinois à la capitale abbasside en 751–62," *T'oung Pao* 26 (1929): 110–12.

82 The ruler of Kish. The text has *ikhrīd*, which is to be understood as Ikshīd: C. E. Bosworth, "Kish," in *EI*[2] s.v. Kish. The title Ikhshīd, Sogdian *xšyδ*, means chief or commander: B. Gharib, Sogd*ian Dictionary: Sogdian-Persian-English* (Tehran: Farhangan, 1995), #10669.

83 Ṭabarī III.79–80.

84 *Kitāb al-Tabaṣṣur* 26.

Chinese felt.[85] Neither of these texts mentions the musk of China, which is well known from other sources, but the geographer Ibn Khurradādhbih (820 or 25–911?) does in his list of Chinese products elsewhere in his book: silk (*ḥarīr*), silk brocade (*firind*), gold brocade (*kīmkhāw*),[86] musk, aloeswood, saddles (*surūj*), sable (*sammūr*), vessels (*ghaḍār*), *ṣīlbanj*,[87] cinnamon (*dārṣīnī*) and galangal (*khūlanjān*).[88] The anonymous 10th century Persian *Ḥudūd al-ʿālam* records gold, silk (*ḥarīr*), silk brocade (*firind*), *khāwkhīr* (presumably a textile), brocade (*dībā*), porcelain (*ghaḍārah*), cinnamon (*dārṣīnī*), *khutū* used for making knife handles,[89] and other marvelous goods, as exports of China.[90] The bulk of these goods are manufactured. Only a few *materia medica et aromatica* came from China, although those that did had considerable importance.

Foremost among the drugs of China was musk, which will be discussed in Chapter 3. We have already noted that rhubarb came from China as well as Central Eurasia. Ibn Khurradādhbih and the other sources mention cinnamon as an export from China. True cinnamon, *Cinnamomum zeylanicum*, is, as the name implies, a native of Sri Lanka and southern India. The similar *Cinnamomum cassia*, called cassia in English but ubiquitously confused with cinnamon, grows in southern China as well as southeast Asia.[91] *Dārṣīnī*, the term for cinnamon used by Ibn Khurradādhbih and the *Kitāb al-Tabaṣṣur bi-l-tijārah*, is Persian for "Chinese wood". The Arabic terminology for cinnamon

---

85 *Kitāb al-Tabaṣṣur* 20, 22.

86 From Chinese *jinhua* 錦花, see B. Laufer, "Loan Words in Tibetan," *T'oung Pao* 17 (1916): 557–8. and his *Sino-Iranica: Chinese Contributions to the History of Civilization in Ancient Iran, with Special Reference to the History of Cultivated Plants*, Field Museum of Natural History Anthropological Series 15:3 (Chicago: Field Museum, 1919; repr. New York: Kraus, 1967), 539.

87 De Goeje suggests that this is the narcotic that the Persians call *gālbang*. It is unclear how far back this substance may be traced, as it is absent from al-Harawī's *al-Abniyah ʿan Ḥaqāʾiq al-Adwiyah*, ed. Aḥmad Bahmanyār, 2nd ed. (Tehran: Dānishgāh-i Tihran, 1975; repr. Tehran: Dānishgāh-i Tihran, 1371/1992), which is the first pharmacological text extant in Classical Persian. The first citation in the *Lughat nāmah* is from the 12th century poet Sūzanī, LN 27.17–8.

88 Ibn Khurradādhbih, *Kitāb al-masālik wa-l-mamālik*, ed. M. J De Goeje (Leiden: Brill, 1889, repr. 1967), 70.

89 See Chapter 4.

90 *Ḥudūd al-ʿālam min al-mashriq ilā al-maghrib*, ed. M. Sutūdah (Tehran: Dānishgāh-i Tihrān, 1983), §9.

91 E. Bretschneider, *Botanicon Sinicum*, 3 vols. (1881–95; repr. Nendeln: Kraus, 1967), 3.443–52; cf. Schafer Golden Peaches 180.

and cassia is as confused as the ancient terminology, which also probably involves plants unrelated to either cinnamon or cassia.[92]

Silk is, of course, the best known of the commodities associated with China in the Early Medieval period.[93] It gave its name to the "Silk Road", *die Seidenstraßen* ("Silk Roads"), and was doubtless of great significance even if it has overshadowed the roles of other commodities. It continued to be exported to Central Asia and the West despite local versions of silk and even after the spread of sericulture because of the fineness and whiteness of the Chinese silk, due partly to the use of white mulberry leaves to feed the worms.[94]

Chinese brocades were certainly silk. The term *firind* is a loanword from Pahlavi *parand*, and could also denote fine linen and woolen brocades.[95] China was well known for its excellent brocades.[96] *Ḥarīr* is the generic Arabic term for silk. China did not merely export silk, for silk was used as a currency and tributes were paid in it. Since Han times, vast quantities of silk entered Central Eurasia in payment to the Xiongnu, as mentioned above. While some of this silk was undoubtedly used by the Xiongnu, much also went further west as gifts exchanged with the potentates of Central Eurasia. As noted above, this process continued into early medieval times, and the Turks and Uyghurs received large quantities of silks as payment for horses, which the Chinese could not produce

92 The cinnamon of the Bible and antiquity seems to have originated in east Africa and Arabia and thus cannot have been *Cinnamomum*, as noted already by Burkill 1.550. See Herodotus 3.111. B. Laufer, *Sino-Iranica* (Chicago: Field Museum, 1919), 541–3. Modern discussion: L. Casson, "Cinnamon and Cassia in the Ancient World," in *Ancient Trade and Society* (Detroit: Wayne State University Press, 1984), 225–46 and P. Crone, *Meccan Trade and the Rise of Islam* (Princeton: Princeton University Press, 1987), 253–63. For references to *dārṣīnī*, see Dietrich DT I.11.

93 For the history of silk technology in China, see D. Kuhn, *Science and Civilisation in China V Part 9: Textile Technology: Spinning and Reeling* (Cambridge: Cambridge University Press, 1988).

94 Sericulture was introduced into Khotan sometime in the early part of the first millenium; there is a legend that a Chinese princess smuggled silkworm eggs there, cf. M. A. Stein, *Ancient Khotan*, 2 v., (Oxford: Oxford University Press, 1907), 229–30 and 259–60, and Hansen 213. On the introduction of sericulture into Byzantium see Procopius VIII.xvii.1–8, who relates that monks coming from Serinda smuggled eggs; Theophanes of Byzantium (in C. Müller, *Fragmenta Historicorum Graecorum t. IV*, Paris, 1868, 210–11), gives the story somewhat differently, stating that a monk smuggled silk-worm eggs from the Seres in a cane.

95 On parand see W. B. Henning, "Two Central Asian Words," *Transactions of the Philological Society* 1945: 153–7 and A. S. Melikian-Chirvani, "Parand and Parniyān Identified: The Royal Silks of Iran from Sasanian to Islamic Times," *BAI* 5 (1991):175–9.

96 Cf. Ibn al-Faqīh, *Kitāb al-Buldān*, ed. Yusūf al-Hāwī (Beirut: ʿĀlam al-Kutub, 1996), 70.

in sufficient quantities domestically. Domestic consumption of these textiles should not be underestimated, but a large quantity of silk will have traveled west to the realms of Central Asia and ultimately into the Middle East.

The vessels mentioned in the *Kitāb al-Tabaṣṣur bi-l-tijārah* are porcelain ceramics, and are attested elsewhere for this period.[97] Porcelain may be distinguished from other types of ceramics by the ring it produces when struck, its white interior, and its translucency. These characteristics were well known in the early medieval Middle East. Chinese vessels were a major influence on the development of many types of Islamic ceramics. Archaeological work has produced fragments of Chinese ceramics from many sites throughout the Middle East.[98] Excavations at Sīrāf and Susa show that Chinese ceramics were readily available in the 9th century even though they were of course vastly outnumbered by Islamic ceramics.[99] More importantly, this evidence also demonstrates that Chinese ceramics were also known by the middle of the 8th century.[100] The Belitung Shipwreck, from the sea near the island of Belitung between Sumatra and Borneo, produced an enormous quantity of Chinese porcelain exported in the 9th century; the vessel which carried it was a dhow from the Middle East.[101] Much of this porcelain trade came by way of the maritime routes of the Indian Ocean; there are only a few examples of sherds of Chinese porcelain from Central Asian sites compared with greater numbers from sites near the maritime routes.[102]

---

97 On this subject see P. Kahle, "Islamische Quellen zum chinesischen Porzellan," ZDMG 83 (1934): 1–45 and "Chinese Porcelain in the lands of Islam," in Paul Kahle, *Opera Minora* (Leiden, 1956), 326–61 (originally published in *Transactions of the Oriental Ceramic Society* 1940–41, London, 1942 and in the *Journal of the Pakistan Historical Society* I (1953): 1–16). For a recent world-historical survey of porcelain, see R. Finlay, *The Pilgrim Art: Cultures of Porcelain in World History* (Berkeley: University of California, 2010).

98 See J. Carswell, "China and the Middle East," *Oriental Art* 45:1 (1999): 2–14, and Carswell in Hourani, 144–6 with further references.

99 The ceramics from Sīrāf are very clearly Chinese, but certainly came through maritime trade, see especially M. Tampoe, *Maritime Trade between China and the West: An Archaeological Study of the Ceramics from Siraf (Persian Gulf), 8th to 15th Centuries A.D.* (Oxford: B.A.R., 1989). See also D. Whitehouse, "Some Chinese and Islamic Pottery from Siraf," in W. Watson, ed., *Pottery and Metalwork in T'ang China* (London, 1970), 35–40.

100 Tampoe 94 and *passim*, and esp. 91–2 on Susa.

101 An exhibition catalog with essays on this important find is R. Krahl, et al., eds., *Shipwrecked: Tang Treasures and Monsoon Winds* (Washington: Smithsonian Institution, 2010).

102 L. Sokolovskaia and A. Rougeulle, "Stratified Finds of Chinese Porcelains from Pre-Mongol Samarkand (Afrasiab)," *BAI* 6 (1992): 95–6.

Porcelain was an expensive product, yet the possessions of wealthy officials and caliphs typically included at least some of it. An anecdote of al-Bīrūnī marvels at the wealth of a man of Rayy who possessed a great many porcelain vessels and implements of different types.[103] In addition to its use as tableware, porcelain was used for storing perfumes and censers. The Fatimid treasury contained porcelain vessels of camphor,[104] and huge porcelain vessels for storing perfume are attested for the early 9th century.[105] Not only was porcelain used for storing perfumes, its use was prescribed for the making of high quality perfumes. The formulas given by the Ibn Mandawayh-Ibn Kaysān tradition as well as some collected by al-Nuwayrī from al-Tamīmī specify the use of porcelain vessels of different types for the preparation of perfumes. Gold and silver vessels are also called for; this indicates the sort of costs associated with the court perfumes. The ingredients themselves were often worth their weight in gold.

The manufacture of paper certainly existed in Central Asia by the 8th century and subsequently spread throughout the Islamic world, but it is probable that paper was imported in addition to the local production.[106] Its introduction to the Islamic world is linked to China, and the story told, that Chinese captured following the battle of Talas (751, actually fought at Aṭlakh) introduced its manufacture at Samarkand, is plausible though the technical skill evinced by Central Asian paper suggests a long development before the arrival of Chinese papermakers.[107] Samarkand was particularly renowned for paper manufacture, and the *Kitāb al-Tabaṣṣur bi-l-tijārah* mentions it as a product of that city as well as China.[108] As mentioned above, the earliest example of an Arabic document on paper has been found at Mt. Mug, along with documents in Chinese and Sogdian.

Saddles, as mentioned by the *Kitāb al-Tabaṣṣur bi-l-tijārah* and Ibn Khurradādhbih, are recorded frequently as products of China in Arabic literature.[109]

103 Quoted by Kahle, "Chinese Porcelain", 342.

104 Kahle, "Schätze", 348–9.

105 Al-Tanūkhī, *Nishwār al-Muḥāḍarah wa akhbār al-mudhākarah*, ed. ʿAbbūd al-Shāljī, 9 vols. ([Beirut: Dār Ṣādir], 1971), 1.289; see also 3.59.

106 See Pelliot 110, Barthold, 237, and Bloom 42–5.

107 Bloom 45.

108 *Kitāb al-Tabaṣṣur* 28.

109 See also Ibn al-Faqīh 62.

These were very likely decorated with elaborate textiles.[110] Chinese lacquerware also seems to have been known, designated as *madhūn*.[111]

Ibn al-Faqīh, as quoted by al-Qazwīnī, mentions that the Chinese have "the manufactured metal (*ḥadīd*, lit. iron) which is called *ṭālīqūn* that is sold for its amalgamation with silver (*yushtarā bi-iḍaʿāfihi fiḍḍatin*)."[112] The exact identity of *ṭālīqūn* is obscure, but it has been lately suggested to have been a mercury amalgam.[113] Other sources on *ṭālīqūn* do not link it with China;[114] although another metal, *khāraṣīnī*, often traced to China, became confused with it. In modern Arabic *khāraṣīnī* means zinc, although this is not certain for medieval times. The metal was best known as the material that constituted Chinese mirrors, which were imported into the Middle East.[115] These were made of high tin bronze.[116] In any case, the signet ring of the caliphs is said to have been made of *ḥadīd ṣīnī* or Chinese iron,[117] and this indicates its status as a precious metal.

China was also known for asbestos,[118] although asbestos is found in the Mediterranean world and Central Asia, and was introduced to China through contact with the West.[119] This mineral was understood by the Arabs as the skin or feathers of an animal called *al-samandal*, "salamander." This is not entirely surprising given the strange nature of this fibrous mineral. The concept of the salamander as an intensely cold animal that can survive fire was borrowed from the Classical world, and the Arabs confused it with the phoenix. Thus in Arabic literature the samandal is sometimes said to be a bird, or a small mammal. Even though asbestos was known (as a mineral, at that) in the Graeco-Roman

110 There is an example of Central Asian cloth cut into the forms of a saddle cover: K. Otavsky, "Stoffe von der Seidenstrasse: Eine neue Sammlungsgruppe in der Abegg-Stiftung," in Otavsky, K., ed., *Entlang der Seidenstrasse. Frühmittelalterliche Kunst zwischen Persien und China in der Abegg-Stiftung* (Riggisberg, 1998), 13–17.

111 Kahle, "Schätze" 343 n. 4.

112 Ibn al-Faqīh 1996, 70.

113 B. Fehér, "Mysterious Alloys in Early Muslim Metallurgy: On the Ṭālīqūn and the Haft-Ǧūš," *The Arabist* 23 (2001): 55–63.

114 Fehér cites accounts from al-Bīrūnī and al-Qazwīnī, neither of which link it with China.

115 A Tang mirror was found at Susa, see R. Ghirshman, "Un miroir T'ang de Suse," *Artibus Asiae* 19 (1956): 230–3; see also J. W. Allan, *Persian Metal Technology 700–1300 AD* (London: Ithaca Press, 1979), 50.

116 Allan 48–52.

117 Ibn Miskawayh, *Kitāb tajārib al-umam*, 3 vols. (Baghdad: al-Muthanna, n.d.), 1.290.

118 Ibn al-Faqīh 1996, 70.

119 B. Laufer, "Asbestos and Salamander: An Essay in Chinese and Hellenistic Folk-Lore," *T'oung Pao* 16 (1915): 299–373, Schafer 199–200.

world, it seems that in the medieval Middle East, it was especially associated with the wonders of the East. The *Kitāb ʿAjāʾib al-Hind* tells of the samandal bird of Wāqwāq (Japan)[120] that can enter a conflagration without catching on fire.[121] Samandal "bird skin" was sent as a gift by the Pāla king Dharmapāla to al-Maʾmūn along with aloeswood, camphor, and other riches.[122]

The geographer al-Idrīsī (1100–c. 1162) gives a long list of the commodities "of China" traded in Aden. He notes that these were brought by the ships of Sind, Hind, and China, and most are not, in fact, products of China, but of different parts of the entire Indian Ocean region, especially from India and Southeast Asia. His list includes iron, silk brocade (*firind*), gold brocade (*kīmkhat*),[123] musk, aloeswood, saddles, porcelain (*ghaḍār*), pepper, long pepper (*al-dār al-fulful*), coconuts (*nārjīl*), *harnuwah*,[124] cardamom (*qāqullah*), cinnamon, galangal, mace, myrobalans (*ihlīlajāt*), ebony, mother of pearl (*dhabl*), camphor, nutmeg, cloves, cubebs, garments made from plants (*al-thiyāb al-muttakhadhah min al-ḥashīsh*) and great velvet garments (*al-thiyāb al-ʿaẓīmah al-makhmal*), elephant tusks, tin (*raṣāṣ qalaʿī*), and other things including bamboo canes and more commodities.[125] It is a dramatic illustration of the scope of Middle Eastern trade with Further Asia. Many of these commodities will be discussed below in the consideration of the contributions of India and Southeast Asia to the Middle Eastern trade with Further Asia. Fifteen of these items are drugs, spices, and aromatics, out of a total of twenty-eight. This is by far the largest category of commodities; specified types of textiles are only four items, and there are two manufactured goods, the saddles and porcelain, two

120 和國 *wa-kʷak* > *kok*(*u*); Christopher Beckwith, personal communication.

121 Buzūrg b. Shāhriyār, *Kitāb ʿAjāʾib al-Hind: Livre des merveilles de l'Inde*, ed. P. A. Van der Lith and trans. L. M. Devic (Leiden: Brill, 1883–6; repr. Tehran: M. H. Asadi's Historical Series, 1966), 172.

122 Abū Bakr Muḥammad al-Khālidī and Abū ʿUthmān b. Hāshim al-Khālidī (called al-Khālidayn), *Kitāb al-Tuḥaf wa-l-hadāyā* (Cairo: Dār al-Maʿārif, 1956), 159–63.

123 This is the spelling given by Idrīsī; it is less correct than the *kīmkhāw* of Ibn Khurradādhbih.

124 Commonly identified with the seed of the aloeswood tree by many medieval texts; it was some sort of pungent seed. Levey, "Ibn Masawaih" 405, calls it "Indian pepper", as does Dietrich Ibn Juljul #10 on 40 ("Indischer Pfeffer") with many references to the relevant Arabic texts, Lev and Amar 555 have "Guinea pepper", which they identify with *Capsicum minimum* (= *C. annuum*), which is the species of the bell and chili peppers of the Americas. By Guinea pepper one might expect such West African plants as *Piper guineense* (Guinea pepper, false cubeb, or Ashanti pepper) or *Aframomum melegueta* (grains of paradise or melegueta pepper, also called Guinea pepper). Also see Schmucker 519–20.

125 Al-Idrīsī, *Kitāb Nuzhat al-mushtāq fī ikhtirāq al-āfāq*, 9 vols. (Naples: Istituto Universitario Orientale di Napoli, 1970–84), 1.54.

metals, and five miscellaneous commodities, including such raw materials as ebony, ivory elephant tusks, mother of pearl, bamboo, and coconuts.

India and the Indian Ocean

The early medieval Arabs knew two designations for India: Sind, the land around the Indus River, now Pakistan, was territory the Arabs had conquered. Hind was the land beyond that, and it included not only India, but what we understand as Southeast Asia.[126] Hind was also used by extension for the entire territory of the East, including Sind and even Makrān, but excluding China and the Turks. During the early medieval period Indian influence was extensive in the lands of Southeast Asia, and to a casual observer it must have seemed that they were indeed lands of India. So perhaps it is artificial to divide the commodities of the Indian subcontinent from Southeast Asia, for Islamic geography certainly did not make such a division, but it reflects the geographical origin of the goods, and the more extensive expeditions necessary to acquire the goods of the Indonesian Archipelago at the source.

Like China and Central Asia, India exported manufactured goods alongside natural products. India exported so many different natural products, though, that they tend to obscure the important manufactured goods. A great variety of natural products were imported from India: animal, vegetable, and mineral. The *Kitāb al-Tabaṣṣur bi-l-tijārah* lists tigers, leopards, elephants, leopard pelts, rubies, white sandalwood, ebony, and Indian nuts (presumably coconuts).[127] These are but a few of the imports from India, and many well known commodities are not included.

The *Ḥudūd al-ʿālam* begins its list of the products of India with perfumes, and specifies musk first of all.[128] Musk was not produced in India, but was transshipped there from the Himalayas, as will be discussed below. The author of the *Ḥudūd al-ʿālam* also counts among the notable products of India elephants, rhinoceroses (*karg*),[129] peacocks, *k.r.k.rī* (unknown, presumably a

126 A. Wink, *Al-Hind: the making of the Indo-Islamic world. Volume 1. Early Medieval India and the Expansion of Islam, 7th–11th Centuries* (Leiden: Brill, 1996; repr. 2002).

127 *Kitāb al-Tabaṣṣur* 25–6. Cf. also Ibn Khurradādhbih 70 for elephants.

128 *Ḥudūd al-ʿālam* §10.

129 Cf. also *Ḥudūd al-ʿālam* §10.1 which notes that rhinoceroses are abundant in Qāmarūn (Assam). See also al-Thaʿālibī 214 and al-Bīrūnī, *Kitāb al-Bīrūnī fī taḥqīq mā li-l-Hind*, ed. E. Sachau (Haydarābād: Dāʾirat al-Maʿārif, 1958), 163.

bird),[130] parrots, and cuckoos (*shārak*).[131] The animals of India always elicited special interest.[132] We have already noted tigers, leopards, elephants, and leopard pelts; Arabic *namr* "leopard" can probably also denote similar big cats such as the snow leopard. The pelts from all of these big cats were very valuable, and this has contributed to the destruction of many of the animals. Live parrots were especially fascinating because of their ability to "talk". The white gerfalcons and black goshawks of India were also esteemed as hunting birds in the Middle East.[133] In addition to the animal products ivory,[134] horn, and feathers, India also exported honey.[135] Lac was an important dyestuff, derived from the lac insect.[136]

There were also maritime products of animal origin from India. Pearl diving occurred in the waters around Sri Lanka,[137] although the most famous pearls in the Middle East came from the waters around the Arabian Peninsula. White conch shells (*sapīd muhra*) are mentioned by the *Ḥudūd al-ʿālam*, which also gives the Indian name *sanbak*, restored by Minorsky to **shank*, corresponding to Sanskrit *śaṅkha*. Conch shells were blown like trumpets (*būq*), and indeed this corresponds to the practice mentioned in Indian epic poetry.[138] *Shank* in the meaning of a conch blown as an instrument is attested in the so-called *Akhbār al-Ṣīn wa-l-Hind* of 851.[139]

---

130 Perhaps *kuraṅkara*, the Indian crane *Ardea sibirica*?

131 Many of these animals are also linked to India (and thus Southeast Asia as well) by al-Jāḥiẓ, *Kitāb al-Ḥayawān*, ed. ʿAbd al-Salām Muḥammad Hārūn, 8 vols. (Cairo: Muṣṭafā al-Bābī al-Ḥalabī, 1966), 7.170. He also mentions the tiger (*babr*) and the Sindī chicken; I do not know what the latter is, but it is presumably the same as the Hindī chicken in al-Thaʿālibī 214; he has a similar list of animals that may be compared to the lists of al-Jāḥiẓ and the *Ḥudūd al-ʿālam*. See also the references in M. M. Ahsan, *Social Life under the Abbasids 170–289 AH, 786–902 AD* (London: Longman, 1979), 79 n. 18.

132 Bīrūnī, *Hind* 162–4.

133 *Kitāb al-Tabaṣṣur* 35. See also Ahsan 216–20.

134 E.g., Thaʿālibī 215.

135 *Ḥudūd al-ʿālam* §10.39.

136 It is attested many times in the Cairo Geniza documents, see the index to S. D. Goitein and M. A. Friedman, *India Traders of the Middle Ages: Documents from the Cairo Geniza: "India Book"* (Leiden: Brill, 2008), for examples. R. A. Donkin, "The Insect Dyes of Western and West-Central Asia," *Anthropos* 72 (1977): 864–5.

137 Ibn Khurradādhbih 64; al-Sīrāfī 80; R. A. Donkin, *Beyond Price: Pearls and Pearl-Fishing: Origins to the Age of Discoveries* (Philadelphia: American Philosophical Society, 1998), 129–31.

138 *Ḥudūd al-ʿālam* §10.7.

139 Sīrāfī 19–20.

Ambergris is a famous product of the Indian Ocean; it is derived from a secretion of the digestive system of the sperm whale.[140] Ambergris is the second most important aromatic in early medieval Islamicate perfumery after musk.[141] Ambergris is still greatly valued in luxury perfumery as it makes an unparalleled enhancer to scents, fixing them with a velvety richness that cannot be obtained by any other means. Yet it is rarely used because the substance is rare and the source of supply is literally dependent on the winds and tides. In addition, use of ambergris is illegal in some countries because the sperm whale is endangered. Synthetic alternatives are usually employed today. Most ambergris is recovered from beaches where it has washed ashore after being excreted by the whale; in the days of whaling some was taken from animals that were slaughtered, but not all sperm whales produce ambergris.

Ambergris is also a problematic case in the study of Islamicate aromatics. Starting in Umayyad times, Arabic poetry mentions *ʿanbar al-hind* "Indian ambergris".[142] Yet the use of ambergris is poorly attested in India. Pliny the Elder claims that the Indians had something that is believed to be ambergris that was valued more than frankincense by them, and Nicias, quoted by him, refers to amber arising from the sea there,[143] but no reflection of this is

140 S. Arctander, *Perfume and Flavor Materials of Natural Origin* (Elisabeth, N.J.: privately published, 1960), 55–8; K. Yamada, *A Short History of Ambergris by the Arabs and Chinese in the Indian Ocean*, 2 parts (Osaka: Kinki University, 1955–6); K. H. Dannenfeldt, "Ambergris: The Search for its Origin," *Isis* 73:3 (1982): 382–97; P. Borschberg, "Der asiatische Ambra-Handel während der frühen Neuzeit (15. bis 18. Jahrhundert)," in J. M. dos Santos Alves, et al., eds., *Mirabilia Asiatica: Productos raros no comércio marítimo*, vol. 2 (Wiesbaden: Harrassowitz, 2005), 167–201. For ambergris in China, see J. F. So, "Scented Trails: Amber as Aromatic in Medieval China," *JRAS* (2013): 85–101.

141 J. Ruska and M. Plessner, "ʿAnbar," *EI*² s.v.; al-Kindī [attr.], *Kitāb Kīmiyāʾ al-ʿIṭr wa-t-Taṣʿīdāt: Buch über die Chemie des Parfüms und die Destillationen. Ein Beitrag zur Geschichte der arabischen Parfümchemie und Drogenkunde aus dem 9. Jahrh. P.C.*, ed. and trans. with commentary by K. Garbers, (Leipzig, 1948), #5 on 168–72 and Schönig 47–50; E. Lev and A. Amar, *Practical Materia Medica of the Medieval Eastern Mediterranean According to the Cairo Genizah* (Leiden: Brill, 2008), 331–3. Newid 12–60 has an overview focused especially on Persian literature.

142 E.g., al-ʿArjī (d. 738) quoted by al-Sarī al-Raffāʾ, *Al-muḥibb wa-l-maḥbūb wa-l-mashmūm wa-l-mashrūb*, ed. Miṣbāḥ Ghalāwinjī, 4 vols. (Damascus: Majmaʿ al-Lughah al-ʿArabiyyah, 1986–7), 3.173 (#326).

143 Pliny the Elder, *Natural History*, ed. and trans. H. Rackham, et al. 10 vols. (Loeb Classical Library. Cambridge, MA: Harvard University Press, 1962), XXXVII.XI.36, and J. André and J. Filliozat, *L'Inde vue de Rome: Textes latins de l'Antiquité relatifs à l'Inde* (Paris: Les Belles Lettres, 1986), 108–11 and n. 211 on 369–70. See also on Greek and Latin references to ambergris in A. Dalby, "Some Byzantine Aromatics," in L. Brubaker and K. Linardou,

found in indigenous sources. The use of ambergris in India appears to begin in the early medieval period, and even then the word used for the substance is derived from the Near East: Sanskrit *ambara*, from either Pahlavi *ambar* or directly from Arabic *ʿanbar*.[144] While there are unclear words in early Sanskrit literature on aromatics, it appears that none of these were linked by later writers with *ambara*. By 1300, however, another series of words comes to denote ambergris: *agnijāra, agnijā, agnigarbha*, and *agniniryāsa*; none seem to be attested for the early medieval period or earlier.[145] Thus it seems that the use of ambergris in India was promoted by the Arabs and the trade of early medieval times.

In this situation, the explanation of the "Indian ambergris" of Arabic literature must be that it refers to ambergris derived from the Indian Ocean (*baḥr al-hind*) and brought through the India trade. Southern Iraq is often called *arḍ al-hind* "the land of India" in Arabic literature supposedly because there was a sizable Indian population there in Sasanian times.[146] It was also known as *farj al-hind* "the opening of India" because ships set sail from there.[147] While the region of Makrān, now southern Iran and western Pakistan, was considered the frontier of India and almost a part of it along with Sind and Hind and the lands beyond,[148] the entire sea was associated with India.[149] Arabic accounts of Indian Ocean trade point out how ambergris could be obtained on various islands in the Indian Ocean. The ambergris obtained from al-Shiḥr on the southern coast of Arabia was especially regarded, but ambergris could

eds., *Eat, Drink, and be Merry (Luke 12:19)- Food and Wine in Byantium* (Aldershot: Ashgate, 2007), 51–3.

144 P. K. Gode, "History of Ambergris in India Between about A.D. 700 and 1900," in Gode, *Studies in Indian Cultural History Volume 1* (Hoshiarpur: Vishveshvaranand Vedic Research Institute, 1961), 9–14 and J. McHugh, *Sandalwood and Carrion: Smell in Indian Religion and Culture* (Oxford: Oxford University Press, 2012), 165. For the Pahlavi, see D. N. Mackenzie, *A Concise Pahlavi Dictionary* (London: Oxford University Press, 1971), 8.

145 *Dhanvantarinighaṇṭu; Rājanighaṇṭu*; cf. André and Filliozat 370 n. 211. Gode does not mention any of these later sources in his paper on ambergris in India.

146 E.g., Ṭabarī I.2378; see J. C. Wilkinson, "Arab-Persian Land Relationships in Late Sasanid Oman," *PSAS* 6 (1972): 41; Crone 47 n. 155 and J.-F. Salles, "Fines Indiae, Ardh el-Hind: Recherches sur le devenir de la mer Érythrée," in *The Roman and Byzantine Army in the East, Proceedings of a colloqium* [sic] *held at the Jagiellonian University, Krakow, in September 1992* (Krakow: Jagiellonian University, 1994), 182–4.

147 E.g., Ṭabarī I. 2223.

148 E.g., Wink 132–3.

149 Cf. Ṭabarī 1.2023 and the above references.

come from any coast, from Africa to Southeast Asia being the best known to the Arabs.

India was especially famous for its numerous spices and medicinal and aromatic plants.[150] Many of these transcended the obscurity of the pharmacopoeias by their renown and became familiar in daily life in medicine, perfume, or cookery. As will be discussed below, it was believed that Adam took with him leaves from the Garden after the expulsion, and when he descended to earth in Sri Lanka these leaves became the origin of the sweet smelling herbs of India.

Spikenard, *sunbul*, was often called *sunbul al-hind*[151] to distinguish it from varieties of valerian found elsewhere that were also designated *sunbul*. Ibn Māsawayh notes that it is an herb brought from India and Tibet.[152] There are several plants that are called spikenard even today: *Nardostachys jatamansi* (synonym *N. grandiflora*) is the most famous, and it is found in the Himalayas from Uttar Pradesh into western China.[153] The root of this herbaceous perennial was greatly valued as a perfume; scented oils were prepared from it, and it was used in incense. Its root is often confused with *Valeriana jatamansi*, which is more widely spread in the same region and has a similar scent. Both were likely designated as *sunbul al-hind* by the Arabs. Costus was exported from Sind; the Indian variety, *Saussurea costus*, grows in the western Himalayas, and there are relatives extending throughout the high country.[154] Like spikenard, it was valued for its root, which has a distinctive pungent scent reminiscent of orris and violets. In perfumery it is a superb fixative, and it is also used as an incense.

Aloeswood, also called agarwood, eagle wood, and agallocha, is the diseased wood of trees of various species of the genus *Aquilaria*, such as *A. agallocha*,

---

150 E.g., *Ḥudūd al-ʿālam* §10.

151 Al-Bīrūnī, *Kitāb al-Ṣaydanah*, ed. H. M. Saʿīd (Karachi: Hamdard National Foundation, 1973), 237, ed. ʿAbbās Zaryāb (Tehran: Markaz-i Nashr-i Dānishgāhī, 1991), 351–2, and Dietrich DT I.6.

152 Ibn Māsawayh 16, Cf. also al-Thaʿālibī 215.

153 W. Dymock, *Pharmacographia Indica: A History of the Principal Drugs of Vegetable Origin met with in British India* (London: Kegan Paul, Tench & Trübner, 1890–3; repr. in one volume, Karachi: Hamdard, 1972), 2.233–8; E. Brucker, "AI. Nálada= Nardostachys Jatamansi DC. Ein Beitrag zur indischen Pflanzenkunde," *Asiatische Studien* 29 (1975), 131–6; O. Polunin and A. Stainton, *Flowers of the Himalaya* (Oxford: Oxford University Press, 2000), 176–7.

154 Ibn Khurradādhbih 62; see also Garbers #94 on 308–11, Maimonides #338, Schmucker 346–7, Dietrich DT I.13, J. Greppin, "Gk. κόστος: A Fragrant Plant and its Eastern Origin," *Journal of Indo-European Studies* 27 (1999): 395–408; Polunin and Stainton 207, Lev and Amar 157–8.

which grow in South and Southeast Asia.[155] The tree responds to injury or disease by producing an oleo-resin that makes the wood heavy, dark, and aromatic. Aloeswood was of immense importance as an incense in medieval Islamicate perfumery. It retains an immense importance for incense in the Middle East and East Asia today.[156] Aloeswood was generally described as an Indian product in the Islamicate literature,[157] but most of this was transshipped from the eastern parts of India and Southeast Asia. Likewise, the camphor of India in the *Ḥudūd al-ʿālam* must have been transshipped from further east. The *Ḥudūd al-ʿālam* mentions that aloeswood came from Qāmarūn, which is Assam in eastern India and from Mandal.[158] The aloeswood of Cambodia (*Qimār*), mentioned as early as Ibn Māsawayh, is also noted by the *Ḥudūd al-ʿālam*.[159]

It is said that aloeswood was in such short supply in the time of the caliph al-Mutawakkil (r. 847–61) that he sent an envoy to India with presents to request a supply. The envoy, who was apparently the famous courtier Ibn Ḥamdūn al-Nadīm, found the king of India in Lahuwār (Lahore?) to be most hospitable, for he had been raised in exile in Oman and was thus conversant with Arabic as well as Indian culture. His court was rich with imported goods from the Middle East and Iran, including perfumes. "Before him were vessels of gold and silver, and many Iraqi metalworks, all beautiful and filled with camphor, rosewater, ambergris, the compound perfume called *nadd*, and figurines."[160] The *nadd* was likely an import from the Middle East rather than an Indian incense, given the context, and the figurines would have been made from aromatics such as ambergris. At the end of Ibn Ḥamdūn's extended stay the king, whose friend he had become, burned for him a rare aloeswood that was a sliver smaller than half a *dāniq*. It had an indescribably pleasant scent, and it lingered in cloth after repeated washing with soap. Apparently this rare scent was the stuff of legends, for the wood grew in an inaccessible place and only could be had by shooting

155 Arctander 49–50 .

156 A. Dietrich and C. E. Bosworth, "ʿŪd," in *EI*² s.v.; Garbers #122 on 360–4; I. H. Burkill, *A Dictionary of the Economic Products of the Malay Peninsula*, 2 vols. (Kuala Lumpur: Ministry of Agriculture, 1966), 198–206; Schmucker 307–8; Schönig 285–90, Lev and Amar 97–8; D. Jung, "The Cultural Biography of Agarwood- Perfumery in Eastern Asia and the Asian Neighbourhood," *JRAS* (2013): 103–25.

157 Aloeswood types are discussed in the standard texts on aromatics by Ibn Māsawayh, al-Yaʿqūbī, al-Tamīmī, and the Ibn Mandawayh-Ibn Kaysān source. For the general idea of India as a source of aloeswood, see *Ḥudūd al-ʿālam* §10.

158 The *Kitāb al-Tabaṣṣur*, 16, praises Mandalī aloeswood as the best, as does Ibn Māsawayh, see below.

159 Ibn Māsawayh 13; *Ḥudūd al-ʿālam* §10.6.

160 Tanūkhī 3.106. On *nadd*, see below 277–8.

deer that happened to have been there and sometimes were found chewing on bits of it, hence its extreme rarity![161] Al-Mutawakkil received a half of a *raṭl*[162] of it and deemed it worth the expedition after he had censed with it.

Today sandalwood is perhaps a more familiar aromatic in the west than aloeswood. There are numerous varieties of woods called "sandalwood" in English, and the situation is similar for medieval times.[163] The most famous and characteristic is the white sandalwood from *Santalum album*. Sandalwood was known from the 6th century in the Graeco-Roman world.[164] It was well known as an import from India.[165] Sandalwood was often used in incense and in other aromatic preparations, but its fame was less than aloeswood. The confusion between cinnamon and cassia, and the difficulties of the ancient and medieval terminology for them, has been mentioned above. Both were, of course, imports from India in the Islamic age.[166]

One of the most important dyes of the pre-modern world was indigo (*nīl*); it was famed as a product of India.[167] Indigo had been imported by Graeco-Roman times and was used as a medicine and dye as well as a textile dye in later Roman times.[168] Indigo cultivation was practiced in various parts of the Islamic world, including Egypt, Palestine, and southern Arabia, but indigo imported from India was a necessary supplement to the quantities of the dyestuff required in medieval Middle Eastern society. Indian indigo was also famed for its quality. Along with the brazil wood of Southeast Asia, indigo was the most important dyestuff of the medieval Islamic world as seen in commercial documents.[169]

161 Tanūkhī 3.113.

162 See W. Hinz, *Islamische Masse und Gewichte* (Leiden: Brill, 1955), on the different weight of the *raṭl* as well as other weights according to place and time.

163 A handy overview is in McHugh, *Sandalwood and Carrion* 182–5.

164 A. Dietrich, "Ṣandal," *EI*² s.v., Garbers #113 on 342–5, Schmucker 282–3, Dietrich, *Ibn Ǧulǧul* #38 on 53–4, Schönig 261–4, Lev and Amar 476–7, and A. Dalby, "Some Byzantine Aromatics" 56–7, King, "New *Materia Medica*" 514.

165 *Kitāb al-Tabaṣṣur* 25–6; *Ḥudūd al-ʿālam* §10.46.

166 Garbers #92, Schmucker 341–2, Lev and Amar 143–6.

167 Ibn al-Faqīh 1996, 72. On indigo, see J. Balfour-Paul, *Indigo in the Arab World* (Richmond: Curzon Press, 1997) and, concisely, A. Dietrich, "Nīl," *EI* 2nd ed. s.v.

168 Balfour-Paul 11–12.

169 Goitein and Friedmann index s.v. "Indigo", S. D. Goitein, *Studies in Islamic History and Institutions* (Leiden: Brill, 1966), 322, and N. A. Stillman, "The Eleventh Century Merchant House of Ibn ʿAwkal (A Geniza Study)," *JESHO* 16 (1973): 38–9.

The trade in pepper goes back into antiquity; pepper was just as popular in Arab and Persian cuisine as it had been among the Greeks and Romans.[170] Pepper was well known as a product of India in Arabic and Persian literature;[171] the spread of the cultivation of pepper throughout Southeast Asia is a phenomenon of more recent history. The English word "pepper" has come to denote many different plants from around the world, notably the *Capsicums* of the Americas in addition to the Old World peppers. In medieval Islamicate culture there were two important kinds: black pepper (*Piper nigrum*) was called *filfil* or *fulful*, from southern India and long pepper (*Piper longum*), from the north. The white, pink, and green peppercorns found in today's markets are also derived from the *Piper nigrum* gathered at different times and subjected to varied treatments. Pepper was known from India, especially the southwest, known as the Land of Pepper,[172] and from Sri Lanka.[173] Long pepper has a more intense peppery quality, and it was the most expensive in Graeco-Roman times but has fallen out of use today outside of Asia. Long pepper root (*fulfulmūl*) was also used in Islamic times.[174]

Cardamom was also imported from India and Southeast Asia.[175] True cardamom or green cardamom is the seed of *Elettaria cardamomum*, but species of *Amomum* such as *A. subulatum* also produce cardamom. In modern parlance, *Amomum* cardamom is black cardamom or greater cardamom. Islamicate sources also distinguish a greater cardamom (*qāqullah kabīr* or just *qāqullah*) and lesser or smaller cardamom (*hāl* or *hālbuwwā*), but the simple identification of the medieval greater and lesser cardamoms with the modern greater and lesser cardamoms is doubtful. The camphoraceous scent of the cardamoms made them an important part of pharmacology, cuisine, and

170 For pepper and its different varieties, see Dymock 3.166–83; Burkill 2.1766–84; A. Dalby, *Dangerous Tastes: The Story of Spices* (Berkeley: University of California Press, 2000), 89–94.

171 Bīrūnī, *Ṣaydanah*, ed. Saʿīd 29–3; ed. Zaryāb, 465–7; *Ḥudūd al-ʿālam* §10.12, 13, 14. See also I. Löw, *Die Flora der Juden*, 4 vols. (Wien, 1928–34; repr. Hildesheim: Olms, 1967), 3.49–61, Stillman 44–5, and Dietrich DT II.143, with many references.

172 *Akhbār al-Ṣīn wa-l-Hind*, ed. and trans. J. Sauvaget (Paris: Les Belles Lettres, 1948), §56; Sīrāfī 48.

173 Ibn Khurradādhbih 64; some of the other products in this list are not found in Sri Lanka, though.

174 Bīrūnī, *Ṣaydanah*, ed. Saʿīd 294 and ed. Zaryāb 468.

175 On the problems of the identification of the various cardamoms and their history, see Dymock 3.428–37, Garbers #89, Maimonides #116 and #325, Dietrich Drogenhandel #1 on 26–9, Schmucker 518, Dietrich DT III.125 n. 5, Ibn Juljul #9, Dalby 102–6, and Lev and Amar 125–6.

perfumery. They appear widely in perfume formulas; sometimes both *qāqullah* and *hālbuwwā* are called for in the same recipe.[176]

Many drugs came to be imported from India; some were known in Graeco-Roman antiquity, while others were new. Here, a few are mentioned that gained some fame beyond the pharmacopoeias. Medieval Asian pharmacology knew several myrobalans, but three were the most famous: the chebulic (*ihlīlaj*), beleric (*balīlaj*), and emblic (*amlaj*).[177] They were imported from India;[178] all had an extensive medical use there and came to be widely used in Islamicate medicine as well. None were known to the ancient Greeks and Romans. Among the numerous other drugs imported from India were Purging Cassia, also called Indian Laburnum (*khiyār shanbar, Cassia fistula*)[179] and zedoary (*zurunbād, Curcuma zedoaria*).[180]

Tamarind (*Tamarindus indica*) is an Arabic word: *tamr hindī*, which means the Indian date. Tamarind was unknown in Classical times, and its introduction occurred under Islam.[181] It is mentioned as early as the 9th c. by physician ʿAlī b. Rabban al-Ṭabarī.[182] The *Ḥudūd al-ʿālam* mentions the production of tamarind in India.[183]

India, then as now, produced rice that was in demand in the Islamic world, although riziculture was rapidly developing in the Near East.[184] The coconut palm (*nārjīl* or *jawz hindī*) originated in southeast Asia but had spread into India in antiquity.[185] The Arabs considered it an Indian plant, and they noted that it came from Sri Lanka[186] as well as India. Abū Zayd al-Sīrāfī paints a picture of a thriving trade out of Oman to Indian islands seeking the coconut

176 E.g., al-Tamīmī's toilet water formula quoted by al-Nuwayrī 12.121.

177 A. Wayman, "Notes on the Three Myrobalans," *Phi Theta Annual* 5 (1954–5): 63–77 and Schafer, *Golden Peaches* 145–6.

178 *Ḥudūd al-ʿālam* §10.38.

179 Bīrūnī, *Ṣaydanah*, ed. Saʿīd 173 and ed. Zaryāb 240; *Ḥudūd al-ʿālam* §10.34; King, "New *materia medica*" 505.

180 Ibn Juljul #13; King, "New *materia medica*" 506.

181 Ibn Juljul #5; King, "New *materia medica*" 504.

182 ʿAlī b. Rabban al-Ṭabarī, *Firdaws al-Ḥikmah fī al-ṭibb*, ed. M. Z. Siddiqi (Berlin: Sonne, 1928), 382.19; W. Schmucker, *Die pflanzliche und mineralische Materia Medica im Firdaus al-Ḥikma des Ṭabarī* (Bonn: Selbstverlag des Orientalischen Seminars, 1969), 131–2.

183 At *jāb.rs.rī*, an unknown place 10.34.

184 Ibn Khurradādhbih 63; Watson 15–19; Decker 194–7.

185 Watson 55–7; King, "New *materia medica*" 507.

186 Ibn Khurradādhbih 64.

palm for use in shipbuilding and for the coconuts themselves.[187] The *Ḥudūd al-ʿālam* also mentions coconuts as a product of India.[188]

Many fruits of India came to be cultivated in the Middle East in early medieval Islamic times. Of the numerous varieties of citrus, only the citron was known in classical antiquity; it diffused into the Roman Empire from the Near East although it perhaps was only successfully grown in the west by late antiquity.[189] During the early medieval age a revolution in citrus occurred, and the cultivation of several important plants spread out of India into the Middle East. Among these were the sour orange, lemon, and lime.[190] While the banana became known in the classical world thanks to the expedition of Alexander, it was only in Islamic times that the banana became common in the Middle East, even if it may have reached it in earlier centuries.[191]

The spread of sugar was one of the most important consequences of the Islamic expansion with regard to foodways. Sugar cane may have been cultivated in late Sasanian times, but it was only in the Islamic period that sugar began to become a basic ingredient of food.[192] Even so, the sugar of India was still well known. Rock candy (*panīdh*)[193] and sugarcane (*nay shakar*) were specialties.[194]

Besides these herbs, spices, and aromatics, wood was imported from India in larger quantities. Teak (*sāj*)[195] was brought from India for use in shipbuilding and construction.[196] Arabia has few sources of wood, and so teak along with coconut wood and fiber used for rope imported from India were critical to Arabian shipping.[197] Teak beams were used in construction in the Mosque in Meccah,[198] as well as in the building of Baghdād under al-Manṣūr.[199] Al-Bīrūnī noted that teak grew in India and the land of the Zanj: East Africa.[200]

187 Sīrāfī 85–6 and see below.

188 *Ḥudūd al-ʿālam* §10.14, 39 .

189 Watson 44; King, "New *materia medica*" 514.

190 Watson 45–50.

191 Watson 51.

192 Watson 25–30; Ahsan 100–2, Dietrich DT II.65.

193 *Ḥudūd al-ʿālam* §10.39; Ibn Ḥawqal 325.

194 *Ḥudūd al-ʿālam* §10.52.

195 The name is ultimately derived from Indic *sāka*. On teak, Ibn Juljul #40; Ibn al-Bayṭār 3.2; King, "New *materia medica*" 510.

196 Ibn Khurradādhbih 67.

197 Hourani 89–91.

198 Ṭabarī III.989.

199 Ṭabarī III.319.

200 Bīrūnī, *Ṣaydanah*, ed. Saʿīd 216 and ed. Zaryāb 327.

Bamboo was also brought from India;[201] it eventually came to be cultivated in the Middle East. It was used, for example, in making the shafts of lances and spears.[202] Domestic items including stools, chests, and censers could also be made from bamboo.[203] The *ṭabāshīr* (Eng. tabasheer), crystalized sap, found within bamboo had many applications in Islamicate medicine.[204] Ebony (*Diospyros ebenum*) has already been noted as a famous import from India.[205]

India was especially famed as a land of precious stones; this reputation stretches back into antiquity. The *Ḥudūd al-ʿālam* lists jewels (*gawhar*),[206] pearls (*marwārīd*), corundum (*yāqūt*), diamonds, coral, and pearls (*durr*).[207] *Yāqūt*, like the English corundum, can denote different members of this family. Red *yāqūt* is thus ruby, and yellow and blue would be sapphire.[208] The *Kitāb al-Tabaṣṣur bi-l-Tijārah* mentions that the best corundum is the red *bahramānī*.[209] The standard name for the blue sapphire is *al-yāqūt al-asmānjūnī*, which is Persian "sky-colored *yāqūt*". The source for *yāqūt* was Sri Lanka.[210] Sri Lanka was an especially famous source of precious stones.[211] *Ballūr*, beryl, came from Sri Lanka,[212] along with diamond (*almās*).[213] Less glamorous than these precious minerals was whetstone (*sunbādhaj*) "with which gems are polished"; it came from Sri Lanka, India, and Nubia.[214] India also exported tutty (Ar. *tūtiyā*) which had various applications in metalworking and medicine.[215]

The *Ḥudūd al-ʿālam* mentions the production of gold of Qāmarūn (Assam).[216] Yet the Islamic world seems to have contributed far more gold and silver to India than it ever would have received, for there was an imbalance of trade in

201 *Ḥudūd al-ʿālam* 10.14; Ibn Khurradādhbih 62.

202 E.g., Ṭabarī III.586.

203 S. D. Goitein, *A Mediterranean Society: The Jewish Communities of the Arab World as Portrayed in the Documents of the Cairo Geniza*, 6 vols. (Berkeley: University of California Press, 1967–93; repr. 1999), 4.138, 386, 388–9.

204 Ibn al-Bayṭār 3.94–5, Ibn Juljul #18.

205 See also Ibn al-Bayṭār 1.8.

206 *Ḥudūd al-ʿālam* §10.26 .

207 *Ḥudūd al-ʿālam* §10.

208 Numerous colors are described in Bīrūnī, *Jamāhir* 74–88.

209 On this term, see Bīrūnī, *Jamāhir* 35.

210 *Kitāb al-Tabaṣṣur* 13; Bīrūnī, *Jamāhir* 38, 43–4; Ibn al-Faqīh 1996, 72.

211 Sīrāfī 80–1.

212 Ibn Khurradādhbih 64.

213 Ibn al-Faqīh 1996, 72.

214 Ibn Khurradādhbih 64; Bīrūnī, *Jamāhir* 102–4.

215 E.g., Thaʿālibī 215; Ibn al-Bayṭār 1.143.

216 *Ḥudūd al-ʿālam* §10.1.

the commodities of India versus the commodities of the Middle East exported there.[217] This trade imbalance was made up with precious metals. However, Indian metalworking was well known and significant. Some of the most famous swords in the Islamic world were imported from India; they are celebrated in poetry.[218] Swords made in the Indian fashion (*muhannad*) in the Middle East ranked highly as well.[219]

As in modern times, India exported textiles. Ibn Khurradādhbih notes among the exports of India garments made from "grass" (*ḥashīsh*)—probably hemp—and garments of cotton velvet.[220] The *Ḥudūd al-ʿālam* also lists garments (*jāma*). These are sometimes specified as velvets (*makhmal*)[221] and once as fabrics for making turbans (*shāra*). These garments could be plain (*sādha*) or figured (*munaqqash*).[222] It was known that cotton garments were worn in India.[223] Cotton garments have been a famous export of India since antiquity, and even though the spread of cotton culture was greatly furthered in the Islamic Middle East, cotton stuffs were still imported.[224] Coarse silk was also brought from India.[225] Shoes or sandals (*naʿlayn*) produced in Kanbāya (Cambay) were "exported to all the lands of the world."[226] Precious shoes of a type called *kanbātī* were also brought from Manṣūrah in Sind.[227]

Southeast Asia

The products of Southeast Asia were of great importance to the pharmacists and perfumers of the medieval Middle Eastern world. The lands of Southeast Asia, consisting of modern Myanmar, Thailand, Cambodia, Laos, and Vietnam, and Malaysia and the vast archipelago of Indonesia, were, at the beginning of the early medieval period, heavily influenced by Indian culture. These cultures

---

217 Wink 63–4.

218 See R. G. Hoyland and B. Gilmour, *Medieval Islamic Swords and Swordmaking: Kindi's treatise "On swords and their kinds"* (Oxford: Gibb Memorial Trust, 2006).

219 H. Kennedy, *The Armies of the Caliphs: Military and Society in the Early Islamic State* (London: Routledge, 2001), 173.

220 Ibn Khurradādhbih 70, cf. also 67.

221 *Ḥudūd al-ʿālam* §10. 27, 38. See also Thaʿālibī 215.

222 *Ḥudūd al-ʿālam* §10.38 .

223 *Ḥudūd al-ʿālam* §10.8.

224 Watson 31–41.

225 Ahsan 33.

226 *Ḥudūd al-ʿālam* §10.14 and Masʿūdī 1.135 (§269). See also Ahsan 74 and n. 376.

227 Muqaddasī 481.

acquired their scripts from India, along with religion, Buddhism or Hinduism, and cultural values.[228] The importance of Indian culture in this region continues even today, although maritime trade brought continual influence from the Middle East, leading to the Islamization of most of the Indonesian Archipelago and the Malay Peninsula by the time of the European voyages of the Renaissance. The early medieval Arabic sources give a fairly detailed, although difficult to interpret, body of information on Southeast Asia.[229] This literature revolves around the aromatic products of these lands such as camphor, aloeswood, cloves, and nutmeg, and their relative qualities and means of procurement. For the medieval Middle East, Southeast Asia was seen as an extension of India, and not as a region in its own right, so the division of Southeast Asian products from those of India in this survey based on Arabic and Persian literature is somewhat artificial.

In Roman times, Greek and Near Eastern traders had come to exploit many of the riches of India, but many of the products of East India and Southeast Asia remained unknown until the deepening of this trade in Late Antiquity. Thus Dioscorides knew of aloeswood, yet his information is exceptional, and the expansion of interest in the aloeswood of Southeast Asia dates to Islamic times. Likewise, camphor, nutmeg and mace, and cloves remained unknown until Late Antiquity. Their introduction into *materia medica et aromatica* may have occurred slightly before Islam, but it was the physicians of the Islamic period, Christian, Jewish, and Muslim, who elaborated upon their use and made them the familiar substances they are today.

Direct sailing from the Middle East brought Muslim merchants into direct contact with the people of the Southeast Asian littoral, who mediated the trade with the interior lands where spices and aromatics were produced. This was known to the Arabs. Ibn Kaysān, for example, notes that no one reached the live aloeswood tree.[230] While his legend about the inaccessible mountains behind which aloeswood grows are greatly exaggerated, they reflect the feeling of Middle Eastern merchants on the futility of reaching the source. Similar stories could be listed for other substances; the point is that Middle Eastern merchants acquired their commodities in the port cities of Southeast Asia, not at the sources inland; that trade seems to have been entirely in local hands.

The sources do not dwell on animal products at a length comparable to their discussions of Indian animals. Some of the Indian animals must surely have

228 Wink, esp. 337–42.

229 G. R. Tibbetts, *A Study of the Arabic Texts containing Material on South-East Asia* (Leiden: Brill, 1979).

230 Ibn Kaysān 196.

also been known from Southeast Asia as well. Some are specifically mentioned for Southeast Asian places; for example, Cambodia (Qimār) exported elephant tusks.[231] More famous was the Southeast Asian rhinoceros. Rāmī was known for its rhinoceroses, the horns of which were extremely valuable.[232] The eastern seas also produced ambergris; Ibn al-Faqīh, among others, mentions it from Shalāhīṭ.[233] Tin was also mined in Kalah,[234] and other areas produced gold and silver.[235] But the number of botanical products brought from Southeast Asia is much higher than the animal or mineral.

Many drugs, spices, and aromatics came from Southeast Asia. Pepper was known to the Arabs from India, not Southeast Asia, where its cultivation has greatly developed in the past few centuries. Another kind of pepper known in the early medieval world was the cubeb (*kabābah* [Ar. "rounded [berry]"] *Piper cubeba*), native to Southeast Asia[236] and a familiar drug and seasoning in the medieval Middle East.[237] It is attested as early as the mid-9th century.[238] Galangal (*khūlanjān, Alpinia galanga* or *A. officinarum*), now commonly used as a seasoning, was also unknown in Classical times.[239] Galangal is native to Southeast Asia but is now cultivated in India; it is unclear when this began.[240] In any case, for the medieval Middle Easterners, it was imported from Further Asia.

After aloeswood, camphor must surely be the most famous botanical export of Southeast Asia in the medieval Islamic world. Today camphor is often derived from *Cinnamomum camphora*, a tree of eastern Asia introduced later into India and Southeast Asia.[241] Camphor is distilled from the wood of this tree; this process apparently developed in China. The most famous camphor of the early medieval world, however, was from the *Dryobalanops aromatica*,

---

231 *Ḥudūd al-ʿālam* §10.6.

232 Ibn Khurradādhbih 65 and Idrīsī 1.75.

233 Ibn al-Faqīh 1996, 72 and Ibn Rustah, *Kitāb al-ʿAlāq al-nafīsah*, ed. M. G. De Goeje (Leiden, 1892; repr. Leiden: Brill, 1967), 138.

234 Ibn Khurradādhbih 66.

235 Masʿūdī 1.130 (§256).

236 Burkill 2.1773–4.

237 Ibn Rustah 138, Ibn Juljul #11, Garbers #52 on 238–41, and Lev and Amar 393.

238 Ṭabarī, *Firdaws* 398, Schmucker 379, and Löw 3.62.

239 Ibn Juljul #8, Ibn al-Bayṭār 2.79–80, and Burkill 2.1323–34.

240 G. Watt, *The Commercial Products of India* (London: John Murray, 1908), 60, notes that it appears in India as early as Marco Polo.

241 A. Dietrich, "Kāfūr," *EI* 2nd ed. s.v., Garbers #54 on 242–6, Schmucker 372–3, Schönig "Camphor," in *EQ* s.v., Lev and Amar 123–5. See also Burkill 1.553–5 and R. A. Donkin, *Dragon's Brain Perfume An Historical Geography of Camphor* (Leiden: Brill, 1999).

which grows in Southeast Asia. This camphor was extracted from the wood directly as whitish crystals that were often further washed in milk to increase their whiteness. The most famous camphor of all came from Fanṣūr, which was in northern Sumatra.[242] Camphor was also well known from Zābaj, which originally referring to the Śailendras of Java, but later applied to the entire Śrivijaya thalassocracy, centered in eastern central Sumatra.[243] Camphor had numerous applications in medicine because of its cooling, soothing quality. It was understood as the antithesis of musk in humoral theory, balancing out musk's heat with its cold. It was also widely used in perfumery. The trade in camphor is mentioned in the Geniza documents; in Fusṭāṭ c. 1130 it cost a hundred *dīnārs* per *mann*, and Fanṣūrī camphor was eighty a *mann* in late 12th century Aden.[244] This extremely high cost makes it one of the most expensive substances of the time by weight.

Much of the aloeswood attributed to "India" was of course from Southeast Asia. Some was produced in eastern India, but the most famous came from further East, especially from Indochina. The most famous varieties were from Qimār, Khmer, i.e. Cambodia, and from Ṣanf, which was Champa, further to the east.[245] Sandalwood, as noted above, originated in Southeast Asia but was introduced into India at an early date. Along with sandalwood from India, Arabs knew that it also came from Southeast Asian regions such as Shalāhiṭ[246] and Zābaj.[247]

The most famous of the Indian spices in the European Age of Discovery were familiar in Islamic times as well: nutmeg (*jawzbuwwā, Myristica fragrans*),[248] mace (*basbāsah*), and cloves (*qaranful*).[249] Nutmeg and mace are both products of the same plant. The nutmeg is the interior seed and the mace is part of its covering. It was generally understood that they were from the same substance, indicating that nutmeg and mace were not always separated but imported as a whole. These plants all grow in the Moluccas, the celebrated "Spice Islands" east of Borneo, and nutmegs only upon the tiny islands of the Banda archipelago to the south. Nutmegs were unknown in classical antiquity,

242 *Ḥudūd al-ʿālam* 10.4.

243 Ibn Khurradādhbih 65.

244 Goitein and Friedmann 288, 505.

245 *Ḥudūd al-ʿālam* §10.2.

246 Ibn Khurradādhbih 66.

247 Ibn al-Faqīh 1996, 72.

248 On nutmeg and mace, see Garbers #33 and #17, Maimonides #38, Schmucker 148 and 112–13, Dietrich Ibn Juljul #10, Schönig 79–80, and Lev and Amar 456–7.

249 E. Ashtor "Ḳaranful" in *EI* 2nd ed. s.v., Garbers #90, Schmucker 343–4, Lev and Amar 151–3, and Dalby, *Dangerous Tastes* 50–5.

and their trade seems to have particularly flourished only in Islamic times. Cloves (*Caryophyllus aromaticus*) were known to the Romans and also to the pre-Islamic Arabs.[250] The exact sources of nutmeg and cloves seem to have been unknown; they are mentioned as products of Shalāhiṭ, i.e. the Strait of Malacca, and those were surely transshipped.[251] Likewise, Ibn al-Faqīh notes cloves, sandalwood, camphor, and nutmegs as products of Zābaj;[252] again, the cloves and nutmegs are products transshipped through the Śrivijayan Empire.

Brazil or sappan wood dye (Ar. *baqqam*) was an important textile dye, producing a red color.[253] It is derived from *Caesalpinia sappan*, a small prickly tree.[254] While various species of *Caesalpinia* occur throughout the tropics of the Old and New World, brazil wood was especially a product of Southeast Asia,[255] even though *C. sappan* is found as far west as India. Ibn Khurradādhbih notes that it was produced in Rāmī.[256] He also notes that its roots were useful against poisons, and that sailors used it against snakebite. Ibn al-Faqīh mentions it from the southern part of Shalāhiṭ.[257] The application of the name Brazil to the eastern portion of South America is sometimes said to be due to the fact that brazil wood was found there by early explorers,[258] but apparently the concept of Brazil as an island in the Atlantic predates the Age of Discovery.[259] Bamboo, familiar from India, was also brought from Rāmī.[260]

Most of the great Old World agricultural plants of modern times were cultivated in the Islamic world. Quite a few of these plants had their ultimate origins in Southeast Asia and their cultivation subsequently spread into India and elsewhere in the Indian Ocean world. In Abbasid times many of them were identified as products of Southeast Asia. For example, Ibn Khurradādhbih

---

250 Ibn Juljul #12 and King, "New *materia medica*" 513. Cf. Imru' al-Qays, *Dīwān*, ed. Yāsīn al-Ayyūbī (Beirut, 1998), 46.

251 Ibn Khurradādhbih 66.

252 Ibn al-Faqīh 1996, 72.

253 P. Pelliot, *Notes on Marco Polo*, ed. L. Hambis, 3 vols. (Paris: Imprimerie Nationale, 1959–73), 1.103–4; see Goitein and Friedmann 260,261 for a Genizah reference to the trade in brazil wood, and see Stillman 39–41.

254 Burkill 1.389–97.

255 Cf. Sīrāfī 89.

256 Ibn Khurradādhbih 65.

257 Ibn al-Faqīh 1996, 72.

258 E.g., Burkill 1.390.

259 References in H. Yule, *The Travels of Marco Polo*, ed. H. Cordier, 3rd. ed, 2 vols. (London, 1929; repr. New York: Dover, 1993), 2.380–1, and Pelliot, *Notes* 1.103.

260 Ibn Khurradādhbih 65.

knew that Bālūs produced sugarcane, coconuts, bananas, and rice,[261] and likewise that Jāba grew coconuts, bananas, and sugarcane.[262]

Place of Origin, "Brand", and Rank

The geographical knowledge that passed with commodities into the Islamic world was not merely informational baggage. It was an integral part of the identity of the commodity in question and necessary both for its successful identification as well as its appreciation.

The historian Abu Jaʿfar al-Ṭabarī (c. 839–923) describes a pot of the compound perfume unguent *ghāliyah*[263] offered to the caliph Hārūn al-Rashīd as follows (the passage is in a comedic context):

> There is no other like it; as for its musk, it is from the navels[264] of the finest Tibetan "dogs",[265] as for its ambergris, it is ambergris from the coast of Aden, and as for its ben oil, it is from so-and-so Madanī, famous for the quality of his work, and as for its compounding, it is by a man in Baṣrah well known for his skill in compounding it.[266]

Later, a witness to the scene condemns the man describing the perfume for behaving like a salesman; the caliph, of course, has no need to worry about such trivia since the finest things of the world are already his. Two techniques are thus used to distinguish this pot of *ghāliyah*: reference to the skill of the craftsmen who produced it, and reference to the source of the ingredients.

In early medieval Islamicate society names for substances such as "Tibetan musk",[267] "Fanṣūrī camphor," and "Shiḥrī ambergris" functioned not only as simple geographic distinctions of origin, but as specific types, somewhat analogous to "brand names" or "controlled appellations". Thus the polymath writer

---

261 Ibn Khurradādhbih 66.

262 Ibn Khurradādhbih 66.

263 *Ghāliyah* formulas are discussed below, 278–80 and 287–90.

264 *Surar*. The musk pod, or structure of the deer which produces musk, was thought to be its navel. See below, Chapter 4.

265 The word *kilāb*, to my knowledge, is nowhere else in Islamic literature applied to the animal that produces musk. Therefore, I suspect that the choice of word is in keeping with the comedic nature of the passage.

266 Ṭabarī III.744–5.

267 On which see A. King, "Tibetan Musk and Medieval Arab Perfumery," in A. Akasoy, ed al., eds., *Islam and Tibet: Interactions along the Musk Routes* (Farnham: Ashgate, 2011), 145–61.

al-Thaʿālibī (961–1038) quotes the writer and diplomat Abū Dulaf (fl. 10th century) in a remarkable exchange listing the most luxurious commodities available, ranging from clothing and textiles, to foods and perfumes. He catalogues the best aromatics as follows: "musk of Tibet, aloeswood of India, ambergris of al-Shiḥr, camphor of Fanṣūr, citron of Ṭabaristān, orange of Baṣrah, narcissus of Jurjān, waterlily of al-Sīrawān, rose of Jūr, wallflower[268] of Baghdād, saffron of Qumm, [and] basil of Samarkand."[269] This selection is representative of the entire passage: in every case, the best of each product is specified according to its place of origin. Everyone knew that the best linens were from Egypt, the finest silks from China, and that the best apples came from Syria or Lebanon.

A major preoccupation of medieval Islamic society was ranking and ordering the universe, both in material and human terms. As far as people are concerned, there is a practical reason for this: the importance of tracing genealogy and the concomitant importance of establishing the validity of potential authorities on Islamic tradition, hence their arrangement in books according to *ṭabaqāt* "ranks" or "stages".[270] Likewise, commodities were ranked as well and in a distinctly qualitative sense. A glance at the surviving early Arabic commercial literature demonstrates this clearly.[271] The descriptor used for indicating relative quality was usually point of origin. Precious stones, textiles, fruits, perfumes—all were typically ranked according to their place of origin rather than their intrinsic qualities, since these were believed to be characteristic of their geographical origin (and indeed, sometimes were). Those qualities had become so intertwined with their point of origin that the place of production was the chief means of distinguishing them. Let us take aromatic aloeswood, derived from diseased trees of the genus *Aquilaria*, as an example. The physician Ibn Māsawayh ranked the types of aloeswood as follows: the best came from the region of "Mandal" in India, the second best from "Samandrūn" in India, the third, from the Khmers (Qimār), the fourth from Champa (Ṣanf), and so on.

The first two of these lands are scarcely known outside of their connection with aloeswood. The references in Ibn Māsawayh and other pharmacologists are the earliest excluding a verse attributed to al-ʿUjayr, an Umayyad poet, mentioning *mandalī* aloeswood:

---

268 *Manthūr*, a somewhat unusual synonym for *khīrī* "wallflower": Ibn al-Bayṭār 4.167.

269 Thaʿālibī 238–9.

270 See C. Gilliot, "Ṭabaḳāt," *EI* 2nd ed. s.v.

271 Jaʿfar b. ʿAlī al-Dimashqī (11th–12th c.), *Kitāb al-Ishārah ilā mahāsin al-tijārah* (Cairo: Maṭbaʿat al-Muʿayyad, 1318/1900–1) [see esp. 19–21 on aromatics] and the *Kitāb al-Tabaṣṣur*.

idhā mashat nādā bi-mā fī thiyābihā
dhakiyyu l-shadhā wa-l-mandaliyyu—l-muṭayyaru

When she walks, the pungent musk (*dhakiyyu l-shadhā*) and soaring Mandalī [aloeswood] announce what is inside her clothing.[272]

The poetry anthologist who preserved this verse, al-Sarī al-Raffāʾ, comments: "*mandalī*: attributed to Mandal, one of the villages of Hind." Later writers who mention this place usually connect it with aloeswood. The geographer Yāqūt (1179–1229)[273] says Mandal is "a land in India from which excellent aloeswood is brought which is called Mandalī aloeswood."[274] In fact, there is considerable confusion on the exact location of Mandal among the Islamicate writers, for the Sanskrit word *maṇḍala* does not refer to a specific place in India but rather in this context means "district" or "province" in the generic sense. Nevertheless, for the purposes of the geographers and *adībs*, mandal was both a "place" that produced aloeswood as well as effectively a brand in the 9th century. It is thus not surprising that the expert pharmacologists after al-Yaʿqūbī do not mention it; doubtless this term was replaced by a more precise term for its origin that cannot yet be linked with one of the previously unattested varieties of aloeswood mentioned by the Ibn Mandawayh-Ibn Kaysān tradition, al-Tamīmī, or al-Bīrūnī.

The branding of commodities by geographic origin still applies in certain areas. Aloeswood is still highly prized as an incense in the Middle East and East Asia and is sold according to its place of origination.[275] The "controlled appellation of origin" (*appellation d'origine contrôlée*) in France or "Protected Designation of Origin" (PDO) of the European Union is utilized in viticulture, cheese production, and other fields to protect the identity of certain recognized types of goods. The principal difference between the modern systems and the geographic branding of the medieval Islamic world is that there was no legal force to these designations as there is today. Naturally, this method of organization appears in the specialized works devoted to aromatics and perfumery along with the more general commercial handbooks.

272 Al-Sarī al-Raffāʾ 3.165–7 and LA 14.525.
273 On whom, see below.
274 Yāqūt, *Muʿjam al-Buldān*, 5 vols. (Beirut: Dār Ṣādir, n.d.), 5.209.
275 Searching "aloeswood" on eBay clearly illustrates this point.

The Impact of Commodity Knowledge in *Adab*

Geographical literature developed in the Islamic world at about the time the scientific literature on aromatics and perfumery first appeared in the context of pharmacological literature. The earliest works extant, however, are slightly later, dating to the very late 9th century and early 10th. Several distinctive genres with different and specific aims are present within early medieval geographical literature. The works of the mathematical geographers, such as al-Khwārazmī (fl. early 9th century), attempted to continue Ptolemy's cartographic project by establishing precise coordinate systems for the world. The Ptolemaic system underlies the work of other geographers, to be sure, but the remaining works of geography are much more human in their orientation.[276] Most geographers wrote not for scientific curiosity, but to provide an assessment of the world as part of the *adab* or polite learning expected of those with any pretensions of culture.[277] Here the focus is on the particularities of the different regions. This focus could be obtained either by focusing on the different lands by region, or, in the case of the genre of "Roads and Kingdoms" (*masālik wa mamālik*) founded by Ibn Khurradādhbih, by following the routes of communication among the lands. Many Islamic geographers (especially the "Balkhī School") chose to focus their efforts exclusively upon the *Dār al-Islām*. Thus the marvelously detailed and descriptive works of Ibn Ḥawqal (943–88) and al-Muqaddasī (c. 945 or 6–1000) are focused on the lands of Islam and contain only minimal and incidental reference to anything Further Asian. But the works of others, such as al-Iṣṭakhrī (fl. mid 10th c.),[278] Ibn al-Faqīh (fl. early 10th c.), and Ibn Rustah (10th century), as well as the Persian *Ḥudūd al-ʿālam* of 982–3 attempt to discuss the entire world, if briefly in some cases.

A major concern of the geographers was keeping their works relevant and interesting to the reader. We must consider the readers of the geography texts, for they had certain expectations for the works they read. Most were not geographers but courtiers, aristocrats, merchants, and the learned classes seeking information not only of practical importance, but information to entertain, edify, and teach them a little about the world along the way. For geographers who dealt with the lands of Further Asia, this meant inclusion of a variety of information, ranging from relatively factual reports on commercial or social conditions, to entertaining marvels and tales. But for relevance, the inclusion

---

276 This orientation has been explored in great detail in the work of A. Miquel, *La géographie humaine du monde musulman jusqu'au milieu du 11e siècle*, 4 vols. (Paris: Mouton, 1967–88).

277 Miquel 1.35–7, etc.

278 Al-Iṣṭakhrī, *Kitāb Masālik al-mamālik*, ed. M. J. De Goeje (Leiden: Brill, 1870; repr. 1967).

of information on the commodities a gentleman of some means, the target audience for this writing, might desire was especially important. This is one reason why the geographers evince a strong fascination with the famous and characteristic products of the regions they describe. Their readers bought, sold, and used these products, or at least aspired to. Ibn al-Faqīh includes in his discussion of Khurāsān, the Central Asian part of the Islamic world, a catalogue of products including fruits, stones, and animals that mentions the products among the "Turks": "... and among the Turks, sable and fennec, and in Tibet, Tibetan musk and Tibetan shields ...".[279] The whole list would be at home in a work of *adab* such as al-Thaʿālibī's, and this is precisely the point: most of the geographical works must be seen in the context of *adab*, the heritage of knowledge about history, society, and the world to which the educated aspired. Part of what made a shield valuable was that it was Tibetan; likewise Tibetan musk was usually considered the best variety.

Here is a portion from the surviving version of Ibn Khurradādhbih's geography giving an itinerary in Southeast Asia that shows the important role played by the aromatic aloeswood in a geographical work:

> [Sailing] from Māʾiṭ (in the Riau Archipelago?)[280] on the left, to the island of Tiyūmah (Pulau Tioman);[281] upon it is Indian (*Hindī*) aloeswood[282] and camphor. From it to Khmer (Qimār) is a distance of five days, and in Qimār is Khmer (*Qimārī*) aloeswood and rice. From Khmer to Champa (*Ṣanf*) upon the coast is a distance of three days, and in it is Champan (*Ṣanfī*) aloeswood. It is better than the Khmer because it sinks in water due to its excellence and weight ...[283]

As can be clearly seen, Ibn Khurradādhbih has emphasized how these places in a region far beyond the pale of Islam and beyond the experience of his readers tie in with the varieties of aloeswood. All of these varieties were already mentioned in Ibn Māsawayh's work that we have mentioned above and reflect

279 Ibn al-Faqīh 1885, 255.
280 Tibbetts 147–9.
281 Tibbetts 136–7.
282 The use of "Indian" in this generic sense referring to Southeast Asia is due to the fact that Southeast Asia was reckoned a part of India by the Islamic geographers, and also to emphasize that the *ʿūd* being discussed is aloeswood, for the Arabic term can also simply mean "wood".
283 Ibn Khurradādhbih 68.

the 9th century aloeswood market.[284] Ibn Khurradādhbih has even inserted a comment about the relative merit of Ṣanfī aloeswood over the Qimārī.[285] In fact, all good aloeswood sinks in water due to the density of aromatic resin within it.[286] It is especially noteworthy that this bit of information survives in the version of Ibn Khurradādhbih still extant, which is abridged; obviously the editor considered information on aloeswood key information. Al-Idrīsī's account of this area is clearly dependent on Ibn Khurradādhbih and includes more detail especially on commodities and trade, among other things, that probably reflect the original text.[287]

Later accounts, like al-Idrīsī's, freely borrowed from the writings of the earlier geographers. Some were more scientific about it; al-Idrīsī fits his information into a geographic and cartographic scheme derived ultimately from Greek geography.[288] Others, such as Yāqūt and Zakariyyā' b. Muḥammad al-Qazwīnī (c. 1203–83),[289] were motivated by different reasons. In their cases especially, information on geography, place, and commodity was seen as part of the intellectual heritage of *adab*. Seldom did they trouble to update their information; they did not read the writings of later writers who could have supplemented and corrected their accounts. (In the case of the pharmacologists, this neglect is not surprising; their literature was highly specialized.) And while these later geographical works are mines of information (and do, in their way, have a lot to say about how that information was perceived in their times), they became increasingly obsolete as practical guides to the world, being useful instead as guides to a literature based upon that world. Within that context, however, we may see once more the commodities that helped to make up the Islamicate understanding of Asia.

---

284 Ibn Māsawayh 13 and Levey 401.

285 This differs from Ibn Māsawayh's sequence, which is Indian of the Mandalī type, Indian of the Samandrūn type, Qimārī, and then Ṣanfī.

286 See, e.g., Schafer, Golden Peaches 163.

287 Idrīsī 1.82–3.

288 We will leave aside the question of al-Idrīsī's pharmacopoeia, the *Kitāb al-Jāmiʿ li-ṣifāt ashtāt al-nabāt wa-ḍurū anwāʿ al-mufradāt* as much more study of this work is needed. Two manuscripts have been published in facsimile, in a three volume edition edited by Fuat Sezgin (Frankfurt am Main: Institute for the History of Arabic-Islamic Science, 1995); these manuscripts have significant variations which has rendered a critical edition unlikely.

289 Author of the cosmographical *'Ajā'ib al-makhlūqāt* and the geographical *Āthār al-Bilād* (both ed. F. Wüstenfeld, Göttingen, 1848–9), highly popular texts in the later medieval Islamic world.

The *Muʿjam al-buldān* of Yāqūt is a geographical lexicon that digests information from earlier geographers, giving data on pronunciation of the place names and particulars on these places from within both the world of Islam and beyond. Not surprisingly, Yāqūt's accounts also usually feature commodities just as earlier geographical literature had done, often with considerable detail. Under his entry on Tibet, for example, Yāqūt quotes an extensive passage from al-Masʿūdī on Tibetan musk and its production.[290] Likewise, the musk of China is duly noted.[291] We have already noted that Mandal makes it into Yāqūt's book because of the fame of Mandalī aloeswood.

This practice of introducing information on commodities in a geographical account is not confined to geographers per se. When a writer such as the historian al-Masʿūdī introduces a place in the geographical discussions in his *Murūj al-dhahab*, he very often includes a mention of a special product to remind the reader what the importance of that land is. For example, in his listing of the seven seas, al-Masʿūdī inserts a comment on the Sea of Ṣanf (Champa), the South China Sea, noting that the lands there produce Ṣanfī aloeswood,[292] which was one of the most desirable varieties.[293] Likewise, in al-Masʿūdī's discussion of the Arabian Sea (the Sea of Lārawī) he notes that it has a little ambergris and includes a digression on how ambergris is produced; he does not fail to point out that the best ambergris is from al-Shiḥr in Yemen and al-Zābaj (Indonesia). Abū Zayd al-Sīrāfī introduces his discussion of the land of Qimār (Khmer) by reminding the reader that this is the region from which Qimārī aloeswood comes.[294] This is not particularly surprising given the commercial orientation of his work, but this practice also extends to writers without any particular interest in Asia. When discussing elephants, the Iraqi *qāḍī* and anecdotist al-Tanūkhī (940–94) mentions the elephants of Ṣanf, and he too includes a note about the Ṣanfī aloeswood.[295] In other words, Ṣanf meant aloeswood for the educated reader. The commodity was tied to the geographical identity of the place.

When an author treats a subject pertaining to one of these exotic lands, beyond the pale of the *Dār al-Islām*, it is common to place a comment on a

290 Yāqūt 2.11.

291 Yāqūt 3.442; here he follows Abū Dulaf's suspect account of his travels in China- more likely a work he cobbled together from other sources. See C. E. Bosworth, "Abū Dulaf," *EIr* s.v.

292 Masʿūdī 1.177 (§361).

293 Even today, aloeswood of Vietnam commands a high price among aloeswood.

294 Sīrāfī 68.

295 Tanūkhī, *Nishwār* 8.208.

particular specialty or specialties of the land in question, even when that is not the subject of the discussion. We have already noted how al-Sarī al-Raffāʾ made sure his reader know just what Mandalī aloeswood was. This bit of information serves to orient the reader within the sea of knowledge by providing a context. It did not take a particularly learned scholar to have heard of ambergris of Shiḥr or aloeswood of Qimār; even if such substances were too expensive to afford, they were as much a part of contemporary culture as a bottle of Dom Pérignon or beluga caviar is to ours—a symbol of luxury to which one aspires.

Conclusion

Surveying the commodities of Asia available in the early medieval Islamic world, one is struck by their diversity as well as by the relatively good knowledge possessed within early medieval Islamicate society of the origins of these goods. They reflect many types of commodities ranging from natural products to manufactured goods. The number of *materia medica et aromatica* mentioned in Arabic and Persian literature is particularly striking; no serious attempt has been made in this survey to go beyond the notices of geographical, commercial, and historical works into the even richer catalogues of these substances preserved in the pharmacopoeias. Were this done exhaustively, the lists would be even more extensive. They reflect the great importance of plant and animal substances used in Islamicate society for purposes such as medicine, perfumery, and dyeing. The survey of commodities mentioned in the business documents of the Cairo Geniza follows this pattern: they form the largest single category of commodities.[296] This has been tentatively attributed to the popularity of the professions of perfumer, druggist, apothecary, and dyer among the Jews. That may well be true, but the survey of commodities above shows that these sorts of goods were familiar to the literati as commodities of Asia and were singled out for mention as important beyond a narrow circle of cognoscenti.

Of the major aromatics in early Islamicate civilization, musk, aloeswood, costus, spikenard, cloves, nutmeg, mace, and camphor were imported from beyond the Middle East and the *Dār al-Islām* exclusively. Some important aromatics could be found closer at hand. Ambergris was found throughout the Indian Ocean, and so some came from the southern Arabian coasts, but this uncommon substance so essential for luxury perfumery was purchased also from wherever it was available. Saffron was mostly obtained from Iran;

296 Goiten and Friedmann 15–16; Goitein, *Letters* 17, 175. See also Lev and Amar for a survey.

labdanum, the sticky fragrant resin of the rockrose (*Cistus creticus* and other species of *Cistus*), was produced in the Mediterranean lands.[297] Onycha (*aẓfār al-ṭīb*), the operculum of mollusks used in incense throughout medieval Eurasia, came from the Red Sea and Indian Ocean.[298] Balsam, valued since antiquity, was available from Egypt.[299] The frankincense and myrrh of Arabia had their uses in medicine, but they are not mentioned at all in the literature on luxury perfumery nor celebrated in poetry, indicating that they were thoroughly without any prestige value, though they were surely used by the common folk.[300]

Finally, the degree to which these commodities penetrated the Middle East should be mentioned. As is well known, the backbone of all pre-modern societies was agrarian, and the farmers in the rural parts of the Middle East can scarcely have afforded most of these exotic goods. But Islamicate civilization had a burgeoning urban class.[301] The roots of the urban civilization stretched back into the cities of the pre-Islamic Near East, but a great impetus to the creation of this culture was the settlement of numerous Arabs in urban areas following the conquests. These Arabs, former members of the army, needed to be settled to channel their efforts in directions other than the destabilization of the nascent state. They were paid from the *dīwān*, the register of the army, according to their rank in the conquests. The money with which they were paid came from the booty acquired during the conquests, and especially included a vast quantity of dethesaurized precious metal that had been languishing in the previous centuries.[302] These factors resulted in the creation of an almost instant monied urban class. Most of these Arabs invested their money in

---

297 *Cistus creticus* or *C. ladanifer*. See Garbers #60 on 250–2, Schmucker 424–5, Dietrich DT I.66, WKAS "*lādhan*" s.v., W. Müller, "Namen von Aromata im antiken Südarabien," in A. Avanzini, ed., *Profumi d'Arabia* (Rome: Bretschneider, 1997), 204–5, Lev and Amar 194–5.

298 See Garbers #10 on 178–9, Lev and Amar 215, H. Nawata, "An Exported Item from Bāḍiʿ on the Western Red Sea Coast in the Eighth Century: Historical and Ethnographical Studies on Operculum as Incense and Perfume," *Papers of the XIIIth International Conference of Ethiopian Studies*, ed. K. Fukui, et al. Kyoto, 1997, vol. 1, 307–25, and J. McHugh, "Blattes de Byzance in India: Mollusk Opercula and the History of Perfumery," *JRAS* (2013): 53–67.

299 M. Milwright, "The balsam of Maṭariyya: an exploration of a medieval panacea," *BSOAS* 66:2 (2003): 193–209.

300 See King, "Importance of Imported Aromatics."

301 S. D. Goitein, "The Rise of the Middle-Eastern Bourgeoisie in Early Islamic Times," *Journal of World History* 3 (1957): 583–604; repr. in S. D. Goitein, *Studies in Islamic History and Institutions* (Leiden: Brill, 1966), 217–41.

302 Cf. Wink 33–6, 173–5.

commerce and industry, and the economy soared. These people, with access to money, formed the basis of the consumer class of the early medieval Middle East. To be sure, vast amounts of luxury goods also flowed to the caliphal court and local seats of government as well.

The result of this process was a great interest in what we would call the "good life". Quite a lot of ink was put to paper recording just how the beneficiaries of the wealth of the early medieval Islamic world should live.[303] The literature upon which the above survey is based is mostly a result of that: works written for the edification of an urban, literate elite. In these books they could learn about the world from the aspect that mattered to them: in terms of what commodities were available to their own households. The sources are full of rankings of different types of goods, with information on how to appraise them as well as suggestions as to their uses. One of the preoccupations of this culture, as we shall see, was perfumery, and this resulted in a great deal of interest in musk, that most precious of the aromatics.

303 Perhaps the most famous is the *Laṭāʾif al-Maʿārif* of al-Thaʿālibī, but even earlier are significant works by Ibn Qutaybah, among others.

CHAPTER 3

History of Musk and the Musk Trade: From Further Asia to the Near East

Introduction

The time when musk was first discovered in the lands of the musk deer habitat in Further Asia is unknown. Undoubtedly the animals were hunted for food as soon as people lived in proximity to them, but perhaps they were not the most desired quarry, as their meat is said to have a distinctive taste.[1] The Tamang of Nepal, proficient collectors of musk, considered the meat hard to digest and prone to cause diarrhea.[2] The hair of the musk deer was used to stuff cushions and mattresses during the 19th century in Kham.[3] In general, we know in modern times that the carcasses of the animals hunted for their musk were often abandoned.

No one can say where the aromatic properties of musk were discovered. It could have been anywhere within the range of the animals. The first person who learned that musk could, in small amounts, have such a profoundly different effect on a perfumed compound than its strong smell would indicate must have been someone who already experimented with some form of aromatic compounds. As the preparation of these is ancient the world over, that cannot help narrow the field.

The only guidance in this matter comes from the references to musk in the literatures of the peoples who lived in proximity to the mountainous habitat of the musk deer. The earliest body of written literature to describe musk with any certainty is from China, where musk is clearly and well attested by the Later Han dynasty (25–220 CE). A few centuries pass before there is an

---

1 Nevertheless, they were certainly eaten: the Lolos in Yunnan "esteemed" their meat, see "Fragrans", "How Musk is Made," *The China Review* 9:4 (1881): 253 and see below. The meat is also discussed in Chinese pharmacological literature, see below.

2 C. Jest, "Valeurs d'Échange en Himalaya et au Tibet: Le amber et le musc," in *De la voûte céleste au terroir, du jardin au foyer: mosaïque sociographique: textes offerts à Lucien Bernot* (Paris: Éditions de l'École des hautes études en sciences sociales, 1987), 232.

3 Evariste-Régis Huc and Joseph Gabet, *Travels in Tartary, Thibet and China, 1844–1846*, trans. W. Hazlitt and intr. P. Pelliot. 2 vols. (London: Routledge, 1928; repr. 2 vols. in 1, New York: Dover, 1987), 2.372, 374.

 | DOI 10.1163/9789004336315_005

explosion of source material from India, Central Eurasia, and the Near East and Mediterranean.

For most of the highlands of Central Eurasia, early written literature is extremely difficult to come by. The peoples of Central Eurasia borrowed their writing systems from their neighbors in the Near East, India, and China over the centuries. Of course the borrowing of scripts signals that important cultural contacts were occurring, and these cultural contacts had a bearing on such things as the uses of aromatics. Except for a very few finds, the vast bulk of indigenous writing recovered from Central Eurasia dates to the second half of the first millennium CE and later. Most of this literature is from the Buddhist canon, but there is a precious handful of business documents, registers, and letters. Materials useful for the history of aromatics in this corpus are scattered and fragmentary, but even so some attestations of musk can be found.

Musk in China

Musk deer are found throughout the forested, mountainous parts of China, Northeast Asia, Siberia and Korea, and south into Vietnam and Burma. The use of musk is reliably attested in China earlier than anywhere else. The earliest extant Chinese dictionary, the *Erya* (爾雅) cites the musk deer, calling it *shefu* (麝父).[4] This work dates to perhaps the third century BCE. Musk is also attested in the Later Han dictionary, the *Shuowen jiezi* (說文解字).[5]

Perhaps the earliest discussion of musk in existence was contained in the *Shennong bencao* (神農本草), a pharmacological work dating apparently to the Later Han and purporting to be the work of the mythical sage Shennong. Despite the presence of musk in the *Shennong bencao*, musk is not attested among the medical manuscripts from Mawangdui. These manuscripts from a Former Han (202 BCE–9 CE) site include religious texts and also some important early medical works, one of which calls for 247 different drugs in its 170 prescriptions.[6]

4 *Erya* 爾雅, recension of Guo Pu 郭璞 (*Bai bu cong shu ji cheng* 百部叢書集成 edition), lower scroll, 29b–30a.

5 See Li Shizhen 李時珍, *Bencao gangmu* 本草綱目 (*Guoxue jiben congshu* edition, Taipei, 1968), 25.51.38. (Abbr. BCGM).

6 The so-called Wushier pingfang. See D. Harper, *Early Chinese Medical Literature: The Mawangdui Medical Manuscripts* (London: Kegan Paul, 1998); P. Unschuld, *Medicine in China: A History of Pharmaceutics* (Berkeley: University of California Press, 1986), 14–15.

The *Shennong bencao* was the foundation upon which the system of Chinese pharmacology was constructed. It survives among the later pharmacological works built upon it. The text was "reconstructed" from the pharmacological literature by the Qing scholars Sun Xinyan 孫星衍 (1753–1818) and his nephew Sun Pingyi 孫馮翼.[7] Their text says:

> Musk. Its taste is acrid and warm. It masters and drives out bad vapors (*eqi* 惡氣) and kills demons and spirits (*gujingwu* 鬼精物). It cures warm malaria (*wennüe* 溫瘧) and *gu* (蠱) poison,[8] epilepsy, and drives out the three worms. If used over a long period of time it gets rid of evil (*xie* 邪). It is good for awakening from sleep without having dreamed, and insomnia. It originates from river valleys.[9]

Tao Hongjing 陶弘景 (452–536) was an early pharmacologist who gave a detailed account of musk as a supplement to the account in the *Shennong bencao*. It is quoted in the *Bencao gangmu* 本草綱目 of Li Shizhen 李時珍 (1518–1593):[10]

> The musk animal has the appearance of a water deer, but is smaller. It is black in color, and always eats cypress leaves (*baiye* 柏葉), and it also eats snakes. Its aromatic substance (*xiang* 香) is just inside the skin in front of the penis; a separate membrane sack (*modai* 膜袋) contains it. In the fifth month one can obtain the musk; occasionally it has snake skin and bones in it. Now people use the sloughed skin (*shetuo* 蛇蛻) of snakes to wrap it, saying that it completes (*mi* 彌) the fragrance. So these are used together. In the summer the musk deer eats many snakes and insects (*shechong* 蛇蟲) until the winter, when the pod fills. In the spring it develops an acute pain in its navel (*qi* 臍) and scrapes it off with its own claws; it is covered over among its excrement. They always do it at one place which does not shift. Once someone happened on one of these places and got the musk. It was as much as one peck (*dou* 斗) and five pints (*sheng* 升). Its fragrance greatly surpasses that [of musk] taken from

---

7 Unschuld, *Medicine in China* 196–7.

8 I.e. virulent magical infection. *Gu* was produced by placing venomous creatures together in a vessel; the one that survived became *gu*. See M. Strickmann, *Chinese Magical Medicine*, ed. B. Faure (Stanford: Stanford University Press, 2002), 27–8.

9 *Shennong bencao* 神農本草, *Congshu jicheng* edition. (Shanghai: Shangwu yinshu guan, 1937), 1.45.

10 BCGM 51.38–9.

> the killed [animals]. Former writers say that it is made from sperm and urine coagulating, but this is certainly not so. That which is produced by the Qiang[11] 羌 and the Yi 夷 is mostly genuine and good. That from the Man[12] 蠻 of Suijun 隨郡, Yiyang 義陽, and Jinxi 晉溪 is inferior. That which is produced by Yizhou 益州 is flat in shape and is still wrapped in its skin and membranes. There is much adulteration. Every genuine musk pod is divided into three or four pods or the like, and the blood and membranes are scraped and taken, along with several other things, and the four are wrapped with the skin of the foot or knee and traded. Those who trade them also adulterate them again. Those people say that the only way to determine whether it is genuine is by breaking a piece open and examining it; that which has hair enclosed in it is superior. To get musk which is certainly genuine now one can only get the animal alive and watch its removal.

The other early source on musk is the *Mingyi bielu* 名醫別錄, which is quoted in later sources. It is apparently a little later than Tao Hongjing's work.[13] It elaborates on the medical uses of musk: "It heals all evil influences and spirits, blows from the malevolent,[14] sudden pain in the heart and stomach, rapid swelling, and constipation, and wind poison. It removes moles from the face, and leucoma from the eyes. In a woman's difficult delivery, it promotes miscarriage. It allows one to communicate with Gods and Transcendents (*shenxian* 神仙)."[15] This text also adds some information about the geographical origin of musk: "it originates from Tai (臺) to Yizhou (益州) and Yongzhou (雍州) in the mountains. In the spring one can collect it. That which comes from Yizhou is best."[16]

The account of musk in the *Xinxiu bencao* 新修本草, the major pharmacological work of the Tang dynasty, is derived entirely, it seems, from these earlier sources.[17] It preserves a number of old readings that are changed in the later versions of these texts quoted in pharmacological literature. Unfortunately,

---

11 Peoples who inhabited the area of modern Gansu, Ningxia and Northeastern Tibet.

12 The aboriginal peoples of southern China.

13 Unschuld, *Medicine in China* 32.

14 Cf. P. Unschuld, *Introductory Readings in Classical Chinese Medicine* (Dordrecht: Kluwer, 1988): 395.

15 Quoted in BCGM 51.39.

16 Quoted by *Shennong bencao* 1:45.

17 The account is found in Chapter 15. Edition consulted: *Shinshū honzō. Zankan* (Tokyo: Meiji Shoin, 1983), 15:87.

the text contains many lacunae and requires a detailed textual study which cannot be undertaken here.

Musk had a reputation for driving away bad influences from at least as early as the Han dynasty. This is repeated in many later books.[18] A demonifuge incense recorded in the *Danjing yaojue* 丹經要訣 included musk along with many ingredients that were poisonous and would produce an intensely irritating smoke to humans and, hopefully, demons.[19]

As it was said to drive away evil influences, musk facilitated a positive relationship to the realms of the immortals. One work quoted by the *Taiping Yulan* 太平御覽 says that musk feeds the soul (*po* 魄)[20] and is also fragrant.[21] Musk was used in Daoist elixers as well.[22] The aroma of musk was also associated with the Transcendents (*xian* 仙); a poem by the Tang dynasty poet He Ning 和凝 describes the Transcendent Who Presides over the River as having the smoke of musk incense in her presence.[23]

In the secular world, musk perfume was associated with women and sex.[24] Women perfumed their bodies with musk; the lingering scent upon a lover was a reminder of previous sex.[25] Musk was also used as a cosmetic; ladies used it to produce beauty marks on their faces. Such a mark in a moon shape (*sheyue* 麝月) is attested as early as the 6th century.[26] Men used musk as well;

---

18 Some quotations in *Taiping yulan* 太平御覽 (Beijing: Zhonghua Shuju, 1960), 981:.5b.

19 N. Sivin, *Chinese Alchemy: Preliminary Studies* (Cambridge, MA: Harvard University Press, 1968), 208–9; Text 27A.

20 Given as 柏 in the text, cf. T. Morohashi, *Dai Kanwa Jiten*, 13 vols. (Tōkyō: Taishūkan Shoten, 1955–60), #14617.4 on v.6 pg. 259 or 5971.

21 *Taiping yulan* 981:5b.

22 Sivin 174–5, text 12A-B ("Cyclically Transformed Elixir") and 188–9, text 18A–B ("Purple Essence Elixir").

23 *Quan Tangshi* 全唐詩, 8 vols. (Taibei: Fu xing shu ju, 1967), vol. 8 p. 5138; S. Cahill, "Sex and the Supernatural in Medieval China," *JAOS* 105:2 (1985): 203.

24 E.g. Wen Tingyun 溫庭筠 (813?–870) in *Quan Tangshi* vol. 6, p. 3504 (description of the singing-girl Jingyuan 靜婉).

25 J. R. Hightower, "Yüan Chen and 'The Story of Ying-Ying,'" *HJAS* 33 (1973): 101, quoting a Tang dynasty tale.

26 J. R. Hightower, "Some Characteristics of Parallel Prose," in *Studia Serica Bernhard Karlgren Dedicata* (Copenhagen: Munksgaard, 1959), 88 (text) and 81 (trans.) quoting Xu Ling 徐陵 (507–83), *Yutai xinyong* 玉臺新詠, preface, line 38. The making of moons and star shaped beauty spots is also attested in late 17th century England, cf. E. Rimmel, *The Book of Perfumes* (London: Chapman and Hall, 1865; repr. n.p.: Elibron, 2005), 207–8, with an illustration from a contemporary source.

a prince of the 8th century placed aloeswood and musk in his mouth so that his breath would be aromatic during his conversation.[27]

Some Daoist adepts carried musk in the mountains to protect against snakebite; since the musk deer was believed to eat snakes, it was considered an appropriate prophylactic.[28] Modern researchers have confirmed that musk has antivenom properties,[29] but this cannot be because the vegetarian musk deer eats snakes. The *Man* 蠻, the aboriginal peoples of Southern China, ate the meat of the musk deer constantly; it was said that they did not have to fear snakebite because of it.[30] And in Eastern Nepal in the 20th century, the belief that musk deer ate snakes and that musk could prevent snakebite was known.[31] Musk compounds were also thought to ward off chiggers and other biting insects.[32]

In addition to these medical and cosmetic uses, musk was also extensively used in incense. The tremendous importance of incense in Chinese and East Asian culture goes without saying; it is burned in temples, shrines, and in the home or palace in vast quantities.[33] During the Tang period the burning of incense was an essential component of Buddhist ritual.[34] Incense was also offered unburned along with flowers, tea, medicines, etc. for later use[35] and

27 E. Schafer, *The Golden Peaches of Samarkand: A Study of T'ang Exotics* (Berkeley: University of California Press, 1963), 157.

28 Ge Hong 葛洪, *Bao puzi* 抱朴子 (Taibei: Zhongguo zixue mingzhu jicheng, 1978), 17.8b–9a (334–5).

29 R. B. Arora, S. D. S. Seth, and P. Somani, "Effectiveness of Musk (Kasturi), an indigenous drug, against Echis carinatus (the Saw-Scaled Viper) Envenomation," *Life Sciences* 9 (1962): 453–7.

30 BCGM 51: 41, citing Meng Xian.

31 Jest 232.

32 Cf. Zakariyyā b. Muḥammad al-Qazwīnī, *'Ajā'ib al-Makhlūqāt* 326, where censing with shavings of the musk deer "horn" was said to drive away pests (*hawāmm*); of course, the musk deer does not have a horn.

33 See J. Needham and Lu Gwei-djen, *Science and Civilisation in China Volume 5. Chemistry and Chemical Technology: Part II. Spagyrical Discovery and Invention: Magisteries of Gold and Immortality* (Cambridge: Cambridge University Press, 1974), 128–54. Even today, the shops and sellers of incense near a major temple in the Peoples' Republic of China are numerous, and large braziers are often provided before the shrines for incense burning even if incense may not be burned inside the shrines owing to the concerns for historic preservation.

34 E. O. Reischauer, trans., *Ennin's Diary: The Record of a Pilgrimage to China in Search of the Law* (New York: Ronald Press, 1955), 62–3, 219, etc.

35 Ennin 221.

was even used as a plaster for the bases of ritual platforms.[36] The patrons who offered incense to Buddhist monasteries and temples included government officials.[37] The use of incense in China dates back to the beginning of the Common Era, if not a little earlier. An incense formula of the 1st century CE has been preserved in the *Mozhuang manlu* 墨莊漫錄 of Zhang Bangji 張邦基.[38] It originates from a work called the *Hangong xiangfang* 漢宮香方, the "Recipes for Fragrances of the Han Palace." This work gives the following ingredients for incense: aloeswood (*chenxiang* 沉香), costus (*guangmuxiang* 廣木香), clove (*dingxiang* 丁香), camphor (*longnaoxiang* 龍腦香), musk, rock honey, and rice gruel. In the proportions of the formula, musk is about 5%, meaning it was used to enhance the scent but not make it overwhelmingly musky.

Musk also appears in Chinese incense compounds of the Tang period. One was prepared for the Huadu Monastery 化度寺.[39] It consisted of aloeswood, white sandalwood 白檀香, storax 蘇合香, onycha 甲香, camphor 龍腦香, and musk. The ingredients were crushed, sieved and mixed with honey. A formula called "The Prince of Shu's Method for Perfuming Imperial Clothing" (蜀王薰御衣法) calls for cloves, *jian* fragrance 籛香,[40] aloeswood, sandalwood, musk, and onycha.[41] The *Bencao gangmu* quotes an incense formula transmitted by Su Song 蘇頌 (1020–1101) for an onycha-based incense paste that included aloeswood and musk.[42] Kou Zongshi 寇宗奭 (c. 1116) is also quoted: he says that onycha is good for governing incense compounds; used together with aloeswood, sandalwood, camphor, and musk it enhances them (*youjia* 尤佳).[43] In general, then, musk was used in incense in small quantities for the purpose of enhancing the scent.

Incense was a preoccupation for the literati, to judge by the number of works that survive on the subject. The gift of incense among the literati was celebrated in poetry.[44] The observation that incense burned at a steady rate

36 Ennin 228.

37 Ennin 233.

38 Ch. 2 p. 17b, 18a; quoted by Needham and Lu 135.

39 The recipe is in the *Xiang Pu* 香譜of Chu Hong 芻撰, 2:28. See also Schafer, *Golden Peaches* 159.

40 Unidentified. Morohashi #44540, where it is said to be identical with *jian* (#15755); both are described as aromatic woods.

41 *Xiang Pu* 2:28.

42 BCGM 46: 41–2, in the article on onycha.

43 BCGM 46: 42.

44 S. Sargent, "'Huang T'ing-chien's 'Incense of Awareness': Poems of Exchange, Poems of Enlightenment," *JAOS* 121:1 (2001): 60–71.

led to the development of many types of incense clocks.[45] These had a long and distinguished history in East Asia. The literati also appreciated musk in another form: musk was mixed with fine inks for calligraphy.[46]

This interest in incense was inherited by the Japanese, who furthered its development as an art form. A form of entertainment for the nobility was burning incense, called *kiki-kō*, literally, "listening to incense"; this is still practiced in the traditional art of *kōdō*, the way of incense.[47] Games were constructed around guessing the ingredients and makers of different blends.[48] This is attested at least as early as the late 10th-early 11th century, for Murasaki Shikibu describes just such a gathering for the identification of incenses that were prepared in competition in the Tale of Genji.[49] Indeed, evoking the moods of the Tale of Genji became the basis for rarified incense games. Rare ingredients were prized and noteworthy pieces of resinous aloeswood (*jinkō*) were named and gained great fame.[50] Musk, along with numerous botanical materials and onycha, was a basic ingredient of the perfumer's palette.[51]

Japan had to acquire musk by trade from the mainland because the musk deer is not native to Japan. The Japanese (*wokou* 倭國) are said to have presented sable furs (貂皮), ginseng and other things to the Chinese, requesting them to present fine *sheng*-instruments (細笙) and musk.[52] Musk is the very first substance specified in the list of sixty medicaments presented to the

45 Schafer 160; S. A. Bedini, *The Trail of Time: Time measurement with incense in East Asia* (Cambridge: Cambridge University Press, 1994).

46 B. Laufer, "History of Ink in China" in F. Wiborg, *Printing Ink: A History* (New York: Harper & Brothers, 1926), 17; Li He in *Quan Tangshi* vol. 4, p. 2336.

47 See A. Gatten, "A Wisp of Smoke: Scent and Character in the Tale of Genji," *Monumenta Nipponica* 32 (1977): 35–48; K. M. Anderson, *Kōdō: the Way of Incense* (MA Thesis, Indiana University, Bloomingon, 1984); D. Pybus, *Kodo: The Way of Incense* (Rutland: Tuttle, 2001); Needham and Lu 143–4; S. Horiguchi and D. Jung, "Kōdō: Its Spiritual and Game Elements and Its Interrelations with the Japanese Literary Arts," *JRAS* (2013): 69–84.

48 See Anderson, *passim*, Pybus 42–65, and Horiguchi and Jung for procedures for these games.

49 Murasaki Shikibu, *The Tale of Genji*, trans. Arthur Waley (New York: Modern Library 1960), 591–6.

50 Anderson 10.

51 See Gatten 37, Needham and Lu 144.

52 *Taiping Yulan* vol. 4 p. 4344 (=981.4b). Cf. J. Kuwabara, "On P'u Shou-kêng 蒲壽庚, a Man of the Western Regions, who was the Superintendent of the Trading Ships' Office in Ch'üan-chou 泉州 towards the End of the Sung dynasty, together with a General Sketch of Trade of the Arabs in China during the T'ang and Sung Eras," *MRDTB* 2 (1928): 75.

Shōsō-in by Empress Kōmyō in 756; the remains of this musk survived at least into the twentieth century.[53]

A pill made of fragrances that perfumed the body and clothing is recorded in the 10th century *Ishimpō* of Yasuyori Tamba; it includes a small amount of musk along with other aromatics.[54] The ingredients were powdered and mixed with honey to make pills, which were dissolved on the tongue. After five days of using it, the recipe says, the body becomes fragrant, and after ten, the clothing and bed covers. After twenty-five days water used for washing the hands and face will be perfumed after being poured upon the ground, and after a month, "even a child carried in your arms will be fragrant."

The musk deer is found in the mountainous parts of Vietnam, and musk was used by the medieval peoples of the area. The Cham anointed themselves with musk twice daily.[55] In Northeast Upper Burma musk was exchanged among the tribes in tribute during the 19th century.[56] The natives of Annam anointed their bodies with camphor and musk and censed their clothes with aromatic woods; they imported at least some of their musk via the Indian Ocean trade.[57]

Musk in India

It is difficult to date the time at which musk begins to appear in Indian literature. Since the usual Sanskrit word for musk, *kastūrī*, is a loanword from Greek, the use of musk must have been in some way promoted and furthered by the Greeks, presumably as a substitute for the castoreum used in Greek medicine that the Greeks could not get in India, as beavers were unknown there. Or perhaps the Indians began to use the substance as a substitute for Greek imported castoreum. It has been suggested that the use of musk began in India during the Gupta period (320–550 CE),[58] and this dating harmonizes with the spread

53 K. Kimura, "Ancient Drugs Preserved in the Shōsōin," *Occasional Papers of the Kansai Asiatic Society* 1 (February, 1954): 4,6.

54 Tamba Yasuyori, *Ishimpō: The Essentials of Medicine in Ancient China and Japan*, trans. E. C. H. Hsia, et al., 2 vols. (Leiden: Brill, 1986), 2.31–2.

55 Jiu Tangshu 197: 5269; Xin Tangshu 222b: 6297; Schafer 72.

56 J. T. Walker, "Expeditions among the Kachin Tribes on the North-East Frontier of Upper Burma," *Proceedings of the Royal Geographical Society* (1892): 165.

57 F. Hirth and W. Rockhill, *Chau Ju-kua: His work on the Chinese and Arab Trade in the Twelfth and Thirteenth Centuries, entitled Chu-fan chih*, 2 vols. (St. Petersburg, 1911), 47, 49.

58 P. V. Sharma, *Vāgbhaṭa-vivecana: Vāgbhaṭa kā sarvāṅgīṇa samīkṣātmaka adhyaya* (Vārāṇasī, 1968) cited by C. J. Meulenbeld, *A History of Indian Medical Literature*, 3 vols. in 5 (Groningen: Egbert Forsten, 2002), IB 703 n. 726. Sharma's earliest sources

of musk to the west but does not explain how it acquired a Greek name. It seems likely that musk was known but not used extensively enough to make it into the literature until later.

No attestation of musk has yet been found in any of the truly ancient Indic manuscripts recovered in Afghanistan or western China or in an ancient inscription; this leaves us with Indian literary texts. Indian literature is notoriously difficult to date, and it was subject to the usual interpolations found regularly in the copying of ancient texts. This is also the case in Sanskrit medical literature, the basic works of which, such as the *Carakasaṃhitā* and the *Suśrutasaṃhitā*, evidently date to the first half of the first millennium CE.

The Gupta poet Kālidāsa, who seems to have flourished during the period of Candragupta II and Kumāragupta I (375–455), is perhaps the first poet to mention musk, if not the earliest writer to mention musk in India. Kālidāsa does not, however, use the term *kastūrī* but rather *nābigandha* "navel perfume" for musk. In his *Meghadūta*, "Cloud Messenger", he says while describing the journey of the cloud that is the subject of the poem:

āsīnānāṃ surabhitaśilaṃ nābhigandhair mṛgāṇāṃ
tasyā eva prabhavaṃ acalaṃ prāpya gauraṃ tuṣāraiḥ
vakṣyasyadhvaśramavinayane tasya śṛṅge niṣaṇṇaḥ
śobhāṃ śubhrāṃ trinayanavṛṣotkhātapaṅkopameyāṃ[59]

Having reached the mountain, the source of that very one (i.e. the Ganges), brilliant with snows, its crag sweetly perfumed with the musk perfume (*nābhigandha*) from deer sitting, you will possess, resting upon its peak to remove weariness of the road, a splendid beauty comparable to the mud of the uneven ground from the bull of the Three-Eyed One.

Kālidāsa also mentions musk in the *Kumārasaṃbhava*:

sa kṛttivāsāstapase yatātmā gaṅgāpravāhokṣitadevadāru
prasthaṃ himādrermṛganābhigandhi kiṃcidgṛṇatkiṃnaramadhyuvāsa[60]

---

are the *Harṣacarita* of Bāṇa and the *Bṛhatsaṃhitā* of Varāhamihira, both discussed below. Meulenbeld points out that Sharma did not notice the mention of *kastūrī* in the *Carakasaṃhitā*; that passage is also discussed below, see Meulenbeld IIB 108 n. 157, but he did not supply a reference there.

59 Kālidāsa, *Meghadūta*, ed. M. R. Kale (Delhi: Motilal Banarsidass, 1991), verse 55.

60 Kālidāsa, *Kumārasaṃbhava*, ed. Suryakanta (New Delhi: Sahitya Akademi, 1962), I.55 on p. 12.

> The Hide-Clad Controlled-Mind One dwelled for ascetic power (*tapas*) on a certain mountain of the Himalayas where the stream of the Ganges sprinkled the deodars (*devadāru*), where there was perfume of musk, where the Kinnaras sang.

It is significant that in both passages Kālidāsa associates musk with the mountainous world of the Himalayas, which he treated as a remote and almost mystical place of solitary seclusion.[61] Kālidāsa's use of the imagery of musk shows that musk must have been generally known in his time, at least among his courtly audience.

Perhaps the earliest work that includes a chapter on the making of perfumes including musk (if not the earliest work on perfume making in India itself) is the *Bṛhatsaṃhitā* of Varāhamihira, dating to the early 6th century.[62] By this time musk was already a significant ingredient in Indian perfumery as Varāhamihira himself shows.[63] It includes some formulas for scented water, hair oil, and various perfumes, especially scented powders and incense. The basic ingredients of all of the scents described by Varāhamihira are from aromatic plants. Some of the plants regularly called for include aloeswood, cassia, costus, patra, sandalwood, vetiver, and gum guggulu (bdellium), which made a base for incense pastes.

A formula called the *kopacchada* or "lid for anger," intended to be calming, was made of four parts each of sugar, benzoin (*śaila*), *Cyperus rotundus* (*musta*), two parts each of pine [resin] (*śrī*), resin of *Vatica robusta* (*sarja*), onycha, and guggulu.[64] It was mixed with camphor powder and rolled into a ball. Camphor is used for its cooling properties in this particular compound.[65]

---

61 On the value that an exotic cachet brings to aromatics in Indian literature, see J. McHugh, "The Incense Trees from the Land of Emeralds: The Exotic Material Culture of Kāmaśāstra," *Journal of Indian Philosophy* 39 (2011): 94.

62 Varāhamihira, *Bṛhatsaṃhitā*, ed. and trans. M. Ramakrishna Bhat, 2 vols. (Delhi: Motilal Banarsidass, 1981–2). See also McHugh, "Incense Trees" 66–7 and his *Sandalwood and Carrion: Smell in Indian Religion and Culture* (Oxford: Oxford University Press, 2012), 108–11.

63 Musk is also attested among the cosmetics and perfumes in the 6th century glossary *Amarakośa*, see McHugh, "Incense Trees" 77.

64 Varāhamihira 77: 12 (pg. 710).

65 On camphor in India, besides the work of R. A. Donkin, *Dragon's Brain Perfume: An Historical Geography of Camphor* (Leiden: Brill, 1999), see L. Sternbach, "Camphor in India," *Vishveshvaranand Indological Journal* 12 (1974): 425–67.

TABLE 3.1 *The sixteen different substances from which gandhārṇava "perfume-ocean" was made*

Ghana (*Cyperus rotundus*)[a]	*Vālaka* (?)[b]	*Śaibya* (Benzoin)	*Karpura* (Camphor)
Uśīra (Vetiver)	Nāgapuṣpa (?)[c]	Vyāghranakha (?)[d]	*Spṛkkā* (*Trigonella corniculata*?)[e]
Aguru (Aloeswood)	*Madanaka* (?)[f]	*Nakha* (Onycha)	*Tagara* (*Tabernaemontana divaricata*)[g]
Dhanyā (Belleric myrobalan)	*Karcūra* (Turmeric)	*Coraka* (?)[h]	*Candana* (Sandalwood)

a Waku Hakuryū, *Bukkyō Shokubutsu Jiten* (Tōkyō: Kokusho Kankōkai, 1979), #356; Dymock 3.552–3.
b Monier-Williams 946b has a kind of *Andropogon*; Bhat has *Aporosa lindieyana* [sic] (= *Aporosa cardiosperma*).
c Bhat has *Mesua ferrea*. Several plants are identified with the "serpent flower," including jack-in-the-pulpit (*Arisaema* spp.).
d Lit. "tiger-nail". Perhaps some form of onycha? Bhat has "cuttle fish bone".
e Waku #376. J. Nobel "Das Zauberbad der Göttin Sarasvatī," in *Beiträge zur indischen Philologie und Altertumskunde. Walther Schubring zum 70. Geburtstag dargebracht von der deutschen Indologie* (Hamburg: Cram, De Gruyter & Co., 1951), 128; McHugh, *Sandalwood* 268 n. 27. Bhat has *Bryonopsis laciniosa* (=*Cayaponia laciniosa*).
f Monier-Williams 778c "*Artemisia indica*".
g Waku #284; Nobel 132; Dymock 2.413–3.
h Monier-Williams 400c "*Trigonella corniculata*" and a kind of perfume. Same as *taskara* which is *Trigonella corniculata*: Meulenbeld Ia 356 and Ib 462 n. 732 and Waku #376.

Varāhamihira called for musk in its role as an enhancer of scent in compounds. He describes a scented powder made of equal amounts of cassia, vetiver (*uśīra*),[66] and patra, plus half the amount of cardamom; it is reinforced with musk (*mṛga*) and camphor.[67] A compound perfume called *gandhārṇava* "perfume-ocean" was prepared from sixteen different substances, almost all of plant origin, in varying proportions.[68]

66 Also called *khus khus* today; incense scented with it is still available. See Dymock 3.571–3.
67 Varāhamihira 77: 12 (pg. 710).
68 See McHugh, *Sandalwood* 109–10, with references on the mathematical implications of combinatory perfume formulas.

Each ingredient was separately fumigated with resins and onycha and then mixed (this mixing is called *bodha*, lit. "awaking, arousing"[69]) with musk and camphor.[70] This mixing and reinforcing a compound with musk and camphor is an act of enhancing the perfume, comparable to the Arabic verb *fataqa*, "to enhance a perfume". *The gandhārṇava* scented powder could be prepared for the scenting of the mouth. In that case it was made from the coral tree (*pārijāta*, *Erythrina variegata*) flower and the sixteen ingredients of the *gandhārṇava* and then reinforced with nutmeg, musk, and camphor and mixed with mango juice and honey.[71]

Varāhamihira's work is a small fragment of the literature on perfumes that apparently existed in medieval India. This was the science of *gandhaśāstra*.[72] There are at least two extant Sanskrit perfumer's handbooks. One is the *Gandhasāra* of Gaṅgādhara (latter half of the 12th century). The other is the anonymous *Gandhavāda* (c. 1450 CE).[73] Both date to the later medieval period; they have not been translated, and, given their late date, have not been used here.

Another work that contains numerous formulas for aromatic compounds is the *Mānasōllāsa* composed by the Western Chāḷukyan king Sōmēśvara III in 1131 CE (the year 1052 of the Śaka Era).[74] The *Mānasōllāsa* is a long, prescriptive account of life at the royal court that includes a wealth of information on social life, science and technology. Sōmēśvara's applications of aromatics are governed by the understanding of their humoral properties, for example, camphor being cold and musk being hot. Musk is thus called for in preparations used in times in which the body required warming, not cooling. A 12th century

69 *Bodha*, in later times, means "reviving the scent of a perfume with the help of aromatic ingredients acting as reviving agents," P. K. Gode "Indian Science of Cosmetics and Perfumery," in *Studies in Indian Cultural History* vol. 1 (Hoshiarpur: Vishveshvaranand Vedic Research Institute, 1961), 7.

70 Varāhamihira 77:16 on pg. 711.

71 Varāhamihira 77:27 on p. 715.

72 See McHugh "Incense Trees" and *Sandalwood* 249–51 for the key texts.

73 P. K. Gode, "Studies in the History of Indian Cosmetics and Perfumery: A Critical Analysis of a Rare Manuscript of *Gandhavāda* and its Marathi Commentary (Between c. A.D. 1350 and 1550)," *Studies in Indian Cultural History* vol. 1. (Hoshiarpur: Vishveshvaranand Vedic Research Institute, 1961), 43–52. Cf. R. T. Vyas, *Gaṅgādhara's* Gandhasāra *and an Unknown Author's* Gandhavāda (*with Marathi Commentary*) (Vadodara: Oriental Institute, 1989), 13.

74 See P. Arundhati, *Royal Life in* Mānasōllāsa (Delhi: Sundeep Prakashan, 1994); McHugh, "Incense Trees" 71–4 and *Sandalwood* 163–4.

Chinese source mentions that the natives of Malabar perfumed themselves with musk and camphor.[75]

Sōmēśvara gives a recipe for an unguent (*lēpana*) called *puṁliṅga* containing musk and used exclusively in the winter months. In this recipe he describes the best musk: "the best *kastūri* removed from a young and energetic musk deer should be fresh, wet, sticky, round in shape, and does not become ash, if burnt; tastes *tikta* [bitter and pungent—A.K.] and *kaṭu* [pungent and hot—A.K.] and is light in weight. Such a best variety should be used by the king, as it gives excellent fragrance. It is mixed with water and applied on the body of the king."[76] For the betel compound used by the king, which was recommended after a refreshing bath, musk was used along with several other sweet smelling ingredients calculated to emit the smell of the *sahakāra* mango blossom.[77]

Incense is also mentioned in the *Mānasōllāsa*.[78] Sōmēśvara describes three methods of censing and gives their ingredients. The *chūrṇadhūpa* was the first of these, and the basis of the other two. It was an incense that contained numerous ingredients, including musk, that were powdered, mixed with water and boiled until the water evaporated. Then it could be burned as incense with camphor, and it was considered to be the best incense.[79] The other methods of censing extended this compound with other ingredients to make lumps of incense or wicks.

Sōmēśvara mentions several kinds of drinking waters. Among these are some that are perfumed. The type called *piṇḍāvāsa* water was perfumed with a little musk and other aromatics.[80]

The *Agnimahāpurāṇam* contains some formulas for aromatic substances. Like all of the *Purāṇas* it is extremely difficult to date, but the core of the text probably belongs to the first millennium CE. The conduct of the king with his women is discussed in chapter 224, and it is in this context that perfumes are mentioned. First the text clarifies that there are eight processes of making the body free from bad smells: cleansing or washing, gargling, vomiting, adorning the body with flowers and garlands, heating, burning incense sticks,

75 Hirth and Rockhill 87.

76 Arundhati 90–1.

77 Arundhati 89, citing 3.4.V.961–79. Use of musk in betel compounds is also mentioned by the 16th c. traveler Jan Huyghen Linschoten, *The Voyage to the East Indes*, ed. A. C. Burnell and P. A Tiele, 2 vols. (London: Hakluyt Society, 1885), 2.67.

78 McHugh, *Sandalwood* 164.

79 Arundhati 137.

80 Arundhati 131.

fumigation, and using scents and perfumes. To purify something by washing, water containing the leaves of scented trees was used. If these were unavailable, the cleaning could be done with water with musk dissolved in it.

Twenty-four different substances are specified for fumigation; they were to be powdered and moistened with the juice of the Śāla tree to be made into incense sticks. Another method for making incense was to mix these substances with honey to form a paste. These ingredients included many components of incense familiar from the Chinese formulas such as onycha (*nakha*), sandalwood, aloeswood, and camphor, along with other ingredients. But musk is not listed among them. Musk (*mṛgadarpa*) is, instead, used to saturate oil into which these were dipped to increase male potency when applied before the bath.[81] The *Agnipurāṇa* also suggests the use of musk, among other substances, in pills to freshen the mouth. One of the other aromatics is a plant called *latākastūrī*, "musk-creeper." Some identify this plant with *Hibiscus abelmoschus* (ambrette).[82]

As suggested by the *Mānasōllāsa*, different combinations of ingredients were used in different seasons because of the heating or cooling qualities of the aromatics. Musk was considered a warming ingredient, so it was especially favored in compounds for the winter. The contrast in perfumes between the summer and winter is reflected in a description of winter from Addahamāṇu's *Saṃneharāsayu*, the date of which is unknown:

> Camphor and sandalwood are not ground by the servants to mix with wax in adorning the lower lip and cheeks:
>
> Avoiding sandalwood, the body is smeared with saffron; champak[83] oil is used, with musk.[84]

---

81 *Agnimahāpurāṇam*, ed. and trans. M. N. Dutt. 2 vols. (Delhi: Parimal Publications, 2001), 29.

82 Monier-Williams 896a.

83 *Magnolia champaca* (= *Michelia champaca*): Watt 821, Dymock 1.42–3, K. M. Nadkarni, *Indian Materia Medica*. 3rd ed. by A. K. Nadkarni, 2 vols. (Bombay: Popular Book Depot, 1955), 1.794–5, and P. K. Gode, "Studies in the History of Indian Cosmetics and Perfumery-The Campaka oil and its Manufacture (Between A.D. 500 and 1850)," in his *Studies in Indian Cultural History*, vol. 1 (Hoshiarpur: Vishveshvaranand Vedic Research Institute, 1961), 57–67.

84 Quoted in A. K. Warder, *Indian Kāvya Literature Volume VII: The Wheel of Time*. 2 parts, paged continuously (Delhi: Motilal Banarsidass 1989), part 2, 547. I have not seen the Sanskrit.

These verses clearly show that cooling perfumes like sandalwood and especially camphor were preferred in the summer, while their warming counterparts were used in the winter.

Sanskrit poetry is particularly rich in the imagery of scent. Many words for perfumes appear, reflecting a large variety of natural substances. The most abundantly attested aromatic is the sandalwood-paste used as a cosmetic. We also find frequently the oil of the champak flower. While musk is well attested in Sanskrit poetry, it is hardly as common as sandalwood, champak, or many other aromatics. But sometimes Sanskrit poetry can elucidate many aspects of the use of musk in aromatics and cosmetics. A poem by Vīryamitra in the 11th century anthology of Vidyākara called *Subhāṣitaratnakoṣa* attests to the practice of women using musk paste as a cosmetic:

> *dūrvāśyāmo jayati pulakaireṣa kāntaḥ kapolaḥ*
> *kastūrībhiḥ kimiha likhito drāviḍaḥ patrabhaṅgaḥ*[85]
>
> This beloved cheek, black as *dūrvā* grass with its erect hairs, is victorious. Why here [should there be] Dravidian streaks sketched with musk (*kastūrī*)?

The poet says that the framing of the beloved's face by bristling black hairs makes any use of black musk cosmetic superfluous. In Sanskrit poetry a sign of erotic excitement is horripilation, or the standing of the hair on end; it is in no way a negative image. Ingalls comments that painting the cheeks or breasts with musk or aloeswood paste is common in Sanskrit poetry;[86] a survey of his references shows that musk is much less common than aloeswood in the *Subhāṣitaratnakoṣa*, but it certainly was used. Another verse, that contains some untranslatable punning, compares the musk patterns on the face of a beautiful girl with the features of the moon:

> *smitajyotsnāliptaṃ mṛgamadamasīpatrahariṇaṃ*
> *mukhaṃ tanmugdhāyā harati hariṇāṅkasya laḍitam*[87]

---

85 Vidyākara, *Subhāṣitaratnakoṣa*, ed. D. D. Kosambi and V. V. Gokhale (Cambridge, MA: Harvard University Press, 1957), 75, #407. Cf. also the trans. of D. H. H. Ingalls, *An Anthology of Sanskrit Court Poetry* (Cambridge, MA: Harvard University Press, 1965), 169.

86 Ingalls 498 note to #389.

87 Vidyākara #389. See also Jayadeva, *Gītagovinda*, ed. and trans. B. S. Miller (New York: Columbia University Press, 1977; repr. Delhi: Motilal Banarsidass, 1984), 12.16.

> The innocent girl's pale face, anointed with a moonlight smile and a pattern of musk powder, takes that which is desired of the deer-marked moon.

In this contest of beauty, the poet suggests, the girl is the clear victor. Sōmēśvara also mentions that musk was used upon the faces of the ladies of the harem.[88] Furthermore, the *Brahmavaivartapurāṇa* mentions this practice of making marks on the forehead with musk, along with other aromatics.[89] Musk was mixed with kohl and used to anoint the eyes at marriage.[90]

The 11th–12th century Kashmīrī poet Bilhaṇa's *Caurapañcāśikā* is a series of verses recollecting his beloved. It includes three verses which describe her use of musk among her cosmetics and perfumes:

> *adyāpi tāṁ masṛṇacandanapaṅkamiśrakastūrikāparimalottha-visarpigandhāṃanyonyacañcupuṭacumbanalagnapakṣmayugmābhirāma-nayanāṃ śayane smarāmi*[91]

> Even now I remember her in bed, perfume of glistening sandalwood paste mixed with musk diffusing and spreading its scent, her lovely eye with paired eyelashes touching and kissing like the closing of a bird's beak.

> *adyāpi tāṃ nidhuvane madhudigdhamugdhalīḍhādharāṃ kṛśatanuṃ capalāyatākṣīm kāśmīrapaṅkamṛganābhikṛtāṅgarāgāṃ karpūrapūgaparipūrṇamukhīṃ smarāmi*[92]

> Even now I remember her during sex, licking innocently her honey-smeared lower lip, her slim figure and unsteady long eye, limbs dyed with saffron ointment made with musk (or, with unguent made from saffron ointment made with musk), her mouth filled with camphor and betel.

---

88 Arundhati 102.

89 A. J. Rawal, "Society and Socio-Economic Life in the Brahmavaivartapurāṇa," *Purāṇa* 15:1 (1973): 56.

90 Rawal 56.

91 Bilhaṇa, *Caurapañcāśikā* (Varanasi: Chawkhamba Sanskrit Series, 1971), 8.

92 Bilhaṇa 9.

Musk was used in other cosmetic preparations as well. In the 19th century the Jains anointed the spot where the hair of a novice was removed with camphor, musk, sandalwood, saffron, and sugar.[93] The Jains also prepared a sacrificial powder called Vasakshepa that consisted of sandalwood, saffron, camphor, and musk.[94] In short, by the middle of the 1st millennium CE, musk was widely used in Indian perfumery and cosmetics, to judge by the evidence of works on those subjects and poetry. In medicine, by contrast, musk seems to have taken a long time to become established in the widely regarded position it has today.

Sanskrit medical literature refers to a vast array of *materia medica* used in prescriptions. The earliest medical texts do not make much use of musk at all; evidently, in medicine musk also came into use somewhat more extensively during the latter half of the first millennium CE. In modern times, musk has attained a great importance in traditional Indian medicine,[95] but this importance seems to have developed within the last millennium, for the oldest texts make little use of musk.

The oldest medical text that mentions musk is the *Carakasaṃhitā*. Like other Indian medical works, the *Carakasaṃhitā* devotes much time to the manners and methods of a healthy lifestyle, and in this context interesting information about the use of aromatics can be gleaned. Before eating, one should apply "sacred scents," though unfortunately these are not specified.[96] Sandalwood paste is recommended for removing the foul odor of the body,[97] while aloeswood (*agaru*) removes its coldness.[98] Musk is not specified in any of the discussions of the use of scent. In the catalogue of deer in the *Carakasaṃhitā*, the musk deer is not clearly mentioned either. The word *ṛṣya*, which has been translated as musk deer, is apparently an antelope.[99] The one formula in the *Carakasaṃhitā* that includes musk is found in the twenty-eighth chapter. It is a formula for *balātaila*, a medicinal oil used for treatment of the *vātika* diseases.[100] It includes several dozen ingredients along with the musk. This formula is very similar to the formula for the *balātaila* found in Vāgbhaṭa's

---

93 E. Bender, "An Early Nineteenth Century Study of the Jains," *JAOS* 96:1 (1976): 116.

94 Dymock 1.199.

95 Nadkarni 1.196–205.

96 *Carakasaṃhitā*, ed. and trans. R. K. Sharma and V. B. Dash, 7 vols. (Varanasi: Chowkhamba Sanskrit Series Office, repr. 2004–5); *Sūtrasthāna* VIII.20 (Vol 1.175–6).

97 *Carakasaṃhitā Sūtrasthāna* XXV.74 (Vol. 1.431); see McHugh, *Sandalwood* 188 on medicinal properties of sandalwood.

98 *Carakasaṃhitā Sūtrasthāna* XXV.75 (Vol. 1.431).

99 *Carakasaṃhitā Sūtrasthāna* XXVII F.16 (Vol. 1.502); cf. Monier-Williams 226c.

100 *Carakasaṃhitā Cikitsāsthāna* XXVIII.148–57 (Vol. 5.61–3).

Aṣṭāṅgahṛdayasaṃhitā, which also includes musk,[101] and may be an interpolation in the *Carakasaṃhitā*.

It has been suggested that the *medaḥpuccha* of Suśruta refers to musk.[102] J. C. Wright proposed that this word is a Sanskritization of the Prakrit *miyapuccha* found in the *Maṇipaticarita*, which has been dated to after 700 CE, although Wright argues it is earlier. This *miyapuccha* has been understood by the editor of the text to refer to a fat-tailed sheep. Since this term refers to something precious, aphrodisiac, and housed within the royal palace, musk seems a better interpretation. Wright suggests an underlying **mṛgapuccha* "musk gland" Prakritized into *miyapuccha*. Suśruta's *medaḥpucchaka*, appearing in a list of domestic animals,[103] would then be a Sanskritization of it. While the *Harṣacarita* (see below) seems to refer to the keeping of live musk deer, the difficulty modern zookeepers have had keeping musk deer suggests that if ancient Indians did keep musk deer in captivity, it was not for long, and they certainly did not domesticate them. Nevertheless, Thubten Jigme Norbu (1922–2008) describes having a baby musk deer for a pet while a child in Amdo, a native climate for the animal, and notes that his grandmother had kept one previously.[104]

Among the early medical writings, it is in the works of Vāgbhaṭa that musk begins to appear with more frequency. Unfortunately, the precise authorship and date of the two works attributed to Vāgbhaṭa, the *Aṣṭāṅgasaṃgraha* and the *Aṣṭāṅgahṛdayasaṃhitā*, are complex problems; both can, in the broadest terms, be attributed to the early medieval period, about the 7th or 8th century.

It has been suggested that the *Aṣṭāṅgasaṃgraha* may be the first Indian medical work that extends the use of musk.[105] But many of its few references to musk are paralleled in the *Aṣṭāṅgahṛdayasaṃhitā*, and indeed, one appears in the *balātaila* formula found in the *Carakasaṃhitā*, as noted above.

101 Vāgbhaṭa, *Aṣṭāṅgahṛdayasaṃhitā*, ed. R. P. Das and R. E. Emmerick (Groningen: Egbert Forsten, 1998); *Cikitsāsthāna* 21.77. Cf. Meulenbeld 1a.436.

102 J. C. Wright, "The Supplement to Ludwig Alsdorf's Kleine Schriften: A Review Article," BSOAS 62 (1999): 533–4.

103 *Suśrutasaṃhitā*, ed. and trans. P. V. Sharma. 3 vols. (Varanasi: Chaukhambha Visvabharati, repr. 2004–5); Sū*trasthāna* 46.85 (Vol. 1.481). It is said to be an aphrodisiac and similar in properties to mutton in line 88 on p. 482.

104 Thubten Jigme Norbu and H. Harrer, *Tibet is My Country: Autobiography of Thubten Jigme Norbu, Brother of the Dalai Lama* (London, 1960; repr. London: Wisdom Publications, 1986), 43–4.

105 Meulenbeld IA 647.

An interesting passage in the *Uttarasthāna* of the *Aṣṭāṅgasaṃgraha* discusses gum guggulu, which is praised. Musk (*darpa*) is held to be inferior to it.[106]

In the *Aṣṭāṅgahṛdayasaṃhitā* there are also a few references to musk. The word *darpa* occurs in the second hemistich of 3.11: *kuṅkumeṇa sadarpeṇa pradigdho 'guru dhūpitaḥ* "smear [the body] with saffron mixed with musk and fumigate with aloeswood."[107] This procedure is part of the regimen of care for the body during the winter. In Chapter 13 of the *Uttarasthāna*, Vāgbhaṭa describes a *cūrṇa* for *kapha* diseases that uses some musk.[108]

Vāgbhaṭa's near contemporary, Ravigupta, in his *Siddhasāra,* apparently does not mention musk at all; it is not found in his chapter listing the various medications. In short, the early medieval period was the time in which musk was beginning to be used in Indian medicine, and not everyone made use of it. Even Vāgbhaṭa makes only a handful of references to it. However, in later centuries musk took up a more secure position in Indian medicine and is better attested. A work that is perhaps a century or more later than Vāgbhaṭa, the *Haramekhalā* of Māhuka, written in Prakrit and dating to the 9th century, makes more use of musk and even contains formulas for imitations.[109] The fifth part of this book discusses fragrant compounds.[110] Some other later texts attesting musk are the *Cikitsāsaṃgraha* of Cakrapāṇi (11th century)[111] and the so-called *Jīvakapustaka*, which is preserved in Sanskrit and a Khotanese translation, presumably dating before the year 1000.[112] A later work, the *Nighaṇṭu* of Madanapāla (14th century), gives synonyms for musk and its properties: pungent in taste, heavy, and hot in potency. It promotes semen and alleviates *kapha* (phlegm) and *vāyu* (wind). It cures feelings of cold, poisoning, vomiting, edema (*śopha*), and bad odor from the body (*daurgandhya*).[113]

One of the earliest securely dated texts that mentions musk is the *Harṣacarita* of Bāṇabhaṭṭa. This long poem is a celebration of the reign of the king Harṣa (ruled in the early 7th century CE); Bāṇa was his contemporary. It is written in an extremely elaborate and densely descriptive language. In Book 7, there is a long list of very valuable and elaborate presents offered to

106 *Aṣṭāṅgasaṃgraha* 49.258; See Meulenbeld IB 703 n. 726.

107 *Aṣṭāṅgahṛdayasaṃhitā*, ed. Emmerick and Das, 5.

108 *Aṣṭāṅgahṛdayasaṃhitā Uttarasthāna* 13.25. Cf. Meulenbeld IA.449.

109 Meulenbeld IIA 134; McHugh 2011: 68, McHugh, *Sandalwood* 195–6.

110 Meulenbeld IIA 132, 134 and IIB 151 n. 341.

111 Meulenbeld IIA 90.

112 Meulenbeld IIA 126. See below 114.

113 V. B. Dash and K. K. Gupta, *Materia Medica of Ayurveda based on Madanapala's Nighaṇṭu* (New Delhi: Health Harmony, 1991; repr. 2001), 162–3.

the king. It includes musk twice: first in the form of musk pods (*kastūrīkākośa*) and, later, as apparently live musk deer perfuming the space around them (*parimalāmoditakakubhaśca kastūrīkākuraṅgān*).[114] This long catalogue of carefully described gifts includes jewels, textiles, and other valuable items. Some of the other aromatics mentioned include aloeswood, aloeswood oil, black aloeswood dark as kohl, and camphor. Musk is listed among the perfumes and drugs in an inscription from Karnataka of about 1150 CE: "cardamoms, cloves, bdellium, sandal, camphor, musk, saffron, malegaja, and other perfumes and drugs."[115] This inscription shows that the trade in musk was well developed at the time and profitable. It records that the customs dues on "camphor, musk, saffron, sandal, pearls and all such commodities sold by weight" was 2 *kāṇi* per *pon*.[116]

Indian culture evidently maintained a high opinion of musk. While it is not as abundantly used in proverbial expressions as sandalwood, which features in dozens of proverbs, those that include it speak of it in a very positive way. For example:

> *mṛganabhisamā prītirna tu gopāyyate kvacit*
> *āvṛtāpi punastasya gandhaḥ sarvatra gacchati*[117]
>
> Friendship is like musk; it cannot ever be hidden. Even when it is covered, that scent still goes everywhere.

A more melancholy aphorism states:

> *dhātastāt tavaiva dūṣaṇamidaṃ yannāma kastūrikā*
> *kāntarāntaracāriṇaṃ tṛṇabhujāṁ nābhau kṛtā mṛtyave*[118]

114 *Harṣacarita*, ed. P. V. Kane (Bombay, 1918; repr. Delhi: Banarsidass, 1997), 117. The description of the gifts begins on 116. Trans. E. B. Cowell and F. W. Thomas, *The Harṣacarita of Bāṇa* (Repr. Delhi: Motilal Banarsidass, 1993), 214; they have incorrectly translated the first instance "scent bags of musk oxen". See also McHugh, "Incense Trees" 84 and *Sandalwood* 170.

115 B. Lewis Rice, *Epigraphia Carnatica Vol. VII. Inscriptions in the Shimoga District (Part I)* (Bangalore: Mysore Government Central Press, 1902), 86.

116 Lewis Rice 87.

117 O. Böhtlingk, *Indische Sprüche*, 3 vols. (St. Petersburg, 1870–3; repr. Osnabrück: Otto Zeller, 1966), #7597.

118 Böhtlingk #3140.

> Oh Creator, is this thus your sin, that indeed musk made in the navels of those who wander in the wilderness and eat grass makes them die?[119]

The poet plays on the fact that wandering in the wilderness eating plants makes men notable ascetics and holy men, but it causes the death of the musk deer, something the poet finds difficult to accept. The concept that one should disregard the origin of something good is often likened to the production of musk from deer:

> *gauravaṁ labhate lōke nīcajātistu sadgunaiḥ*
> *saurabhyātkasya nābhīṣṭā kastūrī mṛganābhijā*[120]
>
> One of low birth in the world gains importance by his good qualities;
> For whom is musk which is derived from the musk deer navel not dear because of its fragrance?

Musk has an excellent scent because of its innate characteristics, not because of its elaborate preparation:

> *yadi santi guṇāḥ puṁsāṁ vikasantyeva te svayam*
> *nahi kastūrikāmodaḥ śapathena vibhāvyate*[121]
>
> If there are qualities in men, they blossom by themselves.
> Certainly the fragrance of musk is not manifested by an ordeal.

Finally, an epigram comments that only those of discernment are capable of appreciating musk and thus those who cannot recognize its qualities are not fit to be superior people:

> *api tyaktāsi kastūri pāmaraiḥ paṅkaśaṅkayā*
> *alaṁ khedena bhūpālāḥ kiṁ na santi mahītale*[122]
>
> Oh musk, you are avoided by fools who think you are mud.
> Enough hardship! Why are they not kings on the surface of the earth?

---

119 Böhtlingk #3140.

120 Böhtlingk #2208.

121 Böhtlingk #5237.

122 Böhtlingk #560.

India, with its abundant indigenous aromatics, never gave musk the sort of importance it gave to sandalwood, aloeswood, or various other aromatics, especially the floral scents. Given the great importance of aromatics in Indian life, in incense for worship, or in perfumed cosmetics, it is perhaps surprising that musk did not attain a greater degree of fame. But musk was certainly known and used, especially for its property of enhancing other scents. Interest in the scent of musk only increased. Perhaps influenced by Islamicate culture, musk, along with civet,[123] came into much greater use after the 9th century or so. Ambergris, called *ambar*, which is borrowed from Persian or perhaps Arabic, also came into wider use after this time while it is unattested for the time of Varāhamihira. By the 17th century, the popularity of musk was well-established. Chardin wrote: "The Indians make great Account of this Aromatick Drug, and esteem it, as well for its Use, as for the great Demand there is for it. They use it in their Perfumes, in their Medicinal Epithems, and Confections, and in all Preparations which they are accustom'd to make, in order to awaken the Passions of Love, and confirm the Vigour of the Body."[124]

Indian culture had a profound and long-term impact on the cultures of Southeast Asia and Central Eurasia. At Ta-prohm in Cambodia, there is an inscription dated to 1186 CE carved at the order of Jayavarman VII that describes the pious foundations he endowed. Among the numerous items catalogued in this inscription are aromatics including camphor, *taruska* (*turuṣka*, frankincense?), sandalwood, musk (*kastūrī*), and other aromatics.[125] The terms for musk used in Southeast Asia are often derived from India. For example, Antonio Pigafetta, who sailed with Magellan's expedition, records that the word for musk in the Moluccas was *castori*.[126]

With the spread of Buddhism, Indian medicine made continuous progress into Tibet and Central Eurasia. As all of our sources for the history of musk in these areas date to periods after contact with India, it is often impossible to separate out the threads of truly indigenous uses of musk from those prescribed by Indian culture and medicine.

---

123 J. McHugh, "The Disputed Civets and the Complexion of the God," *JAOS* 132:2 (2012): 249–51, for early Indian civet.

124 John Chardin, *Travels in Persia 1673–1677* (London, 1927; repr. New York: Dover, 1988), 151.

125 G. Coedès, "La stèle de Ta-prohm," *Bulletin de l'École Français d'Extreme-Orient* 6 (1906): 78–9, etc.

126 A. Pigafetta, *Magellan's voyage: a narrative of the first circumnavigation*, trans. R. A. Skelton (New Haven: Yale, 1969; repr. New York: Dover, 1994), 174.

Musk in Tibet

Tibet and the Himalayas were home to the most famous musk in the medieval tradition. It can be presumed that the knowledge of musk and its exploitation goes back very far in time in Tibet. The sources, however, are extremely limited for the earliest period. With the formation of the Tibetan empire, Tibet gained a powerful centralized state that engaged in world trade.[127] The earliest inscriptions on stone in the Tibetan language appear at the same time. Unfortunately, these inscriptions are all political in nature and do not include references to aromatics. The earliest documents on paper in Tibetan that we have were written a little later; these were excavated in modern times from sites in the Tarim Basin and Gansu. These documents are also of disappointingly limited use for the tracing of the use of musk by the Tibetans. In the business documents that exist, no mention of musk or the musk trade has been found so far. An Old Tibetan medical fragment paralleling a text in the Kanjur was recovered from Turfan in the early 20th century; it contains various remedies but does not mention musk.[128] Two references to musk have been found in IOL Tib. 756, a manuscript from Dunhuang perhaps from the time of the Tibetan Empire.[129] Both passages refer to the use of musk in treating wounds; in one case, it helps stanch bleeding, in the other, it helps dry swollen wounds.

Tibetan traditional literature provides much more information on musk; all of this literature comes from the later (post-imperial) Buddhist tradition. There is an extensive corpus of medical literature that contains much information on *materia medica*. But the Tibetan tradition is so permeated by India that it is difficult to separate peculiarly Tibetan uses of musk from those deriving from Indian usage. Musk is one of the five scents (*dri lṅa*) used in Tibetan Buddhist practice: white and red sandalwood, camphor, saffron, and musk.[130] In this form it was used extensively in ritual practice.

One of the oldest and most important of the Tibetan medical works is known as the *Rgyud bźi*, the Four Tantras. It is the foundation on which later

---

127 C. I. Beckwith, "Tibet and the Early Medieval Florissance in Eurasia: A Preliminary Note on the Economic History of the Tibetan Empire," CAJ 21 (1977): 89–104.

128 G. Kara, "An Old Tibetan Fragment on Healing from the Sutra of the Thousand-Eyed and Thousand-Handed Great Compassionate Bodhisattva Avalokiteśvara in the Berlin Turfan Collection," in D. Durkin-Meisterernst, et al., eds., *Turfan Revisited- The First Century of Research into the Arts and Cultures of the Silk Road* (Berlin: Reimer, 2004), 141–6.

129 See A. Akasoy and R. Yoeli-Tlalim, "Along the Musk Routes: Exchanges between Tibet and the Islamic World," *Asian Medicine* 3 (2007): 231.

130 S. Beyer, *The Cult of Tārā: Magic and Ritual in Tibet* (Berkeley: University of California, 1973; repr. 1978), 290.

Tibetan medical literature is based, and its relationship to Indian medical literature is complex and unclear.[131] Musk is considered one of the "super-potent" substances.[132] In the section on medicines within the second tantra is a short list of the properties of musk: "it cleanses poison, worms, the kidney, liver, and plague (*gnyan-nad*)."[133] This is elaborated in Sdesrid Saṅs-rgyas rgyamtsho's commentary on the *Rgyud bźi*, the *Baiḍūr sṅonpo*.[134] He notes that the musk deer has twenty names[135] and that musk comes from the musk vesicle of the musk deer. He regards grain musk as superior and adds that it cleanses disorders of the eyes and channels, diseases caused by *nāgas* (snake spirits) and other spirits, and retention of urine. The uses of musk thus cluster according to several themes:[136] treatment of poison, humoral disorders, as an anthelmintic, and for eye diseases. Many of these are paralleled in the Islamicate medical tradition.[137] As the *Rgyud bźi* stated, musk was especially regarded for its ability to treat poisons, especially poison from snakebite.[138] Musk's effectiveness at treating snakebite is also attested within the Chinese, Nepalese, and Islamicate traditions. Musk was used in Tibetan (as in Islamicate) versions of the theriac, a panacea of Greek origin which was adapted and elaborated through the ages.[139]

In more modern times, musk was used in numerous medical applications. A specialist in Tibetan medicine in 20th century Kathmandu included musk in 27 out of his 86 preparations, though musk was often used in smaller quantities than called for because of its high cost.[140] Different parts of the musk deer, including hoof, bones, musk, and also the hair, were used against fever in

---

131 On the date and relationship see Meulenbeld IA 656 with citations of the literature.

132 Akasoy and Yoel-Tlalim 231–2.

133 *Rgyud bźi, A Reproduction of a set of prints from the 18th century Zuṅ-Cu Ze Blocks from the Collection of Prof. Raghu Vira* (Leh: S. W. Tashigangpa, 1975), II.19c.

134 *Baiḍūr sṅonpo* (Leh, 1974), 95c.

135 Some synonyms are listed in Vaidya Bhawan Dash, *Materia Medica of Tibetan Medicine* (Delhi: Sri Satguru, 1994), 415.

136 Following Akasoy and Yoeli-Tlalim 232–3.

137 Akasoy and Yoeli Tlalim 233.

138 Akasoy and Yoeli-Tlalim 232. 'Jam dpal rdo rje f.39v and 117r.

139 Akasoy and Yoeli-Tlalim 234–5. See also C. I. Beckwith, "The Introduction of Greek Medicine into Tibet in the Seventh and Eighth Centuries," *JAOS* 99 (1979): 297–313, and D. Martin, "Greek and Islamic Medicines' Historical Contact with Tibet: A Reassessment in View of Recently Available but Relatively Early Sources on Tibetan Medical Eclecticism," in A. Akasoy, et al., eds., *Islam and Tibet: Interactions along the Musk Routes* (Farnham: Ashgate, 2011), 117–43.

140 Jest 233–4.

the early 20th century.[141] Musk was used in a charm against *nāgas* and against clawing animals.[142]

The great Saskya Paṇḍita (1182–1251) wrote, among his other works, a book called the *Subhāṣitaratnanidhi*. This work is a collection of aphorisms, many of which use material from daily life to illustrate wisdom. Aromatics appear quite frequently. The most frequent seems to be sandalwood, often mentioned as a wood or because it can be burned. Musk is mentioned only once:[143]

> (30) *legs-bshad byispadag laskyaṅ mkhasparnamskyi yoṅssu lendri źim 'byung na ridagskyi lteba*[144] *laskyang glartsi len*
>
> The wise wholly accept a wise remark, even from ignorant people; if a sweet scent arises, one accepts musk even from the navel of a wild animal.

This aphorism is comparable to Indian aphorisms that suggest that musk is valued regardless of its somewhat disturbing origin, but so far an exact Sanskrit parallel has not been traced.[145]

The Tibetan translation of Vāgbhaṭa mentions the use of musk in a cosmetic procedure:

> *gur-gum glartsi daṅ bcaspas lus bsku akaruyis bdug*[146]
>
> Smear the body with saffron together with musk and fumigate with aloeswood.

141 W. L. Hildburgh, "Notes on some Tibetan and Bhutia Amulets and Folk-Medicines, and a few Nepalese Amulets," *Journal of the Royal Anthropological Institute of Great Britain and Ireland* 39 (1909): 394.

142 L. A. Waddell, "Some Ancient Indian Charms, from the Tibetan," *Journal of the Anthropological Institute of Great Britain and Ireland* 24 (1895): 42, 43.

143 Sa Skya Paṇḍita, *Subhāṣitaratnanidhi*. Tibetan and Mongolian texts edited and translated by J. E. Bosson, A Treasury of Aphoristic Jewels (Bloomington: Indiana University, 1969), I.30.

144 Bosson gives this as *ldeba* and *lteba* is in some manuscripts.

145 I can find none exactly like it in Böhtlingk's collection, but compare #2208 on vol. 1, p. 432: "One of low birth in the world gains importance by his good qualities; For whom is musk which is derived from the musk deer navel not dear because of its fragrance?".

146 C. Vogel, *Aṣṭāṅgahṛdayasaṃhitā. The First Five Chapters of its Tibetan Version* (Wiesbaden, 1965), 3.11.

Of course, the most important use of musk was likely in incense, as in China. Incense is used in great abundance in ritual practice.[147] Traditional Tibetan incense makes extensive use of juniper leaves; a paste for making incense sticks is basically formed of ground juniper leaves, saffron, sandalwood, aloeswood, musk, and other ingredients.[148] Berthold Laufer located an incense recipe in the *Tanjur*.[149] It is called the "Garland of Jewels" (Tibetan *rinpoce'i phreṅba*, Sanskrit *ratnamālā*). The text is stated to be a Tibetan translation of a Sanskrit text called the *dhūpayogaratnamālānāma*, which does not appear to be extant. In any case, it is simply a short recipe. It cannot be dated precisely, but must be earlier than the compilation of the *Tanjur* in the 14th century:

> A half ounce of musk (*ridags lteba'i dri*, lit. "deer navel scent")
> Two ounces of camphor (*gabur*)
> A half ounce of saffron from Kashmir which must be pure (*khacheyi gurgumdag ni sraṅ phyed yin*)
> One puts an ounce of *nagi* upon it
> Just sixteen ounces of valerian (*spaṅ spos*)
> Four ounces of what is called *sila* (Skt. *sillakī*)
> Two (ounces) of cumin (?) come into it
> And an equal amount of gum guggulu
> Having put it into a pile it is beaten to a powder
> Combine it well with the gum guggulu
> Take for it the measure of its thickness of charcoal.

While this is surely an original Indian formula, as it claims, it shows the kind of incense being made in Tibet. The basis of the paste was the gum guggulu, which was used for this purpose in India as well.

Earlier incense formulas are found in the Tibetan translations of the *Kālacakratantra*. The Sanskrit original of this text dates to the first half of the 11th century and is accompanied by a commentary known as the *Vimalaprabhā*. In the text itself musk is included in a list of 25 substances used for making incense that were to be stored in a special compartmented

147 W. Zwalf, *The Heritage of Tibet* (London: British Museum, 1981), 80 and N. Bazin, "Fragrant Ritual Offerings in the Art of Tibetan Buddhism," *JRAS* (2013): 179–207.

148 Bazin 33,34.

149 B. Laufer, "Indisches Recept zur Herstellung von Räucherwerk," *Verhandlungen der Berliner Gesellschaft für Anthropologie, Ethnologie und Urgeschichte* (July 18, 1896): 394–8.

box according to the commentary.[150] Musk seems to have been a basic ingredient in the incenses prepared according to these directions, but it was also used in a small quantity: the text notes that musk was used in a single portion in the composition of incenses, which could include a total of 15 portions and up. The incense formulas of the *Kālacakratantra* were often prepared with ground ingredients mixed with sugar and/or honey to make a paste. The role of musk was to enhance the incense. Interestingly, the *Kālacakratantra* also calls for civet (*pūti*) in likewise small quantites[151] and the Vimalaprabhā commentary includes ambergris (*ambara*) as a substitution for saffron in the *Kālacakratantra*.[152] The *Vimalaprabhā* also describes an oil for bathing or anointing that could be scented with musk.[153]

The incense made in India and Tibet, as attested from these formulas and the formulas preserved in Indian sources, was extremely different from the incense formulas in the Islamicate Middle East.[154] The Arabic formulas are based on the mixing of large proportions of ambergris, which was little used in India, with musk, as discussed in the next chapter. Indian formulas made much more use of resins and woods; ironically, some of these resins, such as frankincense and perhaps some bdellium (gum guggulu)[155] were imported from the Near East, where they became most uncommon in the wake of the new imports from the East.

Musk in Central Asia

Apart from China, musk is next attested in Central Asia, in the famous "Sogdian Ancient Letters" as noted in Chapter 1.[156] These documents are generally

---

150 V. A. Wallace, trans., *The Kālacakratantra: The Chapter on the Individual together with the Vimalaprabhā* (New York: American Institute of Buddhist Studies, 2004), 189–92.

151 Wallace 201.

152 Wallace 190.

153 Wallace 205.

154 A. King, "Tibetan Musk and Medieval Arab Perfumery," in A. Akasoy, et al., eds., *Islam and Tibet: Interactions along the Musk Routes* (Farnham: Ashgate, 2011), 145–61.

155 The West Asian bdellium is also known as *anxi xiang* 安息香 "Parthian Fragrance" in Chinese. See K. Yamada, *A Study on the Introduction of An-hsi-hsiang in China and that of Gum Benzoin in Europe*, 2 parts (Kinki University, 1954–5).

156 Originally published by H. Reichelt, *Die soghdischen Handschriftenreste des Britischen Museums II. Teil* (Heidelberg: Winter, 1931). See now F. Grenet, N. Sims-Williams, and É. de la Vaissière, "The Sogdian Letter V," *BAI* 12 (1998): 91–104 and N. Sims-Williams, "The Sogdian Ancient Letter II," in *Philologica et linguistica: historia, pluralitas, universitas:*

believed to date to the early 4th century. Despite this early and promising start, the materials for the study of musk in Central Asia are extremely incomplete. It is impossible to say very much about how musk was used and regarded by the peoples of Central Asia because of this insufficiency of written material.

The Sogdian Ancient Letter II was written and posted to Samarkand by a Sogdian merchant who was active in the towns on the western frontiers of China; luckily for posterity it never arrived and remained in a cache with the other Sogdian Ancient Letters in the Dunhuang *limes*. Nanai-vandak had an understanding of events in the heart of China even if communications had become disrupted by the time of his letter; the letter's reference to the sack of Luoyang by the "Huns" enabled the dating of this letter, and hence the whole corpus, to shortly after 311. This particular sack of Luoyang fits the other evidence for the dating of the letters better than the other sieges of Luoyang in 190 and 535.[157]

In Sogdian Ancient Letter II the merchant Nanai-vandak records a testament of his goods: "And Wan-razmak sent to Dunhuang for me 32 (vesicles of) musk (*yxsyh*) belonging to Takut so that he might deliver them to you. When they are handed over, you should make five shares, and therefrom Takhsīch-vandak should take three shares, and Pēsakk (should take) one share, and you (should take) one share."[158] No word such as "vesicle" is present in the original. The translator has supplied it because to an English speaker it seems a logical way for musk to have been counted here.[159] Taking as an average weight of the musk in pods, de La Vaissière calculates that this amounted to 0.8 kg of pure musk. He further notes that in Turfan in 743, musk cost 110 Chinese cash (equal to 3.43 silver coins) per *fen* (0.41 g). That gives a cost of 8.4 silver coins per gram of musk, making it extremely expensive. If one assumes that the price of musk had remained constant, the value of this shipment of musk would have been 27 kg of silver. There is no evidence available for the price of musk in 311, and that is centuries from 743. But the great value of this musk can be shown, as

Festschrift für Helmut Humbach zum 80. Geburtstag, (Trier: Wissenschaftlicher Verlag, 2001), 267–80.

157 F. Grenet and N. Sims-Williams, "The Historical Context of the Sogdian Ancient Letters," *Transition Periods in Iranian History* (Paris, 1987), 101–22 and É. de la Vaissière, *Sogdian Traders: A History*, trans. J. Ward (Leiden: Brill, 2005), 45–6.

158 Trans. N. Sims-Williams, "The Sogdian Ancient Letter II," in *Philologica et linguistica: historia, pluralitas, universitas: Festschrift für Helmut Humbach zum 80. Geburtstag*, (Trier: Wissenschaftlicher Verlag, 2001), 267–80.

159 Cf. also N. Sims-Williams, "On the Plural and Dual in Sogdian," *BSOAS* 42 (1979): 341.

de La Vaissière notes, by the fact that three fifths of this musk was to form the entire inheritance of Takhsīch-vandak, his son.[160]

The Sogdian Ancient Letter II shows that Sogdian traders in China, who were exporting goods to Samarkand, were trading in large quantities of valuable musk in the early 4th century. This musk probably originated in China, but could have also come from northeastern Tibet. This is around the time musk begins to appear in Indian literature, as discussed below, and a bit before it must have reached the Near East; it is of course not known for how long the Sogdians had been trading in musk. This suggests that the Sogdians played an important role in the development of the trade in musk and its adoption outside of the lands of its origin. The Sogdians continued to trade as long as their mercantile empire survived, and then their commerce merged with the commerce of Islamic Khurāsān as Islam spread through Transoxiana during the 8th century. In Islamic times there was a type of musk called *al-misk al-sughdī*. This term refers to musk imported by the Sogdians from Tibet, and it was usually considered the best musk.

As mentioned in Chapter 1, the Khotanese word for musk is *yausa*,[161] which is probably connected to the Sogdian words *yγs* (*yaγs*) and *yγš* (*yaγš*). The so-called *Jīvakapustaka*, which exists in a corrupt Sanskrit version and a Khotanese translation, calls for musk in just one of its formulas. This particular formula includes a number of Indian ingredients: lavanga, camphor, cassia or cinnamon bark (*tvak*), sandalwood, pepper, etc., along with musk. A paste was made from them with sesame oil, cow's fat, water, and mother's milk. The resulting substance was put into the nose and was said to benefit eye diseases, pains in the eyebrow and ear, and partial blindness (*timira*) and migraine (*ardhāvabhedaka*).[162]

A native Khotanese composition is a *deśana* in the name of a prince named Tcūṃ-Ttehi. It cannot be precisely dated, but presumably comes from the last few centuries of the first millenium CE. It is a personal testimony of Buddhist faith and includes a description of the prince's offering:

---

160 All of these calculations are found in É. de la Vaissière, *Sogdian Traders: A History*, trans. J. Ward (Leiden: Brill, 2005), 53.

161 H. W. Bailey, *A Dictionary of Khotan Saka* (Cambridge: Cambridge University Press, 1979), 343.

162 S. Konow, *A Medical Text in Khotanese: CH.ii 003 of the India Office Library* (Oslo: Dybwad, 1941), #57, 62–3. Bailey, *Indo-Scythian Studies being Khotanese Texts Volume 1*, reprinted in one volume with volumes 2 and 3 (Cambridge: Cambridge University Press: 1969), 1.178 (= 97v4).

> The *kauṇḍana-sāraga* sandal[wood]-perfume, the exquisite musk perfume, *uragasāra* sandal[wood], saffron, camphor and the rest, the white and red sandal[wood], the royal perfumes, according to my vow may I rain down as rain upon the Buddhas. The gifts of jewels, fine pearls, strewn incense, musk, with flowers of all colors,—with them all seats are strewn, the *cintāmaṇi* jewel, pure and clear, the *sūryakānta, indragopaka* gems, the *padmarāga* jewels, the noble *candrakānta*. May I thus ripen myself with the devotion of reverence.[163]

While many of the precious materials mentioned by Tcūṃ-Ttehi are Indian in origin, the prominence of musk seems unusual for a source in India. Presumably in Khotan, on an important trade route from China into Central Asia, musk was a highly regarded perfume since it is mentioned in such a luxurious context.

As mentioned in Chapter 1, Old Turkic has the complicated terminology *kin yıpar* and just *yıpar* for musk. Many of the attestations of these words appear in documents that are fragmentary and difficult to interpret. As in Tibet, the extent of Indian influence makes it hard to distinguish between peculiarly Turkic uses of musk and those borrowed from Indian, especially Buddhist, culture. And as far as Buddhist culture is concerned, perhaps the vast bulk of the extant Buddhist writings in Old Turkic are translations not from Indian originals, but from Chinese versions. The Uyghur *Suvarṇabhāsottamasūtra*, discussed in Chapter 1, included musk, as in the Chinese version, but that is not found in the Sanskrit original.

Paul Pelliot recovered several commercial documents in Uyghur from Dunhuang. These documents, along with some others,[164] form the corpus of Uyghur business documents.[165] They are of great importance for the history of the Eurasian overland commerce. Unfortunately, the dating of all of these documents has been disputed. The corpus of Uyghur contracts from East Turkestan

163 Trans. H. W. Bailey, "The Profession of Prince Tcūṃ-Ttehi," in E. Bender, ed., *Indological Studies in Honor of W. Norman Brown* (New Haven: American Oriental Society, 1962), 19–20. Cf. R. E. Emmerick, *Guide to the Literature of Khotan*, 2nd ed. (Tokyo: International Institute for Buddhist Studies, 1992), 37.

164 J. Oda, et al., eds., *Sammlung uigurischer Kontrakte* 3 vols. (Osaka: Osaka University Press, 1993). Older edition of many of these documents is W. Radloff [=V. Radlov], *Uigurische Sprachdenkmäler* (1928; repr. Osnabrück: Biblio Verlag, 1972).

165 J. R. Hamilton, *Manuscrits ouïgours du Ixe–Xe siècle de Touen-houang*, 2 vols. (Paris: Peeters, 1986) contains documents from Paris and London.

may date to the Mongol period (13th–14th centuries)[166] or earlier.[167] For the Pelliot Uyghur business documents, the dating is similarly disputed. Their editor, James Hamilton, favored a date from around the 10th century,[168] but he has been challenged by Marcel Erdal, who argues that some of the documents are likely from the Mongol period.[169] Hamilton's dating seems persuasive for the particular documents discussed here, and these can be provisionally placed in the 10th century.

The first of these commercial documents is a legal document drawn up by an unknown official in an inquiry about the disposition of goods belonging to a deceased merchant.[170] This man was named Oγšaγu, described as a merchant with a caravan (*atlıγ sart*), and he had come to the city of Khotan (called Odon in Uyghur) but died. His goods had been taken by his younger brother's son. Later, a man named Ču came to claim the goods; his relationship to Oγšaγu is not known. The remainder of this fragmentary document lists some of Oγšaγu's goods and their locations. A certain Ču Xayšın had received *bir täk yıpar* each from the son of Yartaš and the son of Oγšaγu. This phrase means "just one musk," presumably, "one musk pod." But in the next item description, a measure is used to describe the musk, and the quantity involved is much more than a single pod. A certain Qutluγ Arslan El-Tiräk took twenty *zatır* of musk. The word *zatır* is derived from Sogdian *st'yr*, which ultimately goes back to the Greek στατήρ.[171] The twenty staters are probably the unit of weight rather than the cost of the musk; in the fourth century the stater had remained at close to its ancient Greek measurement of 16 grams,[172] but in the Uyghur documents the *satır* is equal to the Chinese *liang* 兩,[173] about 37.3 grams under the Song.[174]

166 L. V. Clark, *Introduction to the Uyghur Civil Documents of East Turkestan (13th–14th CC.)*. PhD. Dissertation. Indiana University, Bloomington, 1975.

167 T. Moriyasu, "Notes on Uighur Documents," *Memoirs of the Research Department of the Toyo Bunko* 53 (1995): 68–9.

168 Hamilton esp. xi-xii. See also Hamilton, "On the dating of the Old Turkish Manuscripts from Tunhuang," in *Turfan, Khotan und Dunhuang. Vorträge der Tagung 'Annemarie v. Gabain und die Turfanforschung'* (Berlin: Akademie Verlag, 1996), 135–46, and G. Kara's review of Hamilton's book in *Acta Orientalia Academiae Scientiarum Hungaricae* 43 (1989): 126.

169 M. Erdal, "Uigurica from Dunhuang," *BSOAS* 51 (1988): 252.

170 Hamilton #18, 103–5 and pl. on 300.

171 Hamilton 105.

172 V. A. Livshits and V. G. Lukonin, "Srednepersidskie i sogdiiskie nadpisi na serebrianikh sosudakh," *Vestnik Drevnei Istorii* 3 (1964): 155–76.

173 F. W. K. Müller, "Uigurische Glossen," *Ostasiatische Zeitschrift* 8 (1919–20): 319–21.

174 So Sivin 74. It ranged, according to Müller, from 35 to 37 ¾ grams.

This would produce a total weight of about 746 grams of musk, representing anywhere from 25 to 50 pods or so (given that the musk pod can contain from about 15 to 30 grams of musk). This is a large quantity of musk and would have been extremely valuable. Since musk deer are not native to the area of Khotan, this musk must have been either imported—from China, or Tibet—for local use or as merchandise destined for the West, for the Islamic world.

The second Uyghur document that includes a mention of musk is an account written in Shazhou 沙州 (spelled *š'čyw* in Uyghur), which had its prefectural capital at Dunhuang 敦煌, where the documents were found.[175] This document describes the location of numerous goods ranging from silverware to textiles. In the fourth line it states: "four *patrı* of musk remain with Mir Yegän." *Patrı* is a loanword from Indic *pātra*, "bowl," which, in the spelling *patır* or *batır* is used for different bowls in Uyghur literature, from bowls used in medical prescriptions to the begging bowl of the Buddhist monk.[176] One can imagine that a bowl of musk also represented a fairly sizable quantity of the substance, but it is impossible to be more precise in the absence of information on how much the bowls contained.

There are several medical documents in Uyghur. They all appear to be heavily influenced by the Indian medical tradition; indeed, some fragments have been identified with specific Indian medical works. Musk is definitely mentioned six times in the Old Turkic medical texts.[177] None of the formulas in which musk is required have been identified with an Indian prototype, and it is likely that at least some of them represent the indigenous medical knowledge of the Uyghur. One formula runs: "If a woman's ulcer becomes itchy (*kayu tiši kartı kicinür bolsar*)[178] she should eat the flesh of a swallow (*karlıγac*) or she should drink wine (*bor*) mixed together with saffron, millet flour (*konak meni*),[179] and musk, and it will be beneficial."[180] Another formula with ingredients mixed with wine and drunk calls for *sıpar*.[181] Possibly this *sıpar* is an

175 Hamilton #34 165–9 & pl. on 323.

176 G. Clauson, *An Etymological Dictionary of Pre-Thirteenth Century Turkish* (Oxford: Oxford University Press, 1972), 307a with references.

177 One of these, G. R. Rachmati, "Zur Heilkunde der Uiguren II," *Sitzungsberichte der preussischen Akademie der Wissenschaften zu Berlin, Phil.-hist. Kl.* 1932: 5.6 on 432, is very fragmentary.

178 Clauson's reading, *Etymological Dictionary* 698a.

179 Clauson, *Etymological Dictionary* 637a.

180 G. R. Rachmati, "Zur Heilkunde der Uiguren," *Sitzungsberichte der preussischen Akademie der Wissenschaften zu Berlin, Phil.-hist. Kl.* 1930: l.93–5 on 458–9.

181 Rachmati "Heilkunde" l.67 on 456–7.

error for *yıpar*.[182] Musk is used in a remedy for headache: "one takes sesame oil mixed with musk warmed a little bit, and if it is put in the nose, the pain goes away."[183] Musk is called for in a remedy used if the ear buzzes (*kulkak tigileser*)[184] and also in a remedy for the suppuration of the ear that would also get rid of ear pain.[185] Finally, mixed with sugar and sal ammoniac (*čatır*), musk was applied to the teeth as a remedy for tooth pain.[186]

Musk is mentioned in the religious literature of the Turks as well. An example of a religious text referring to musk is a Manichaean hymn by Aprin-čor Tegin, describing the soul's longing for its heavenly alter ego, which is described as *kın yıpar yidlıγım* "my musk-scented one."[187]

It is unfortunately impossible to say that many of the occurrences of the word *yıpar* in Uyghur and Old Turkic literature refer to musk rather than to perfume in general. Bang and von Gabain translated *atıng keng yatıldı yıd yıpar täg* with the *yıd yıpar* as musk, thus "your name is spread wide like the scent of musk."[188] But this appears in a divination manual based on the Chinese hexagrams familiar from the *Yijing*, so perhaps the comparison of renown to the scent of musk in this text is Chinese in origin.[189] Even so, it is likely that the Uyghurs had such a concept since it is so widespread.[190] However, *yıd yıpar* can simply mean perfume.

Under the so-called Karakhanid Empire the first Islamic literature in Turkic appeared. Perhaps the greatest monument of this literature is the *Dīwān Lughāt al-Turk* of Maḥmūd al-Kāshgharī, the earliest dictionary of Turkish, written between 1072 and 1077.[191] Kāshgharī wrote his book in Arabic as a guide to the languages spoken by the Turks; he dedicated it to the Caliph al-Muqtadī (r. 1075–94). He distinguishes between several dialects and gives specimens of

182 Clauson, *Etymological Dictionary* 879a.

183 Rachmati "Heilkunde" l.158–60 on 460–1.

184 Rachmati, "Heilkunde ... II," 1.92–4 on 410–11.

185 Rachmati "Heilkunde ... II" 1.107–8 on 410–11.

186 Rachmati "Heilkunde ... II" 2.33 on 416–7.

187 A. von Le Coq, Türkische Manichaica II," *Abhandlungen der preussischen Akademie der Wissenschaften* 1919 No. 3 (Berlin, 1919): 8 lines 14–15 (= T.M. 419r).

188 W. Bang and A. von Gabain, "Türkische Turfan-Texte." *Sitzungsberichte der preussischen Akademie der Wissenschaften zu Berlin, Phil.-hist. Kl.* (1929): l. 146 on 251.

189 Although this exact comparison does not appear in the Yijing. The hexagram in question is *xun* 巽, number 57.

190 Alternatively, they could have gotten it from India!

191 Maḥmūd al-Kāshgharī, *Maḥmūd al-Kāšγarī: Compendium of the Turkic Dialects* (Dīwān Luγāt at-Turk), trans. R. Dankoff and J. Kelly, 3 vols. (Cambridge: Harvard, 1982–5), 1.23 on the date.

their scripts. Another important feature of the *Dīwān Lughāt al-Turk* is that Kāshgharī quotes many proverbs and passages from cycles of epic poetry that are otherwise unknown now.

Kāshgharī reveals that a complex vocabulary was used for musk. He directly identifies the Turkic *yıpar* with Arabic misk.[192] He also specifies that the Turkic term *kin yıpār* equals the Arabic *nāfijat misk*, meaning musk pod.[193] But Kāshgharī also discusses the words associated with musk. For example, the verb *sač-* is used for the scattering of musk: *män yıpār sačtım* "I scattered musk," glossed as *nathartu misk*.[194] *Sač-* is also used for sprinkling water (*rashsha*). For the diffusion of the aroma of musk, Turkic used *bur-* as in *yıpar burdı* "the musk fragrance spread (*fāḥat*)."[195] For anointing with musk Turkic used the verb *yuqtur*, as in on *anıg tonıŋa yıpar yuqturdı* "he had his garment rubbed (*alṭakha, amassa*) with musk."[196] Musk is quoted under the verb *yıdhla-* "to smell": *ol yıpar yıdhladı* "he smelled (*tashammama*) the musk."[197] The expression *yıparlıgh är* means "a man who owns musk."[198]

Kāshgharī quotes several times from an ode to spring that is no longer extant. One passage is given under the verb *sučul-* "to be scattered":

yaghmur yaghıp	(when) the raindrops are scattered
sačıldı	and the flowers are brought forth (from the earth)
türlüg čečäk sučuldı	and the pearl (and coral) shells have opened (meaning the
yinčü qapı ačıldı	flowers)
čından yıpar	then (the scent of) sandalwood and musk are kneaded
yoghrušur	together (and their fragrances spread).[199]

The connection of musk with the scent of springtime is quite common in Persian literature as well; even so, it is likely these lines represent an authentic Turkic composition. Kāshgharī also gives a proverb under the entry for *yıparlıgh käsürgü*, "a bag with musk": *yıparlıgh käsürgüdin yıpar ketsä yıdhi qalır* "when the musk goes from the bag its fragrance does not depart [lit. is left

192 Kāshgharī 2.161 (=456).

193 Kāshgharī 1.171 (=171).

194 Kāshgharī 1.389 (=265).

195 Kāshgharī 1.390 (= 266), cf. 2.243 (= 524) under *būr-* is given *yipar burdi* "the musk-fragrance spread (*fāḥat*)".

196 Kāshgharī 2.196 (= 484).

197 Kāshgharī 2.307 (= 579).

198 Kāshgharī 2.174 (= 466).

199 Kāshgharī 2.25 (=329–30); cf. also 3.307–8.

behind].” Kāshgharī noted: “This is coined about one who bestows a favor, and the trace of his favor remains with him even though the favor itself is gone, so that something of it can still be found with him.”[200]

Finally, Kāshgharī gives an insight into the views of the Turks on the goodness of the musk scent: *yond äti yıpār* “Horse flesh (exudes) musk.” He said: “this means that when it is cooked and left to cool there exudes from it a good odor.”[201]

The other great work of Karakhanid literature that survives is the *Kutadgu Bilig* (1069 CE), a mirror for princes. It is written in an allusive and allegorical style reminiscent of Persian works in the same genre; there are Persian calques in its Turkic language. But on the whole it seems to preserve many details about life in the Karakhanid realm. Musk appears twice in this book. One occurrence of it is in a familiar aphorism comparing wisdom to musk: “Musk and wisdom are of the same sort: neither one can be kept hidden. If you try to hide musk, its scent gives it away; and if you conceal wisdom, it nevertheless continues to regulate your tongue.” (*yıparlı biligli tengi bir yangı tutup kizlese bolmaz özde öngi yıpar kizlese sen yıdı belgürer bilig kizlese sen tilig ülgüler.*)[202] This sort of aphorism is found not only in Sanskrit literature, but in Persian and Arabic, so there is the possibility that it is a common motif. Musk also appears in a description of the springtime that could also come straight from a Persian ode:

The brown earth filled with musk as the camphor melted away.	*yağız yir yıpar toldı kafur kitip*
***	
Brown earth wrapped a veil of green silk over her face;	*yağız yir yaşıl torku yüzke badı*
the Cathay caravan spread out its Chinese wares.	*hıtay arkısı yadtı tavgaç edi*
***	
Countless kinds of flowers opened with a smile,	*tümen tü çiçekler yazıldı küle*
and the world filled with the fragrance of musk and camphor.	*yıpar toldı kafur ajun yıd bile*

200 Kāshgharī 2.173 (=465).

201 Kāshgharī 2.149 (=447).

202 Yūsuf Khāṣṣ Ḥājib, *Kutadgu Bilig, Vol. 1*, ed. R. Arat, 2nd ed. (Ankara: Türk Dil Kurumu Yayınları, 1979), lines 311–12 on 46; R. Dankoff's trans. is quoted, *Yūsuf Khāṣṣ Ḥājib. Wisdom of Royal Glory: Kutadgu Bilig* (Chicago: University of Chicago Press, 1983), 50.

Zephyr sprang up carrying the scent of cloves	*saba yili koptı karanfil yıdın*
and all the world was suffused with musk.[203]	*ajun barça bütrü yıpar burdı kin*

Aromatics in the Persian World

As the linguistic evidence given in Chapter One shows, the words for musk found in the languages of the Near East and Europe derive from Persian. The Iranian world has had a longstanding history of interest in perfumery and scent.[204] Pliny the Elder, in his work *Natural History*, summarized the knowledge that the Romans had of aromatics during the first century CE. He mused that the use of *unguentum* (aromatic preparations) was an innovation, because he considered that aromatics did not exist at the time of the Trojan War, when offerings to the gods did not include incense. Pliny credits the development of aromatics to the Persians (*unguentum Persarum gentis esse debet*) and dates their introduction to the Greeks to the expedition of Alexander. He states that a chest of perfumes owned by the last Achaemenid king Darius was captured along with his camp.[205] Thus, the story goes, the Greeks (and ultimately Romans) came to know the pleasures of perfumes. Whether or not this is the case, it is significant that the use of aromatics was particularly associated with the ancient Persians.

Some measure of the legacy of Persian aromatics can be found in Arabic, where numerous terms for aromatic substances are attributed to Persian, sometimes through the intermediation of Aramaic, e.g. *wardah* "scented rose",[206]

203 Lines 64–71 on Arat 24; Dankoff's trans. 41. Clauson translates "the sweet smell (of flowers) has filled the brown earth," *Etymological Dictionary* 879.

204 See S. Anṣārī, *Tārīkh-i ʿIṭr dar Īrān* (Tehrān: Wizārat-i Farhang, 1381/2002–3) and M. A. Newid, *Aromata in der iranischen Kultur unter besonderer Berücksichtigung der persischen Dichtung* (Wiesbaden: Reichert, 2010) for an overview.

205 Pliny, *Natural History*, XIII.I. See S. Lilja, *The Treatment of Odours in the Poetry of Antiquity. Commentationes Humanarum Litterarum* 49 (Helsinki: Societas Scientiarum Fennica, 1972), 31–2.

206 A. Jeffrey, *The Foreign Vocabulary of the Qur'ān* (Baroda: Oriental Institute, 1938), 287 and A. Asbaghi, *Persische Lehnwörter im Arabischen* (Wiesbaden: Harrassowitz, 1988), 271. The philologist al-Jawālīqī, *al-Muʿarrab min al-kalām al-aʿjamī* (Tehran, 1966), 344, reported that it was not considered Arabic originally.

kāfūr "camphor",[207] and, of course, *misk* "musk." The Arabic word for ambergris, *ʿanbar*, resembles the Pahlavi *ambar*,[208] though it is perhaps more likely that both the Arabic and Pahlavi words are from Somali.[209] Even in medieval times many of these words, all of which except for *ʿanbar* occur in the Qurʾān, were thought to be loanwords. The scented world of the first few centuries of Islam was strikingly different from that known to the occupants of the Classical Mediterranean; many new substances, particularly imported substances such as musk, camphor, and ambergris, had come into prominence in perfumery.[210] Persia under the Parthian and Sasanian empires appears to have played an important role in this transformation, so it is important to assess the knowledge and use of aromatics by those empires.

Iranian tradition holds that Jamshīd, the legendary ancient king who introduced many amenities of civilized life to the world, initiated the use of aromatics. We find this concept in the *Shāhnāmah* of Firdawsī (early 11th century), the great Persian epic of kings, where the use of aromatics such as ben (*bān*), camphor (*kāfūr*), pure musk (*mushk-i nāb*), aloeswood (*ʿūd*), ambergris (*ʿanbar*) and "luminous rosewater" (*roshan gulāb*) are attributed to his civilizing influence.[211] This canon of aromatics is not free from the influence of the Islamic period, for aloeswood is called by its Arabic name. Three out of five of these aromatics only become important in the Late Antique period; camphor and musk were unknown in antiquity, and aloeswood was barely known.[212]

207 Jeffrey 246–7 and Asbaghi 225. Jawālīqī 285–6 quoted Ibn Durayd's opinion that it was not Arabic. The immediate source of the Persian word is Sanskrit *karpūra*, which has deeper roots in Draviḍian and Austronesian; cf. R. A. Donkin, *Dragon's Brain Perfume* (Leiden: Brill, 1999), 80, M. Mayrhofer, *Kurzgefaßtes etymologisches Wörterbuch des Altindischen*, 4 vols., Heidelberg: Winter, 1956–1980), 1.175, and W. Mahdi, "Linguistic and Philological Data Towards a Chronology of Austronesian Activity in India and Sri Lanka," in R. Blench and M. Spriggs, eds., *Archaeology and Language, vol.4: Language Change and Cultural Transformation* (London: Routledge, 1999), 216–20.

208 Cf. Asbaghi 204; Gignoux, Lexique, 20.

209 L. Reinisch, *Die Somali-Sprache II. Wörterbuch* (Wien: Hölder, 1902), 59a; Reinisch's etymology is followed by G. Ferrand, "Review of R. Campbell Thompson, The Assyrian Herbal," *JA* (1925:1): 172 and noted by P. Pelliot, *Notes on Marco Polo* (Paris: Imprimerie National, 1959–73), 1.33.

210 Z. Amar and E. Lev, "Trends in the Use of Perfumes and Incense in the near East after the Muslim Conquests," *JRAS* (2013): 28–30 emphasize the role of Islam in the promotion of the new aromatics.

211 Firdawsī, *Shāhnāmah*, ed. Djalal Khalegi-Motlagh, 8 vols. (New York: Bibliotheca Persica, 1988–2008), 1.43 l. 42. See also Newid 12.

212 Dioscorides I.22 knew of the difference between the medicinal aloe and aloeswood, but his report is unique. See J. Greppin, "The Various Aloës in Ancient Times," *Journal of Indo-European Studies* 16 (1988): 33–48.

Ben oil was produced in the Middle East and was frequently used in the compounding of perfumes, while rosewater is certainly the most famous aromatic product of Iran to this day.

The use of incense goes back into antiquity in Iran. The Zoroastrian high liturgy called the *Yasna*, which is performed by Zoroastrian priests for the benefit of the community, includes the burning of fragrant wood as part of the ceremony. This is called the *khūsh-bōy*, and today sandalwood and frankincense are used.[213] Perfuming of the individual was also practiced. Before his ordeal, the righteous man Wirāz, the subject of the Pahlavi *Ardā Wirāz Nāmag*, prepared by washing his head and dressing himself in a new garment and then perfuming himself with sweet scent (*bōy xwaš bōyēnīd*).[214] Good scent was important symbolically in religious belief as well; after death the soul meets his deeds personified as a woman. If he was good, the woman is beautiful, and if he was bad, the woman is ugly in proportion to his bad deeds. The *Ardā Wirāz Nāmag* says:

> On the third [day] at dawn the soul of the righteous went about among the sweet-smelling plants (*urwar bōy xwaš*), and to him that scent seemed more pleasant than all the sweet smells that had come into his nostrils [when he was among] the living, and that scented breeze came from the southern direction, from the region of God. Then [came forward] his own religion and his own deeds in the form of a girl in appearance, well grown, that is, she was grown in virtue, with prominent breasts [that is] her breasts swelling out, whose fingers were long, whose body was as brilliant as [her] appearance was most pleasing and [her] looks most fitting.[215]

While this personification of the religion and good deeds is not explicitly said to have possessed a sweet scent, it is implied. In the case of the evil person, the soul appears as follows:

> Then a cold stinking wind (*wād-ē sard ī gandag*) came to meet him. He had not experienced a fouler wind than that in the world. And in that wind he saw his own religion and deeds [in the form of] a naked whore, rotten, filthy, with crooked knees, with projecting buttocks, with

213 Dastur F. M. Kotwal and J. W. Boyd, *A Persian Offering. The Yasna: A Zoroastrian High Liturgy* (Paris: Association pour l'avancement des études iraniennes, 1991,) 138 and *passim*.

214 *Ardā Wirāz Nāmag: The Iranian 'Divina Commedia'*, ed. and trans. F. Vahman (London: Scandinavian Institute of Asian Studies, 1986), 5.12; trans. 193.

215 Vahman's trans.; text 8–9; trans. 194–5.

> unlimited [number of] spots, i.e. spots joined [to] spots, resembling the most hideous reptile, most filthy and most stinking.[216]

This imagery of scent is paralleled in Christian literature, where it may be due to Iranian influence, and in the Qur'ān and other early Islamic literature, where the role of good scent in religion reaches an even greater importance.[217] Those developments will be discussed in the last chapter.

Under the Achaemenids (559–330 BCE) incense must have played a major role. The Arabians provided the Achaemenid court an annual tribute of a thousand talents of frankincense.[218] Philostratus reports that the palace of the king of the "Medes" possessed the fragrance of frankincense and myrrh.[219] Achaemenid incense burners have been excavated and they are also represented in Achaemenid art. A famous frieze from the Treasury Room at Persepolis depicts courtiers standing before the king. The gesture of the courtiers indicates the solemnity of the occasion. The scene is quite stark and is free from extraneous details that would distract the viewer from the ceremony being represented. However, between the figure of the seated king and the courtier are two objects: incense burners. This identification is secure because similar pieces have been found from around the Achaemenid world.[220] These conical incense burners have been suggested as a prototype for a major type of Chinese incense burner as well.[221]

---

216 Vahman's trans.; text 21–2; trans. 201.

217 The account of Ardā Wirāz is paralleled in Islamic literature; see al-Ghazālī, *Iḥyā' 'Ulūm al-Dīn*, 5 vols. (Beirut: Dār al-Kutub al-'Ilmiyyah, 2005), 4.663–4. The pious deeds of the believer are described as beautiful in face, with a good scent (*ṭayyib al-rīḥ*), and with beautiful garments.

218 Herodotus, *The Persian Wars*, ed. and trans. A. D. Godley. Rev. ed. 4 vols., Loeb Classical Libary (Cambridge, MA: Harvard University Press, 1926), III.97.

219 Philostratus, *Imagines* II.31 *Themistocles*, ed. and trans. A. Fairbanks, Loeb Classical Library (Cambridge, MA: Harvard University Press, 1931). Thanks to Cynthia King for pointing out this reference.

220 In general, see B. Goldman, "Persian Domed Turibula," *Studia Iranica* 20 (1991): 179–188, which provides a list of examples, and A. S. Melikian-Chirvani, "The International Achaemenid Style," *BAI* 7 (1993): 112–13. There are two similar audience scene reliefs from Persepolis; they were originally located on the staircase friezes of the Apadana. For a full discussion of these and other scenes involving incense burners, see L. Allen, "Le roi imaginaire: An Audience with the Achaemenid King," in *Imaginary Kings: Royal Images in the Ancient Near East, Greece and Rome, Oriens et Occidens 11*, ed. by O. Hekster and R. Fowler (Stuttgart: Steiner, 2005), 39–62.

221 See J. Rawson, "The Chinese Hill Censer, boshan lu: A Note on Origins, Influences and Meanings," *Arts Asiatiques* 61 (2006): 75–86.

The first supposedly Iranian incense formula comes from classical antiquity. Pliny the Elder provides us with the formula for what he calls the Parthian royal unguent: ben, costus, amomum, Syrian cinnamon, cardamom, spikenard, marum,[222] myrrh, cassia, styrax, labdanum, opobalsam, Syrian calamus, Syrian rush, oenanthe, malabathrum, serichatum, henna fruit, aspalathus,[223] panacea, saffron, cyperus, marjoram, lotus, honey, and wine.[224] This is a total of twenty-six individual ingredients. While some of these ingredients, such as costus and malabathrum, were imports from Further Asia, particularly India, the majority are not. Many are from Arabia and the Near East. Aromatics such as musk and camphor, which are well known from the medieval period, are completely absent. One question which we cannot easily answer is whether this recipe is authentic; that is, is a true Parthian formula represented here? All of these ingredients are also aromatics well known in Classical Antiquity to the Greeks and Romans, and it is entirely possible that it is a formula produced by them and attributed to the Parthian royal court.[225] Some of the ingredients in Pliny's Parthian royal incense are indeed Near Eastern, and there is no way to prove whether the unguent formula is authentic or not.

The aromatics of highest standing in the early medieval Islamic world are quite different from the aromatics called for in Pliny's formula. There we find musk, camphor, ambergris, aloeswood, cloves, sandalwood—all aromatics with very sketchy attestations in Classical Antiquity, if they are attested at all. Moreover, the words for these substances in Arabic are often considered to be Persian loanwords, and many have been attested and identified as loanwords since early Islamic times. Therefore, we must suspect that a revolution in the field of aromatics occurred in Sasanian times, with new aromatics of Further Asian origin flowing into the west and minimizing the importance of the favored aromatics of Parthian and Graeco-Roman times.

---

222 Dioscorides, *De Materia Medica*, ed. M. Wellmann. 3 vols. (Berlin: Weidmann, 1958), III.49.

223 Dioscorides I.19.

224 Pliny XIII.II.18.

225 For example, Apicius gives in his cookbook two "Parthian" recipes, which both include characteristically Iranian ingredients: VI.8.3 for "Parthian chicken" and VIII.6.10 for "Parthian kid or lamb"; see *Apicius: A Critical Edition with an Introduction and an English Translation of the Latin Recipe Text* Apicius, ed. and trans. C. Grocock and S. Grainger (Totnes, Devon: Prospect Books, 2006). Unfortunately, the Parthianness of these dishes may stem only from the use of characteristically Parthian spices as main flavoring, cf. A. C. Gunter, "The Art of Eating and Drinking in Ancient Iran," *Asian Art* 1:2 (1988): 36.

Musk and Aromatics in Sasanian Persia

There are two major bodies of source material for aromatics in the Sasanian Empire. The first consists of writings about the Sasanians preserved mainly in the Arabic, Syriac, and Greek texts. The second consists of Zoroastrian Pahlavi literature. Neither corpus is, strictly speaking, primary source material. The primary source material extant from the Sasanians consists of inscriptions and documents on leather and papyrus, and none of these so far discovered include any information on the aromatics used in the Sasanian Empire.

We may regard portions of the extant literature in Pahlavi as texts which were known in the Sasanian period, but by no means all, and it is difficult to date the elements in these texts, all of which were redacted in the Islamic period. Much of the literature itself dates from long after the Islamic conquest of Iran. Therefore, it is very hard to determine the philological history of the terminology for aromatics that appears in Pahlavi literature.[226]

Most of this literature is religious. There are frequent references to the use of incense in worship, but these often do not specify exactly what was burned. One must bear in mind that the sorts of works we have are not representative of the scope of Sasanian literature because they are skewed in favor of the religious texts and exclude many important secular works that are believed to have existed. Given the nature of the extant literature, it is perhaps surprising that there are as many references to musk in it as there are. Under these circumstances, it seems likely that musk was not just known in Sasanian times but was familiar and important as an aromatic. The facts that musk was being traded by the Sogdians, northeastern neighbors of the Sasanians in the early 4th century, and that it was known in India shortly after this time strongly support the early presence of musk in Sasanian Persia as well.

The Pahlavi word *mušk* appears in several texts. Unfortunately, all of these are works that took their final form long after the fall of the Sasanian Empire in Islamic times. The notice of the musk animals contained in the *Bundahišn* was mentioned in Chapter One, where the Persian origin of the "musk" word family was discussed. The *Bundahišn* certainly attests to the knowledge of musk at the time it was written. Since scholars consider it to be a composition of Sasanian origin with additions made during Islamic times, we may suppose on the basis of this text alone that the Sasanians were acquainted with musk because the zoological chapters are not explicitly from the Islamic period and

226 On the general reconstructions of plant names from the Iranian tradition, see especially the many studies of P. Gignoux, who makes full use of the data from loanwords and later texts.

their structure seems to have been fixed—and that fixed conception included a category of musk animals, as we have seen. However, there are more Pahlavi texts that mention musk.

The poem *Draxt-ī Āsūrīg* is perhaps the next most important of the Pahlavi texts that attest to the existence of musk. This poem, besides being a rare specimen of Middle Persian poetry, has many Parthian elements in its language; this is not necessarily a guarantee that the source is archaic, as Parthian culture continued in Sasanian times.[227] The subject of the *Draxt-ī Āsūrīg* is a contest between a date palm tree and a goat as to who is the more useful. Among the boasts of the goat, who wins the contest, is the following:

> *hambān az man karēnd　wāzārgānān wasnād*
> *kē nān ud pōst ud panīr　harwīn rōyn-xwardīg*
> *kāpūr ud mušk syā　ud xaz tuxārīg*[228]
>
> They make skin bags for the merchants from me
> Bread, skin, and cheese and all the sweetmeats
> Camphor and black musk and the Tukhārī martens.

This is the text as it is in the standard edition by Nawwābī. Leaving aside the question of how the author thought camphor derived from the goat, the text of the third line has been doubted by Henning, who notes that Ṭukhāristān was not famous for martens. Observing that the sequence read *xaz* "marten" can also be read as the ideograph AZ standing for *buz* "goat," he suggested a possible emendation of the last part of the line to *mušk syā* [*čē*] *buz tuxārīg* "black musk of the Tokharian goat."[229] It seems reasonable that the small musk deer that lives among the mountains could be described as a goat, at least poetically.[230]

---

227 For example, early Sasanian inscriptions were often bilingual in Parthian, and the corpus of Manichaean literature includes many texts in Parthian. The role of the Parthian aristocracy in Sasanian times is stressed by P. Pourshariati, *Decline and Fall of the Sasanian Empire: The Sasanian-Parthian Confederacy and the Arab Conquest of Iran* (London: Tauris, 2008).

228 Lines 77–9; pg. 78–81 in Māhyār Nawwābī's edition, *Manẓūmah-i Drakht-i Āsūrīg* (Tehrān: Intishārāt-i Bunyād-i Farhang-i Īrān, 1346/1967).

229 W. B. Henning, "A Pahlavi Poem," BSOAS 13 (1950): 644 n. 11.

230 A passage of al-Idrīsī, *Kitāb Nuzhat al-mushtāq fī ikhtirāq al-āfāq*, 9 vols. (Naples: Istituto Universitario Orientale di Napoli, 1970–84), 2.204, possibly deriving from al-Jayhānī, classes musk deer among the goats. It is discussed in the 6th chapter. Ananias of Širak, *The Geography of Ananias of Širak (Ašxarhacʿoycʿ): The Long and Short Recensions*, trans. R. H. Hewsen (Wiesbaden: Reichert, 1992), 75, states that the musk animal resembles

But Henning also noted that musk is not associated with Ṭukhāristān either. The Armenian geography attributed to the 7th century writer Ananias of Širak, however, does mention musk from *Xupi-Tuxarstan*, which is surely to be identified with Ṭukhāristān.[231] Ṭukhāristān was on the trade routes by which furs arrived from Tibet, and they had many different varieties of furs according to the *Ḥudūd al-ʿālam*. So it is possible that somehow marten fur had become associated with Ṭukhāristān. The musk deer, however, still lives in the mountainous northeast of Afghanistan[232] and might have extended further west into the rugged mountains of Ṭukhāristān; even so, the situation of Ṭukhāristān allows the same possibility of the importation of musk from Tibet as for marten fur. Neither reading can thus be dismissed from possibility on the grounds that the subject is improbable, but since the text has "Tukhārī martens" it is best to keep to that reading,[233] as appealing as an early Pahlavi reference to the musk trade through Ṭukhāristān would be.

The Zoroastrian writer Zādsparam, who lived in the late 9th century, wrote a book called *Wizīdagīhā-ī Zādspram*, the "Selections of Zādspram." In the section on eschatology describing the bodies of the resurrected faithful he wrote: *ušān āmēzišn andar tan hu-bōytar hēnd kū mušk ud ambar ud kapur*:[234] "The humors in their bodies will be more fragrant than musk, ambergris, or camphor." The idea of the resurrected bodies of the faithful having a good fragrance could be pre-Islamic (and thus in turn could have influenced Islamic ideas of the afterlife), but this particular collection of aromatics, unattested in other Pahlavi texts on the resurrected body, probably shows the influence of Islamic conceptions of the afterlife, and the latter could be the source of the former.

Another late work is a Pazand text called the *Ayādgār-ī Jāmāspīg*, a work based on the Pahlavi *Jāmāsp Nāmag* that exists in fragments. Musk is

---

a four-month old goat except for the teeth, which are said to resemble those of a fox. Tomé Pires, *The Summa Oriental of Tomé Pires and the Book of Francisco Rodrigues*, trans. A. Cortesão (London: Hakluyt Society, 1944), 1.96 says that "musk comes from animals such as goats".

231 Ananias of Širak 74 and note 35 on 236 plus addendum on 346 that discusses the connections of this term with the Tokharians; it is much simpler to regard it as an attestation of the region of Ṭukhāristān.

232 S. Ostrowski, et al. "Musk deer *Moschus cupreus* persist in the eastern forests of Afghanistan," *Oryx* (2014): 1–6.

233 As has been done by C. Brunner in his translation, "The Fable of the Babylonian Tree," *JNES* 39 (1980): 293 and cf. his note on this line on 300.

234 Zādspram, *Wizīdagīhā-ī Zādspram*, ed. and trans. P. Gignoux and A. Tafazzoli, *Anthologie de Zādspram*, Paris: Association pour l'avancement des études iraniennes, 1993), 35:51 on 310 (text) and 138–9 (transcription and translation).

mentioned in connection with China: *ud čēnastān šahrhā-ī wuzurg was-ī zar was mušk was gōhr* ... "And China has large cities, much gold, much musk, and many jewels ..."[235]

There is a catalogue of scents in the well-known story of Khusraw and the page that consists of a page (*rēdag*) being questioned by the Sasanian king Khusraw about the finer things in life, such as the best fruit, entertainment, etc.[236] This text is extant both in Pahlavi and in an Arabic version that is preserved in the *Tārīkh ghurar al-siyar* of al-Thaʿālibī (d. 1038); there are many great differences between these two versions. In the Pahlavi version Khusraw inquires about which flower or herb (*sprahm*) is the best scented (*hubōytar*), and as expected, the scents listed in the Pahlavi version are all single floral and herbal scents that are likened to specific offices and groups of people.[237] For example, the first scent mentioned by the page in the Pahlavi version is the jasmine: "the flower of the jasmine (*yāsimīn*) is the best scented, because its scent resembles the scent of the lords (*xwadāyān*)."[238] The page lists twenty-four different individual flowers and herbs.

In the Arabic version the question is different: Khusraw first inquires about which are the sweetest of scents (*aṭyab al-mashmūmāt*), with no requirement that the response be limited to floral scents, and after the page's short listing of these, Khusraw asks a second question, about the scent of the Garden (*rāʾiḥat al-jannah*). As we might expect from the second question, we find aromatics that are well-known in Islamic times in both answers. As it turns out, the page's answer to the first question is that sweet basil (*shāhisfaram*) censed with *nadd* with rosewater sprinkled upon it is the best, followed by violet with ambergris and waterlily (*nīlūfar*) with musk and fragrant bean (*fūl al-bāqilāʾ*) with camphor. These three pairs could be ingredients in a single compound perfume or, as is more likely, three individual pairs of compound perfumes. The Arabic version is thus different from the Pahlavi version not only in the specific list of aromatics, but also because they are linked in compound perfumes and no

---

235 H.W. Bailey, "Iranian Studies," *BSOS* 6:4 (1932): 948; I have modified his romanization.

236 S. Azarnouche, Husraw ī Kawādān ud Rēdag-ē: *Khosrow fils de Kawād et un page* (Paris: Association pour l'Avancement des Études Iraniennes, 2013) and the older edition of D. Monchi-Zadeh, "Xusrōv Kavātān ut Rētak," *Monumentum Georg Morgenstierne* 2, Leiden: Brill, 1982 (= AI 22), 47–92.

237 *Kāpūr*, camphor, has been read in §76, comparing it to the scent of the priests by Unvala, followed by Donkin, 108, Azarnouche 156, and A. King, "The new *materia medica* of the Islamicate tradition: the pre-Islamic context," *JAOS* 135, no. 3 (2015): 512. Monchi-Zadeh suggests that *kāčūr* "safflower" is more consistent with the context. In light of the religious associations of camphor, this reason seems insufficient.

238 *Husraw ī Kawādān ud Rēdag-ē* §69.

comparisons with offices are given. Likewise, the page's answer to the second question, about the scent of the Garden, consists of a combination of aromatics: "When you combine in scent Khusrawānī wine, Syrian apple, Persian rose, the sweet basil of Samarkand, the citron of Ṭabaristān, the narcissus of Maskā,[239] the violet of Iṣfahān, the saffron of Qumm and Bawan, the waterlily of Shīrāz, tripartite perfume[240] with Indian aloeswood, Tibetan musk, and Shiḥrī ambergris, the scent of the Garden that is promised to the faithful is not lacking."[241] Many of these scents, if not all, are very typical of the scents extolled in medieval Islamicate culture as the most excellent. This is certainly true of the tripartite perfume made with the most highly regarded forms of aloeswood, musk, and ambergris.

It is thus likely that the Pahlavi version without the aromatics prized in Islamic times preserves something closer to the text as it might have been known in Sasanian times than the Arabic version, which is really a free adaptation rather than a translation and thus cannot be considered a source for Sasanian aromatics.

Besides this doubtful Arabic reference to Sasanian musk, it is attested in Chinese sources that the Persians used musk. There are multiple versions of this brief statement:

> *Jiu Tangshu* 198: 5311: 其事神以麝香和蘇塗鬚點額及耳鼻用以為敬拜必交股 "When they worship, they smear musk together with oil-bearing plants on their beards and dot their foreheads, reaching up to their ears and noses, and they use this to be respectful, and in paying obeisance must pay their shares."[242]

> *Xin Tangshu* 221b: 6258: 祠夕以麝揉蘇澤彨顏鼻耳 "On the night of sacrifice, they anoint their beards, face and nose with musk rubbed with oil-bearing plants."

> *Taiping Yulan* 981: 4b: 唐書曰波斯國人皆以麝香如蘇塗點額及於耳鼻用以為敬 "The Tangshu says that the people of Persia all use musk-like

---

239 A town in Kirmān, see Yāqūt, *Muʿjam al-buldān*, 5 vols. (Beirut: Dār Ṣādir, n.d.), 5.128.

240 Al-Thaʿālibī, *Tārīkh ghurar al-siyar*, ed. and trans. H. Zotenberg (Paris: Imprimerie Nationale, 1900; repr. Tehran: Maktabat al-Asadī, 1963), 709. Zotenberg supplies the word *nadd* before *al-muthallath* but it is not apparently in the manuscript.

241 Thaʿālibī, *Tārīkh*, 709.

242 The latter part of this sentence is unclear, and C. I. Beckwith suggests it may be a separate sentence.

> oil-bearing plants to smear their beards and dot their foreheads reaching up to their ears and noses, and they use this to be respectful."

Without entering into the question of which of these texts best preserves the original version of this statement, we can say for certain that it indicates that the Sasanian Persians used musk. The *Taiping yulan* has surely substituted 如 for the 和 of the *Jiu Tangshu*, changing the meaning from musk together with oil-bearing plants to musk-like oil bearing plants.

Arabic literature maintains the idea that the Sasanians were familiar with and used musk. A recipe for a scented powder including musk said to have been used by Kisrā ("Khusraw", a generic name for a Sasanian monarch in Arabic) is mentioned in the *Firdaws al-ḥikmah* of ʿAlī b. Rabban al-Ṭabarī; that the formula is authentic cannot be proved, and it may be a later invention.[243] It does illustrate that the later Sasanians were credited with these substances in Abbasid times, as does the catalogue of scents in the Arabic version of the story of Khusraw and the Page. Kisrā is also said to have gotten scents forwarded to him from Yemen; his governor Bādhām sent to him Yamanī clothing, musk, and ambergris.[244] As we have seen, musk is attested in a document from pre-Islamic Yemen, so this story has a credible ring to it. The *Kitāb al-Tāj* of pseudo-al-Jāḥiẓ records that the Sasanian kings received tribute from their grandees according to what they specialized in. People specializing in musk, as well as ambergris, are specifically mentioned.[245] And al-Bīrūnī, describing the treasury of the Sasanians at Ctesiphon, called *Bahār-i Khurram*, notes that it contained a hundred baskets each with a thousand pods of musk, as well as a hundred bags of camphor.[246]

It is probable that Classical Persian literature preserves information on pre-Islamic uses of aromatics. Unfortunately, this literature was written at a time in which Arab-Islamic culture had attained such an importance that the suspicion of influence from that tradition upon the Persian literature cannot be dismissed.[247] For example, in Firdawsī's *Shāhnāmah*, women are associated

243 ʿAlī b. Rabban al-Ṭabarī, *Firdaws al-ḥikmah fī al-ṭibb*, ed. M. Z. Siddiqi (Berlin: Sonne, 1928), 611.

244 Al-Iṣbahānī, *Kitāb al-Aghānī*, 24 vols. (Cairo: Dār al-Kutub, 1927–74), 17.318.

245 Al-Jāḥiẓ (attr.), *Kitāb al-Tāj*, ed. Ahmed Zeki Pacha (Cairo: Imprimerie Nationale, 1914), 146–7.

246 Al-Bīrūnī, *Kitāb al-Jamāhir fī maʿrifat al-jawāhir*, ed. F. Krenkow (Haydarābād: Dāʾirat al-Maʿārif, 1355/1936), 71–2.

247 The *Farhangnāmah*, Newid, and B. Grami, "Perfumery Plant Materials As Reflected in Early Persian Poetry," *JRAS* (2013): 39–52, contain extensive quotations and references to Classical Persian poetry's use of aromatics.

with scents such as musk and ambergris.[248] Aromatics such as saffron and aloeswood also appear, but the aloeswood is referred to under the Arabic name *ʿūd*, and the saffron with *zaʿfarān* instead of the Pahlavi *kurkum*. Suspicions of Arabic influence become certain when the Arabic compound perfume *ʿabīr* is mentioned. So, Classical Persian literature cannot be used as an unqualified source for pre-Islamic aromatics, but it surely is an important source for the aromatics used in medieval Islamic Persian culture and thus figures in the discussion of Islamicate views and uses of musk in what follows.

The Westward Spread of Musk in Late Antiquity

It thus seems very likely that musk was introduced into the Near East through the Iranian world. The next question is how the Persians became acquainted with musk. To address this subject, we must discuss the history of musk in the Near East and the Mediterranean world. As shown above, the existing Iranian sources are of little help, so it is to the Near Eastern and classical texts we must turn.

The first traces of musk in the Near East are earlier than Jerome, who is regarded as the earliest classical source. One is an Arabic business document on wood from Yemen that mentions musk and is broadly paleographically dated to the first three centuries CE.[249] This document records the receipt of musk by a group of three men and asks the recipient the cost so that payment could be sent. It gives no indication what sort of value was attached to the musk. Musk is mentioned in both the Babylonian and Jerusalem Talmuds.[250] Both Talmuds, in the discussion of reciting benedictions over censers, mention that one uses a different blessing for musk. Both attribute the statement to Rav Ḥisda, who was active in approximately the period 250–290 in Babylonia.[251] Over incense one was to recite the blessing "He Who gave pleasant fragrance in spice wood," but for musk it was "He Who gave pleasant fragrance in all kinds

248 F. Vahman, "A Beautiful Girl," in *Papers in Honour of Professor Mary Boyce*, vol. 2 (Leiden: Brill, 1985), 668, catalogued them but translated *ʿanbar* as "amber" and *ʿabīr* as "ambergris."

249 J. Ryckmans, W. W. Müller, and Y. M. Abdallah, *Textes du Yémen antique inscrits sur bois* (Louvain-la-Neuve: Institut Orientaliste, 1994), 65–6 and 102–3.

250 Jerusalem Talmud Tractate *Berakhot* 6.6; Babylonian Talmud Tractate *Berakhot* 43a. See also D. A. Green, *The Aroma of Righteousness: Scent and Seduction in Rabbinic Life and Literature* (University Park: Pennsylvania State University Press, 2011), 47.

251 H. W. Guggenheimer, *The Jerusalem Talmud: First Order: Zeraïm, Tractate* Berakhot (Berlin: De Gruyter, 2000), 99.

of spices."[252] The Babylonian Talmud adds that this is because musk is from an animal.[253] In both places the word used is a form borrowed from Persian *mušk* rather than the *mūr* used in later Hebrew. The Arabic form is *ʾmskn*, which is an independent borrowing from Persian.[254] The Arabic document is likely to be roughly contemporary with the Talmuds, from the late 3rd century CE. These sources are thus earlier than the oldest Indian evidence for musk in Sanskrit literature.

Musk is completely absent from classical Greek and Latin literature until Late Antiquity.[255] It is striking that such an excellent source as the *Periplus Maris Erythraei*, with its apparently extensive body of information on the trade with the Indian Ocean lands, does not mention it.[256] The *Periplus* also does not mention the other aromatics that became important during the Sasanian period—ambergris, camphor, aloeswood, and cloves. The likely explanation for the absence of musk is that the exploitation of musk in India had not yet begun; this is consistent with the fact that the earliest traces of musk in Indian literature date from the Gupta period (320–550 CE). The Indians, it will be recalled, applied the Greek word *kastūrī*, meaning castoreum, which was used in Greek medicine, to musk, indicating that they originally used musk as a substitute for castoreum.

The absence of knowledge about musk also kept it from inclusion in Classical medical literature; the inclusion of musk in the Classical medical canon later received in the west shows the influence of Sasanian and Islamicate medicine, in which Classical Greek medicine was synthesized with the new discoveries of late antiquity and the early medieval period.[257]

---

252 Guggenheimer 500–1.

253 L. Goldschmidt, *Der babylonische Talmud mit Einschluss der vollstaendigen Mišnah*, vol. 1 (Berlin: Benjamin Harz, 1925), 153.

254 But cf. W. Müller, "Namen von Aromata im antiken Südarabien," in A. Avanzini, ed., *Profumi d'Arabia* (Rome: Bretschneider, 1997), 209–10.

255 I shall leave aside M. Brust's (*Die indischen und iranischen Lehnwörter im Griechischen* (Innsbruck: Instituten für Sprachen und Literaturen, 2005), 467–8] idea that Dioscorides' μόσχος (II.75.2) and μόσχειος (II.76.8 and 16 and II.77.1) represent "musk" and "musky" rather than their usual meaning "calf" and "of a calf". This seems too early, and does not fit well with the context of the text, which discusses fat and marrow. I am at a loss to imagine what real substance "Moschusfett" might be.

256 L. Casson, *Periplus Maris Erythraei* (Princeton: Princeton University Press, 1989). The collection of Latin sources on India, J. André and J. Filliozat, *L'Inde vue de Rome: Textes latins de l'Antiquité relatifs à l'Inde*, Paris, 1986), includes no mention of musk.

257 A. Dalby, *Dangerous Tastes: The Story of Spices* (Berkeley: University of California Press, 2000), 46–7, has a brief overview of the history of musk.

The first attestation of the term for musk in the Graeco-Roman west is apparently *muscus* in Jerome's (d. 419) *Adversus Iovinianum*. The sentence goes, *odoris autem suavitas, et diversa thymiamata, et amomum, et cyphi, oenanthe, muscus et peregrini muris pellicula, quod dissolutis et amatoribus conveniat, nemo nisi dissolutae negat.*[258] "No one save dissolute women denies that the dissolute and licentious find pleasure in sweet fragrances and various incenses, and *amomum*,[259] *kuphi*,[260] *oenanthe*[261] and musk, which is the skin of a foreign rodent." None of these substances, though they are likely unfamiliar to modern people, is uncommon in ancient literature except for musk. Jerome seems to have been running off a list of aromatics and then to have added in musk, giving his comment about its origin. It seems, then, to have been known beyond the circle of traders and doctors at the time and was used by other people. It is striking and so far out of place, because musk does not appear with any frequency in classical literature until the 6th century, so there is a risk that this passage is an interpolation from a later period. The notion that musk is the skin of a foreign rodent, however, is consistent with an early medieval Syriac source where musk is counted by skins,[262] and early Arabic sources also have the *fa'rat al-misk*, as discussed in Chapter 4, so Jerome's idea is not a Latin folk etymology but clearly derives from Near Eastern tradition.

Cosmas Indicopleustes, an Alexandrian Greek of the early 6th century CE, traveled and traded in the Red Sea and perhaps the Indian Ocean.[263] He wrote a book, the *Christian Topography*, containing much important information on the sea trade with India and Ceylon, even though its main purpose was the refutation of Ptolemaic cosmology. By his time musk was evidently well known in Byzantine culture because he mentions it as a trade good.[264] Three

---

258 *Adversus Jovinianum* II.8. The text is in J. P. Migne, *Patrologia Latina* (Paris, 1883), 23.311a.

259 Two Indian species can be understood under this term: *Amomum aromaticum* and *Amomum cardamomum*, André and Filliozat 361 n. 165.

260 Gk. κῦφι, "an Egyptian compound incense," LSJ 1015a; it is first mentioned in Dioskorides I.25, where a recipe is given. There are a variety of formulations for it, cf. J. Scarborough, "Early Byzantine Pharmacology," *Dumbarton Oaks Papers* 38 (1984; =*Symposium on Byzantine Medicine*, ed. J. Scarborough): 229–32. A long discussion of it is in L. Manniche, *Sacred Luxuries: Fragrance, Aromatherapy & Cosmetics in Ancient Egypt* (Ithaca: Cornell University Press, 1999), 47–57.

261 Obtained from flowers of the wild grape vine, see Pliny XII.LXI.132–3.

262 Discussed in Chapter 1, p. 31.

263 See B. Baldwin and A. Cutler, "Kosmas Indikopleustes" in *Oxford Dictionary of Byzantium*, vol. 2, 1151–2.

264 Cosmas' passage on musk is mentioned by, among others, E. H. Warmington, *The Commerce between the Roman Empire and India*, 2nd ed. (Cambridge: Cambridge University Press,

manuscripts of his work are illustrated, and there is an illustration of what purports to be the musk-animal, looking more like some type of dog or fox.[265] His description of the musk producing animal is: "... the small animal which is musk (μόσχος, *móskhos*): the Indians call it, in their dialect, καστοῦρι (*kastoûri*). Chasing after it, they shoot it with a bow, and after having bound the navel, where the blood gathers, they cut it off. This is the aromatic part, that which we call musk; the rest of the body they abandon."[266] The animal is not identified as a rodent here, and Cosmas is also nearly correct about its Indian name, *kastūrī* being musk and the *kastūrikā* the musk deer. The idea that musk is produced from blood in the navel of the musk deer continued in Arabic writings; those will be discussed in Chapter 6.

The physician Aetius of Amida (d. 578) used musk along with camphor in a prescription in Book 16 of his *Tetrabiblon*; the camphor was to be included if enough was available.[267] This indicates the rarity of camphor but says little about musk.[268] Book 16 of Aetius' work includes numerous other references to musk in gynecological prescriptions. However, musk also appears in the first book in a nard preparation for use in the church.[269] Alexander Trallianus

---

1974; repr. New Delhi: Munshiram Manoharlal, 1995), 161. See also A. McCabe, "Imported *materia medica*, 4th–12th centuries, and Byzantine pharmacology," in M. M. Mango, ed. *Byzantine Trade, 4th–12th Centuries: The Archaeology of Local, Regional, and Intercultural Exchange* (Farnham: Ashgate, 2009), 273–92, 282. On Byzantine trade with Asia in general, see M. M. Mango, "Byzantine Maritime Trade with the East," *ARAM* 8 (1996): 139–63. Her statement on 148 implying that musk was a well known import to the Roman Empire is not correct.

265 Indeed, Linschoten describes the musk deer as "beastes like Foxes, or little Dogs," 2.94. See also Juan Gonzalez de Mendoza, *The History of the Great and Mighty Kingdom of China*, trans. R. Parke, ed. G. T. Staunton (London: Hakluyt Society, 1854), 2.285, who says that the musk animal is "like vnto a litle dogge."

266 W. Wolska-Conus, *Cosmas Indicopleustes: Topographie chrétienne* (Paris: Les Éditions du Cerf, 1973), xi.6 on 3.324–5. See also P. Lindegger, *Griechische und römische Quellen zum Peripheren Tibet*, 3 vols. (Zürich: Tibetan Monastic Institute, 1979–93), 2.87 and 193.

267 Aetius Book 16, in *Gynaekologie des Aëtios*, ed. S. Zervos (Leipzig: Fock, 1901), 163, ll. 10–11. See also J. Scarborough, "Early Byzantine Pharmacology," *Dumbarton Oaks Papers* 38 (1984): 224–6. There is also a Latin translation of Book 16 from the Renaissance, and it has been translated into English: J. V. Ricci, *Aetios of Amida: The Gynaecology and Obstetrics of the VIth Century, A.D.* (Philadelphia: Blackiston, 1950). It seems to correspond fairly closely to the Greek. Books 1–8 are edited by A. Olivieri, *Aetii Amideni Libri Medicinales*, 2 vols. (Berlin: In aedibus Academiae Litterarum, 1935–50).

268 Donkin 106.

269 Aetius, *Libri medicinales*, I.131 on 1.66. Cf. G. Harig, "Von den arabischen Quellen des Simeon Seth," *Medizinhistorisches Journal* 2 (1967): 256 and n. 204 on 266.

(c. 525–c. 605) called for a tiny bit of musk in one of his prescriptions for gout.[270] These scanty attestations prove merely that while musk was known, it was either quite uncommon or not often used in medicine, or both. Paul of Aegina (7th century) does not discuss musk in his descriptions of simples. One of his formulas, however, does. This formula is for a malagma called *baion*; the notation that pepper and musk could be added is said to come from another manuscript in the text and is probably an interpolation.[271] He also mentions that musk ointment, made of musk and oil, was one of the ointments used only by women.[272] The *Book of the Eparch*, probably dating to the late 9th or early 10th century, mentions musk as well as ambergris among the aromatics sold by perfumers.[273] Only much later does the 11th century Jewish physician Simeon Seth give the first detailed discussion of musk in Byzantine literature.[274] This unprecedented discussion is influenced by Arabic accounts of musk and mentions that the best musk came from Tibet (Τουπάτ, *Toupát*), which was located east of Khurāsān (Χοράση, *Khorásē*).[275]

From these early Greek, Latin, Arabic and Hebrew sources we find words derived from Persian/Pahlavi *mušk* used for musk, rather than words used in India such as *darpa* or *kastūrī*. Cosmas Indicopleustes is even aware that the Indians had a different terminology for musk. This indicates that the introduction of musk into the Graeco-Roman world occurred outside of the classical sea trade with India, which declined in Late Antiquity under pressure from the Sasanians.

Sasanian maritime trade became very important in Late Antiquity.[276] The roots of the trade with India go back at least into Parthian times and are attested

270 Alexander of Tralles, *Alexander von Tralles. Original-Text und Übersetzung nebst einer einleitenden Abhandlung. Ein Beitrag zur Geschichte der Medicin*, ed. and trans. T. Puschmann, 2 vols. (Vienna: Braumüller, 1878–9), 2.64–5, and F. Burnet, trans., *Oeuvres médicales d'Alexandre de Tralles*, vol. 4 (Paris: Geuthner, 1937), 249.

271 Paul of Aegina, ed. I. L. Heiberg, 2 vols. (Leipzig: Teubner, 1921–4), VII.18.8 on 2.370; Trans. I. Berendes, *Paulos von Aegina. Des besten Arztes Sieben Bücher* (Leiden: Brill, 1914), 810.

272 Paul of Aegina VII.20.3 on 2.382; Berendes 819.

273 *The Book of the Eparch*: I. Dujčev, ed., *The Book of the Eparch* (London: Variorum, 1970), containing Greek text ed. by J. Nicole on 41 and Eng. trans. by E. H. Freshfield on 250.

274 *Simeonis Sethi syntagma de alimentorum facultatibus*, ed. B. Langkavel (Leipzig: Teubner, 1868), 66–7, and trans. M. Brunet, *Siméon Seth, médecin de l'empereur Michel Doucas, sa vie- son œuvre* (Bordeaux: Delmas, 1939), 77–8.

275 See also Harig 256.

276 A. D. H. Bivar, "Trade between China and the Near East in the Sasanian and Early Muslim Periods," in W. Watson, ed., *Pottery and Metalwork in T'ang China* (London: Percival David Foundation of Chinese Art, 1970), 1–11, D. Whitehouse and A. Williamson, "Sasanian

archaeologically in the form of Indian pottery of the first three centuries CE found on the Bushire Peninsula.[277] Ardashīr I, the first Sasanian king, pursued a policy of consolidating the Sasanian hold over the lands along the Persian Gulf.[278] By the time of Justinian (527–65) much of the Byzantine commerce in the Indian Ocean was handled by the Ethiopians, and they supposedly had difficulty acquiring silk from India because Persian merchant ships would occupy the harbors of India and buy entire cargoes of silk coming, presumably, from China itself by sea.[279] The Ethiopians, urged on by the Byzantines, expanded their influence in the Yemen, where they were contested by the Sasanians. Under Khusraw I (531–79), the Sasanians expelled the Ethiopians sometime in the early 570's and established their power over the *Bāb al-Mandab*, the gateway between the Red Sea and the Indian Ocean.[280] This gave the Sasanians enormous influence in the Indian Ocean trade. One important beneficiary of Sasanian trade in the Indian Ocean was the Nestorian church, which expanded using the routes the Sasanians were opening.[281]

We must thus attribute the spread of musk consumption to the Sasanian Empire, but it is not certain where the Sasanians became acquainted with

---

Maritime Trade," *Iran* 11 (1973): 29–49, D. Whitehouse, "Sasanian Maritime Activity," in J. Reade, ed., *The Indian Ocean in Antiquity* (London, 1995), 339–48, and M. Morony, "Economic Boundaries? Late Antiquity and Early Islam," *JESHO* 47 (2004): 184–8.

277 Whitehouse and Williamson 38–9.

278 V. Fiorani Piacentini, "Merchants- Merchandise and Military Power in the Persian Gulf (Sūriyānj/Shahriyāj-Sīrāf)," *Atti della Accademia Nazionale dei Lincei, Memorie*, Series IX, 3:2 (1992): 136–7, 146.

279 Procopius, *History of the Wars*, ed. and trans. H. B. Dewing, 5 vols., Loeb Classical Library (New York: Macmillan, 1914–1928), I.20.12, and T. Power, *The Red Sea from Byzantium to the Caliphate: AD 500–1000* (Cairo: American Univ. in Cairo Press), 79–80. Despite the importance of middlemen such as the Ethiopians, the range and quantity of trade goods between the Byzantines and Further Asia was substantial, see Mango.

280 See J. Harmatta, "The Struggle for the Possession of South Arabia between Aksūm and the Sāsānians," *IV Congresso internazionale di studi etiopici (Roma, 10–15 aprile 1972)* (Rome: Accademia nazionale dei Lincei, 1974), 95–106, I. Shahid, *Byzantium and the Arabs in the Sixth Century Volume 1 Part 1: Political and Military History* (Washington: Dumbarton Oaks, 1995), 364–72, and his *Byzantium and the Arabs in the Sixth Century Volume 2 Part 2: Economic, Social, and Cultural History* (Washington: Dumbarton Oaks, 2009), 52–7, and Power 81–86.

281 B. C. Colless, "The Traders of the Pearl," *Abr-Nahrain* 9 (1969–70): 17–38, 10 (1970–1): 102–21, 11 (1971): 1–21, 13 (1972–3): 115–35, 14 (1973): 1–16, 15 (1974–5): 6–17, and 18 (1978–9): 1–18. This series traces the expansion of the Nestorians in the Indian Ocean lands and Southeast Asia into nearly modern times. See also Colless, "Persian Merchants and Missionaries in Medieval Malaya," *JMBRAS* 42 (1969): 10–47.

musk. There are three different avenues for the introduction of musk to the Sasanians. The Sasanian maritime trade with India is the best known. The roots of this trade go back far in Sasanian history, and during the 5th century Sasanian power extended to a city known for its trade in musk. The Islamic historian al-Ṭabarī reports that Bahrām Gōr (420–38) received Daybul, Makrān, and the adjoining parts of Sind through his marriage to a local princess.[282] Cosmas Indicopleustes tells us that in his time, the middle of the 6th century, musk, κοστάριν (costus?),[283] and spikenard existed in Sind, called Σινδοῦ (*Sindoū*) by him.[284] Actually, Daybul was a major trade emporium that received goods from the Himalayas and Pamirs in the north for transshipment by sea; none of these items originated there. It so happens that musk is explicitly mentioned passing through Daybul in Islamic times as well, so it can be safely assumed that the city was a major emporium for the musk trade.

But even earlier than this, the Sasanians had also extended their power into Bactria and the East. During the third century the Sasanians overthrew the Kushan empire and soon created the so-called Kushano-Sasanian state in Bactria.[285] The geographical extent of this state was such that it brought Persians close to the mountainous areas of the Pamirs where musk deer were—and still are—to be found. It is possible that they discovered the use of musk there as well. The Pamir language Parachi uses the Persian word for musk rather than an indigenous one, along with the word *tātār*, also from Persian *tātārī*, referring to Tatar musk; this was borrowed much later, of course.[286] The Kushano-Sasanian state faced invasion by the Chionites in the middle of the 4th century. During the 5th century this area passed under the control of the Hephthalites, who forged a powerful empire. Eventually the Hephthalites gained control over Sogdiana as well, but under them the commercial empire

282 Ṭabarī I.868.

283 Cf. note 15.5 on 346. Certainly not "castor" as in de la Vaissière, *Sogdian Traders* 86, presumably following McCrindle's translation, which I have not seen. Also the "androstachys" there is a typo for "Nardostachys."

284 Cosmas XI.15 on 346–7. See below on later attestations of the musk trade in Sind.

285 The exact chronology of the Kushano-Sasanians is disputed. See A. D. H. Bivar, "The History of Eastern Iran," in *Cambridge History of Iran Volume 3 Part 1* (Cambridge: Cambridge University Press, 1983), 209–13, J. Cribb, "Numismatic Evidence for Kushano-Sasanian Chronology," *Studia Iranica* 19 (1990): 151–93, and A. Nikitin, "Note on the Chronology of the Kushano-Sasanian Kingdom," in *Coins, Art, and Chronology* (Vienna: Akademie der Wissenschaften, 1999), 259–63.

286 G. Morgenstierne, *Indo-Iranian Frontier Languages* (Oslo, 1929–38) 1.274 and 2.39*, 1.296.

of the Sogdians flourished.[287] At this time a great expansion in items of western origin—coins, glassware, and metalwork—found in Eastern Central Eurasia and China occurred, indicating a great expansion of trade.[288]

The Sogdians themselves are the third route by which musk could have reached the Sasanians.[289] Positioned between China, the nomadic empires of the Hephthalites and then the Turks, and Sasanian Persia, the Sogdians were ideally situated to dominate the trade of Central Eurasia. The Sogdians used their position in the Turk Empire to further the trade with China. In 545/6 the emperor of Western Wei sent an ambassador from Jiuquan 酒泉 in Gansu to *Tümen/Bumın, who would become the first kaghan of the Turk Empire. This ambassador was a Hu 胡, here meaning a Sogdian, named An Napantuo 安諾槃陁.[290] The town of Jiuquan was a center for Sogdian merchants and is attested as such in the Sogdian Ancient Letters.[291] This first Turkic contact with China was motivated by a desire to trade with the Chinese for silk, and the Sogdians, who were merchants *par excellence*, served as intermediaries.[292] Likewise, the first Turk embassy to the Byzantines in 568/9 was organized by the Sogdians, who had previously sought to trade raw silk with the Persians but failed. They turned to the Byzantine Empire, with help from their Turk overlords.[293] These examples show the great scope of the Sogdian trade network in the 6th century. It was a commercial empire built on trade in goods like silk and musk.

As mentioned above, the Sogdians traded in musk from the time of the Ancient Letters, which are believed to date to shortly after 311 CE; in these letters musk is attested as traveling to Samarkand from the Sogdian merchants

---

287 De la Vaissière 110–12.

288 F. Thierry, "Sur les monnaies sassanides trouvées en Chine," in R. Gyselen, ed., *Circulation des monnaies, des marchandises et des biens* (Bures-sur-Yvette: Groupe pour l'étude de la civilisation du Moyen-Orient, 1993), 115–19. See also R. N. Frye, "Sasanian-Central Asian Trade Relations," *BAI* 7 (1993): 73–7. There is a large literature on archaeological discoveries of western objects in China. For an overview, emphasizing the importance of these goods in Chinese culture, see E. R. Knauer, *The Camel's Load in Life and Death* (Zürich: Acanthus, 1998).

289 De La Vaissière.

290 *Zhoushu* (Beijing: Zhonghua shuju), 50: 908.

291 N. Sims-Williams and F. Grenet, "The Historical Context of the Sogdian Ancient Letters," in *Transition Periods in Iranian History* (Paris, 1987), 108.

292 Cf. de la Vaissière 205.

293 Menander Protector, *The History of Menander the Guardsman*, ed. and trans. R. C. Blockley (Liverpool: Cairns, 1985), 10.1, 1–6.

in China, where musk was well known by that time. During the 5th and 6th centuries, when Sogdian trade was flourishing, musk became well known in Western Asia. Therefore, it is quite likely that the Sasanians could have become acquainted with musk through the Sogdians, although they did not adopt the Sogdian word for musk. It is certain that the later Sogdians traded musk to the west as well. In Islamic times the term *al-misk al-sughdī* referred not to musk from Sogdiana, but musk transshipped from Tibet by Sogdian merchants.[294]

There are no grounds to specify through which of these routes musk reached the Persians first. Musk could have arrived simultaneously through more than one of them. The Sogdians were of course active in the trade through Sind as well, coming down through the Karakorum.[295] Even if the Persians first encountered musk in Sind, it may have been through the Sogdians. Something more than simple Persian adoption of musk from India is likely, especially since musk was probably little used in India prior to the 4th or 5th century and the Persians did not adopt the Indian word for musk.

Musk reached the Arabs through the Sasanians.[296] Not only did Arabs visit the court of the Sasanian King of Kings, the Arabs along the Sasanian frontier had their own Persianate states that acted as buffers between the Sasanian and Byzantine Empires. Many famous poets were associated with the court at Ḥīra. One of them, Maymūn b. Qays al-Aʿshā, was later reported to have used many Persian words: "al-Aʿshā used to visit the court of the kings of Persia, and because of that, Persian words abound in his poetry."[297] These words reflected the luxurious court life of the poets' patrons. Many of the Persian words used in the Qur'ān, including *misk*, are also found in pre-Islamic Arabic poetry; they appeared in the Qur'ān because they had a great prestige value from their use in court life.

---

294 Ibn Māsawayh 9.

295 See N. Sims-Williams, "The Sogdian Merchants in China and India," in A. Cadonna and L. Lanciotti, eds., *Cina e Iran da Alessandro Magno alla dinastia Tang* (Firenze: Olschki, 1996), 52–6.

296 For the botanical heritage of Sasanian times in Islamicate medicine, see King, "New *materia medica*".

297 Ibn Quṭaybah, *Kitāb al-Shiʿr wa-l-shuʿarā'* (Leiden: Brill, 1902), 137, and C. E. Bosworth, "Iran and the Arabs before Islam," in *The Cambridge History of Iran 3:1 The Seleucid, Parthian and Sasanian Periods* (Cambridge: Cambridge University Press, 1983), 610.

It is in this body of pre-Islamic poetry that musk first commonly appears in Arabic; a verse from al-Aʿshā including musk is quoted in Chapter 4. The terminology employed by early poets such as Imruʾ al-Qays and ʿAntarah already uses complex series of words to refer to musk, such as *misk, faʾrat al-misk*, and simply *faʾrah*. This corpus includes a number of loanwords from Persian that may be attributed to the Arab contacts with the Sasanian world, either directly or through the intermediation of Syriac culture. In the case of *misk*, had the Arabs borrowed directly from Syriac *muškā*, one might expect a word closer to the Syriac.

Arabs may have acquired musk from merchants in the cities on the edge of the Sasanian Empire. But musk certainly also reached the Arabs by sea through the network of sea trade established by the Sasanians that extended to the eastern and southern shores of the Arabian Peninsula.[298] The Sasanians were present in the market fairs of the south. The market of Aden, held on the first of Ramaḍān, was taxed by the Persian government, and this was a market from which "aromatics (*al-ṭīb*) were carried to the rest of the regions".[299]

The port city of Dārīn was famous for the importing of musk. Dārīn was located on Tarut, an island in the Persian Gulf near Baḥrayn[300] and was an important port city during late antiquity and the early medieval period; it had its own Christian community as well.[301] It must eventually have been the most important source of musk for the Arab peoples, because the musk of Dārīn is frequently attested in early Arabic poetry, while the perfumers who dealt in aromatics traded through Dārīn came to be called Dārī.[302] Al-Bīrūnī

298 Fiorani Piacentini 141–2.

299 Al-Yaʿqūbī, *Tārīkh*, ed. M. T. Houtsma, 2 vols. (Leiden: Brill, 1883; repr. 1969), 1.314.

300 D. T. Potts, *The Arabian Gulf in Antiquity vol. 2* (Oxford: Clarendon Press, 1990), 212–16. Cf. LA 13.186.

301 See J.-M. Fiey, "Diocèses syriens orientaux du Golfe Persique," in *Mémorial Mgr. Gabriel Khouri-Sarkis* (Louvain: Impr. orientaliste, 1969), 213–5 and map between 180 and 181.

302 Al-Bīrūnī, *Kitāb al-Ṣaydanah*, ed. and trans. H. M. Said (Karachi: Hamdard National Foundation, 1973), 4, and ed. ʿAbbās Zaryāb (Tehran: Markaz-i Nashr-i Dānishgāhī, 1991), 5, quotes a version of the ḥadīth of the good companion (discussed in Chapter 4) which has a *dārī* as the possessor of the good scent. See also M. Meyerhof," Das Vorwort zur Drogenkunde des Bērūnī," *Quellen und Studien zur Geschichte der Naturwissenschaften und der Medizin* 3:3 (1932): text 5 and trans. 30. Musk of *Dārīn* is also mentioned by al-Maʿarrī, *Risālat al-ghufrān*, ed. ʿĀʾisha ʿAbd al-Raḥman, 5th ed. (Cairo: Dār al-Maʿārif, 1969), 222.

(973–c. 1050) wrote concerning the Arabic term *dārī*, which had come to mean "perfumer":[303]

> Dārīn was a port in old times for ships laden with perfumes (*ʿiṭr*) and aromatics (*ṭīb*); then the perfumers [having put in at Dārīn] distributed them among the Bedouins and people of that sort like the Quraysh who were specialists with skill in mixing and compounding them, and in testing them, or like the people of Yemen with the making of oils. Because of this perfumers became known among the Arabs as Dārī, attributed to that port just as perfume is also attributed to it.[304]

The concept of Dārīn as a source of aromatics appears in poetry. A pair of verses by al-Nābighah al-Jaʿdī (d. c. 683) read:

raḥīqan ʿirāqiyyan wa rayṭan yamāniyyan
  wa-muʿtabiṭan min miski dārīna adhfarā
bi-aṣdāfi hindiyyayni zabba liḥāhuma
  yabīʿāni fī dārīna miskan wa-ʿanbarā[305]

Fine wine of Iraq, cloth of Yemen
  and pure, pungent musk of Dārīn
With shells of the two Indians who have hairy beards
  who traffic in Dārīn musk and ambergris.

These verses indicate that Indians sailed to Dārīn to sell musk, but this may refer to people involved in the India trade if not to actual Indians, who were certainly present in the Gulf, though. He also said:

ulqiya fīhi filjāni min miski dā-
  rīna wa-filjun min fulfulin ḍarimi[306]

Two measures of Dārīn musk were cast into it, and a measure of burning pepper.

---

303 Cf. also Ibn Khallikān, *Wafāt al-aʿyān*, ed. I. ʿAbbās, 4th ed., 8 vols. (Beirut: Dār Ṣādir, 2005), 3.41.

304 Bīrūnī, *Ṣaydanah*, ed. Saʿīd 4, ed. Zaryāb 5, Meyerhof text 5, trans. 29–30.

305 Bīrūnī, *Ṣaydanah*, ed. Sa ʿīd 4, ed. Zaryāb 6, Meyerhof text 5 and trans. 30.

306 LA 13.186; Bīrūnī, *Ṣaydanah*, ed. Saʿīd 5, ed. Zaryāb 6, Meyerhof text 6 and trans. 32.

Kuthayyir (c. 660–723) has:

masāʾiḥu fawday raʾsihi musbaghillatun
jarā misku dārīna l-aḥammu khilālahā[307]

The locks of the temples of his head were loosened, the blackest musk of Dārīn flowed among them.

Kuthayyir also has:

ufīda ʿalayhā l-misku ḥattā kaʾannahā
laṭīmatu dāriyyin tafattaqa fāruhā[308]

Musk was used upon her, until it was as if she was a Dārī's perfume-box with a pod split open.

Hammām b. Ghālib al-Farazdaq (c. 640–728) has a verse:

ka-ʾanna tarīkatan min māʾi muznin
wa-dāriyya al-dhakiyyi min al-mudāmi[309]

Like the water left from the rain of the clouds and the pungent Dārī [musk] from the wine.

Ibn al-Rūmī (836–896):

niʿālu kanbāyata wa-l-ʿanbaru
wa-misku dārīnikum al-adhfaru[310]

Sandals of Cambay and ambergris, and your most potent musk of Dārīn.

---

307 LA 13.186; al-Sarī al-Raffāʾ 3.156 (#282); Bīrūnī, *Ṣaydanah*, ed. Saʿīd 4, ed. Zaryāb 6, Meyerhof text 6 and trans. 30; Kuthayyir, *Sharḥ Dīwān Kuthayyir*, ed. H. Peres. 2 vols. (Paris: Paul Geuthner, 1928–30), 2.52–2.

308 LA 13.186; Kuthayyir, *Dīwān*, 1.93.

309 LA 13.186; al-Sarī al-Raffāʾ 3.154 (#276) with variant.

310 Bīrūnī, *Ṣaydanah*, ed. Saʿīd 5, ed. Zaryāb 6, ed. Meyerhof text 6 and trans. 31.

All of these examples show that the Arabs were most familiar with musk imported through the Persian Gulf, and not by the Indian Ocean through Yemen, although that route certainly existed in later times. Taken at face value, al-Bīrūnī implies even that the people of Yemen acquired their musk through Dārīn; there is no indication whether they got it through transshipment by sea or overland. But the Bedouin certainly must have acquired their musk by overland trade from the Arabian coast adjoining Tarut. Bīrūnī's reference to the Quraysh as perfumers is interesting also, as that would provide a context for the familiarity of Muḥammad and the Meccans with musk.[311] While the role of Mecca in Eurasian commerce has undoubtedly been exaggerated,[312] the city itself was a market for goods, especially aromatics.[313] Abū Ṭālib, the uncle of Muḥammad who raised him, is said to have sold perfumes (*ʿiṭr*).[314] The trade in musk was well known in the early days of Islam. ʿUmar b. al-Khaṭṭāb, the second caliph (r. 634–44) is credited with the following aphorism: "If I were a merchant, I would choose nothing but musk. Even if its profit (*ribḥ*) escaped me, its scent (*rīḥ*) would not."[315]

The trade through eastern Arabia seems to have declined during Islamic times. By the time the geographers were describing the trade routes, the great ports were elsewhere, such as Ubullah at the head of the Gulf, Sīrāf on the Persian side, opposite Baḥrayn, and Ṣuḥār in Oman. The area around Baḥrayn, on the other hand, had destabilized,[316] and musk of Dārīn had become a matter fit for the investigations of antiquarians.

---

311 On the Quraysh as dealers in aromatics, see P. Crone, *Meccan Trade and the Rise of Islam* (Princeton: Princeton University Press, 1987), 95–6.

312 Crone. See G. Heck, "Arabia without Spices: An Alternative Hypothesis," *JAOS* 123:3 (2002): 547–76 for a review of the matter, emphasizing the importance of local industry and commerce within the Arabian economy.

313 Some references have been collected by Heck 571.

314 Al-Thaʿālibī, *Laṭāʾif al-maʿārif* (Cairo: Dār Iḥyāʾ al-Kutub al-ʿArabiyyah ʿĪsā al-Bābī al-Ḥalabī wa-Shurakah, n.d.), 127; Ibn Rustah, *Kitāb al-ʿAlāq al-nafīsah*, ed. M. G. De Goeje (Leiden, 1892; repr. Leiden: Brill, 1967), 215.

315 Al-Ghuzūlī, *Maṭāliʿ al-budūr fī manāzil al-surūr*, 2 vols. bound as one (Cairo: Maktabat al-Thaqāfah al-Dīniyyah, n.d. [2000]), 1.75. A variation, perhaps the original, has *ʿiṭr* instead of musk: al-Zamakhsharī, *Rabīʿ al-Abrār wa Fuṣūṣ al-Akhbār*, eds. ʿAbd al-Majīd Diyāb and Ramaḍān ʿAbd al-Tawwāb, 2 vols. (Cairo: al-Hayʾah al-Miṣriyyah al-ʿĀmmah li-l-Kitāb, 1992–2001), 2.214 and Ibn Abī al-Ḥadīd 19.342.

316 See J. C. Wilkinson, "Ṣuḥār in the Early Islamic Period," *South Asian Archaeology 1977* (Naples: Istituto universitario orientale, 1979), 895.

Conclusion

Musk was thus well known and used in the lands where the musk deer lived. All the cultures surveyed made use of musk in perfumery and had medicinal uses for it as well. Each of the cultural areas has its own particular terminology for musk, indicating that the substance came to be employed in each of these regions independently, even if the first use of musk cannot be determined because of a lack of sources. Only in China do we have sources antedating the attested use of musk in the Near East, allowing us to determine approximately when its utilization began.

The Chinese employment of musk in medicine and spiritual medicine is surely autochthonous. The coming of Buddhism also promoted the burning of incense, no doubt enhancing the importance of musk, which was already used by the Chinese in incense formulas. The precise origins of the use of musk in India are obscure, since no reliable attestations of musk can be traced earlier than the Gupta period. But from its very beginnings in India musk became an important perfume ingredient and gradually entered the repertory of *materia medica* as well. In India and Tibet, as in China, musk was used in small quantities in a blend based on other ingredients to enhance them. The scent of musk was not to dominate blends. This is consistent with the use of musk in European perfumery over the past two or three centuries.

In the case of Central Eurasia, one cannot separate indigenous uses of musk from the influence of India. Presumably if we had sources that antedated the spread of Buddhism and Indic cultural influence, we could find out how musk was seen by the inhabitants of Tibet and Central Eurasia. Some clues that musk was important among the Turks are found in the wide ranging uses of musk in the remains of Old Turkic literature. But even here, the influence of India may be possible, and, in the case of Karakhanid works such as the *Dīwān Lughāt al-Turk* of Kāshgharī and especially the *Kutadgu Bilig* there is probably Persian influence as well.

For the earliest period of the use of musk in the Near East and the Mediterranean, Sasanian Persia is never far from the action. The Arabs must have encountered musk in the courts of the Persians on the Mesopotamian frontier, and then it started to come to the Arabian Peninsula directly through a trading network created by the Sasanians. Cosmas Indicopleustes and Procopius mention the Persian ships in the Indian Ocean transporting the goods of India, and by these same routes the Byzantines must have been able to acquire musk, since Cosmas is well informed about the trade. Few records of the trade between the Near East and Further Asia are preserved; merchants may keep records, but they are often kept private to prevent their competition from learning their trade secrets. And over the centuries such documents tend to be lost and neglected. In the exceptional cases where records are available, in Ibn Khurradādhbih and the Cairo Geniza, they show a flourishing Jewish trade with Asia that included the importation of musk. We know that Arabs also had such trading networks through the Indian Ocean. Overland, into Khurāsān, the Central Asian musk trade was in the hands of the Sogdians, who had a trading system stretching far into China. The Sogdians have the distinction of being the first people known to have sent musk to the west, and their importance is reflected in the term *al-misk-al-sughdī*, used for the best musk in the early 9th century.

CHAPTER 4

Islamicate Knowledge of Musk and Musk Producing Lands

Arabic and Persian literature—Muslim, Christian, and Jewish—of the Islamic period preserves a great deal of information about musk, as befits a commodity of such great cultural importance. There was a sophisticated terminology for musk and the musk pod, as well as for the perfumes which were made from it. The former are discussed below, while the latter will be discussed in Chapter Six. The mere existence of such a complex vocabulary associated with musk illustrates its significance. The types of musk were determined especially by their geographic origin rather than by any assessment of their incidental qualities, and the detailed body of information that survives is analyzed in the last part of this chapter. We have seen how geographical information intertwined with the identity of commodities as part of connoisseurship. The point of origin and story of its genesis was a key part of the appreciation of musk.

Arabic Terminology Relating to Musk

As noted in Chapter One, the Arabic word *misk* "musk" is closely related to the words for musk used throughout the western end of Eurasia and the Middle East. Arabic philologists generally acknowledged that it was a loanword originally from Persian, as is indeed the case. While musk is usually just denoted *misk*, there are a number of special terms associated with musk as well as synonyms that appear throughout Arabic literature. Some of the more important of these are surveyed below.

1. *Faʾrat al-misk*

Apart from the word *misk*, early Arabic poetry frequently applies the term *faʾrat al-misk* or simply *faʾrah* (literally, "mouse") to musk. A pre-Islamic example of the term *faʾrah* used by itself for musk is found in a famous line from the *Muʿallaqah* of ʿAntarah:

> *wa-kaʾanna faʾrata tājirin bi-qasīmatin*
> *sabaqat ʿawāriḍahā ʿilayka min al-fami*[1]

1 *Sharḥ dīwān ʿAntara*, ed. Ibrāhīm al-Zayn (Beirut: Dār al-Najjāḥ, 1964), 205.

 | DOI 10.1163/9789004336315_006

As if a merchant's musk in a perfume box
comes to you from her mouth before her teeth.

A variant quoted in the *Kitāb al-Nabāt* of al-Dīnawarī has:

wa-kaʾanna rayyā faʾratin hindiyyatin
sabaqat ʿawāriḍahā ilayka min al-fami[2]

As if the aroma of the Indian Mouse
comes to you from her mouth before her teeth.

The term *faʾrat al-misk* was used in Islamic poetry as well; here is a line of the Umayyad poet Dhū al-Rummah:

taḥuffu bi-turbi l-rawḍi min kulli jānibin
nasīmun ka-faʾri l-miski ḥīna tufattiḥu[3]

Surrounding the earth of the meadow from every direction
was a breeze like a musk pod when it is opened.

The confusion over the term was resolved by the Basran polymath al-Jāḥiẓ (c. 776–868 or 9) in his *Kitāb al-Ḥayawān*: "I asked one of the perfumers of the Muʿtazilites about *faʾrat al-misk* and he said: 'it is not like a mouse (*faʾrah*) but a small gazelle (*khishf*) is more similar to it.' Then he told me about the business of musk and how it is produced."[4] This was also known to al-Dīnawarī (d. late 9th century), who wrote in his *Kitāb al-Nabāt* while commenting on the use of the term in a verse: "*al-faʾr* is the plural of *faʾrah* and he means by it the *faʾrat al-misk*. It is its pods (*nawāfijuhu*) which are in it. They are called *al-faʾr* but are not mice, but rather they are the navels (*surar*) of the musk gazelles (*ẓibāʾ al-misk*)."[5]

As people got a better idea of the origin of musk, *faʾrah* was replaced in most literature by the term *nāfijah*, which is discussed in the next section. In poetry,

2 Abū Ḥanīfa al-Dīnawarī, *Kitāb al-Nabāt: The Book of Plants: Part of the Monograph Section*, ed. Bernhard Lewin (Wiesbaden: Steiner, 1974), 192.

3 Dhū al-Rummah, *The Dīwān of Ghailān ibn ʿUqbah known as Dhu ʾr-Rummah*, ed. C. H. H. Macartney (Cambridge: Cambridge University Press, 1919), #10 l. 24 on p. 83.

4 Al-Jāḥiẓ, *Kitāb al-Ḥayawān*, ed. ʿAbd al-Salām Muḥammad Hārūn, 8 vols. (Cairo: Muṣṭafā al-Bābī al-Ḥalabī, 1966), 5.304; cf. 7.210.

5 Dīnawarī, *Monograph*, 194.

however, *fa'rah* survived as an archaism, and in early medical literature *fa'rat al-misk* is still encountered.[6] Al-Damīrī, following the lexicographer al-Jawharī, prefers to spell it *fārah* and derives the word from the verb *fāra/yafūru* "to boil, to be roused", no doubt due to the heating and exciting properties of musk in pharmaceutical thought.[7]

In addition, the term *fa'rat al-misk* does seem to be used in several places to describe the actual musk shrew that lives in the lands around the Indian Ocean and beyond. Al-Jāḥiẓ mentions that there is a kind of black rat (*juradh*) sometimes found in peoples' houses, presumably in Iraq or Arabia, that produces a musky scent. He explicitly states that these are not *faʿrat al-misk*, for he says that the latter occur only in Khurāsān, here meaning the east in general, and resemble small gazelles.[8] His black rats are surely musk shrews. It is probably the musk shrew that Ibn al-Faqīh refers to as *fa'r al-misk* in the following place: "The *fa'r al-misk* is brought sometimes from Sind to Zābaj. Civet is sweeter in scent than musk when the female produces musk. When one enters a house the scent of musk diffuses from it, and when you touch it with your hand, your hand becomes redolent."[9] It is also possible that Ibn al-Faqīh refers to the civet cat under the term *fa'r al-misk* here. Ibn al-Faqīh later mentions that Tibet has musk, but he does not say what it comes from; in any case, he does not say it comes from *fa'rat al-misk*.[10] Each instance must be evaluated according to its context.

There seems to have been a persistent suspicion about musk and mice due to the existence of the musk shrew and the unusual archaic nomenclature despite al-Jāḥiẓ's clarity on the matter. Al-Dīnawarī writes:

6 E.g., Pseudo-al-Kindī, *Kitāb Kīmiyā' al-ʿiṭr wa-t-taṣʿīdāt: Buch über die Chemie des Parfüms und die Destillationen. Ein Beitrag zur Geschichte der arabischen Parfümchemie und Drogenkunde aus dem 9. Jahrh. P.C.*, ed. and trans. with commentary by K. Garbers (Leipzig, 1948), #63 on p. 41 and *The Medical Formulary or Aqrābādhīn of al-Kindī*, ed. and trans. M. Levey (Madison: University of Wisconsin, 1966), 16, etc. and Levey's entries numbers 217 and 218. The 14th century Persian writer Ḥamdullāh al-Mustawfī Qazwīnī, *The Zoological Section of the Nuzhatu-l-Qulūb*, ed. and trans. J. Stephenson (London: Royal Asiatic Society, 1928), text 30 and trans. 21, also mentions it and claims its musk is even better than deer musk by perhaps ten times! He probably refers to a musk pod.

7 Al-Damīrī, *Ḥayāt al-ḥayawān*, 2 vols. (Būlāq, 1284/1867; repr. Frankfurt am Main: Institute for the History of Arabic-Islamic Science, 2001), 2.236.

8 Jāḥiẓ, *Ḥayawān*, 7.211.

9 Ibn al-Faqīh al-Hamadhānī, *Mukhtaṣar Kitāb al-Buldān*, ed. M. J. De Goeje (Leiden: Brill, 1885; repr. 1967), 11. Cf. also Dīnawarī, *Monograph*, 194–5.

10 Ibn al-Faqīh 255.

> In the region of India there is a mouse which is exported to the land of the Arabs sometimes and I had become used to it circulating in houses and entering between garments. Nothing that it came into contact with or entered between, nor anything that it defecated upon or urinated upon did not fail to diffuse a scent. Merchants import its excrement and people buy it and put it into bags (*ṣurar*) and place it between their garments and they are perfumed.[11]

Al-Dīnawarī is clearly describing the musk shrew in the first part of this passage. But the conclusion is most interesting: that the excrement of the musk shrew was collected and put into *ṣurar*, a term used to describe musk vesicles, as we will see.

One late account of musk from al-Damīrī's (14th century) zoological book *Ḥāyat al-ḥayawān* shows that the detailed knowledge of musk arrived at by the scholars of the 9th–11th centuries did not successfully displace all of the earlier ideas about the origins of musk. Al-Damīrī's work is based on older Arabic literary and especially religious sources. Most notably, he continues to mention the *fa'rat al-misk* and has this creature listed in his account of mice, based on the *Kitāb al-Ḥayawān* of al-Jāḥiẓ, along with other types of mice. Al-Damīrī gives an account of *fa'rat al-misk* under the word *fa'r* and here, following al-Jāḥiẓ, he clearly distinguishes between the musk pod and the musk shrew.[12] In his account of the gazelles (*ẓibā'*) he discusses the musk deer proper, and here he also attempts to distinguish between the musk produced by the "mouse" and the gazelle; the former is impure and unclean (rodents are impure in Islamic law), while the other is lawful.[13] Musk from the gazelle and the vesicle was considered lawful as long as it was removed from the animal while it was alive, as one would expect. Al-Damīrī quotes a certain al-Maḥāmilī, author of a *Kitāb al-Lubāb*, who says that "musk from the gazelle is pure, that is the musk taken from the gazelle." Al-Damīrī comments:

> By this he warned against Tibetan musk taken from the mouse which will be discussed in the chapter of the letter *fā'*, God the Exalted willing. It is impure, as can be inferred from the prohibition against eating it; if it could be eaten, then its musk would join the musk of the gazelle. The perfumers call Tibetan musk Turkish musk, and it is the best musk according

11 Dīnawarī, *Monograph*, 194–5.
12 Damīrī 2.226–7.
13 Damīrī 2.127.

> to them and the highest in value, but it is necessary to guard against using it because of its impurity.[14]

It is most striking that al-Damīrī regards Tibetan musk as derived from a mouse: he has thus created two arbitrary categories of genuine musk, one lawful and one prohibited, from the supposed origin of it.

2. *Nāfijah*

The term *nāfijah* refers to the musk vesicle; it is the most commonly used word in Classical Arabic scientific literature for it. It can denote the pod in its full state as well as the actual skin of the pod without the musk. In some cases in perfumery literature one suspects that these vesicles were made of other types of skin that were prepared to be filled with adulterated or imitated musk. The medieval writers believed that it was originally the navel (*surrah*) of the animal. The term *nāfijah* derives from the Persian word for navel, which is *nāfag* in Pahlavi;[15] it later became *nāfah* in Classical Persian where it was used by itself or with the word *mushk* to mean the musk pod, e.g. Minūchihrī (early 11th century):

nāfah-yi mushkast harch ān bi-ngari dar bustān
dānah-yi durrast harch ān bi-ngari dar jūybār[16]

Wherever you look in the garden is a musk pod;
wherever you look in the stream is a seed pearl.

The foreign origin of this word was well known to Arabic scholars. Al-Jawālīqī wrote: "*al-nāfijah* is musk. It is Persian (*ʿajamiyyah*) arabicized."[17] The word *nāfijah* appears sometimes in poetry in the sense of musk although it is nowhere near as frequently encountered there as *faʾrat al-misk* and its variations. An anonymous *rajaz* verse says:

kaʾanna ḥashwa l-qurṭi wa-l-damāliji
nāfijatun min aṭyabi l-nawāfiji[18]

---

14 Damīrī 2.127.
15 *Mušk-i nāfag* is attested in the *Bundahišn*, see above, Chapter 1.
16 Quoted LN vol. 30 part 1, 198.
17 Al-Jawālīqī, *al-Muʿarrab min al-kalām al-aʿjamī ʿalā ḥurūf al-muʿjam*, ed. by Abū al-Ashbāl Aḥmad Muḥammad Shākir (Tehrān, 1966), 341.
18 Al-Sarī al-Raffāʾ, *Al-muḥibb wa-l-maḥbūb wa-l-mashmūm wa-l-mashrūb*, ed. Miṣbāḥ Ghalāwinjī, 4 vols. (Damascus: Majmaʿ al-Lughah al-ʿArabiyyah, 1986–7), 3.156 (#283).

As if the filling of the earring and bracelets
was a pod of the most fragrant musk pods.

3. *Shiyāf* and *akrāsh*

Musk grains are denoted by the term *shiyāf*, which usually denotes an eye medication. These *shiyāf* are described as fine (*diqāq*) and coarse (*jilāl*) by Ibn Mandawayh-Ibn Kaysān.[19] The application of the term *shiyāf* by the Ibn Mandawayh-Ibn Kaysān tradition explains the mysterious phrase *bayna al-jilāl wa-l-diqāq* used to describe musk in Ibn Māsawayh's *Kitāb Jawāhir al-ṭīb al-mufradah*; he must be referring to these particles.[20] Likewise, al-Yaʿqūbī (d. 284/897) describes the best musk: "the dimensions [of its grains] are between coarse grains and fine grains" (*wa maqādīruhu wasaṭan bayna al-jilāl wa-l-diqāq*).[21] This makes sense if we assume the "dimensions" refers to the grains of musk.

In literature on the manufacture of perfumes the terms *kirsh/akrāsh* are encountered. Usually this word denotes the stomach, especially of ruminant animals, but in literature on the manufacture of perfumes it denotes a part of the musk pod. Since the root *k-r-sh* is also associated with the idea of being crinkled and convoluted, a provisional hypothesis is that this term denotes membranes or structures enfolded within the vesicle among the musk, or perhaps it denotes the inner layer of the vesicle itself, or perhaps simply the empty pod. The word *karish* also can denote a perfume container, in the sense of a pouch probably originally made of intestine used to store aromatics.[22] Since *nāfijah* must denote something different than *akrāsh* in, for example, Sahlān b. Kaysān's work (he uses both terms and the *nāfijah* most certainly refers to the pod), it is difficult to accept that it could be simply a synonym for the pod. Likewise, Ibn Wāfid distinguishes the compound musk perfume

---

19 Ibn Mandawayh 225 and Ibn Kaysān 188. Cf. Frederick Markham, *Shooting in the Himalayas: A Journal of Sporting Adventures and Travel in Chinese Tartary, Ladac, Thibet, Cashmere, &c.* (London: Richard Bentley, 1854), 88: "The musk itself is in grains, from the size of a small bullet to small shot …"

20 Ibn Māsawayh 10. M. Levey, "Ibn Māsawaih and His Treatise on Simple Aromatic Substances," *Journal of the History of Medicine* 16 (1961): 399, translated this "between the large and small" with no indication of the precise reference.

21 Quoted by al-Qalqashandī in *Ṣubḥ al-Aʿshā*, 14 vols. (Cairo: Al-Mu'assasah al-Miṣriyyah al-ʿĀmmah li-l-Ta'lif, 1964), 2.120. The text has *al-riqāq*, but this should be corrected to *al-diqāq*, as found in all the other versions of this statement.

22 E.g., al-Iṣbahānī, *Kitāb al-Aghānī* (Cairo: Dār al-Kutub), 9.125 where a *karish* filled with ambergris is given as a gift.

sukk made from *julūd* "skins", which he defines as musk pods (*nawāfij al-misk*) from *sukk al-akrāsh*, defined as *sukk* made with chopped musk pods and kneaded with *rāmik*.[23] Sahlān b. Kaysān regards the *akrāsh* as an impurity to be removed from musk during its processing, along with the *luqaṭ*, "remains," presumably referring to other interior parts of the musk vesicle that are not musk.[24]

4. *Ṣiwār*

Ṣiwār (also voweled *ṣuwār*) is another term used for the musk pod. It can also refer to the scent of musk. The *Lisān al-ʿArab* states the term is Persian, but this does not seem to be the case.[25] Bashshār b. Burd (c. 714–c. 784) used it in a well-known verse:

> *idhā nafakha l-ṣiwāru dhakartu suʿadā*
> *wa-adhkuruhā idhā lāḥa l-ṣiwāru*[26]

> When the musk pod diffuses its scent, I remember Suʿadā,
> and I remember her when the oryx come into view.

Here Bashshār plays on two senses of the word *ṣiwār*: that of a herd of oryx and that of the musk pod.[27] This term seems to go back to pre-Islamic times because it is found (in the plural form *aṣwirah*) in a frequently cited verse by al-Aʿshā (before 565–after c. 629) from his poem included by some among the *Muʿallaqāt*:

---

23 Ibn Wāfid, *Kitāb al-adwiya al-mufrada* (*Libro de los medicamentos simples*), ed. and trans. Luisa Fernanda Aguirre de Cárcer, 2 vols. (Madrid: Consejo superior de investigaciones científicas agencia española de cooperación internacional, 1995), 2.118 for text and 1.172–3 for trans. Cárcer notes that the reading of the word I have given as *akrāsh* is unclear in the manuscript (she gives variants including *akrās* in her n. 51.2 on page 172), and she has read it *akhrās*. The parallel in Ibn al-Bayṭār, *al-Jāmiʿ li-mufradāt al-adwiyah wa-l-aghdhiyah*, 4 vols. (Būlāq, n.d.), 3.24, however, has *akrāsh* and this reading also harmonizes with Ibn Mandawayh-Ibn Kaysān, etc.

24 Ibn Kaysān 198. See below for the entire preparation procedure.

25 LA 4.548,549.

26 Al-Sarī al-Raffāʾ 3.155 (#280). Cf. LA 4.548 without attribution and with Laylā replacing Suʿadā. Another verse by Bashshār using *ṣiwār* is quoted in Chapter 6.

27 E. W. Lane, *An Arabic-English Lexicon*, 8 vols. (repr. Beirut: Libraire du Liban, 1968), 4.1745b.

idhā taqūmu yaḍūʿu l-misku aṣwiratan
wa-l-zanbaqu al-wardu min ardānihā shamilu[28]

When she rises, the scent of musk emanates from the musk pods
and jasmine and rose from her cuffs pervades [the air].

5. *Laṭīmah*

The term *laṭīmah* has a variety of senses. It appears that it originally denoted a caravan and came to mean the expensive goods carried by a caravan.[29] It also denoted by extension a perfume box or vessel. In later times it certainly referred to musk, probably in the pod. The Egyptian al-Qalqashandī (1355–1418), quoting an unknown source on the pre-Islamic market fairs of the Arabs, notes that Arabs would purchase *laṭāʾima wa-anwāʿa l-ṭībi* in Aden;[30] here the meaning is obviously perfumes and we can suspect that it meant musk pods specifically, brought over the Indian Ocean. It appears in a verse of Bashshār:

fa-qāla a-ʿaṭṭārun thawā fī riḥālinā
wa-laysa bi-mawmātin tubāʿu l-laṭāʾimu[31]

He said, did a perfumer stay in our camp?
But there is no sale of musk pods (*laṭāʾim*) in the wide desert.

It is impossible to know whether the *laṭāʾim* in this verse refer to perfumes in general or to musk. A verse of Dhū al-Rummah (d. 735–6) shows clearly that it can mean musk pods, or a unit of musk:

ka-annahū baytu ʿaṭṭārin yuḍamminuhū
laṭāʾima l-miski yaḥwīhā wa-tuntahabu[32]

As if [the sand-dune] was the abode of a perfumer which he fills with musk pods; he collects them and they are sold off.

28 Al-Aʿshā, *Gedichte von Abû Baṣîr Maimûn ibn Qais al-Aʿšâ*, ed. R. Geyer (London: Gibb Memorial Series, 1928), #6 line 13 on 43. Cf. Dīnawarī, *Monograph*, 190; LA 4.549.

29 S. Fraenkel, *Die aramäischen Fremdwörter im Arabischen* (Leiden, 1886; repr. Hildesheim: Olms, 1962), 176.

30 Qalqashandī 1.411.

31 Al-Sarī al-Raffāʾ 3.155 (#278).

32 Dhū al-Rummah # 1 l.78 on p. 20.

Imru' al-Qays has the line:[33]

idhā qāmatā taḍawwaʿa l-misku minhumā
bi-rāʾiḥatin min al-laṭīmatin wa-l-quṭur[34]

When they rose, the fragrance of musk emanated from them with an aroma from the musk pod and of incense.

6. *Shadhā/ al-shadhw/ al-shadhā*
The verbal root can mean "to perfume oneself with musk." It is probably a development from the noun *shadhan*, which originally meant "pungency" but came to be applied to certain strong smelling or tasting substances such as salt and musk. *Al-shadhā* can also denote small pieces of aloeswood.[35] *Al-shadhw* is the usual form of the noun in the sense of musk. We have already met this term in a verse of al-ʿUjayr quoted in Chapter 2. The word *shadhī*, from the same root, can also mean musk-colored or black.[36]

7. *Rāmik* and *Sukk*
The *Lisān al-ʿArab* quotes two verses involving musk using the term *al-shadhw* for musk and another substance called *rāmik*:

inna laka l-faḍla ʿalā ṣuḥbatī
wa-l-misku qad yastaṣḥibu l-rāmikā
ḥattā yaẓalla l-shadhwu min lawnihi
aswada maḍnūnan bihi ḥālikā[37]

Truly you have too much excellence for my companionship,
but musk sometimes keeps company with *rāmik*
So that the musk (*al-shadhw*) continues to be pitch black
in its color and used sparingly.

Rāmik referred originally to a substance made with gallnut that had a black color; it is often mentioned as a substance used to extend or adulterate musk.

33 Another variation is quoted in Dīnawarī, *Monograph*, 197, 215; see below.
34 Imru' al-Qays, *Dīwān*, ed. Yāsīn al-Ayyūbī (Beirut: al-Maktab al-Islāmī, 1998), Line 8 on p. 272.
35 LA 14.525.
36 LA 14.525 with a verse.
37 LA 14.525. The first verse is also quoted by Lane 1159.

The use of the word *al-shadhw* allowed the poet to avoid repeating the word *misk* in both lines. The point of these verses is that musk, which is the best of the aromatics, makes a fine compound with the less valuable *rāmik*, just as the *mamdūḥ* or object of praise exceeds the level of the poet but is complemented by him.

Rāmik was a main ingredient in a compound perfume and perfume ingredient called *sukk* or *sukk* of musk. In this preparation, *rāmik* and other substances provided a way to extend the usefulness of a small quantity of musk. Since one of the valued characteristics of musk was its dark color, it is not surprising that the black *rāmik* would prove a useful adulterant. *Sukk* is mentioned in pre-Islamic poetry and is said to have been used by Muḥammad as well.[38] It was an important preparation throughout the early medieval period.[39]

8. *al-misk al-adhfar*

The adjective *adhfar* is commonly used to describe the odor of musk. The phrase *al-misk al-adhfar* means "pungent smelling musk." It appears in poetry as early as Imru' al-Qays.[40] *Adhfar* denotes any sort of pungent smell associated, it seems, with either musk or the human body, for example, with strong-smelling sweat. Musk has a strong smell that can resemble the acrid smell produced by the human body, especially from the axillae, as we have seen. *Adhfar* is not used with any other aromatic than musk.[41] This term appears in ḥadīths and so has a wide usage in both Arabic and Persian Islamic literatures.[42] Musk is also described as *dhafir*, pungent.[43]

---

38 Ibn Sa'd, *al-Ṭabaqāt al-kubrā*, 9 vols. (Beirut: Dār Ṣādir, 1960), 1.399 and Ibn Qayyim al-Jawziyyah, *al-Ṭibb al-Nabawī* (Cairo: Dār al-Turāth, 1978), 331.

39 R. Gottheil, "Fragment on Pharmacy from the Cairo Genizah," *JRAS* (1935): 123–44 contains two *sukk* formulas from Egypt. The standard perfume formularies—Ibn Mandawayh, Ibn Kaysān, etc.- contain *sukk* formulas.

40 Imru' al-Qays, *Dīwān*, 181.

41 LA 4. 307.

42 Abī 'Abdallāh al-Bukhārī, *Ṣaḥīḥ Abī 'Abdallāh al-Bukhārī bi-sharḥ al-Kirmānī*, 25 vols. (Cairo: Al-Maṭba'ah al-Bahiyyah al-Miṣriyyah, 1933–62), 9.451. See also S. Anṣārī, *Tārīkh-i 'Iṭr dar Īrān* (Tehran: Wizārat-i Farhang wa Irshād-i Islāmī, 1381/2002–3), 52–3, and *Farhangnāmah-i Adab-i Fārsī*, vol. 2 of *Dānishnāmah-i Adab-i Fārsī* (Tehran: Mu'assasah-i Farhangī wa Intishārāt-i Dānishnāmah, 1996–), 257b for some Persian examples.

43 E.g., 'Umar b. Abī Rabī'ah, *Dīwān* (Cairo: al-Hay'ah al-Miṣriyyah al-'Āmmah li-l-Kitāb, 1978), 72.

Persian Terminology for Musk

As noted above, many Arabic terms relating to musk have passed into Persian, although the Persians still preserve the distinctive form of their original pronunciation of musk itself, *mushk*, alongside an Arabic-influenced *mishk*. Islamic Persian literature began to flourish during the 10th century. The whole lexical resources of Arabic were essentially made available to writers in the newly developing classical Persian language. This meant the borrowing of special terminology from Arabic, although the core terminology for musk and its pod, *nāfijah*, was Persian originally.

The musk deer is usually called *āhū-yi mushk* "musk deer" in Persian; another name is *nāfah-bāf* "the weaver of the musk pod". The term *mushk* for the substance musk is ubiquitous and appears in a great many compounds and expressions. *Nāfah* is also used extensively in Persian in the meaning of musk. The best musk is called *mushk-i nāb* "pure musk"[44] and also *mushk-i sārā*[45] alongside the *muskh-i adhfar* of Arabic origin.

The number of words and phrases based on *mushk* in Persian is very great, reflecting the high value placed on musk as a paragon of aromatics. Some will be discussed in Chapter 7 among the symbolic meanings of musk. In poetry *mushk* or the adjectival *mushkīn* are used to describe many things, especially parts of the body of the beloved and things of the color black. Various kinds of hair, ranging from eyebrows to head hair to facial hair, especially the newly growing facial hair of the young male beloved (*khaṭṭ*), are described as musky.[46] The musky mole (*khāl*) on the face of the beloved is another common image.[47]

Sources of Musk: Middle Eastern Knowledge of the Geography of the Musk Producing Lands and the Origins of Musk

The many ways in which musk was used in medieval Middle Eastern society and the many references to it in literature show that a large amount of musk must have been imported. Information on the origins of musk is preserved in a variety of sources. The majority of this information goes back to the 9th and

44 *Farhangnāmah* 261a.

45 *Farhangnāmah* 260a.

46 Numerous examples are among the phrases employing musk in the *Farhangnāmah* 257–60 and M. A. Newid, *Aromata in der iranischen Kultur unter besonderer Berücksichtigung der persischen Dichtung* (Wiesbaden: Reichert, 2010), 72–81.

47 *Farhangnāmah* 259a; Newid 78–9.

10th centuries (a golden age of investigation about Further Asia) and stems from a handful of writers. Two general and somewhat overlapping categories can be discerned among these sources: writers interested primarily in geography and those interested in pharmacology. Those who were concerned especially with geography typically evince a considerable interest in commerce and the goods of the lands they discuss. Knowledge of the specialties of exotic lands was a branch of *adab*, and within this broader arena of *adab* writers such as al-Jāḥiẓ and al-Thaʿālibī also took an interest in musk as one of the substances consumed by polite society. The pharmacologists had their own purpose for assessing the medicinal value of musk, but they also frequently provide more general accounts of the musk deer and details of musk production and trade. The material on musk in these works ranges from the *mirabilia* of travelers' yarns to more scientific accounts.

In addition, what frequently survives are later works that incorporate evidence from earlier writers, whose works are frequently lost; only the quotations from them survive. In many cases it is impossible to determine the precise authorship of these quotations, so some speculation is required. In the following, an attempt has been made to arrange material by its point of origin rather than the source in which it was preserved.

It is only during the 9th century that extensive information on musk appears; of course, this is also the time of the earliest Arabic scientific works. It can be assumed that there was information commonly understood about musk prior to this time. In this chapter, we will focus upon the accounts of the musk deer and the origins of musk, and the places in which it lived. Accounts of the medicinal properties of musk will be left to Chapter 6.

TABLE 4.1 *Major early Arabic sources on musk*

Author/Date	Book	Genre
Ibn Māsawayh (777–857)	*Kitāb Jawāhir al-ṭīb al-mufradah*	Perfumery
Al-Jāḥiẓ (c. 776–868 or 9)	*Kitāb al-Ḥayawān*	*Adab*, zoology
Ps.-al-Jāḥiẓ (late 9th c.?)	*Kitāb al-Tabaṣṣur bi-l-tijārah*	*Adab*, commerce
Al-Yaʿqūbī (d. 897)	Quotations from an unknown work in al-Nuwayrī	

Author/Date	Book	Genre
Al-Jayhānī (early 10th c.)	*Kitāb al-Masālik wa-l-mamālik*	Geography
Abū Zayd al-Sīrafī (early 10th c.)	*Akhbār al-Ṣīn wa-l-Hind*	Geography
Al-Masʿūdī (d. 956)	*Murūj al-dhahab*	History/Geography
Al-Tamīmī (d. 980)	*Ṭīb al-ʿarūs*	Perfumery
Ibn Kaysān (d. 990)	*Mukhtaṣar fī al-ṭīb*	Perfumery
Ibn Mandawayh (writing before 995)	*Risālah fī uṣūl al-ṭīb*	Perfumery
Ibn Sīnā (d. 1037)	*al-Qānūn fī al-ṭibb*	Medicine
Al-Bīrūnī (d. c. 1050)	*Kitāb al-Ṣaydanah*	Pharmacopoeia
Ibn Bādīs (d. 1061)	*ʿUmdat al-kuttāb*	Bookmaking

Ibn Māsawayh (777–857)

The Christian physician Ibn Māsawayh is the earliest writer known to have dealt with the different types of musk in a systematic way. His *Kitāb Jawāhir al-ṭīb al-mufradah* says:[48]

> The varieties differ in their relative superiority. The best of them is the Sogdian; it is what arrives in Sogdiana from Tibet, and then it is carried to the horizons overland (lit. by back, *ʿalā al-ẓahr*). Next is the Indian; it is what arrives in India from Tibet, and then to al-Daybul. It is then transported by sea; it is inferior to the first type because of its transport by sea. Next is the Chinese; it is inferior to the Indian because of the length of its being kept at sea. Perhaps, moreover, they differ because of the difference in grazing land at their origin because these also compete for superiority. The best of it derives from [the animal] which grazed on the plant called *al-kandasah*; it is in Tibet and Kashmir or in one of them. The next best derives from [the animals] which grazed on the spikenard (*sunbul*) which is used by the perfumers, and it is found in Tibet. The inferior derives from animals which grazed on the plant whose root is called bitter (*murr*) and the scent of that plant and its root is the scent of musk except that musk is stronger and more redolent than it.

48 Ibn Māsawayh, 9–10; cf. Levey's translation 399.

> The best musk is in scent and appearance like an apple; its scent resembles the scent of a good Syrian or Lebanese apple. Its color is predominantly yellow and it has [particles of a size] between the large and small ones; the next best is more strongly black than it and it competes with it in odor but it is not quite its equal. Then there is what is even blacker and it is the worst.[49]

Ibn Māsawayh preferred musk that was not greatly aged, when it was still somewhat yellow in color and granular.

Ibn Māsawayh's account of musk seems to have been widely known, but it was evidently through a version of it in the work of someone else, probably a certain Muḥammad b. Aḥmad b. al-ʿAbbās al-Miskī ("the musk-specialist"), since Ibn Māsawayh is not quoted explicitly in the later sources' accounts of musk. Al-Miskī's dates are unknown, but he evidently lived during the early 10th century or earlier because his work was used by Muḥammad b. Aḥmad b. Saʿīd al-Tamīmī (d. 980) in his book *Ṭīb al-ʿarūs wa rayḥān al-nufūs*. We will look at a passage quoted from al-Miskī in the section on al-Tamīmī.

Al-Jāḥiẓ (c. 776–868 or 9)

As discussed above, it was the Baṣran littérateur Abū ʿUthmān ʿAmr b. Baḥr al-Jāḥiẓ who ascertained that the *faʾrat al-misk* was not a mouse at all, but part of a small gazelle (*khishf*). However, in the same work in which he makes this statement, the *Kitāb al-Ḥayawān*, a few pages before it, he says the following:

> Among the types of *al-faʾr* is *faʾrat al-misk*. It is a small animal (*duwaybbah*) which lives in the region of Tibet. It is hunted for its pods (*nawāfij*) and navels (*surar*). When [a hunter] catches it he binds its navel with a strong ligature and its navel is suspended and its blood collects in it. When it has become firm he slaughters it—and how many have eaten it!—and when it dies he cuts out the navel which he had bound while the "mouse" was alive. Then he buries it in barley until that congested blood transforms there, solid after its death, into pungent musk. After that the stench departs from the blood.[50]

This inconsistency could be due to the heterogeneous character of the material assembled by al-Jāḥiẓ, or perhaps he did not consider it a contradiction to

---

49 Ibn Māsawayh 10.

50 Jāḥiẓ, *Ḥayawān*, 5.301. This passage is quoted by Ibn Abī al-Ḥadīd, *Sharḥ nahj al-balāghah*, 20 vols. (Cairo: ʿĪsā al-Bābī al-Ḥalabī, 1960–4), 19.345 and by Damīrī 2.236–7.

his later statement and did not alter it; it is not clear how small an animal must be to be a *duwaybbah*, as opposed to a *dābbah*, which often denotes an animal large enough to be ridden. Given the modern sense of the word *duwaybbah* it is doubtful if even a musk shrew would be described with that term. Other isolated Arabic authors have likened the musk deer to the rabbit and the dog.[51] Perhaps al-Jāḥiẓ did not believe what the Muʿtazilite perfumer told him, that musk was derived from an animal more like a small gazelle, yet he later emphasizes that the *fa'rat al-misk* is from a small gazelle.[52] But it is more likely that he was confusing the musk deer and the musk shrew—which can be found in the Middle East and presumably was eaten—with the musk deer. Nevertheless, in the text he later has a proper mention of the musk shrew, which he calls a *juradh* or rat, as we have seen earlier. He notes that it is not the *fa'rat al-misk*, that it has an odor which resembles musk, and that it collects shiny objects.[53] In any case, the description of the preparation of the musk pod by binding the "navel" and allowing the blood to collect within is similar to Cosmas Indicopleustes' early account and certainly applies only to the musk deer.

Another work of approximately the same period is the *Kitāb al-tabaṣṣur bi-l-tijārah*, which has traditionally been attributed to al-Jāḥiẓ. The *Kitāb al-Tabaṣṣur bi-l-tijārah* is a guidebook to the goods traded by merchants. It discusses specific categories of items, such as gemstones, textiles, and perfumes, and also discusses the specialties of particular regions and lands. In the section on musk, the writer gives the qualities desired in that substance: "The best musk is the dry, light-colored (*al-fātiḥ*) Tibetan; the inferior (*ardāhu*) is *al-buddī*.... The best musk is that which is light in weight and fragrant."[54] This *al-buddī* probably refers to musk that was used in Indian temples upon Buddhist statues (*budd*); al-Masʿūdī described it, and it will be discussed below.

Al-Yaʿqūbī (d. 897)

One final 9th century source is Aḥmad b. Abī Yaʿqūb, called al-Yaʿqūbī. Later writers quote from him a series of extensive passages about aromatics of the

---

51 Rabbit: Abū Jaʿfar Aḥmad b. Ibrāhīm al-Jazzār, *al-I'timād fī al-adwiyah al-ʿarabiyyah*, ed. Idwār al-Qashsh (Beirut: Sharikat al-Maṭbuʿat li-l-Tawziʿ wa-l-Nashr, 1998), 61, and dog: Ṭabarī III.744.

52 Jāḥiẓ, *Ḥayawān*, 7.210, 211.

53 Jāḥiẓ, *Ḥayawān*, 7.211.

54 *Kitāb al-Tabaṣṣur bi-l-tijārah fī waṣf ma yustaẓraf fī al-buldān min al-amtiʿah al-rafīʿah wa-l-aʿlāq al-nafīsah wa-l-jawāhir al-thamīnah*, ed. Ḥasan Ḥusnī ʿAbd al-Wahhāb al-Tūnisī (Cairo: Maktabat al-Khānjī, 1994), 17. See also the trans. of Ch. Pellat, "Ǧāhiẓiana, I. Le Kitāb al-Tabaṣṣur bi-l-Tiǧāra attribué à Ǧāḥiẓ," *Arabica* 1 (1954): 157.

east, including musk, which do not appear in his extant works.[55] The authenticity of parts of Yaʿqūbī's account of the varieties of musk, which includes several unknown toponyms, has been doubted,[56] but since there is no concrete reason to reject portions of it, it may provisionally be accepted as a whole. An unknown work is just as likely to be early as later. Much of it evidently reached later writers through the book *Jayb al-ʿarūs wa rayḥān al-nufūs* of the 10th c. al-Tamīmī, the relevant portion of which is also lost, that formed the basis of the accounts of musk and other aromatics in the encyclopaedias of al-Nuwayrī and al-Qalqashandī. Even if it is thus not a genuine work of the 9th century, the passages attributed to al-Yaʿqūbī would be almost certainly of the 10th.

Al-Tamīmī, quoted by al-Nuwayrī, gives the following from al-Yaʿqūbī:

> Aḥmad b. Abī Yaʿqūb *mawlā* of the Banī al-ʿAbbās said, "A group of those learned in the source of musk told me that its sources are in the land of Tibet and other known lands. The traders set up a structure resembling a minaret (*manār*) the length of a cubit. When this animal from whose navel musk originates comes, it rubs its navel on that minaret and the navel falls off there. The traders come to it in a time of the year which they know, and they gather it freely (*mubāḥan*). When they bring it into Tibet, one tenth of its value is taken from them. People say that this animal was created by God the Exalted as a source of musk; it produces it every year. It is the bloody excess (*faḍlun damawiyyun*) which collects from its body in its navel every year at a certain time in the way that matters (*mawādd*) flow to the organs. When swelling and enlargement arise in its navel, it makes it sick and causes pain until it matures. When it ripens and is finished, it scratches it with its hooves and it falls off in those deserts and steppes. Then the traders go out to it and they collect it." He (al-Tamīmī) said, "That which was said in the chapter on musk is correct".[57]

Al-Tamīmī concludes by noting the accuracy of this information, meaning that it was regarded as true by an authority of the century after al-Yaʿqūbī's time.[58] Al-Nuwayrī also provides a description of the musk deer in the zoological

---

55 Cf. M. Ullmann, *Die Medizin im Islam* (Leiden: Brill, 1970), 315.

56 M. J. De Goeje in al-Yaʿqūbī, *Kitāb al-Buldān*, ed. M. J. De Goeje (Leiden: Brill, 1892), 366.

57 Nuwayrī 12.4; Yaʿqūbī 364–5. Yaʿqūbī's account has been translated in the context of the translation of al-Nuwayrī's chapter by E. Wiedemann, "Über von den Arabern benutzte Drogen," in *Aufsätze zur arabischen Wissenschaftsgeschichte*, vol. 2 (Hildesheim: Olms, 1970), 240–4; see also Newid 68–9.

58 It is perhaps also the basis of Ibn al-Jazzār's terse notice of the musk deer, 61.

section of his work, which appears ahead of the section on aromatics. While not credited to al-Yaʿqūbī, it is very likely either taken from or influenced by his work because it appears together with the above account of the production of musk in another later source, as we shall see.

> The musk gazelle is among those added to this type [the gazelle]. Its color is black, and it has two upright white canines protruding from its mouth in its lower jaw upright like the tusks of a pig; each of them is less than a span with the appearance of an elephant's tusk. This gazelle is in the land of Tibet and in Hind. It is said that it travels from Tibet to Hind after grazing on the grass of Tibet, which is not aromatic, and casts down musk in Hind which is bad because it is produced from that fodder. Then it grazes on the aromatic grass of Hind, produces musk from it, and comes to the land of Tibet and casts it down in it and that is better than what it casts down in the land of Hind. We will mention, God willing, the information on musk in detail in its chapter in the last section of plants in the supplementary part.[59]

This account of the musk deer and the biological process responsible for musk influenced the brief discussion of the musk deer in the anonymous Egyptian *Book of Curiosities* (*Kitāb Gharāʾib al-funūn wa mulaḥ al-ʿuyūn*), c. 1020–50[60] and the discussion of musk deer by al-Marwazī (possibly taken from al-Jayhānī, see below). The account in the *Mabāhij al-fikar wa-manāhij al-ʿibar* of Muḥammad b. Ibrāhīm al-Waṭwāṭ (1234–1319) combines both the above account of the biological process of the production of musk with the description of the musk deer from al-Nuwayrī, probably reflecting the organization of the original source, which should be al-Yaʿqūbī:

> Attached to this type [the gazelle] is the musk gazelle; its color is black and it resembles the preceding in shape, the fineness of its legs, the separation of its hooves, the configuration of its horns and their curvature except that each one of them has two upright white canines protruding from its mouth in its lower jaw upright like the tusks of a pig; each of them is less than a span with the appearance of an elephant's tusk. It is in Tibet and Hind. It is said that the gazelles travel from Tibet to Hind after grazing on the grass of Tibet, which is not aromatic, and casts down musk

59 Nuwayrī 9.333.

60 *An Eleventh-Century Egyptian Guide to the Universe: The Book of Curiosities*, ed. and trans. Y. Rapoport and E. Savage-Smith (Leiden: Brill, 2014), text 45 and trans. 523.

> in Hind which is bad. It grazes on the aromatic grass of Hind, produces musk from it, and comes to the land of Tibet and there casts it down and that is good quality. Musk is the bloody excess which collects from its body in its navel at a certain time of the year in the way that matters flow to the organs. God the Exalted created these navels as a source for musk and it fruits each year in the manner of a tree which gives its harvest each season. When this blood reaches the navel it swells and enlarges, and the gazelle becomes ill on account of it and is in pain until it matures. When it has ripened, it scratches it with its hooves and wallows in the dust so that it falls off in those deserts and steppes. Then the traders go out and they collect it. It is said that the people of Tibet put up stakes (*awtād*[61]) in the steppes that it scratches upon when its navel hurts it, then it breaks away and falls off. When it falls off from the gazelle, that improves it and makes it well again. Thereupon the gazelle returns to the pasturage and sources of water. There are some that die from the intensity of the pain.[62]

A passage of al-Yaʿqūbī quoted in the *Ṣubḥ al-Aʿshā* of al-Qalqashandī (1355–1418) gives the characteristics of musk:

> Aḥmad b. Yaʿqūb said, "The best musk in scent and appearance is what is apple-like, its scent resembling the scent of a Lebanese apple. Its color is predominantly yellow, and the dimensions [of its grains] are between coarse grains and fine grains.[63] Then comes what is more strongly black than it, though it approximates it in appearance. Then comes what is more strongly black than that, and it is the least in value and price." He said, "It reached me from the merchants of India that there are two other types of musk which are made from plants of the land. One of them does not spoil even with long storage, but the other spoils with long storage".[64]

It is not clear if the last part of this is indeed from al-Yaʿqūbī—it is probably from al-Tamīmī—but the first part is, and it is consistent with what Ibn Māsawayh, who was at least a generation before al-Yaʿqūbī, said. Perhaps al-Yaʿqūbī was using the work of Ibn Māsawayh, either directly or through someone else.

---

61 This term is also used by Ibn al-Jazzār and the *Book of Curiosities*.

62 Muḥammad b. Ibrāhīm al-Waṭwāṭ, *Mabāhij al-fikar wa-manāhij al-ʿibar*, ed. ʿAbd al-Razzaq Aḥmad al-Ḥarbī (Beirut: al-Dār al-ʿArabiyyah li-l-Mawsūʿāt, 2000), 278.

63 The edition has *al-riqāq* instead of *al-diqāq*, with the same meaning.

64 Qalqashandī 2.120.

The other extensive quotation on musk from al-Yaʿqūbī preserved in the *Nihāyat al-Arab* concerns the different kinds of musk:

> Aḥmad b. Abı Yaʿqūb said, "The best musk is the Tibetan, then after it the Sogdian, and after the Sogdian, the Chinese. The best of the Chinese is what is brought from Khānfū, a great city which is the port of China at which ships of Muslim merchants disembark. Then it is carried by sea to the Straits of Hormuz (*al-Zaqāq*) and when it comes near to the land of Ubullah, its scent arises and the merchants cannot hide it from the tax collectors. When it is brought out of the ships its scent improves, and the smell of the sea leaves it.
>
> Then comes Indian musk; it is what comes from Tibet to India, is then carried to Daybul, and then shipped by sea. It is inferior to the first.
>
> After the Indian comes the musk of *Q.nbār*, which is good quality musk except that it is beneath the Tibetan in value, essence (*jawhar*), color and scent. It is brought from the land which is called *Q.nbār* between China and Tibet; sometimes they misrepresent it and attribute it to Tibet."
>
> He said, "Following it in quality is the Tughuzghuzī musk. It is heavy (*razīn*) musk which tends towards black. It is brought from the land of the Tughuzghuz Turks; it is imported by merchants and they misrepresent it, but it does not have the essence or color [appropriate to good musk]; it takes crushing slowly, and is not free from coarseness.
>
> Following it in quality is *Q.ṣārī* musk; it is brought from the land called *Q.sār* between India and China." He said, "It might be grouped with the Chinese except that it is below it in value, essence, and scent."
>
> He said, "Kirghiz[65] musk is musk which is similar to the Tibetan and resembles it, and is yellow, attractive, and strong of scent.
>
> After it is the *ʿ.ṣmārī* musk; it is the weakest type of all musk and the lowest in value. One may remove from a musk pod which weighs an *ūqiyyah* a single *dirham* of musk.
>
> Then comes *Jabalī* ("mountain") musk; it is what is brought from the direction of the land of Sind, from the land of Mūltān. It has large pods and attractive color; however, it has a weak scent".[66]

---

65 *Jirjīzī* in the manuscripts.

66 Nuwayrī 12.11–3; Yaʿqūbī 365–6; G. Ferrand, *Relations de voyages et textes géographiques arabes, persans et turks relatifs à l'Extrême-Orient du VIIIe au XVIIIe sieclès*, 2 vols. (Paris: Leroux, 1913–14), 1.50, translates the beginning of this section.

This list comprises the following types of musk: 1. Tibetan, 2. Sogdian, 3. Chinese, 4. Indian, 5. *Q.nbār*, 6. Tughuzghuzī, 7. *Q.ṣārī*, 8. Kirghiz, 9. *ʿ.ṣmārī*, 10. *Jabalī*. Al-Qalqashandī gives the same list in the same order, except that he reverses the order of the last two types, placing Jabalī musk before *ʿ.ṣmārī* musk.[67] Presumably al-Qalqashandī's order is correct because in al-Nuwayrī's version *ʿ.ṣmārī* musk is described as the weakest type of musk of all.

Al-Jayhānī (Early 10th Century)

The lost geography *Kitāb al-masālik wa-l-mamālik* of al-Jayhānī probably was a family work, added to by successive members of the illustrious family of Sāmānid viziers.[68] This factor of its composition makes its fragments difficult to use in a precise sense, but we may well hope that the data quoted by our later writers reflects the state of knowledge in the 10th century Sāmānid realm in a general way. Only one fragment quoted by al-Marwazī in his *Ṭabāʾiʿ al-ḥayawān* (c. CE 1120) touching on musk is directly attributed to al-Jayhānī:

> Al-Jayhānī mentioned in his book *al-Masālik wa-l-mamālik* that a traveler passing from Shazhou (*Sājū*) to China sees on his right side a mountain which has musk animals on it, and the bulls (*thīrān*) from the tails of which fly-whisks (*al-madhābb*) and the heads of banners are made.[69]

Passing from Shazhou into China one would be travelling to the southeast, and south of the route are the mountainous parts of Gansu and the eastern parts of the Tibetan plateau; in the more moist parts of these mountains there are apparently still musk deer today.[70] The "bulls" from the tails of which fly-whisks were made refers to yaks; yak tails were indeed an important trade good from Tibet; they were used also in making medieval military standards. Al-Marwazī's account of the musk deer, which is probably taken from al-Jayhānī, is as follows:

---

67 Qalqashandī 2.120–1.

68 See C. Pellat, "al-Djayhānī" in *EI*² supplement s.v. But see J.-C. Ducène, "Al-Ǧayhānī: fragments (Extraits du *K. al-masālik wa l-mamālik* d'al-Bakrī," *Der Islam* 75 (1998): 260–1.

69 Al-Jayhānī quoted on *Ṭabāʾiʿ al-ḥayawān*, Folio 75a, in V. Minorsky's ed. of al-Marwazī, *Ṭabāʾiʿ al-ḥayawān*, publ. as *Sharaf al-Zaman Tahir Marvazi on China, the Turks and India* (London: Royal Asiatic Society, 1942), text *51 and trans. 91–2.

70 Lanzhou and Longyou sent musk as tribute to the court. See also the map of musk deer production in China in E. H. Schafer and B. E. Wallacker, "Local Tribute Products of the T'ang Dynasty," *Journal of Oriental Studies* 4 (1957–8), pl. 5 and 225. The *Ḥudūd al-ʿālam* mentions musk deer and yaks in the mountains near Suzhou (*sawkjū*).

> As for the musk attributed to Tibet, it is the best type of musk and the most fragrant (*adhkā*) in scent. It is the navel of an animal that resembles the large gazelles. It becomes excited at a certain time of the year and black blood gathers in its navel, flooding (*yafīḍu*) into it from the rest of its body. The swelling (*waram*) and the pain in its head and all of its body intensifies. So it comes to places in those deserts in which are its accustomed wallows (*al-marāghah*), abstaining from grazing and water until its swollen navel falls off from the abundance of blood, and sometimes its horns fall off also. Some of them die there and some of them recover; they return to the grazing land. Navels from them collect in those wallows. As the years pass over them all the blood solidifies and dries and becomes transformed into musk. Then the youths of Tibet go out to those deserts during the rainy season and they might find in a wallow thousands of vesicles. They gather the sound ones. Perhaps their endeavor [to find musk] might fail.[71]

This account is similar to the account of al-Yaʿqūbī, although it is worded differently enough that it is unlikely to be directly based on it.

Abū Zayd al-Sīrafī (Early 10th Century) and al-Masʿūdī (c. 896–956)

Abū Zayd al-Sīrāfī was the editor of the *Akhbār al-Ṣīn wa-l-Hind*, which had been written long before his time in 851.[72] The text known as the *Akhbār al-Ṣīn wa-l-Hind* is extant as the first part of the complete work in al-Sīrāfī's edition. Al-Sīrāfī also appended a supplemental chapter to the *Akhbār* that has not received as much attention as the *Akhbār* itself. Al-Sīrāfī was from Sīrāf, on the Iranian coast of the Persian Gulf, which was one of the most important trading emporiums of its time.[73] He did not apparently travel in the East, but he had access to the sailors and merchants who did. Al-Nuwayrī, who quotes from his account of the musk deer, describes him as "experienced in the land of China, its ocean, its routes and kingdoms".[74] He was a friend of the famed historian

---

71 Marwazī Book 8 sec. 43, text *17 and trans. 28–9.

72 The *Akhbār aṣ-Ṣīn wa-l-Hind*, shorn of al-Sīrāfī's additions, is edited and translated by J. Sauvaget, *Akhbār aṣ-Ṣīn wa l-Hind: Relation de la Chine et de l'Inde* (Paris: Les Belles Lettres, 1948).

73 A. Lamb, "A Visit to Sīrāf, an ancient port on the Persian Gulf," *JMBRAS* 37 (1964), 1–19, D. Whitehouse and A. Williamson, "Sasanian Maritime Trade," *Iran* 11 (1973): 33–5, and D. Whitehouse, *Siraf: history, topography and environment* (Oxford: Oxbow Books, 2009). On one of the wealthy merchants of Sīrāf, see S. M. Stern, "Rāmisht of Sīrāf, a Merchant Millionaire of the Twelfth Century," *JRAS* (1967): 10–4.

74 Nuwayrī 12.10.

Abū al-Ḥasan ʿAlī b. Ḥusayn al-Masʿūdī. His account of musk is his supplement to the *Akhbār al-Ṣīn wa-l-Hind*, which lacked an account of musk. This account is paralleled in part by al-Masʿūdī's account of the musk deer.

Al-Masʿūdī's *Kitāb Murūj al-dhahab wa-madāʾin al-jawhar* includes much important material about Further Asia and a more extended discussion of musk than al-Sīrāfī's.[75] Al-Masʿūdī is the source quoted for the description of the musk deer in the influential *Kitāb al-Jāmiʿ li-mufradāt* of Ibn al-Bayṭār (d. 1248) and his account was used widely by later writers.[76] Al-Masʿūdī himself was a noted traveler and seems to have visited many areas of Iran, East Africa, the Caspian region, and western India. His work is particularly noteworthy because of his great curiosity and the wealth of detail about many regions of the world; he was, in effect, a sort of Muslim Herodotus. Identifying al-Masʿūdī's sources is important before trying to assess his material. Al-Masʿūdı's account of the musk deer is very similar to that of Abū Zayd al-Sīrāfī. The two apparently met,[77] so this similarity is not surprising. Nevertheless, al-Masʿūdī's account is worded differently, even though the meaning is generally identical, so it is unlikely that he used the written work of Abū Zayd al-Sīrāfī as we have it as a source. The two writers may have exchanged information orally, information which must have come from merchants who traveled in India.

---

75 On al-Masʿūdī, see A. M. H. Shboul, *Al-Masʿūdī & His World: A Muslim Humanist and his Interest in non-Muslims* (London, 1979).

76 Ibn al-Bayṭār 4.155–6. It became a standard account of the musk deer, and is used by, among others, Abū ʿUbayd al-Bakrī (d. 1094), *al-Masālik wa-l-mamālik*, ed. A. P. Van Leeuwen and A. Ferre, 2 vols. (Tunis: al-Dār al-ʿArabiyyah li-l-Kitāb, 1992) 1.270, al-Muẓaffar Yūsuf b. ʿUmar b. ʿAlī al-Ghassānī (2nd half of the 13th century), *Kitāb al-Muʿtamad fī al-adwiyah al-mufradah*, ed. Muṣṭafā Saqqā, 3rd ed. (Beirut: Dār al-Maʿrifah, 1975), 495–6, and Abū Bakr b. ʿAbdallāh b. al-Dawādārī (fl. early 14th c.), *Kanz al-durar wa-jāmiʿ al-ghurar*, vol. 1, ed. B. Radtke (Wiesbaden: Franz Steiner, 1982), 182–3. Much of Zakariyyāʾ b. Muḥammad al-Qazwīnī's account of musk in the *ʿAjāʾib al-makhluqāt*, ed. T. Wüstenfeld in *Zakarija ben Muhammed ben Mahmud el-Cazwini's Kosmographie*, 2 vols. (Göttingen, 1848–9; repr. in 1 vol.: Wiesbaden: Martin Sändig, 1967), 383, is probably abridged from Masʿūdī or a source derived from him, although Qazwīnī adds some interesting information about how different parts of the musk deer were used.

77 Al-Masʿūdī, *Murūj al-dhahab wa-maʿādin al-jawhar*, ed. B. de Meynard and P. de Courteille, revised by C. Pellat, 5 vols. (Beirut: Manshūrāt al-Jāmiʿah al-Lubnāniyyah, 1966–74), 1. 172 (§351). Cf. Shboul 155.

Al-Sīrāfī[78]

The land which has the Chinese musk gazelles and Tibet is one land with no distinction among it. The people of China carry off the gazelles which are close to them and the people of Tibet those which are close to them.

Rather Tibetan musk is superior to Chinese musk for two reasons. First, the musk gazelle which is in the territory of Tibet has grazing lands of spikenard (*sunbul al-ṭīb*) while those which are near the land of China have grazing land of other herbs (*ḥashāʾish*).

The other reason is that the people of Tibet leave the pod in its original condition while the people of China adulterate the pods which come to them. Their routes are also over the sea and moisture[80] clings to them. If the people of China left the musk in the pods and placed them into vessels (*barānī*) and secured them, it would arrive in the land of the Arabs like the Tibetan in its quality.

Al-Masʿūdī[79]

The land which has musk gazelles (*ẓibāʾ*) is in Tibet and China, and it is a continuous piece of land.

Instead, Tibetan musk is superior to Chinese musk in two respects. First, the gazelles of Tibet graze on spikenard and various types of aromatic plants (*anwāʿ al-afāwīh*) while the Chinese gazelles graze on herbs inferior to the aromatic herbs (*ḥashāʾish al-ṭīb*) which we mentioned that Tibetan ones graze on.

The other respect is that the people of Tibet do not bother to remove the musk from the vesicle and they leave it as it is within it. The people of China remove it from the pod and augment it with adulterant blood and other types of adulterants. The Chinese also must be transported over the distance of the seas, as we have described,[81] and is exposed to much moisture and differences of climate. If the Chinese

78 Al-Sīrāfī, *Riḥlah*, ed. ʿAbdallāh al-Ḥabashī (Abu Dhabi: Manshūrāt al-Majmaʿ al-Thaqāfī, 1999), 75–7.

79 Masʿūdī 1.188–9 (§391–4). German translation in Newid 64–5.

80 The text has *al-ʾīdhāʾ*, but the parallel in Masʿūdī has *al-andāʾ*. G. Ferrand, *Voyage du marchand arabe Sulaymân en Inde et en Chine* (Paris: Bossard, 1922), 110, has also translated this "celui-ci s'imprègne d'humidité".

81 Masʿūdī 1.176–83 (§361–379) describes the seas between Iraq and China.

did not adulterate their musk and placed it in vessels of glass with strong covers and pads, then it would arrive in the lands of Islam such as Oman, Persia and Iraq, and other realms, and it would be like the Tibetan in its quality.

Then their accounts present their information in a different order, with al-Masʿudi typically providing much more information.

The best musk of all is that which the gazelle scrapes on the rocks of the mountains; it is matter which forms in its navel as fresh blood collects in the manner in which blood collects as it appears from boils. When it ripens, he rubs it and it torments him so he flees to the stones until he tears a hole in it, and what is inside it flows out. When it comes out from it, it dries and heals over and the substance returns and collects in it as before.

There are men in Tibet who go out in search of it. They have a knowledge of it and when they find it, they pick it up, collect it, put it inside the pods, and carry it to their kings. This is the ultimate musk, when it has ripened in its pod upon its animal. Its excellence over other musk is like the excellence of fruit which has ripened on its tree over the other which has been torn from it before its ripening. Other than this is musk which is hunted with set-up nets (*sharak*) and arrows. The pod might be cut off from the gazelle

The best and most fragrant musk is that which comes out from the gazelle after it has reached the end of its [the musk's] maturation.

To be precise, there is no difference between our gazelles and the musk gazelles in appearance, form, color, or horn, except that the latter can be distinguished by its canine teeth (*anyāb*) which are like the canines of elephants. Each gazelle has protruding from its jaws two vertical, upright, and white canines, about the span of a hand, more or less.

In Tibet and China they set up snares (*ḥabāʾil*) and nets (*ashrāk* and *shibāk*)[82] for them and they trap them. Sometimes they shoot arrows at them and fell them. Then they cut the pods from them. The blood which is in the navel is raw (*khām*) and not matured, and fresh and not ripened. It has a stench for its scent. They leave it for a time until that stinking, offensive odor has left it. It changes because of substances in the air and

82 It is unclear what the distinction between these two kinds of nets might be.

before the musk has matured inside it. Although when it is cut from its gazelle it has a bad odor for a time until it has dried over long days, as it dries it changes until it becomes musk.

The musk gazelle is like the other gazelles among us in shape and color, and the delicacy of its legs, and the cleaving of its hooves, and the configuration of its horns and their curvature, but it has two fine, white canine teeth in its jaws standing up on the face of the gazelle. The length of each of them is the measure of a span or less, with the form of an elephant's tusk. This is the difference between them and the other gazelles.

becomes musk. The manner of this is like the manner of fruit when it is separated from the tree and cut before it is completely ripened on its tree and its matter (*mawādd*) has become ingrained within it.

The best musk is what matured in its vessel and ripened in the navel, became ingrained upon the animal, and is the end result of its matter. To be precise, natural forces push the matter of the blood to his navel. When the blood has matured and ripened within, it bothers the animal, and it causes him to scratch, so he seeks refuge at stones or rocks warmed by the heat of the sun and there rubs himself on them finding relief through that. Then it [the navel]

bursts forth and it [the musk] flows upon those rocks as an abscess or boil bursts when whatever matter has collected in it successively has matured. Thus he finds relief in its expulsion. When what was in its pod (*nāfijah*)–this means the navel, and it is a Persian word–has gone, it heals for a time, but then the matter of the blood rushes into it and collects again as it was before.

The men of Tibet go out towards its grazing lands among those rocks and mountains to find the blood which has become dried on those stones and rocks, the matter having matured, nature having ripened it in the animal, the sun having dried it, and the air having influenced it. Then they take it and that is the best musk. They place it in pods they have brought with them which they took from gazelles which they had trapped and prepared. That is what their kings use and exchange as gifts between them; the merchants carry it rarely from their country. Tibet has many cities and musk is ascribed to every region of it.[83]

Al-Tamīmī (d. 980)

Muḥammad b. Aḥmad b. Saʿīd al-Tamīmī is one of the most important Arabic writers on aromatics; his partially lost book *Jayb al-ʿarūs wa rayḥān al-nufūs* has been mentioned before because he made use of al-Yaʿqūbī's work on aromatics. He is quoted extensively by al-Nuwayrī in his encyclopaedia *Nihāyat al-arab*.[84] The first of these quotations gives a unique piece of information about the geography of the musk producing lands:

> Musk has many types and differing kinds. The best and most excellent of them is the Tibetan which is brought from the place called *dhū sm.t* (Mdosmad); between it and Tibet is a journey of two months, then it is brought to Tibet whence it is transported to Khurāsān. He said: The origin of musk is in a four-footed animal (*bahīmah*) whose form resembles a small gazelle.[85]

Al-Nuwayrī writes:

> We have mentioned the musk gazelle in the third chapter of the second part of the third division. It is in volume nine of this manuscript[86] and there is no point in repeating it. They had mentioned in the description of obtaining musk from this animal information which we will mention. People claim that the gazelle is killed and its navel is taken by the hair upon it and what is in it is fresh blood (*ʿabīṭ*) and sometimes the navel has a lot of blood, sometimes it might be large and broad with little blood.

---

83 Following this passage on musk, Masʿūdī resumes his narrative with a "*qāla al-masʿūdī*", implying that the preceding passage is a quotation.

84 Nuwayrī's entire chapter on musk has been translated by Wiedemann, see above under al-Yaʿqūbī.

85 Nuwayrī 12.2.

86 Nuwayrī 9.333; see the discussion of this passage above in the section on al-Yaʿqūbī, 163.

> It is suspended in the throat of a privy (*ḥalq mustarāḥ*) for a period of forty days, then it is taken out and suspended in another place until it is completely dried and its odor has become strong. Then the pods (*nawāfij*) are placed inside small bags (*mazāwid*) and sewn shut, and they are carried from Tibet to Khurāsān.[87]

The latter part of this paragraph is probably derived from al-Tamīmī, although it likely goes back to an earlier source, as al-Tamīmī also transmitted al-Yaʿqūbī's story of the origin of musk.

Nuwayrī also gives a quotation apparently from al-Tamīmī on the animal's production of musk:

> He said, "They mentioned that it stirs at a certain time of the year and the area of its navel begins to putrefy and thick, black blood collects in it, flowing into it from the rest of its body. Its pain intensifies and it goes to a place in which there is soft soil, which appears like a wallow (*marāghah*) in those steppelands."[88]

Al-Tamīmī is probably the source of al-Nuwayrī's quotation on the three different types of musk. Alternately this text may also be by al-Yaʿqūbī or al-Miskī; it is not clear. The passage runs:

> He said, "It has come to me that the Indian merchants experienced with musk mention that musk is of three types and they do not deviate from that. The first type—it is the most excellent and best—is the true musk of well-known nature. The two other types are imitations: the first of them is made from a dried mixture consisting of plants from their land, and there is of the true musk not a bit. They enjoin its use and purchase from the places of its origin and their neighboring lands and from those who know about it and these are the people of Tibet. The other type is manufactured and they forbid it and its sale and commerce in it, and that is because it changes and is corrupt from the time it originated. It is half the price of good musk."[89]

This passage shows the extent of falsification of musk and demonstrates that some of the imitations of musk were also called musk. Al-Nuwayrī's account

87 Nuwayrī 12.2–3.

88 Nuwayrī 12.5.

89 Nuwayrī 12.14.

of the properties of musk is quite similar to Ibn Māsawayh's and is probably derived from it, perhaps by al-Miskī and then quoted by al-Nuwayrī from al-Tamīmī, who used al-Miskī's book.

> Musk in its natural state is sharp (*ḥādd*),[90] pleasant, and penetrating, good for heart (*fu'ād*) pain, strengthening to the heart (*qalb*). It stops blood flow when a wound is dressed with it. It is added to kohl for the eye and in great pastes (*al-ma'ājīn al-kibār*). Its substitute is castoreum for it is the closest thing to it in nature and action.[91]

Al-Tamīmī is certainly the source of Nuwayrī's note on the musk of Dārīn:

> Muḥammad b. Aḥmad said, As for the musk which is ascribed to Dārīn, it is a kind of Indian musk which merchants bring to Dārīn, an island in Baḥrayn at which the ships of the India merchants land, and it is transported from it to different places. Dārīn is not a source of musk.[92]

Since al-Tamīmī used al-Yaʿqūbī, one is not surprised to find: "He (al-Tamīmī) said, 'The best musk is what had for the pasturage of its gazelles the plant called *al-kdhms* which grows in Tibet and Kashmir or in one of them'".[93] The spelling *kdhms* is probably a corruption of the *kandasah* of al-Yaʿqūbī.

Besides al-Yaʿqūbī, Muḥammad b. Aḥmad b. al-ʿAbbās al-Miskī was another of al-Tamīmī's sources. From him we have the following:

> He (al-Tamīmī) said, what Muḥammad b. al-ʿAbbās al-Miskī said in his book testifies to the truth of that and agrees with it, that the merchants of musk from the people of Sughd mention that musk is the navel of an animal (*dābbah*) in appearance the size of a gazelle. It has a single horn in the center of its head. He said, And from its horn and bone of its forehead they make the sword hilts (*nuṣub*) known as *khutū* sword hilts.[94]

In this passage, al-Miskī has confused the rhinoceros, the horns of which were indeed traded from Southeast Asia, with the antlerless musk deer, combining

90 This is probably an error for *ḥārr* "hot", cf. Ibn Māsawayh 11 (see below, 203).
91 Nuwayrī 12.14–15 .
92 Nuwayrī 12.15.
93 Nuwayri 12.7.
94 Nuwayrī 12.4–5.

the source of two rarities into one animal.[95] The *khutū* is properly ivory derived from the walrus and narwhal, which was called *guduo* 骨咄 in Chinese.[96] Al-Miskī also explodes a notion about the plants grazed by the musk deer:

> Muḥammad b. Aḥmad b. al-ʿAbbās al-Miskī said, some Arab mentioned that the musk animal grazes upon camphor trees and he concluded that from the verse of the poet al-ʿUklī:[97]
>
> *taksū l-mafāriqa l-labbāti dhā arajin min quṣbi muʿtalifi l-kāfūri darrāji*
>
> You clothe the part of hair and upper part of chest with fragrance from a belly (*quṣb*) fed on camphor coming and going.
>
> *Al-quṣb* means the gut (*al-miʿā*); of it is the word of the Prophet, peace be upon him, "I saw ʿAmr b. Luḥayy dragging his gut (*quṣb*) in the fire."[98] Muḥammad b. Aḥmad said, this is a Bedouin opinion, and is not supported by the learned.[99]

Many ideas circulated about the plants grazed by the musk deer, how it was supposed to have consumed fragrant plants in order for the musk to develop its scent. The verse quoted was understood by Ibn Qutaybah (828–89) to refer to musk.[100] This verse is an unusual example of camphor, which was a polar opposite to musk in medieval thought, as the fodder of musk.

95 Zakarīyā b. Muḥammad al-Qazwīnī, *ʿAjāʾib*, 386, also mentions the virtues of the supposed musk deer horn along with its tongue: "As for the properties of its parts, its horn is carved up and censed with to drive away pests (*hawāmm*), and its tongue is dried in the shade and fed to impudent women to stop their impudence."

96 The history of *khutū* has been explored in detail by Berthold Laufer, "Arabic and Chinese Trade in Walrus and Narwhal Ivory," *T'oung Pao* 14 (1913): esp. 318–9. Chinese *guduo* 骨咄 in turn apparently derives from a Tungusic word, see *Liaoshi* 遼史 (Beijing: Zhonghua shuju, 1974), 106: 1549. See also A. King, "Early Islamic Sources on the Kitan Liao: The Role of Trade," *Journal of Song-Yuan Studies* 43 (2013): 263–7.

97 This verse is attributed to al-Rāʿī by Ibn Qutaybah, *Kitāb al-Shiʿr wa-l-shuʿarā* (Leiden: Brill, 1902), 247.

98 Cf. Ibn Isḥāq, *Sīrat Rasūl Allāh. Das Leben Muhammed's*, ed. Ibn Hishām, text ed. F. Wüstenfeld, v. 1 part 1, (Göttingen, 1858, repr. Frankfurt am Main, 1961), 1.51–6. He is credited with introducing the worship of stone idols among the sons of Ismāʿīl.

99 Nuwayrī 12.9–10.

100 Ibn Qutaybah 247 says: "He means musk; it is produced from the belly (*quṣb*) of the musk gazelle and deemed it appropriate for his verse on the woman."

Al-Tamīmī also quotes the following from al-Miskī:

> Muḥammad b. al-ʿAbbās said, "The best musk is the Sogdian, and it is that which the merchants of Khurāsān purchase from Tibet and transport overland (*ʿalā al-ẓahr*) to Khurāsān. Then it is carried from Khurāsān to the distant lands (*al-āfāq*). Then next in quality is the Indian musk; it is what comes from Tibet to the land of India and then is carried to Daybul, and then is transported by sea to Sīrāf, ʿAdan (Aden) and Oman and other districts; it is inferior to the Sogdian. The Chinese musk follows the Indian and it is inferior to it because of the length of time it is kept at sea, and what rottenness of the atmosphere comes in contact with it and for another reason, and that is the difference of pasturage originally."

This quotation gives a clear picture of the structure of the musk trade, showing the priority of musk that was transported overland. It is identical to Ibn Māsawayh's account, which it may have quoted directly. Ibn Māsawayh's *Kitāb Jawāhir al-ṭīb al-mufradah* is apparently not quoted by later writers explicitly;[101] it could be that it was only known through the work of al-Miskī, which was used by al-Tamīmī, who was in turn used especially by al-Nuwayrī and al-Qalqashandī.

Ibn Mandawayh (d.1019) and Ibn Kaysān (d. 990)

Another late 10th century account comes from two closely parallel sources: the *Risālah fī uṣūl al-ṭīb wa-l-murakkabāt al-ʿiṭriyyah* of Abū ʿAlī Aḥmad b. ʿAbd al-Raḥmān b. Mandawayh[102] and the *Mukhtaṣar fī al-ṭīb* of the Egyptian Christian Abū al-Ḥasan Sahlān b. ʿUthmān b. Kaysān.[103] Ibn Mandawayh was a Persian physician who worked in the hospital of the Buyid ʿAḍud al-Dawlah (r. 949–82) in Baghdad.[104] He dedicated his *Risālah* to the Buyid vizier Ibn

---

101 The citations in Ullmann 314 do not correspond to the published text. As it stands, the work only discusses, as its title implies, simple perfumes, while the formulas quoted by al-Nuwayrī from "the book of Ibn Māsawayh" are for compound perfumes. Most likely they come from another work, a formulary of some sort, which has been lost.

102 Ibn Mandawayh 225–9. I have prepared a translation of this text and Ibn Kaysān's which will appear in *Medieval Islamicate Perfumery*. See Ullmann 146 on Ibn Mandawayh.

103 Ibn Kaysān 187–9. Biographical data can be found in P. Sbath's edition of the *Mukhtaṣar fī al-ṭīb*: "Abrégé sur les arômes," *Bulletin de l'Institut d'Égypte* 26 (1943–4): 184–5. Cf. also P. Sbath and C. D. Avierinos, *Sahlān ibn Kaysān et Rashīd al-Dīn Abū Ḥulayqa: Deux traités médicaux* (Cairo: Institut Français d'archéologie orientale, 1953) and Ullmann 315–16.

104 Ullmann 146.

ʿAbbād (938–995). Ibn Kaysān was active during the reign of the Fatimid ʿAzīz Abū Manṣūr Nazār, 975–996, which places his book dedicated to that caliph in the period 975–990. Thus both works were essentially contemporary, and neither writer mentions the other.

Both works in their entireties have almost identical content and are comparable in length. While there are many passages where Ibn Kaysān is more terse, there are also many places where Ibn Mandawayh does not provide the detail given by Ibn Kaysān. Notably, Ibn Kaysān has an account of Tatar musk that is lacking in Ibn Mandawayh, but which probably stems from the common source because Tatar musk is included in the same place in a passage derived from the Ibn Mandawayh-Ibn Kaysān source in an uncredited marginal comment in al-Bīrūnī's *Kitāb al-ṣaydanah*, and it is doubtful that al-Bīrūnī knew about Ibn Kaysān. Thus, of the two possibilities—that one copied from the other, or that both are derived from another common source—the latter seems most likely.

The Ibn Mandawayh-Ibn Kaysān account is quite different from all of the other early surviving accounts; there is no indication as to the sources used beyond Ibn Mandawayh's notice that he read about the identity of the musk deer with the animal that produces *khutū* in "a book"—a passage not paralleled by Ibn Kaysān but similar to a passage quoted by al-Nuwayrī from al-Tamīmī, ultimately going back to Muḥammad b. al-ʿAbbās al-Miskī, which has been discussed above. While the origins of the Ibn Mandawayh-Ibn Kaysān account are obscure, it was an influential one in the eastern Islamic world and is quoted and alluded to in later works in Arabic and Persian.[105]

---

105 See below for the marginal comment in al-Bīrūnī that derives from this account. The accounts of musk by the Persian writers Abū al-Qāsim ʿAbd Allāh b. ʿAlī al-Kāshānī, *ʿArāyis al-jawāhir wa-nafāyis al-aṭāyib*, ed. Īraj Afshār (Tehran: Intishārāt-i Anjuman-i Athār-i Millī, 1966), 250–4, Naṣīr al-Dīn Muḥammad b. Muḥammad al-Ṭūsī, *Tansūkhnāmah-i Īlkhānī*, ed. Mudarris Raḍawī (Tehran: Intishārāt-i Bunyād-i Farhang-i Īrān, 1969), 247–51, and Ismaʿīl b. Ḥasan al-Jurjānī, *Dhakhirah-i Khwārazmshāhī*, ed. Saʿīd Sīrjānī (Tehran: Bunyād-i Farhang-i Īrān, 1976), 156 and trans. Newid 65–7, are also influenced by it. Al-Bīrūnī, *Kitāb al-Ṣaydanah*, ed. H. M. Saʿīd, *Al-Biruni's Book on Pharmacy and Materia Medica* (Karachi: Hamdard National Foundation, 1973), 278, and ed. A. Zaryāb (Tehran: Markaz-i Nashr-i Dānishgāhī, 1991), 444, also tells an uncredited story about aloeswood derived from the Ibn Mandawayh-Ibn Kaysān source (Ibn Mandawayh 235–6, Ibn Kaysān 196).

Ibn Mandawayh

The varieties of musk: The best of them is the Chinese (*al-ṣīnī*), which is only met with rarely. It is known that every pod of it is twenty dirhams' weight, give or take a dirham. The grains (*shiyāf*) are evident in it because of the fineness of its skin. The pod is smooth and does not have tufts of hair upon it. When it is opened, the weight of the skin of the pod is half a dirham or two thirds of a *dirham*. Every perfumer who opens this musk gets a nosebleed from the potency of its scent. Whoever wishes to crush it needs to smell camphor and rosewater during its crushing. One cannot crush more than ten *mithqāl*s in a day. For every perfume which calls for two *mithqāl*s of good musk or rare Tibetan, a third of a *mithqāl* of this musk will suffice. When it is crushed it resembles crushed saffron because of its yellowness.

Ibn Kaysān

Musk. There are different varieties: The best of them is the Chinese. It is the musk which comes from China in pods, and it is very scarce and only appears seldom. It is known that its pod is twenty *waznah* or more than twenty. Since its skin is fine, one can reckon the grains in it.[106] The pod is shaved and does not have tufts of hair upon it. When it is opened, the weight of the pod[107] is half a *dirham* to two thirds of a *dirham*. Every perfumer or person in his presence who opens this musk gets a nosebleed from the potency of its scent. Whoever wishes to crush it must take a small piece of cotton (*quṭnah*), dip it in rosewater and rub a bit of camphor on it and then put it in his nostrils; then he can begin to crush it. One cannot crush more than ten *mithqāl*s in a day, and if one crushes it in the summer and does not follow the procedure I have described, it will be dreadful for him. For every perfume which calls for two *mithqāl*s of Tibetan musk, a third of a *mithqāl* of this musk will suffice. When it is crushed and compared with crushed saffron, one cannot distinguish one from the other because of the intensity of its yellowness.

106 This means that one does not have to open the pod to determine the quality of the musk inside.

107 As Ibn Mandawayh's "skin of the pod" suggests, "pod" here must refer to the pod which has been emptied of musk; the higher the percentage of musk to vesicle in weight, the better the pod.

After it is the Tibetan (*al-tubbatī*). Most of its pods are delicate with little hair. The weight of every pod is several *dirhams*. When you remove the musk from it, some of it is yellow and some of it is black, and in it are fine grains (*diqāq shiyāf*). There is no distinction between the yellow and the black, and between the fine grains and large grains when they are obtained from this variety. The yellow are the freshest and the black are the oldest; blood is red when it is fresh and when it ages, it becomes black.

After it is the Tibetan. Most of its pods are delicate with little hair, and are yellow. Each pod weighs from five to six *dirhams*. When the musk is scattered from it some of it is black and some of it is yellow, and in it are fine grains (*shiyāf diqāq*) and large grains (*shiyāf jilāl*). There is no difference between the black and the yellow and the fine and the large. When this type is formed, the yellow are the freshest and the black are the oldest,[108] just like blood which is red when it is fresh and when it ages, it becomes black.

After it is the *Ṭūsmatī*. It is a type close to the Tibetan; it has comparable strength in making compounds, but it can be distinguished from it by examining the pods. This is that its pod (*faʾrah*) has white hair and may be yellow when stripped. Each pod of it weighs eight *dirhams*, and might reach up to ten *dirhams*.

Then after it is the *Ṭūmsatī*. It is a type close to the Tibetan. There is not much difference between them because its musk is the same color as the Tibetan and has comparable strength in making compounds, but it can be distinguished by examining it. This is that its pod (*nāfijah*) has white hair and may be yellow when stripped. Each pod weighs from eight *dirhams* to ten *dirhams*. A pod of this type came into my hands weighing fifteen *dirhams*. I opened it and what was in it came out in a single piece with a weight reaching ten *dirhams*.

After it is the Nepalese (*al-naybālī*). This type mostly arrives loose (*manthūr*), not secure in the pods. It is a musk of basic strength in making compounds, black colored with many grains (*shiyāf*). For every perfume

Then after it is the Nepalese. Most of what is imported of this type is loose. It is a basic musk in making compounds with a black color. For every perfume which calls for a *mithqāl* of Tibetan musk, a *mithqāl* and a half of

108 Reading *aʿtaq* instead of the editor's *aʿbaq* "most fragrant" for ms. *aʿmaq* "deepest".

which calls for a *mithqāl* of Tibetan musk, half a *mithqāl* of this musk will suffice.

this musk will suffice. Most of what is imported now of this type is in pods which have been stuffed with white lead (*ānuk*).

After it is the Kitan (*al-khiṭāʾī*). There is no other musk among the types of musk which resembles the Chinese except for this musk; its pods are without hair and have a fine skin. Its musk is very much like the Tibetan when it is scattered, except that its strength in making compounds is like the strength of the Nepalese. Most of what is imported comes with the Tibetan; when you find among all of it a pod which is smooth, then you know that it is the Kitan. It is not acceptable to sell it at the same rate as the Tibetan, for it is worth two thirds of its price because of its weakness.

Then after it is the Kitan.[109] Its pods resemble the Chinese in fineness and lack of hair, and its musk resembles the Tibetan except that its strength in making compounds is like the strength[110] of the Nepalese.

Then after it is the Tatar (*tatārī*). Its pods resemble the Tibetan pods and so does its musk, but it is only suitable for selling and it is weak in making compounds.

After it is the Kirghiz (*al-kh.rālkhīrī*). It is a famous type which everyone knows; it is not suitable for any making of compounds at all. It is not used except perhaps in musk pills (*aqrāṣ al-misk*), so it may be added to them.

Next is the Kirghiz (*al-khirkhīzī*). It is the well-known musk which everyone knows and is only useful for making compounds except for *ghāliyah*s, *lakhlakhah*s, and scented powders (*dharāʾir*) only.

109 Sbath's text has *al-kh.ṭāfī* instead of *al-khiṭāʾī*.

110 Reading *quwwah* for Sbath's *fuwwah*, "madder". This correction should also be applied to King, "Early Islamic Sources" 259.

After it is the Oceanic (*al-baḥrī*). It is musk which comes from the region of Oman, having been brought over the sea. This musk is originally good except that it is weakened by the smell of the sea. There is not another type of musk except for this one after the Chinese that is beneficial for constipation (*imsāk*), and this is really one of its types, except that it is not suitable for manufacturing, because when one passes the pestle over it, it becomes like ash (*ramād*) with no scent and no strength. When someone holds a bit of it, its scent lingers a long time. It is strongly black, with a good color and many grains (*shiyāf*).

Then after it is the Oceanic. It is musk which comes from the region of Oman having been brought over the sea; it is originally good except that it is weakened by the smell of the sea. It is beneficial for constipation, and it is not suitable for the making of perfumes at all because when the perfumer passes the pestle over it, it becomes like ash with no scent and no strength. It is strongly black, with an attractive color, and many grains.

After it is the Kashmiri (*al-qashmīrī*). It is the most inferior type of musk, because each pod of it weighs ten *dirhams*, but its musk is merely a *mithqāl*. All of its pods produce musk according to this proportion. It is not suitable for use in making compounds. When it is opened, there are no fine grains (*diqāq*) in it, but rather all of it consists of coarse grains (*shiyāf kibār*). It is really only suitable for perfumers, who sell it as good freshened musk (*miskan ṭariyyan jayyidan*); they anoint the surface of this musk in order to improve its color and scent, and then they sell it. Otherwise, it is not suitable for any sort of preparation.

Then after it is the Kashmiri. It is the most inferior type of musk; each pod of it reaches a weight of ten *dirhams* but only produces a *mithqāl* of musk, and most of it is coarse grains. When a bit of them is crushed, it all comes out as folded skin which is not musk; God the Exalted created it, and it has no adulteration.

I read in a book that musk is from the navel of a beast having the appearance of a bulky

> gazelle which has a single horn in the middle of its forehead, and from the bone of its forehead rounded hilts are made which are called *khutū*.[111]

When one increases its grains by crushing, it all comes out as folded skin (*jild malfūf*) which is not musk. God the Exalted created it, and it has no adulteration.

To set these toponyms out in order, we have: 1) Chinese, 2) Tibetan, 3) *Ṭūsmatī*/*Ṭūmsatī*, 4) Nepalese, 5) Kitan, 6) Tatar [only in Ibn Kaysān's version], 7) Kirghiz, 8) Oceanic, and 9) Kashmiri. We will discuss these places below. Most of these names are paralleled in the work of al-Azdī, the *Ḥikāyat Abī al-Qāsim al-Baghdādī*, which gives a list of types of musk.[112] This work is usually dated to the early 11th century, and so is a little later than Ibn Kaysān.

The Ibn Mandawayh-Ibn Kaysān account also has a discussion of the production of musk that is again textually quite independent of the earlier authors:

Ibn Mandawayh
The origin of musk is blood collected in the navel of the gazelle. The cause of the fine particles is that the blood descends from the veins to the navel, and the man hunts for the gazelle, slaughters it, and continually rubs

Ibn Kaysān
The origin of all musk is blood collected in the navel of the gazelle which it is said to graze on fragrant spikenard (*sunbul al-ṭīb*), and for that reason its blood changes into musk. This is not true, because there

111 This passage is evidently out of place. It probably represents an addition by Ibn Mandawayh to the source text, or a passage added by a copyist.

112 Muḥammad b. Aḥmad Abū al-Muṭahhar al-Azdī, *Ḥikāyat Abī al-Qāsim al-Baghdādī*, ed. A. Mez (Heidelberg: Carl Winter, 1902), 36. This list is part of a speech by Abū al-Qāsim listing the various luxuries absent from Iṣfahān but available in Baghdād; it is an important list for aromatics besides musk. See also E. Wiedemann, "Über Parfüms und Drogen bei den Arabern," in *Aufsätze zur arabischen Wissenschaftsgeschichte* (Hildesheim, Olms, 1970), 2.420–2.

with his hands all of its parts so that more of what blood is in its veins will descend to the navel. When it is cold, and he knows that he cannot make any more blood descend, he cuts out the place and hangs it until a year (*ḥawl*) has passed over it, and the blood has transformed into musk. All of the blood which was present in the navel of the gazelle before it was slaughtered is the fine grained musk (*misk diqāq*), and what descends into it from the veins after it was slaughtered in drops is the grains.

are gazelles which graze on wheat, barley, and grasses and musk comes from them, too. As for the cause of the grains (*shiyāf*) of musk and its fine grains (*diqāq*): it is that a man hunts the gazelle, slaughters it, and continually rubs with his hands all of its parts in order that more of what blood is in its veins will descend to its navel. When he knows that the navel has become filled with blood, he cuts out the place and hangs it until a year has passed over it, and when a year has passed, it has transformed into musk. All the blood which was present in the gazelle's navel before it was slaughtered, when it transforms is fine grain musk, and what descends into it after it was slaughtered from the veins drop by drop becomes grains.

It is said that the fodder of this gazelle is spikenard, and because of this its blood becomes musk, but there are different opinions about this, because there are gazelles which graze upon wheat, barley, and grass and musk is produced by them. Rather, God the Exalted ascribed it to their nature. God the Almighty created the delicacies and marvels of the many animals as benefits for humanity, such as the silk which is used for the softest garments from a worm, the honey which is the most famous of foods from a beast (*dābbah*), the bezoar which is produced from the belly of a wild cow, and the musk which is produced from a gazelle, and the examples are many, every delicacy made by the Creator, Exalted be His Mention: "Such is the decree of the Almighty, the Omniscient."[113]

Ibn Sīnā

Abū ʿAlī al-Ḥusayn b. ʿAbdāllah b. Sīnā (d. 1037), commonly known as Avicenna in the European tradition, was born in Central Asia and spent his life there and in Iran. He gives a brief account of musk in his medical encyclopaedia, *al-Qānūn fī al-ṭibb*:[114]

> Musk. Identification: Musk is the navel of a quadruped like the gazelle, or it is the gazelle itself. It has two white teeth curving in towards the part near the armpit (*unsā*) like two horns. Selection: The best on account of its source is the Tibetan. It is also said that the Chinese is best, followed by the Kirghiz[115] and then that of the Indian Ocean. With regard to its pasturage then comes the "horns"-type[116] which grazes on the two aconites (*al-bahmanayn*)[117] and spikenard (*al-sunbul*), then myrrh. The best with regard to its color and its aroma is yellow *faqāḥā*.

He continues with his discussion of the medical properties of musk, which will be quoted in Chapter 6. It is not clear what Ibn Sīnā's sources were; much of his account does not derive directly from any of the known previous works on musk. His passage on musk is probably the main source for Ibn Jazlah's (d. 1100)

---

113 Qur'ān 6:96, etc. Cf. al-Ghazālī, *Iḥyā' ʿulūm al-dīn*, 5 vols. (Beirut: Dār al-Kutub al-ʿIlmiyyah, 2005), 3.283 and the Conclusion. Compare a passage from an early 14th century Jain text on the lowly origins of precious substances cited by McHugh, *Sandalwood*, 176–7.

114 Ibn Sīnā, *al-Qānūn fī al-Ṭibb* (Dār Sādir reprint of Būlāq ed., n.d.), 1.360 and the critical edition edited by Idwār al-Qashsh (Beirut: Mu'assasat ʿIzz al-Dīn, 1987), 1.593.

115 Text has *al-j.rjīzī* in the Būlāq and *al-j.rjīrī* in al-Qashsh's ed.

116 Something seems wrong with the text here. The Būlāq edition has *f.rūn* while the critical edition has *qurūn* "horns", which doesn't make much sense either. The parallel in Abū ʿAlī Yaḥyā b. ʿĪsā Ibn Jazlah, *Minhāj al-bayān fī mā yastaʿmiluhu al-insān* (Cairo: Dār al-Kutub al-Miṣriyyah, 2010), 779, says "The best of it with respect to the pasturage of its animal is what grazed upon aconites ..."

117 Meaning, red and black aconites, see R. Dozy, *Supplément aux dictionnaires arabes*, 2nd ed., 2 vols. (Paris: Maisonneuve, 1927), 1.123, and W. Schmucker, *Die pflanzliche und mineralische Materia Medica im Firdaus al-Ḥikma des Ṭabarī* (Bonn: Selbstverlag des Orientalischen Seminars, 1969), 122–3.

account of musk.[118] Ibn Sīnā notes that some prefer Chinese musk. It is not known whether Ibn Sīnā had used the Ibn Mandawayh-Ibn Kaysān account, which prioritizes Chinese musk. But it is also notable that Ibn Sīnā, who was active in the eastern part of the Islamic world and thus lived much closer to the sources of Central Eurasian musk, personally preferred the Tibetan.

Al-Bīrūnī (d. c. 1050)

Abū Rayḥān al-Bīrūnī is one of the most renowned Islamic scientists. He was born in Khwarazm and lived most of his life in Central Asia. With the rise of the Ghaznavid Empire, al-Bīrūnī traveled to India and made a detailed study of Indian culture, learning Sanskrit. His book, the *Kitāb fī taḥqīq mā li-l-hind,* is a major work for the Islamic understanding of India and for medieval India in general, but it has little to say about musk. Nevertheless, the experiences al-Bīrūnī gained in his inquisitive life make all his works valuable and of high quality due to his careful method.

Al-Bīrūnī's last work was the *Kitāb al-Ṣaydanah,* or *Book of Pharmacy.*[119] He left it incomplete at his death, and the book may also be lacking the final polish that its distinguished author would have given it in places. The *Kitāb al-Ṣaydanah* is arranged alphabetically and contains detailed accounts of the *materia medica.* It generally includes synonyms in different languages, information on sources and production, types, properties, and therapeutic uses for the substances it covers.

> Musk. In Indian it is *kasturī katūrī,* and in Turkic it is *yıpar,* and in Khwarezmian *akt bnjl.* It is available in Turk and Indian forms, and each has different types. As for the Turk, the best is the Kitan (*Quta'ı*)[120] and after it is the Tibetan. It is intensely fragrant and pungent. After it is the Tatār, and it is like manure,[121] weak in scent, with an offensive smell. After it is the Kirghiz, which is foul, and not pleasant; its scent is like the scent of kewda (*kādhī*) oil. As for the Indian, the best of it is the Nepalese, and it follows the Tibetan, and the color of its musk is black inclining to yellow, and it has a sweeter scent than the others. In pharmaceuticals it

118 Ibn Jazlah 779–80.

119 *Kitāb al-Ṣaydanah,* ed. in Z. V. Togan, *Bīrūnī's Picture of the World* (New Delhi: Archaeological Society of India, 1938), 136, ed. Saʿīd, 345–6, and ed. A. Zaryab, 577–8.

120 Thus Saʿīd. Togan has *Q.tāy,* and Zaryāb *Q.nāy.* See King, "Early Islamic Sources".

121 Following Saʿīd and Zaryāb's *musabbikh.* The word in the ms. is undotted according to Togan, who has *musabbih* "praiseworthy".

is most excellent and pungent. Among the Indian is a [type][122] named *h.t.r.s.rī.*[123] After the Nepalese is the Kashmīrī, and after it is the *Udiyākhī*,[124] and it is close in quality to the Kashmīrī; one can only distinguish it by the ugliness of its appearance, and its musk is quite black, but when it is compared closely, one sees in it something like white sand. A type of black musk comes from the sea, and is brought from China (*Ṣīn*) stripped from its pod in phials (*qawārīr*) and hence is called *qawārīrī*. A book has "The best of it is the Chinese, then the Tibetan, then the Tūmsatī (from Mdosmad), then the Kitan (*Khiṭā'ī*), then the Tatār, then the Khirkhīz, then the Oceanic.

Marginal Comment:[125] The best musk is the Chinese (*al-Ṣīnī*), and it is very scarce and only appears seldom. The weight of one of its pods is twenty *dirhams*, and perhaps a little more or less. Since its skin is fine, one can reckon the grains within it when it is shaved and has no tufts of hair. When it is opened, the weight of the skin is half of a *dirham* to three *dirhams*; one splitting it gets a nosebleed. Its scent lingers for forty years.[126] Next is the Tibetan, then the Ṭūmsatī, then the Nepalese, then the Kitan, then the Tatar, then the Kirghiz, then the Oceanic, then the Kashmiri. It is said of it [the musk animal] that it grazes upon spikenard; this is nonsense because there are gazelles which graze upon wheat and barley and they still produce musk. The hunter hunts it and slaughters it, and does not stop pressing its limbs to force the blood in its veins into its naval before slaughter, and when it is filled, he cuts out the place and suspends until a year has passed over it, and it has transformed into musk. All of the blood which is in the navel before slaughter becomes the fine grains, and what descended into it after slaughter drop by drop becomes the larger grains (*shiyāf*).

---

122 Togan supplies *naw*ʿ, Saʿīd and Zaryāb *mā*.

123 Thus Togan. Saʿīd has *ch.t.r.s.rī* and Zaryāb *j.t.r.s.rī*.

124 Zaryāb's reading; Togan 136 has *al-udhiyākhī*; Saʿīd 345 has *al-uriyākhī* (with Urriyākhī in the trans., 304).

125 This marginal comment is derived from the Ibn Mandawayh-Ibn Kaysān account or its source, as discussed above. There is one addition to the text, as noted below. The marginal comment is not given in Saʿīd's edition, but appears in Togan's and Zaryāb's.

126 This sentence is not in Ibn Mandawayh or Ibn Kaysān, who instead describes the procedure for opening musk pods. It is in al-Kāshānī's version, 250.

Its name in Greek is *mūrūn*,[127] and in Syriac, *muskā*.[128] On its adulteration: with "the black drug" (*siyāh dāwrān*), emblic myrobalan (*amlaj*), Indian pepperwort (*shīṭaraj*), dried acorns (*ballūṭ*), and by Lebanese apple, spikenard, and cloves, then it is supported with musk and it penetrates into it. In the land of Daylam is an herb which is called *mushkrāsh*,[129] its scent cannot be condemned in comparison with the scent of musk.

Ibn Bādīs (1007–1061)

Muʿizz b. Bādīs was ruler of the Zīrid dynasty of Qayrawān in what is now Tunisia. His work, the *ʿUmdat al-Kuttāb*, is a work on bookmaking. It includes many recipes for inks, and in an appendix it discusses ingredients, including musk. Ibn Bādīs writes: "There are different types of musk, and there are five which are well known: the Indian, Bahārī, Tibetan (written *tnbtī*), ʿIrāqī, and *misk al-yad*."[130] He then gives extensive information on their testing.

Types of Musk and their Rankings

In Islamicate literature, musk is usually ranked according to geographic origin and named according to the places it originated. However, there were other considerations in the classification of musk. Musk was often traded in the pod, which contained perhaps 25 grams of musk. Musk was also traded as grain musk by the phial (sg. *qarīrah*, pl. *qawārīr*), so as a type it is not surprising that we should meet the descriptor *qawārīrī* musk. It was more difficult to ascertain the origin of musk without the pod, as pods from specific places were often described as having a particular appearance. In addition, there seems to have been a prejudice against musk in phials because it was more removed from its natural source; in other words, musk in the pod was thought less liable to be tampered with. The Ibn Mandawayh-Ibn Kaysān tradition, which seems to have been particularly quality conscious, dismisses the value of acquiring

127 Greek μύρον has a generic meaning of perfume; LSJ 1155a.

128 Correctly, *muškā*, which is the emphatic state of *mušāk* in Syriac.

129 Zaryāb's reading; Saʿīd and Togan have *miskrāsh*. Zaryāb's note 9 on 579 states that the Persian version also calls it *muskh rāst* "true musk".

130 Ibn Bādīs, *ʿUmdat al-kuttāb wa ʿuddat dhawī al-albāb*, ed. Najīb Māyil al-Harawī (Mashhad: Majmaʿ al-Buḥūth al-Islāmiyyah, 1409), 115, and cf. the trans. by M. Levey, *Medieval Arabic Bookmaking and its Relation to Early Chemistry and Pharmacology*, Transactions of the American Philosophical Society New Series 52:4 (Philadelphia: American Philosophical Society, 1962), 47.

musk in anything other than a pod. Al-Sīrāfī notes "this is the ultimate musk, when it has ripened in its pod upon its animal." His account, however, also mentions that musk gathered as loose scrapings of the musk deer was collected into pods, meaning that one had to be careful of pods, too.

The appearance of "Oceanic" (*baḥrī*) musk among musk types signals an important issue in the Islamicate perception of musk. Beginning with Ibn Māsawayh, a clear preference is enunciated for musk brought by the overland trade over the musk traded by sea. Ibn Māsawayh calls this musk Sogdian musk. The accounts of Abū Zayd al-Sīrāfī and al-Masʿūdī deal with this issue in a head-on fashion, stressing that the humidity of the sea damages musk. This concept is rooted in the humoral theory of medicine, in which musk was characterized as hot and dry, the ocean cold and moist. Thus the sea weakened musk in its most important aspects. Exposure to the elements, especially light, heat, and humidity, is very damaging to the volatile essential oils in aromatic materials, so the Islamicate writers had reason to complain about musk improperly transported. Abū Zayd al-Sīrāfī and al-Masʿūdī note that it would be possible to combat this influence by transporting the musk carefully in sealed waterproof vessels. With its notice of the *baḥrī* or "Oceanic" musk, the Ibn Mandawayh-Ibn Kaysān tradition shows once again that the musk brought via the overland routes was more highly prized than that brought by sea. This source states that it was useful as a laxative, but not in perfumery because it had no scent and strength. This Oceanic musk is also mentioned by al-Azdī.[131] For Ibn Māsawayh, Chinese musk was effectively musk shipped by sea. Of course, Ibn Mandawayh and Ibn Kaysān do not explicitly state that this Chinese musk arrived by sea; their Chinese musk may be Chinese musk shipped overland, distinguished from that shipped by sea, which they called Oceanic. They note that the *baḥrī* musk arrived in Oman from its long voyage over the sea, and then it would have had to be shipped elsewhere in the Middle East.

Ibn Bādīs mentions *ʿIrāqī* musk. There are a few possibilities for the meaning of this term. If taken as actual deer musk, *ʿIrāqī* musk must have been musk exported through Iraq since the musk deer was not native there,[132] unless some obscure toponym has been corrupted into the word *ʿIrāqī*. Another explanation is that Ibn Bādīs' *ʿIrāqī* musk is civet, which could be produced in Iraq, as Ibn Mandawayh and Ibn Kaysān report.[133]

131 Azdī 36.

132 This is the explanation of the name given by the 12th century physician Ibn Jumayʿ, quoted in Ibn al-Bayṭār 2.131. Masʿūdī had already noted musk being imported into Iraq.

133 Ibn Mandawayh 232 and Ibn Kaysān 192–3.

One other type of musk with a name unrelated to a geographical point of origin is the occasionally encountered *buddī* musk. As noted above, the *Kitāb al-Tabaṣṣur bi-l-tijarah* attributed to al-Jāḥiẓ included mention of a type of musk called *buddī*. A passage of al-Masʿūdī, the origin of which is undetermined, elucidates the meaning of this name. Al-Qalqashandī quotes the following from al-Masʿūdī on Indian musk:

> When it is brought to India, the unbelievers of India take it and plaster it upon their idols (*aṣnām*) from year to year, then they exchange it for new, and the custodians of the idols sell it. Because of its being kept a long time upon the idols, its scent weakens.[134]

This passage provides the key to the mysterious *misk al-budd* or *buddī* musk, for the Arabic word *budd*, derived originally from the word Buddha, refers to a Buddhist or Hindu sacred image. Thus these musks were supposedly derived from the religious images of the Indians, and that explained their inferiority and weakness, for the Indians had, in the opinion of the Muslims, desecrated them.

Buddī musk is probably behind the mysterious *misk al-yad* of Ibn Bādīs. The printed edition has *misk al-yad*, but the translator Levey (who was working before the printed edition appeared) has translated "musk of the *bad* plant."[135] There is nothing in Ibn Bādīs' text that requires this item to refer to a plant; later, in his discussion of its testing, he says "it is gathered from upon (*min ʿalā*) the *yad* in the land of India and imported."[136] Levey has read this sentence a little differently, and he has translated it as, "it is fathered [sic] from the *bad* of the Indian country. It is imported." He compares the also unknown plants *badah* or *badad*, the latter mentioned in Ibn al-Bayṭār. Given the low opinion of "musk" made from plants, it is possible that this type is derived from a plant or at least manufactured. But it is better to read the word as *al-budd*, thus changing it to "musk of the Buddha statue", and also agreeing with al-Masʿūdī.

Musk was generally graded and ranked according to point of origin rather than more intrinsic appreciation of its qualities. The qualities of musk were expected to follow from point of origin in a place or among a people, hence the

---

134 Qalqashandī 2.121.

135 Levey, *Medieval Arabic Bookmaking* 47. There is merely a one-dot difference between the b and y in Arabic script and based on the printed edition one assumes that at least some of the manuscripts read *yāʾ*, however, I prefer Levey's reading, as *yad* has no meaning in this context.

136 Ibn Bādīs 116.

statements in the texts that musk from specific areas had specific qualities. In effect, point of origin functioned as a sort of brand-name for musk varieties in the early medieval Islamic world.

TABLE 4.2 *The varieties of musk ranked by excellence*

Ibn Māsawayh[a]	al-Yaʿqūbī[b]	al-Tamīmī[c]	Ibn Mandawayh-Ibn Kaysān[d]
1. Sogdian (*sughdī*) [Tibetan][e]	1. Tibetan (*tubbatı*)	1. Tibetan[f]	1. Chinese
2. Indian (*hindī*)	2. Sogdian		2. Tibetan
3. Chinese (*ṣīnī*)	3. Chinese		3. *Ṭūsmatī*/ *Ṭūmsatī* [from Mdosmad]
	4. Indian		4. From Nepal (*naybālī*)
	5. *Q.nbārī*[g]		5. Kitan (*khiṭāʾī*)[h]
	6. Toquz Oghuz (*tughuzghuzī*)		6. Tatar (*tatārī*) [Only in Ibn Kaysān]
	7. *Q.ṣārī*[i]		7. Kirghiz
	8. Kirghiz (*khirkhīzī*)[j]		8. "Oceanic" (*baḥrī*)[k]
	9. *ʿ.ṣmārī*[l]		9. Kashmiri (*qashmīrī*)
	10. "Mountain" (*jabalī*)		

a Ibn Māsawayh 9–10.
b Nuwayrī 12.11–13; see also De Goeje 325.
c Nuwayrī 12.2.
d Ibn Mandawayh 225–7, Ibn Kaysān 188–9.
e Ibn Māsawayh specifies Tibetan musk imported by overland by the Sogdians.
f Al-Tamīmī himself does not opine on further varieties beyond apparently quoting the list of al-Yaʿqūbī.
g Unidentified; the text says "it is brought from the land called *Q.nbār* between China and Tibet, and sometimes they misrepresent it and attribute it to Tibet", Nuwayrī 12.12.
h Sbath's edition of Ibn Kaysān has *kh.ṭāfī*.
i Unidentified. The text says "… from the land called Qaṣār between India and China (*al-Ṣīn*)," Nuwayrī 12.12.
j Dotted *al-ḥarjīrī* in the text. De Goeje has marked it (sic).
k Musk transported by sea.
l Unidentified. The text gives no indication of its location or other meaning.

TABLE 4.3 *The varieties of musk ranked by excellence according to al-Bīrūnī*

al-Bīrūnī[a] Turk Types:	Indian Types:	Chinese: brought by sea in phials (hence, *qawārīrī*)[b]
1. Turk: Kitan (*Qitāy*)	1. Indian: Nepalese ("it follows the Tibetan")	
2. Turk: Tibetan	[1a. Indian: *H.t.r.s.rī*[c]]	
3. Turk: Tatar	3. Indian: Kashmiri	
4. Turk: Kirghiz	4. Indian: *Udiyākhī*	

a Bīrūnī, *Kitāb al-Ṣaydanah*, ed. Togan 136, Said 345–6, and Zaryab 577–8.
b This is not explicitly ranked, but by being listed after the other varieties it is implied that it is inferior.
c Not explicitly ranked; see the translation above.

The Ibn Mandawayh-Ibn Kaysān tradition's list of musk varieties forms an interesting contrast with the two different lists of musk types given by al-Azdī in the *Ḥikāyat Abī al-Qāsim al-Baghdādī*. This book contains a fictional account of the rake and moocher Abū al-Qāsim in early medieval Islamic high society. In one scene at a gathering he has crashed, he shows off his knowledge of the good life by lamenting the poor amenities available to him at that particular affair. Thus, he recites a litany in overheated rhetoric of items *not* present: a truly worthwhile event would have a full complement of foods and scents. His catalog of missing scents (which includes many significant aromatics besides the musk types) is almost a full page of text long and includes two different lists of musk types.[137]

Al-Azdī's second list of musk varieties is especially unusual, and it includes several types of musk not based on point of origin: "Apple-scented," "Night-black", and the already mentioned "Oceanic musk" and "musk in phials". Ibn Māsawayh had likened the scent of good musk to the scent of a Lebanese apple. The comparison of the scent of musk to apples is common; evidently the apples of Syria and Lebanon were strongly aromatic. The Syrian apple was very highly

137 Azdī, 36–7. This list is part of a speech by Abū al-Qāsim listing the various luxuries absent from Iṣfāhān but available in Baghdād; it is an important list for aromatics besides musk. Cf. Wiedemann, "Über Parfüms und Drogen" 420–2.

TABLE 4.4 *The Ibn Mandawayh-Ibn Kaysān tradition's ranking of musk compared with al-Azdī's rankings*

Ibn Mandawayh- Ibn Kaysān	al-Azdī list 1[a]	al-Azdī list 2[b]
1. Chinese	1. Sogdian	1. Tibetan
2. Tibetan	2. Tibetan	2. "Apple-scented"
3. *Ṭūsmatī/ Ṭūmsatī*	3. *Ṭūmanī*[c]	3. Indian
4. Nepalese	4. Nepalese	4. Chinese
5. Kitan	5. Kirghiz	5. *Wadyā*(ʾ or *n*)*ī*
6. Tatar [only in Ibn Kaysān]	6. Kitan	6. "Night-black" (*samīrī*)
7. Kirghiz	7. "Oceanic"	7. "Oceanic"
8. "Oceanic"	8. Chinese	8. Musk in phials (*qawārīrī*)

a This list refers to the types of musk used to coat Mandalī aloeswood to increase its fragrance.
b This list is his "regular" list of musk types.
c This is a representation of *mdosmad* in Arabic script; see below.

regarded; its excellence was proverbial.[138] Musk was also valued because of its blackness; this is a theme we will discuss in greater detail in Chapter 7.

Toponyms, Ethnonyms, and Sources

Musk of Tibet

There is a near consensus that the best musk was Tibetan, as we have seen. The priority of Tibetan musk can be traced in our earliest detailed source. Ibn Māsawayh stated that the best musk originated in Tibet. According to him, it had come to be known as Sogdian musk on account of the Sogdian intermediaries in the trade who brought musk by the overland route into Khurāsān, just as his Indian musk means musk brought through India. After Tibetan musk transported through Sogdiana, Ibn Māsawayh favored Tibetan musk transported through India, actually through Sind. There was a third path as

138 Cf. al-Washshāʾ, *Kitāb al-Muwashshā*, ed. R. E. Brünnow (Leiden: Brill, 1886), 132, al-Muqaddasī, *Aḥsan al-Taqāsīm fī maʿarifat al-aqālīm*, ed. M. J. De Goeje (Leiden: Brill, 1877; repr. 1967), 181, al-Thaʿālibī, *Laṭāʾif al-maʿārif* (Cairo: Dār Iḥyāʾ al-Kutub al-ʿArabīyah ʿĪsā al-Bābī al-Ḥalabī wa-Shurakah, n.d.), 156, and trans. C. E. Bosworth, *The Book of Curious and Entertaining Information* (Edinburgh: University Press, 1968), 118.

well: the geographer Ibn Ḥawqal noted that much Tibetan musk came through Wakhān.[139]

The early 10th century geographer Ibn al-Faqīh included a section in his work on the specialties of the different regions of the world, and in it he noted musk as a specialty of China and of Tibet, along with Tibetan shields (*diraq*).[140] Ibn Khurradādhbih also mentions that the King of Tibet gave Alexander the Great a gift of musk.[141] This is, of course, completely ahistorical, but it does imply that Ibn Khurradādhbih knew of the musk of Tibet, though curiously he does not otherwise mention musk as a product of Tibet. Likewise, al-Masʿūdī states that the Khāqān of Tibet sent the Sasanian Khusraw Anushīrwān 4000 *mann* of royal treasury (*khazāʾinī*) musk along with characteristic Tibetan armaments.[142] The anonymous Persian geography *Ḥudūd al-ʿālam* of 982 also refers to musk from Tibet.[143]

Tibetan musk usually held the greatest fame among later authorities as well: al-Yaʿqūbī and al-Tamīmī placed it first, and other writers ranked it highly.[144] It was equally famed in literature. The early 10th century Persian poet Muṣʿabī considered musk of Tibet exemplary, along with aloeswood of Qimār and ambergris from Yemen.[145] In the 11th century it was famous enough for the

139 Ibn Ḥawqal, *Kitāb Ṣūrat al-ʿarḍ*, ed. M. G. De Goeje and rev. J. H. Kramers (Leiden: Brill, 1938; repr. 1967), 449, 476.

140 Ibn al-Faqīh, *Mukhtaṣar kitāb al-buldān*, ed. M. J. De Goeje (Leiden: Brill, 1885; repr. 1967), 251, 255.

141 Ibn Khurradādhbih, *Kitāb al-Masālik wa-l-mamālik*, ed. M. J. De Goeje (Leiden: Brill, 1889, repr. 1967), 263. The passage is translated in A. Akasoy, "Tibet in Islamic Geography and Cartography," in *Islam and Tibet: Interactions along the Musk Routes* (Farnham: Ashgate, 2011), 31. The Syriac text has an anecdote in which the king of China gives musk along with other valuables, cf. E. A. W. Budge, ed. and trans., *The History of Alexander the Great, being the Syriac version, edited from five manuscripts, of the Pseudo-Callisthenes* (Cambridge, 1889; repr. Amsterdam: Philo Press, 1976), text 200. Tibet is not mentioned in this text, and that is perhaps a good reason to regard the Syriac translation as genuinely early.

142 Masʿūdī 1.309 (§624).

143 *Ḥudūd al-ʿālam min al-mashriq ilā al-maghrib*, ed. M. Sutūdah (Tehran: Dānishgāh-i Tihrān, 1983), §10:41–43, and V. Minorsky's comments in *The Regions of the World, A Persian Geography 372 A.H.–982 A.D.*, 2nd ed. (London: Gibb Memorial Series, 1970), 248.

144 Al-Zahrāwī, *Kitāb al-Taṣrīf li-man ʿajiza ʿan al-taʾlīf*, facsimile of Süleymaniye Beşirağa collection, ms. 502, ed. F. Sezgin (Frankfurt am Main: Institute for the History of Arabic-Islamic Science, 1986), 2.39 considered Tibetan the best musk; it is the only type he mentions in his description of musk.

145 Quoted in Abū Faḍl Bayhaqī, *Tārīkh-i Bayhaqī*, ed. ʿAlī Akbār Fayyāḍ (Mashhad: Dānishgāh-i Mashhad, 1971), 482.

Persian poet Farrukhī to use it in the peroration of a *qaṣīdah* where the poet invokes the ends of the earth:

> *tā zi kashmīr ṣanm khīzad wa az tubbat mushk hamchu k-az miṣr qaṣab khīzad wa az ṭā'if adīm*[146]
>
> As long as idols come from Kashmir, musk from Tibet, and likewise papyrus from Egypt and leather from Ṭā'if.

The 12th century Jewish traveler Benjamin of Tudela also emphasizes the musk of Tibet. He traveled in the Middle East from 1165–73; he never traveled as far east as Tibet, let alone China, so the parts of his account discussing these regions must derive from other sources, perhaps Jewish merchants who had traveled there. He wrote an account of his journeys in Hebrew. He says: "From there (Samarkand), it is four days to Tibet (ṬWBWT). This is the region (*medînah*) where the musk (*mûr*) is found in the forests (*C'est la région* (medînah) *où l'on trouve le musc* (mûr) *dans les fôrets*)."[147] It is most interesting that Benjamin accurately records that musk comes from forests; the other writers suggest that the musk deer lived in steppeland.

The *Book of Curiosities*' account of musk mentions that those who would gather musk had to take meat with them to throw at dangerous large ants to appease them so they could gain access to the musk.[148] This is a variation of the legend of the gold-digging ants, attested as early as Herodotus (3.102). The many versions of this story in literature from Central Asia, India, and the Graeco-Roman world typically focus on ants that dig up gold and then protect it; this passage substituting musk for the gold may be unique.[149]

---

146 *Farhangnāmah-i Adab-i Fārsī*, 258a. Also quoted by Melikian-Chirvani, "Iran to Tibet," in *Islam and Tibet: Interactions along the Musk Routes* (Farnham: Ashgate, 2011), 93–4.

147 M. Tardieu, "Le Tibet de Samarcande et le pays de Kûsh: mythes et réalités d'Asie centrale chez Benjamin de Tudèle," *Cahiers d'Asie Centrale* 1–2 (1996): 301 and M. N. Adler, trans., *The Itinerary of Benjamin of Tudela* (London, 1907; repr. New York: Philipp Feldheim, n.d.), 59. Page 52 of Adler's trans. has Sinjar Shah ben Shah, Sultan al-Fars al-Kabir, ruling over Tibet, where one finds the animals from which the musk is obtained.

148 *Book of Curiosities* 45 (text) and 523 (trans.).

149 There is a large literature: see P. Lindegger, *Griechische und römische Quellen zum Peripheren Tibet*, 3 vols. (Zürich: Tibetan Monastic Institute, 1979–93) and 2.35–50; K. Karttunen, *India in Early Greek Literature. Studia Orientalia* 65. (Helsinki: Finnish Oriental Society, 1989), 171–80. Rapoport and Savage-Smith 523 n. 31 note Ibn al-Faqīh's story of gold-guarding ants.

China

Ibn Māsawayh, our earliest source, also mentions musk from China. Ibn Kaysān regarded it as the best musk but also lamented its scarcity; Ibn Sīnā was aware of the view that Chinese musk was best.[150] Other writers ranked it highly but were cautious about allegations of Chinese adulteration of musk. Ibn Khurradādhbih also specifically mentions the existence of musk in China. He gives it in a list of Chinese products elsewhere in his book: silk (*ḥarīr*), silk brocade (*firind*), gold brocade (*kīmkhāw*),[151] musk, aloeswood, saddles (*surūj*), sable (*sammūr*), vessels (*ghaḍār*), *ṣīlbanj*,[152] cinnamon (*dār ṣīnī*) and galangal.[153]

China was linked with many rare items in early medieval Arabic literature as we have discussed in Chapter 2. These included perfumes, especially musk, but also cinnamon and other products. China was more famous as a source of manufactured goods; the craftsmanship of the Chinese was considered the most excellent of all peoples. The notion of China as a source of perfumes penetrated legend as well. Shamir and Ḥassān's legendary expedition for the Tubbaʿ of Yemen returned with goods (*amwāl*) obtained in China including different kinds of jewels, perfumes and slaves.[154] The Armenian geographical work formerly ascribed to Moses of Chorene, but now attributed to Ananias of Širak (fl. 7th century),[155] includes a discussion of China and its products, including musk.[156]

---

150 Followers of the Ibn Mandawayh-Ibn Kaysān tradition such as Kāshānī 250 and Jurjānī 156 also regard the Chinese as best.

151 From Chinese *jinhua* 錦花, see B. Laufer, "Loan Words in Tibetan," *T'oung Pao* 17 (1916): 557–8, and his *Sino-Iranica: Chinese Contributions to the History of Civilization in Ancient Iran, with Special Reference to the History of Cultivated Plants*. Field Museum of Natural History Anthropological Series 15:3 (Chicago: Field Museum, 1919; repr. New York: Kraus, 1967), 539.

152 Ibn Khurradādhbih's editor De Goeje suggests that this is the narcotic the Persians call *gālbang*. It is unclear how far back this substance may be traced, as it is absent from al-Harawī's *al-Abniyah ʿan ḥaqāʾiq al-adwiyah*, ed. Aḥmad Bahmanyār, 2nd ed. (Tehran: Dānishgāh-i Tihrān, 1975; repr. Tehran: Dānishgāh-i Tihrān, 1371/1992), which is the first pharmacological text extant in Classical Persian. The first citation in the *Lughat nāmah* is from the 12th century poet Sūzanī, LN 27.17–18.

153 Ibn Khurradādhbih 70.

154 Ṭabarī I.892.

155 On date and authorship, R. H. Hewsen, *The Geography of Ananias of Širak (Ašxarhacʿoycʿ): The Long and Short Recensions*, trans. R. H. Hewsen (Wiesbaden: Reichert, 1992), 7–15; he dates it to the early 7th century ("between 591 and 636" on page 34). See also his Appendix II on pages 272–83 on the life and works of Ananias.

156 Not the musk ox, as translated by Hewsen, 76 and 77. See also G. R. Cardona, "L'India e la Cina secondo l'Ašxarhaċoyċ," in *Armeniaca: Melanges d'études armeniennes* (Venice: Saint Lazare, 1969), 88.

Ibn Baṭṭūṭah's account of a gift sent to the Delhi Sultan Muḥammad b. Tughluq (r. 1325–1351) by the emperor of China includes five *mann* of musk among other characteristic items.[157] The musk trade out of China (in the "Tonquin" musk) remained the most important of the musk trades in the 19th century,[158] and in the 20th and 21st centuries, with continued hunting and the introduction of farmed musk deer, China continues to export musk.

India

Musk was produced only in the Himalayan frontiers of India and eastward and was effectively an import for medieval Indian society.[159] The use of the term Indian (*Hindī*) musk as a type is found in Ibn Māsawayh and al-Ya'qūbī among the specialists in perfumes. Both writers stress that Indian musk is actually from Tibet but was traded through India rather than by the Central Asian routes by the Sogdians, or from China. Al-Ya'qūbī also uniquely refers to Jabalī musk, which he says is musk brought through Mūltān in Sind from the Himalayas. The term *Hindī*, of course, had a wide range of associations with exotica in Islamicate culture, and other aromatics, such as aloeswood, also received this term as a generic marker even if they were produced from lands not strictly "India". The early medieval Islamic geographical conception of South and Southeast Asia distinguished between Sind, the lands of the Indus River Valley, and further India, or Hind, and under the latter term all of Southeast Asia up to the frontiers of China was understood. Tibet, incidentally, was usually classed with the Turk peoples of Central Asia rather than the Indians.

Al-Bīrūnī's account of musk divides the varieties of musk into three categories: Turkish, Indian, and Chinese. Two of these categories, the Turkish and Indian, contain several varieties. Al-Bīrūnī's Indian varieties are: Nepalese, *h.t.r.s.rī*, Kashmiri, and *Udiyākhī*. Each of these types (excluding the completely unknown *h.t.r.s.rī*) is from the northern mountains of the Indian subcontinent.

Musk from Nepal (dotted as *Baytāl* in the manuscript) is attested by the *Ḥudūd al-'ālam*[160] and in Ibn Kaysān's Nepalese (*al-naybālī*) musk; both references are essentially contemporary.[161] *Naybālī* is identical to the form

157 Ibn Baṭṭūṭah, *Riḥlat Ibn Baṭṭūṭah* (Beirut: Dār Ṣādir, 1960), 530.

158 E.g., C. H. Piesse, *Piesse's Art of Perfumery*, 3rd ed. (London: Piesse and Lubin, 1891), 269; Anon. "The Town Warehouses of the East and West India Dock Company," *The Chemist and Druggist* 34 (1889): 452–3.

159 Cf. J. McHugh, *Sandalwood and Carrion: Smell in Indian Religion and Culture* (Oxford: Oxford University Press, 2012), 172–3.

160 *Ḥudūd al-'ālam* §10:41–3 and Minorsky's commentary, 248. Minorsky's reading is now supported by Ibn Mandawayh- Ibn Kaysān's *Naybālī* musk.

161 A. King, "Some 10th century material on Asian Toponymy from Sahlan b. Kaysan," *Archivum Eurasiae Medii Aevi* 16 (2008–9): 122–3.

employed by al-Azdī and al-Bīrūnī. While the marginal comment in al-Bīrūnī's *Kitāb al-Ṣaydanah* evinces knowledge of the tradition of Ibn Kaysān or perhaps Ibn Kaysān himself, it is likely that al-Bīrūnī's knowledge of Nepalese musk was verified through his own experiences in India. Unlike Ibn Mandawayh-Ibn Kaysān, he clearly knows the location of Nepal as can be seen in his detailed presentation of the geography of India in his book on India.[162]

The same can be said for Kashmir, about which al-Bīrūnī writes from his own experiences in India even though he laments that the land was closed to foreigners in his time.[163] Ibn Kaysān is again the first to mention Kashmiri musk explicitly, although Ibn Māsawayh had noted that the plants that were grazed on by the musk deer grew in Tibet and/or Kashmir. One could infer from this that Ibn Māsawayh understood Kashmir as a source of musk, but instead he emphasizes Tibet and China as the original sources. Both the Ibn Mandawayh-Ibn Kaysān tradition and al-Bīrūnī deplore the quality of Kashmiri musk; Ibn Kaysān goes so far as to call it the lowest form of musk.

As to the *Udiyākhī* musk, we propose that name is perhaps a corruption of Udyāna in the Northwest of the Indian subcontinent, in modern Pakistan, i.e. **Udiyānī*.[164] Udyāna is usually identified with the Swat Valley, where musk deer occur. We may also compare the *Wadyā*(*ʾ* or *n*)*ī* musk of al-Azdī. Since the Sanskrit name of Udyāna means "orchard" or "garden", it is possible that this is an epithet for musk, but it is untraced in Sanskrit literature.

Ibn Bādīs mentions an otherwise unattested *bahārī* musk. There was a Bahār in Central Fārs,[165] but it seems most unlikely that such an obscure town in the middle of Fars would be associated with a type of musk. A more plausible guess would be that this Bahār is a form of the Indian toponym *B.lhārī* of the *Ḥudūd al-ʿālam*, which exported musk.[166]

---

162 Al-Bīrūnī, *Kitāb al-Bīrūnī fī taḥqīq mā li-l-Hind*, ed. E. Sachau (Haydarābād: Dāʾirat al-Maʿārif, 1958), 160.

163 Bīrūnī, *Hind*, 165–6.

164 Transcribed *Wuchang* 烏長 (Early Mandarin *ʔɔdriaŋ*, E. G. Pulleyblank, *Lexicon of Reconstructed Pronunciation in Early Middle Chinese, Late Middle Chinese, and Early Mandarin* (Vancouver: UBC Press, 1991), 325 and 50) in Chinese, attested as late as the early 8th century pilgrim Hyech'o 慧超, *Echōō-Gotenjiku-koku-den Kenkyū*, ed. S. Kuwayama (Kyoto: Kyōto Daigaku Jinbun Kagaku Kenkyūjo, 1992), 126.

165 Al-Iṣṭakhrī, *Kitāb Masālik al-Mamālik*, ed. M. J. De Goeje (Leiden: Brill, 1870; repr. 1967), 160, 168, etc.

166 *Ḥudūd al-ʿālam* §10:24. Minorsky suggests possibly *Ballahārī*, the Rāshṭrakūṭas, who did not control land in which musk deer lived but may have transshiped it.

The Tughuzghuz/The Nine Oghuz

Al-Yaʿqūbī is thus the first writer known to have mentioned the musk of the Tughuzghuz. The *Ḥudūd al-ʿālam* also mentions musk from the Tughuzghuz. As discussed in Chapter 1, musk deer are found in the Altai mountains and mountains further east through Mongolia and Siberia. During the first half of the 9th century the Mongolian steppe was under the control of the Uyghur steppe empire, that had assumed power in 744. The Arabs refer to the peoples of this state as the Tughuzghuz, the Arabic transcription of Tokuz Oghuz, the Nine Oghuz, who were the leading clans of the empire. While the Nine Oghuz certainly existed before 744, it is much more likely that the arrival of Tughuzghuzī musk in the Islamic world occurred after the creation of the Uyghur Steppe Empire, which had an interest in profiting from trade. Tughuzghuzī musk probably denoted musk obtained either from the Altai, or from Eastern Siberia or Manchuria shipped through the Uyghur state into Khurāsān. The Tughuzghuz, for the author of the *Ḥudūd al-ʿālam,* were nomadic tribes. These were the tribes that made up the Uyghur steppe empire, some of whom had remained nomadic; their list of products, all of which come from the steppe and northern forests, confirms this. The term Tughuzghuzī musk nevertheless seems to have been a generic term, along with Khirkhīzī musk, because it was used beyond the time limits of the Tokuz Oghuz (Uyghur) steppe empire. Tughuzghuz musk does not appear in any new, non-derivative sources after the 10th century, for new confederations of Turkic and Mongolic peoples formed and played their part in the nomenclature of the musk trade.

The Kirghiz

Attestations of Kirghiz musk are as old as the musk of the Nine Oghuz, if not a little earlier, for Ibn Khurradādhbih mentions musk among the Kirghiz.[167] Other geographers, including al-Iṣṭakhrī (fl. mid-10th c.),[168] Ibn al-Faqīh,[169] and the anonymous author of the *Ḥudūd al-ʿālam* mention it; the latter states that much musk originates with the Kirghiz.[170] The Persian historian Gardīzī

167 Ibn Khurradādhbih 31; the passage lists several steppe peoples including one which has been emended by De Goeje to Arabic *khirkhīz,* the usual Arabic form of the name of the Kirghiz. De Goeje considers the mention of musk an error and suggests that it is a proper name; see also his translation, 23. Even if one considered the existence of musk deer in lands controlled by the Kirghiz problematic, there can be no objection to the possibility of their having acted as intermediaries in the trade. Kirghiz musk will be discussed below.

168 Iṣṭakhrī 288.

169 Ibn al-Faqīh, *Kitāb al-Buldān,* ed. Yusūf al-Hāwī (Beirut: ʿĀlam al-Kutub, 1996), 634.

170 *Ḥudūd al-ʿālam* §14.

(fl. 1049–53) also notes that musk, skins, and *khutū* horn (*shākh-i khutū*) are brought from the region of the Kirghiz.[171] Kirghiz musk appears in most of the literature on musk after Ibn Māsawayh; al-Yaqʿūbī, Ibn Kaysān, and al-Bīrūnī all mention it, making it one of the best known varieties after the Tibetan and Chinese. None of these writers, however, held it in high esteem. Ibn Ḥawqal, by contrast, stated that both Tibetan and Kirghiz musk brought via Transoxiana was superior to all other musk and was exported everywhere.[172]

The Kirghiz were even associated with musk in literature; in the work of the 11th-century Persian poet Minuchihrī, Kirghiz musk is brought by the spring season:

> *zadah yāqūt-i rummānī bi-ṣaḥrāhā bi-khirmanhā*
> *fashāndah mushk-i khirkhīzī bi-bustānhā bi-zanbarhā*[173]

> It studded pomegranate-red rubies against the fields in heaps,
> it spread Khirghiz musk by buckets in the gardens.

This image continues the old association of musk with fertile soil found in Arabic poetry and other literature, which will be discussed in Chapter 7.

As early as the 6th century, the Kirghiz lived north and west of the Altai and Tannu-Ola ranges, in the upper reaches of the Yenisei River in the Minusinsk Basin. From this position they harassed the so-called Second Turk Kaghanate (682–741) and later the Uyghur Kaghanate (744–840) to the southeast on the Mongolian steppe.[174] Eventually the Kirghiz overthrew the Uyghur Kaghanate, but they did not seek to dominate the Mongolian plateau, remaining in their previous territory.[175] Living between the forests of Siberia and the steppelands, the Kirghiz were able to export many goods of the forests, such as furs and musk.

---

171 Gardīzī, *Zayn al-Akhbār*, ed. ʿAbd al-Ḥayy Ḥabībī as *Tārīkh-i Gardīzī* (Tehran: Dunyā-yi Kitāb, 1363/1984), 559; A. P. Martinez, "Gardīzī's Two Chapters on the Turks," *Archivum Eurasiae Medii Aevi* 2 (1982): 128.

172 Ibn Ḥawqal 465.

173 Quoted *Farhangnāmah* 258a.

174 P. B. Golden, *An Introduction to the History of the Turkic Peoples: Ethnogenesis and State-Formation in Medieval and Early Modern Eurasia and the Middle East* (Wiesbaden: Harrassowitz, 1992), 176–83.

175 M. Drompp, "Breaking the Orkhon Tradition: Kirghiz Adherence to the Yenisei Region after A.D. 840," *JAOS* 119:3 (1999): 390–403.

Mdosmad

The place name *Dhū sm.t* given by al-Tamīmī is a transcription of the Old Tibetan *Mdosmad*, referring to northeast Tibet or Amdo, which loosely corresponds to Qinghai in the Peoples' Republic of China. *Mdosmad* appears in a number of variant spellings in Arabic and Persian in our texts: *Dhūsm.t* (Dhūsmat), *Ṭūsm.t* (Ṭūsmat), and *Ṭūm.s.t*. *Mdosmad* is given in the *Ḥudūd al-ʿālam* and al-Bīrūnī's *Qānūn* as *Tūsmat*, with *tāʾ* instead of *ṭāʾ*.[176] Ibn Mandawayh's *Ṭūsmatī* musk and Sahlān b. Kaysān's *Ṭūmsatī* (presupposing a *Ṭūmsat* for the toponym) are placed third after Chinese and Tibetan musk. Ibn Kaysān's is another attempt at rendering the Tibetan name *Mdosmad* into Arabic script, with metathesis of the m and s. We may regard al-Azdī's *Ṭūmanī* musk as another corrupt rendering of *Mdosmad* since it corresponds in position within the list of musk types exactly with Ibn Kaysān's *Ṭūmsatī* musk.[177]

Much of Qinghai province is desert and steppe, which is unsuitable for musk deer, but in the mountainous areas around the Koko Nor (Qinghai Lake) there are forest areas in which musk deer appear today.[178] In addition, musk may

176 *Ḥudūd al-ʿālam* §11:9 and Minorsky's trans. 93 and notes on lxxiii and 259–60. For al-Bīrūnī, see Togan, 50 and Minorsky's comments in his ed. of al-Marwazī, 68 and 91–2. On the identification of these places see L. Petech, "Nota su *Mābd* e *Twsmt*," *Rivista degli Studi Orientali* 25 (1950); repr. in his *Selected Papers on Asian History* (Rome: 1988), 46–7; C. I. Beckwith, "The Location and Population of Tibet according to Early Islamic Sources," *AOH* 43 (1989): 168–9; and H. Stang, "Arabic Sources on the Amdo and a Note on Gesar of Gliṅ," *AOH* 44 (1990): 159–61.

177 The lists do not exactly correspond in order, though item 2 in both lists is Tibetan musk and item 4 is Nepalese musk, with our *Ṭūsmatī*/*Ṭūmsatī* and *Ṭūmanī* musk in between. The types listed are also identical except that Azdī's list has Sogdian musk instead of the Tatar musk of Ibn Kaysān. The corruption of *st* to *n* in rapid handwriting is possible. Among sources dependent on the Ibn Mandawayh-Ibn Kaysān tradition, Jurjānī has *Ṭūmsatī*, Ṭūsī 249 has *Ṭūsm.sī*, and Kāshānī 250 has *Ṭūmsī*.

178 In Zhiduo County, Yushu Tibetan Autonomous Prefecture, in the extreme south of Qinghai Province, which contains a section of the Sanjiangyuan National Nature Reserve, musk deer are present, see "Mammals of Zhiduo County," http://www.plateauperspectives.org/mammals.htm. (Accessed June 16, 2012) and *Qinghai: Sourceland of Three Great Rivers* (Beijing: Foreign Languages Press, 2006), 75, and R. B. Harris, "Conservation Prospects for Musk Deer and other Wildlife in Southern Qinghai, China," *Mountain Research and Development* 11:4 (1991): 353–8. In the northeastern area of Qinghai, in the environs of Kumbum Monastery, musk deer were also present; see Thubten Jigme Norbu and H. Harrer, *Tibet is My Country: Autobiography of Thubten Jigme Norbu, Brother of the Dalai Lama* (London, 1960; repr. London: Wisdom Publications, 1986), 43–4, 69. Norbu's remark that the presence of Buddhist monasteries protected the local musk deer from exploitation is confirmed by Harris 356–7.

have come up the trade route from Sichuan as it did in more modern times.[179] Al-Marwazī, as quoted above, mentioned that al-Jayhānī said, "A traveler passing from Shazhou (*Sājū*)[180] to China sees on his right side a mountain which has musk animals on it. . . ."[181] This would refer to the Nanshan along the northern fringe of Qinghai along the border with Gansu. In the more moist parts of these mountains there are apparently still musk deer today.

The Kitan

Khiṭāʾī musk appears fifth in Ibn Mandawayh's list, corresponding to the *Kh.ṭāfī* of Ibn Kaysān, which is easily emended to *Khiṭāʾī*. *Khiṭāʾī* musk is also included in al-Azdī's list.[182] Al-Bīrūnī was also familiar with musk from the *Qutāy* kingdom, *Qutāy* and *Khiṭāʾ* being two ways of writing the name of the Kitan people in Islamicate sources.[183] The Kitan were a people related to the Mongols who founded the Liao Dynasty in 907 on the eastern Steppes and northern frontier of China.[184] The Kitan themselves go back for centuries earlier; they are mentioned in the Old Turkic inscriptions as an enemy of the Second Turk Kaghanate situated towards the east of the Mongolian Plateau, and long before that in Chinese sources. By the time Ibn Mandawayh and Ibn Kaysān wrote, the Kitan Liao Dynasty controlled the northern part of China and southern Mongolia. Musk deer are found in the mountainous parts of the east into Manchuria and the Korean Peninsula. It is possible that Kitan musk refers to musk from these regions, but it is more likely to have been musk exported from the Kitan Liao Empire in general.[185]

The Tatars

Musk from the Tatar people is mentioned by Ibn Kaysān and al-Bīrūnī. *Tatarī* musk must also have come via the overland route into Khurāsān. In the time of the Old Turkic Runic inscriptions the Tatars seem to have lived east and southeast of Lake Baikal. However, Tatar tribal groups were widespread, as

179 W. W. Rockhill, *Land of the Lamas: Notes of a Journey through China, Mongolia, and Tibet* (London, 1891; repr. New Delhi: Asian Educational Services, 1997) 112, speaks of musk among the goods traded by the Sharba merchants plying the Songpan-Koko Nor/Qaidam route.

180 I.e., the Dunhuang area.

181 Jayhānī quoted on *Ṭabāʾiʿ al-Ḥayawān* Folio 75a; Minorsky/Marwazī 91. Quoted in full above.

182 Azdī 36.

183 For different forms, see King, "Early Islamic Sources".

184 See the references in Chapter 2.

185 More considerations about Kitan musk are in King, "Early Islamic Sources".

a group of them formed part of the Kimek, far to the west.[186] Gardīzī states that the Kimek were actually of Tatar origin. This raises the question of whether Ibn Kaysān's Tatar musk arrived from the territory of the Kimek, to the west of the Kirghiz, thus coming from trade with the Kirghiz or from further north and east, or whether the Tatar musk came from further east, at the northeastern edge of the Mongolian Plateau, home to the Tatars familiar from the early history of the Mongol Empire. The Tatar people became one of the leading constituencies of the Mongol Empire, which led many medieval writers to describe the Mongols themselves (incorrectly) as Tatars or "'Tartars". In later Persian and Chagatay literature Tatar musk became rather well known, perhaps because of the flourishing trade of the Mongol period.[187] Saʿdī says:

ʿūd mīsūzand yā gul mīdamad dar bustān
dūstān yā kārwān-i mushk-i tātār āmadast[188]

Is aloeswood kindled or does a rose blossom in the garden;
have friends come or a caravan of Tatar musk?

Miscellaneous Musks of the Sino-Indian Borderlands

So far we have dealt largely with fairly well known toponyms and ethnonyms. The musk literature, however, includes some toponyms that are attested only once in their respective sources and in later copies from those sources. These refer to places in the Sino-Indian borderlands—an area with poorly understood historical geography. We will take these in turn according to the date of the sources in which they appear.

A place called Mūjah is attested as a source of musk in the *Akhbār al-Ṣīn wa-l-Hind*, the most important Arabic work on the trade routes of the 9th century. The *Akhbār al-Ṣīn wa-l-Hind* has only one notice of musk in connection with the country called Mūjah:[189]

186 Gardīzī 55 and Martinez 120. See also Golden 202.

187 See I. V. Zaytsev, "Tatar Musk," in B. Kellner-Heinkele, et al., eds., *Man and Nature in the Altaic World: Proceedings of the 49th Permanent International Altaistic Conferene, Berlin, July 30–August 4, 2006* (Berlin: Klaus Schwarz, 2012), 479–82, for examples.

188 *Farhangnāmah* 258a.

189 Mūjah is also mentioned by Masʿūdī, 1.206 (§434), who also notes that there is much musk there.

> They are a white people who resemble the Chinese in their clothing. They have much musk. In their land there are white mountains; there are none taller than them. They make war with many kings around them. The musk which is in their country is very excellent.[190]

This land has been identified with the Mūsa of other Arabic sources[191] and with the Mosuo 磨些 (Middle Chinese *ma sa*^{h}[192]) of the Chinese sources, who lived in Northwestern Yunnan.[193] In the 10th century Persian geography *Ḥudūd al-ʿālam*, a place called Mūsa is also said to have produced much musk.[194] This identification has the advantage of according with the Chinese statement that the Mosuo produced musk.[195] Sauvaget identifies it with the Mocha 麼察, located in Yunnan and southwest Szechuan; however, this proposed identification is phonetically unlikely to be correct.[196]

No other source besides al-Yaʿqūbī and his descendants mentions the *Q.nbārī*, *Q.sārī*, *ʿ.ṣmārī*, and *Jabalī* musks. The precise form of these words and especially their vocalization is uncertain; *Jabalī* is assumed to mean "mountain" musk but it could represent a completely different name with different vocalization. *Q.nbār* is said to be between China and Tibet, and *Q.sār* between India and China. There is no information given on the location of *ʿ.ṣmār*.[197] None of these toponyms are found in any other tradition, so whatever they represented, they are probably corrupt now.

In a general sense, it is likely that these toponyms represent places in the Sino-Indian borderlands and the mountainous parts of northern Myanmar, Thailand, Laos, and Vietnam, areas with musk deer. Musk trade is attested in Burma in the 16th century. Tomé Pires says that the people living in the mountains of Capelanguam on the border of Arakan (modern Rakhine, Myanmar)

---

190 Sauvaget, *Akhbār aṣ-Ṣīn wa l-Hind*, §31, Sīrāfī, *Riḥlah*, 37.

191 Marwazī, 12.56: text 37 and trans. 50. Marwazī also notes the presence of musk there. Al-Yaʿqūbī, *Tārīkh*, ed. M. T. Houtsma. 2 vols. (Leiden: Brill, 1883; repr. 1969), 1.106 gives *mūshah*.

192 Pulleyblank, 217, 298.

193 Cf. Minorsky, *Marwazī*, 149–50.

194 *Ḥudūd al-ʿālam* §10:9 and Minorsky, *Regions* 87.

195 Cf. W. Eberhard, *Kultur und Siedlung der Randvölker Chinas*, Supplement to *T'oung Pao* vol. 36 (Leiden: Brill, 1946), 110. See also B. Laufer, *Sino-Iranica: Chinese Contributions to the History of Civilization in Ancient Iran, with Special Reference to the History of Cultivated Plants*, Field Museum of Natural History Anthropological Series 15:3 (Chicago: Field Museum, 1919; repr. New York: Kraus, 1967), 469, on musk from the Nanzhao of Yunnan.

196 Sauvaget 54–5; Christopher Beckwith, personal communication.

197 Newid 68 n. 600 proposes that it is a scribal error for *kashmīrī*.

brought musk, along with rubies, to trade in Ava, the main city of Arakan.[198] Musk trade is also attested in the port of Pegu (modern Bago, Myanmar).[199]

Central Eurasia and the Musk Trade

For al-Bīrūnī, as we have seen, there were three basic categories of musk: Turk, Indian, and Chinese. This division reflects both the cultural orientations in a broad sense of the regions that produced or exported the musk along with the direction from which it came. *Ṣīn*, China, and *Hind*, (greater) India, were two categories of Asian places renowned for their products, as we have seen in Chapter 2. They are grouped together in consideration of the Further Asia trade, for one sailed to China by way of India. Chinese (*ṣīnī*) musk was especially that which arrived by sea, although obviously some would have come by the overland trade routes as well. Indian musk was traded by overland routes from the Himalayas. Some of this musk must have continued along the caravan routes across the Hindu Kush and entered Khurāsān. However, Indian musk most often meant musk which traveled by sea from the ports of India, especially from Daybul in Sind.

But pride of place in quality is consistently given to the musks of Central Eurasia itself, al-Bīrūnī's "Turk" musks, including especially the Tibetan musk. The "Turk" musks arrived from the overland caravan routes. All originated from Central Eurasian peoples, not necessarily of specifically Turkic ethnolinguistic background. As we have seen, many Central Eurasian peoples lent their names to types of musk: the Turkic Tokuz Oghuz (an obsolete designation by al-Bīrūnī's time) and Kirghiz, the Kitans, the Tatars, and the Tibetans. Tibet was understood by the medieval ethnographic tradition to be a land with geographic contiguity with and cultural practices similar to the Turkic peoples of Central Eurasia. Early medieval Tibetan culture belonged to the Central Eurasian Culture Complex, the repertory of cultural commonalities shared among the peoples of Central Eurasia.[200]

---

198 Tomé Pires, *The Summa Oriental of Tomé Pires and the Book of Francisco Rodrigues*, trans. A. Cortesão (London: Hakluyt Society, 1944), 1.96; see also Duarte Barbosa, *The Book of Duarte Barbosa*, trans. M. Longworth Dames (London: Hakluyt Society, 1918–21), 2.159, 160.

199 Pires 1.98.

200 C. I. Beckwith,"The Central Eurasian Culture Complex in the Tibetan Empire," *1000 Jahre Asiatisch-Europäische Begegnung*, ed. R. Erken (Berlin: Peter Lang, 2011), and *Empires of the Silk Road* (Princeton: Princeton University Press, 2009), 127–31, and "A Note on The Heavenly Kings of Ancient and Medieval Central Eurasia," *AEMA* 17 (2010): 7–10.

Other Central Eurasian lands exported musk, although they did not gain the fame and recognition of their own "brand name" of musk. The *Ḥudūd al-ʿālam* also mentions musk as an export from the land of the Tukhs: "from that place comes much musk and various kinds of furs."[201] The Tukhs were a remnant of the Türgish and were one of the tribes which came to make up the state of the Karakhanids. This land was situated in the mountains north of the Chu River in what is now southern Kazakhstan near its border with Kyrgyzstan. Maḥmūd al-Kāshgharī, in his *Dīwān Lughāt al-Turk*, uses the name Qayās for the country of the Tukhs and the Chigil.[202] Geographically this is the extreme southwestern corner of Central Asia in which musk deer can be securely located by historical sources identified so far. No source yet found places musk deer in the Tianshan, and it seems likely that if musk deer ever lived in these mountains, they were not common in medieval times.

The Persian *Zayn al-Akhbār* of Abū Saʿīd ʿAbd al-Ḥayy b. al-Ḍaḥḥāk b. Maḥmūd Gardīzī was dedicated to the Ghaznavid Sultan ʿAbd al-Rashīd b. Maḥmūd (r. 1049–53). This work is also thought to depend largely on Jayhānī for its material on India and Central and East Asia.[203] He describes the Mānabag-lū mountains as a high mountain range in which there are many ermines, squirrels, and musk deer, as well as many trees and abundant game.[204] Gardīzī states that this range is located four days' south of the Kögmän mountains, which are known also from the Old Turkic Runic Inscriptions and are identified with the Tannu Ola range; from this Minorsky concluded that the Mānabag-lū must be the Altai mountains; the descriptions of fauna like ermine and squirrel, apart from the musk deer, accord with this identification.[205]

A general preference is evident in both Arabic and Persian literature for musk brought by the overland trade from Central Eurasia. In the early period this meant especially musk supplied by the Sogdians through Transoxiana that originated in Tibet.[206] The users of musk preferred the quality of Central Eurasian musk both for the alleged superior diet of the deer as well as for the conditions of its transport, far removed from the damaging vapors of the ocean.

---

201 *Ḥudūd al-ʿālam* §17.

202 Maḥmūd al-Kāshgarī, *Maḥmūd al-Kāšγarī: Compendium of the Turkic Dialects*, ed. and trans. R. Dankoff and J. Kelly, 3 vols. (Cambridge: Harvard University Printing Office, 1982–5), 2.238 (=520): "the country of Tukhsi and Čigil."

203 V. M. Minorsky, "Gardizī on India," *BSOAS* 12 (1948): 626.

204 Gardīzī 557 and Martinez 126.

205 Minorsky, *Regions* 196.

206 E.g., Ibn Māsawayh, and cf. Ibn Ḥawqal 465, who specifies both Tibetan and Kirghiz musk brought through Sogdiana.

The importance of Tatar and Kirghiz musk in later Persian poetry is exceeded by the musk of the cities of Kashgar and especially of Khotan.[207] While musk deer are not found in Kashgar or Khotan, these important trade cities came to be associated with the source of musk, for it was there that the caravans heading across the mountains into Khurāsān bringing musk must have originated.[208] Musk destined for the Muslim world from China and the Altai as well as Tibet was obviously gathered in Kāshgar and Khotan and then transshipped. Khotan became proverbial in Persian poetry for its abundance of musk, and there is even a Persian saying "to carry musk to Khotan," equivalent to taking coals to Newcastle.[209] Saʿdī complains of his neglect:

hamānā ki dar fārs inshā-yi man chu mushkast bī qīmat andar khutan[210]

Musk is valueless in Khotan, just as my composition is in Fārs.

The frequency of Central Eurasian toponyms associated with musk in Persian poetry was also probably influenced by the context of many of the poems, for the beloved of a poet like Ḥāfiz was a young Turk with a newly sprouting beard. This is called the *khaṭṭ*, and it is often described as musky. This tradition goes back centuries before Ḥāfiz. Daqīqī (d. 975) says:

dānī ki dil-i man ki fikandah-ast bi-tārāj
ān du khaṭṭ-i mushkīn ki padīd āmadash az ʿāj[211]

Do you know who has plundered my heart?
Those two musky lines which appeared to it from ivory.

---

207 Examples can be found in *Farhangnāmah* 258a. A detailed discussion of musk of Khotan is in Naṣīr al-Dīn Muḥammad b. Muḥammad al-Ṭūsī, *Tansūkhnāmah-i Īlkhānī*, ed. Mudarris Raḍawī (Tehran: Intishārāt-i Bunyād-i Farhang-i Īrān, 1969), 248. Musk of Khotan is also mentioned as superior to Tibetan musk, but less common, by the 14th century Persian writer Ḥamdullāh al-Mustawfī al-Qazwīnī's *Nuzhat al-Qulūb*, ed. and trans. J. Stephenson, *The Zoological Section of the Nuzhatu-l-Qulūb of Ḥamdullāh al-Mustaufī al-Qazwīnī* (London: Royal Asiatic Society, 1928), text 29 and trans. 20.

208 This was true also of Yarkand in the early 17th century, see W. Finch in S. Purchas, *Hakluytus Posthumus, or Purchas His Pilgrimes* (Glasgow: James MacLehose and Sons, 1905), 4.59.

209 S. Anṣārī, *Tārīkh-i ʿIṭr dar Īrān* (Tehran: Wizārat-i Farhang wa Irshād-i Islāmī, 1381/2002–3), 229.

210 *Farhangnāmah* 257a.

211 *Farhangnāmah* 257a.

Another feature of the beloved was the mole, *khāl*, which is often described as musk as well.[212] This tradition also goes back a long time in Persian poetry. The 10th-century poet Rūdakī says:

wa ān zanakhdān bi-sīb mānad rāst
agar az mushk khāl dārad sīb[213]

And that chin truly resembles an apple if an apple has a mole of musk.

The toponyms associated with musk in Persian poetry are, more often than not, taken from the same repertory of cities in Central Eurasia associated with the origins of the Turks who played such a large role in Middle Eastern history.

The musk trade cannot be separated from the commercial system of Central Eurasia. Islamicate writers knew the peoples of Central Eurasia especially because of the impact of Turkic peoples in the Islamic world, first in the military, later as rulers. But the musk trade shows that important commercial relations existed as well—commercial relations so significant that the geographers made sure to mention musk in their descriptions of these lands and so persistent that the names of Central Eurasian peoples became "brand-names" for musk in the apothecaries of far away Iraq and Egypt. The musk trade is one of the most visible aspects of early medieval Islamic relations with Central Eurasia.

The Islamicate Understanding of the Production of Musk

Several of our authors discuss how the musk deer produces musk. The biology of the musk deer was of great interest to Islamicate writers because most perfumes came from plants rather than animals. The idea that an animal could produce one of the most precious of them was strange and fascinating. Even more curious is the fact that ambergris, the second of the principal aromatics, is also of animal origin, although that was less clear to many Islamicate writers.

First, Islamicate authors had to explain the source of the fragrance itself. Most, but not all, writers this traced to the supposed consumption of aromatic plants by the musk deer. Second, just what exactly was musk? Following in the footsteps of pre-Islamic tradition, it was deemed to be blood, but that raised a very important problem since consumption of blood is illicit in Islamic law.

212 Newid 76–7 with more references.

213 *Farhangnāmah* 259a.

Related to the bloody origin of musk is the question of why and how the musk deer produced musk. We will also consider the tales of hunting the musk deer and other peculiarities of the Islamicate accounts of musk. Many medieval ideas about musk will be shown to be erroneous, yet some reflect good information about the hunting of musk deer and the properties of musk.

A Diet for Fragrance

The notion that it is the plants consumed by the animal that produced the aroma of musk is common in Arabic and Persian literature. This trend began with Ibn Māsawayh, who claimed that musk could be classed according to quality on the basis of the fodder of the animal: *kandasah* producing the best musk, spikenard (*sunbul*) the next best, and *murr* the poorest. He also notes that the plants that are grazed on by the musk deer grow in Kashmir or Tibet.

Unfortunately, the plant *kandasah* is difficult to identify; the proper vocalization is not known either. It is also mentioned by al-Yaʿqūbī, quoted by al-Nuwayrī, who also gives the form *k.d.h.m.s* apparently from al-Tamīmī; De Goeje's glossary merely explains the word as a plant that grows in Tibet.[214] *K.n.d.s* or *q.n.d.s* can mean the white hellebore, and this must be why Levey, who gives no explanation for his translation of this rare word, has translated it as hellebore.[215] Another more usual meaning for Arabic *kundus* is the soapwort, Greek στρούθιον,[216] *Saponaria officinalis* and *Gypsophila struthium*.[217] A further level of complexity is added to this problem by the Persian word *gund*, "testicle," as in *gundbīdastar*, "castoreum." This word was borrowed by Arabic in the spelling *qundus*[218] and came to be associated with *kundus*. Ibn Māsawayh explicitly states that *kandasah* is a plant, but this may be an error, and the phrase may reflect an older belief in the association of musk, testicles, and castoreum that had become obscured by his time.

Spikenard has the longest association with musk deer fodder; having started with Ibn Māsawayh, this concept became commonplace in later accounts. That musk deer grazed upon spikenard explained its scent; the various plants called *sunbul* or *sunbul al-ṭīb* "spikenard," such as *Nardostachys jatamansi*, have an animal, musky quality to the scent prepared from their roots. Al-Sīrāfī and al-Masʿūdī noted that the musk-producing land is a single region;

214 M. J. De Goeje, *Bibliotheca Geographorum Arabicorum*, vol. 8, p. xxxvi.

215 *Veratrum album*, see I. Löw, *Die Flora der Juden*, 4 vols., (Vienna, 1928–34; repr. Hildesheim: Olms, 1967), 2.160; 1.641.

216 Dioscorides II.163, Dietrich DT II.147, and Schmucker 363.

217 Cf. Löw 1.335.

218 Ibn al-Bayṭār 4.40 and cf. 86–7.

the differences in the quality of musk stem from either the processing and transportation of the musk, or the plant on which the animals graze. These writers emphasized that the best musk came from animals that grazed upon spikenard and fragrant herbs; this combination was found in Tibet. The Chinese musk deer had to make do with grass only, with no spikenard. Ibn Māsawayh had considered musk produced from animals that grazed on spikenard inferior to that originating from *kandasah*; the intervening accounts make no mention of spikenard, but in the early 10th century from al-Sīrāfī on it becomes ubiquitous in the literature. The grazing of musk deer on spikenard was sufficiently well known for the Persian poet Khāqānī (1106 or 7–1185) to mention it in a verse with word play on the two meanings of Persian *chīn*:

> *chūn mushk-i chīn tu dārī az āhū-yi chīn mapurs*
> *āhū bi-chīn bih ast ki sunbul charā kunad*[219]

> Since you have "musk of China" (i.e. plaited hair), do not ask more from the deer of China. The deer of China is better off grazing on spikenard.

The association of *murr* with musk is first attested by Ibn Māsawayh. His statement that inferior musk is derived from a plant whose origin or root is *murr* raises some questions. Myrrh (*Commiphora* sp.) is not native to any of the musk-producing lands, although it is not known if Ibn Māsawayh knew this. More likely the bitter, dark color of musk was associated with the bitter and dark, strong-smelling myrrh. Medieval Hebrew adopted the word *mūr* for musk.[220]

Ibn Sīnā's account reflects Ibn Māsawayh's tripartite division of musk fodder. Ibn Māsawayh's *kandasah* is replaced by *al-bahmanayn*, the two aconites, which were the red and black aconites, but the remaining two follow Ibn Māsawayh: fodder of spikenard, and then the *murr*. Ibn Sīnā's mention of the aconite in connection with the musk deer evokes the tradition of the *Bundahišn* mentioned in Chapter 1, in which a type of musk animal that eats the *bīš* or aconite is noted.[221] The idea of a *bīš*-eating musk animal is also known from the later Islamic sources Ḥamdullāh al-Mustawfī Qazwīnī in Persian and

---

219 *Farhangnāmah* 258a.

220 Löw 1.310; J. Leibowitz and S. Marcus, *Moses Maimonides on the Causes of Symptoms* (Berkeley: University of California Press, 1974), 227.

221 Kāshānī 251, probably influenced by Ibn Sīnā, mentions the possibility that the musk deer eats spikenard and the two aconites, but no *murr*. Qazwīnī, *ʿAjāʾib* 386 mentions spikenard, the two aconites, and fragrant herbs.

al-Damīrī in Arabic, as cited above. The association of musk with aconite must stem from the musky scent of the aconite root.

Notably, Ibn Mandawayh and Ibn Kaysān do not believe that the diet of the animal makes any difference in the ability of the animal to produce musk. Ibn Kaysān writes: "there are gazelles that graze upon wheat, barley, and grasses, and musk comes from them, too".[222] This passage appears (minus the grasses) in the marginalia to al-Bīrūnī's *Kitāb al-Ṣaydanah*, but he does not comment on the matter himself. The 19th century traveler Frederick Markham also expressed his doubts that the fodder of the musk deer made any difference in its scent.[223]

The belief that the fodder of the musk deer was responsible for the scent of musk persisted, however.[224] Even in the 19th century the travelers Huc and Gabet believed that the musk deer that grazed on "pines, cedars, hollies, and cypresses" were attracted to them by their "strong aromatic perfume".[225] That the Chinese believed the fodder of musk deer contributed to the scent of musk was noted in Chapter 3.

So, why should this idea develop and continue so tenaciously in the literature about the musk deer? The repertory of aromatics employed in ancient times in the Near East and Mediterranean for perfuming was almost all vegetal in origin. The important animal-based perfumes of medieval and early modern times—ambergris, musk, and civet—were either unknown or scarcely known. (And even in Islamic times the very origin of ambergris was a controversial question.) Castoreum was used by the Greeks more as a drug than a perfume and does not appear to have been used in the context of an aromatic preparation at all.[226] Only onycha, the operculum of certain mollusks that could be added to incense, was known, and it had a problematic reception in the Middle East, although it was certainly used in Greek and Roman perfumery.[227] When

---

222 Ibn Kaysān 190.

223 Markham 89.

224 E.g., Juan Gonzalez de Mendoza, *The History of the Great and Mighty Kingdom of China*, trans. R. Parke, ed. G. T. Staunton (London: Hakluyt Society, 1853), 1.16, who attributes the scent to the root called *camarus*.

225 Evariste-Régis Huc and Joseph Gabet, *Travels in Tartary, Thibet and China, 1844–1846*, trans. W. Hazlitt and intr. P. Pelliot, 2 vols. (London: Routledge, 1928; repr. 2 vols. in 1, New York: Dover, 1987), 371–2.

226 See Pliny VIII.47, and Dioscorides II.24 on the uses of castoreum. Cf also S. Lilja, *The Treatment of Odours in the Poetry of Antiquity*, Commentationes Humanarum Litterarum 49 (Helsinki: Societas Scientiarum Fennica, 1972), 164–5.

227 There has been controversy as to whether true onycha was one of the ingredients used in incense in the Jewish tradition, mollusks not being kosher. The Hebrew Bible's incense

musk first came into use, it was regarded as different from plant-based aromatics. As noted in Chapter 3, the Talmud carefully distinguished musk from aromatics of vegetal origin in ritual practice. Thus there was a typological predisposition to look for the causes of good scents in the world of plants rather than the world of animals.

The Bloody Origins of Musk

From the accounts of Ibn Māsawayh and al-Jāḥiẓ it is evident that the production of musk was not well understood in the medieval Middle East. In Ibn Māsawayh we find the notion that the plants upon which the animal grazes help to produce the musk. Ibn Māsawayh does not say that musk is blood, or even related to blood; he is silent on the subject, but al-Jāḥiẓ reports the notion that musk is produced from blood, which would henceforth dominate medieval Arabic and Persian literature on the musk deer. The belief that musk derived from blood was certainly pre-Islamic since Cosmas Indicopleustes mentions it. The identification of musk with blood probably rests in the need to explain what musk was and the fact that musk does have a resemblance to dried blood.[228] In addition, blood was hot in the humoral system just like musk.

The Qur'ān contains explicit prohibition of the consumption of blood in 2:173: "He has made forbidden (*ḥarrama*) to you carrion (*al-maytah*) and blood, and the meat of the pig, and that upon which any other than God has been invoked."[229] Another significant passage is 6:145, which states that no meat is prohibited "except if it be carrion, or blood poured forth, or the meat of a pig..." The "blood poured forth" (*damman masfūḥan*) illustrates that

---

formula for temple worship mentions *sh^eḥelet* (Exodus 30:34), which is translated in the Greek *Septuagint* as *onyx* (Latin *onycha*); the latter most certainly means onycha and is given in that meaning in Brown-Driver-Briggs' Hebrew Lexicon; L. Koehler, *The Hebrew and Aramaic Lexicon of the Old Testament*, rev. ed. by Walter Baumgartner, 5 vols. (Leiden: Brill, 1994–2000), 4.1462, notes the literature and conclude that the meaning is uncertain. See also K. Nielsen, *Incense in Ancient Israel* (Leiden: Brill, 1986), 65–6 for earlier literature (he prefers to leave the question of *sh^eḥelet*'s identity open) and H. J. Abrahams, "Onycha, Ingredient of the Ancient Jewish Incense: An Attempt at Identification," *Economic Botany* 33 (1979): 233–6, who favors labdanum. More references for the history of onycha are in Chapter 3.

228 E.g., the 17th century Jean-Baptiste Tavernier, *Travels in India by Jean-Baptiste Tavernier*, trans. V. Ball and ed. W. Crooke, 2 vols. (London, 1889, repr. New Delhi: Munshiram Manoharlal, 1995), 2.113: "The musk is then [after the removal of the pod from the animal] extracted from the bladder which contains it—it is then like coagulated blood".

229 See also Qur'ān 5:3, with more detail on prohibited slaughtering methods.

any blood removed from an animal not adhering within its meat was illicit.[230] Thus, with no familiarity with Islamic tradition one might assume that musk would be considered illicit because of its origin. The opposite is the case, of course, and the licitness of musk in Sunni Islam is due primarily to the high regard Muḥammad held it in, as we will discuss in Chapter 7.[231] Since the substance was approved by Muḥammad, a theoretical justification reconciling the bloody origin of musk and its licit status had to be reached. Al-Jāḥiẓ's discussion with the Muʿtazilite perfumer continued on to this very question:

> He said, "If the Messenger of God, Peace be upon him, had not perfumed with musk, I would not perfume myself with it. And as for civet,[232] none of it comes near my clothing." I said to him, "How is it that a kid can suck the milk of a sow and its meat is not prohibited? [He said,] "It is because that milk has been changed into meat, and it has left that nature, and that form, and that name, likewise the meat of the impure cow (*jallālah*). Musk is not blood, and vinegar is not wine, and the essence is not prohibited in itself. Rather it is prohibited for its nonessential characteristics and particular reasons. So do not shrink from it when we mention to you that it is retained blood, because it is not. Fire changes into air, and air into water, and the resemblance which is between water and fire is very remote indeed!"[233]

---

230 See al-Ṭabarī, *Jāmiʿ al-bayān ʿan taʾwīl āy al-Qurʾān*, 30 vols. (Cairo: Muṣṭafā al-Bābī al-Ḥalabī, 1954), 8.70–1. Perhaps this concept partially explains the regular emphasis given by writers on musk to the idea that this blood was collected and retained by the musk deer within its vesicle. Cf. Nuwayrī 12.10–11, Ibn Kaysān 190–1, etc.

231 Cf. Damīrī 2.126, where the licitness of musk is grounded firmly in Muḥammad's tradition. He also mentions that some Shiʿites disapproved of musk. This disapproval does not seem to have had much of an effect on the prestige or consumption of musk; musk remained a potent symbol in Shiʿite traditions and in the Twelver Shiʿite Ṣafavid state. Jean Chardin, *Travels in Persia 1673–1677* (London: Argonaut, 1927; repr. New York: Dover, 1988), 152, describes "a great consumption of [musk]" brought by Armenians, Persians, and Pathans". The famed Twelver Ṣafavid scholar al-Majlisī's *Biḥār al-anwār* (Beirut: Muʾassasat al-wafāʾ, 1983), 73.142–3, contains the usual praises of musk in its section on perfumes.

232 This is perhaps the earliest Arabic reference to civet, and it shows that it was known but not as favored as musk. It is not clear when the Arabs became acquainted with civet, but it could be that the use of musk furthered the discovery of substances like civet, which comes from the various species of civet "cats", some of which live in the Near East and East Africa.

233 Jāḥiẓ, *Ḥayawān* 5.304–5. Quoted by Ibn Abī al-Ḥadīd 19.345–6.

This is a very interesting passage because it gives a thoroughgoing philosophical justification for the use of musk: the transformation of blood into musk, something not blood, means that it is licit, just like the meat of an impure cow that suckled from a sow. When the blood in the musk pod dries out and becomes musk, it ceases to be blood. Ibn ʿAbd Rabbih (860–940) sums up the process in his discussion of wine in *al-ʿIqd al-Farīd*: "The parallel to wine in that it is lawful or prohibited according to its nonessential characteristic is musk, which when it is fresh (*ʿabīṭ*) blood is prohibited, and then when it dries its scent comes into being, and it becomes lawful and good."[234] One later source suggests that this process occurs after the navel has been separated from the animal,[235] but usually the maturation of the musk is said to occur before it is separated from the animal, as in al-Sīrāfī and al-Masʿūdī. This is an important point, because if it is blood when it is separated from the animal, it runs a greater risk of being illicit than if it is already musk when removed from the animal according to the Qurʾānic prohibition of blood.

The concept of the transformation of blood into musk became key; as we will see, its lawfulness gave hope to the Sūfīs, who longed for release from the bondage of their own corporeal existence through the miraculous transformatory powers of God. We will return to this subject in Chapter 7. The concept of the bloody origin of musk even appears in poetry. The poet al-Mutanabbī (915–965) says in a panegyric:

faʾin tafuqi l-anāma wa-anta minhum
fa-inna l-miska baʿḍu dami l-ghazāli[236]

If you surpass the human race, and yet you are of them,
well surely musk is part of the blood of the gazelle.

Here the poet marvels at the greatness of the patron, who seems to transcend humankind, but then reminds himself that musk, too, comes from a humble source.

The next problem is, why did blood collect in the musk vesicle to begin with? The view of Cosmas Indicopleustes, our most detailed pre-Islamic source from the west on musk, emphasizes human agency: humans bind the

234 Ibn ʿAbd Rabbih, *al-ʿIqd al-Farīd*, 31 vols. (Beirut: Maktabat Ṣādir, 1951–4), 30.93.

235 Marwazī Book 8 sec. 43, text 17.

236 Abū al-Ṭayyib Aḥmad b. al-Ḥusayn al-Mutanabbī, *Dīwān* (Beirut: Dār Ṣādir, 1958), 268; al-Sarī al-Raffāʾ 3.159 (#292) has *nafuqi* instead of *tafuqi*, but the latter seems to make better sense.

navel, and blood collects there. Al-Jāḥiẓ's account of the hunting of the musk deer states that the hunter binds the navel to collect the blood there, following exactly the procedure described by Cosmas Indicopleustes. No indication is given that the concentration of blood in the navel is a natural phenomenon in this passage or its antecedent. Ibn Mandawayh and Ibn Kaysān also emphasize that the hunter of the musk deer plays a key role in the formation of the musk pod. Ibn Kaysān writes: "... A man hunts the gazelle, slaughters it, and continually rubs with his hands all of its parts in order that more of what blood is in its veins will descend to its navel. When he knows that the navel has become filled with blood, he cuts out the place and hangs it ..." Yet, Islamic sources in general, as we will see, take pains to explain that the collection of blood in the navel is indeed a natural process.

We can first see the new approach to this problem in al-Yaʿqūbī: "[The musk deer] produces [musk] every year. It is the bloody excess (*faḍlun damawiyyun*) which collects from its body in its navel every year at a certain time in the way that matters (*mawādd*) flow to the organs. When swelling and enlargement arise in its navel, it makes it sick and causes pain until it matures. When it ripens and is finished, it scratches it with its hooves and it falls off in those deserts and steppes." For al-Yaʿqūbī, musk production is a natural fact of the musk deer's biology. Indeed, he states: "People say that this animal was created by God the Exalted as a source of musk"; thus the creation of musk within the musk deer is due to God's planning. The concept of how the musk pod is produced has changed from the earlier accounts that emphasize that humans had to bind the navel to cause the blood to collect in it and become musk. From this point in time, it became a common opinion that the production of musk was a natural process that does not require human intervention; it is later likened to the bearing of fruit on trees. There is also a role for human agency in the account of al-Yaʿqūbī. He notes that all people had to do was set up "scratching posts" for the animals, and they would be able to acquire the musk at certain times of the year. This tale, which appears in several other sources influenced by al-Yaʿqūbī discussed above, belongs to the category of the marvels.

In al-Jayhānī's account we likewise find the origin of musk in blood, a concept going far back, coupled with the idea that the production of musk was a natural and periodic process for the musk deer. The suffering imposed on the musk deer by this process has been greatly emphasized by al-Jayhānī, and while other accounts also emphasize the pain associated with the production of musk, none do it with the graphic intensity of the account quoted by al-Marwazī: "The swelling and the pain in its head and all of its body intensifies. So it comes to places in those deserts in which are its accustomed wallows, abstaining from grazing and water until its swollen navel falls off from the abundance of blood, and sometimes its horns fall off also. Some of them die there and some of them recover." Al-Tamīmī's account also emphasizes the suffering of the musk deer and the natural process of musk production as part of its biology.

Abū Zayd al-Sīrāfī (paralleled by his friend al-Masʿūdī) adds a new twist: the idea that the pods themselves do not fall off, but that the musk is excreted from them. "The best musk of all is that which the gazelle scrapes on the rocks of the mountains; it is matter which forms in its navel as fresh blood collects in the manner in which blood collects as it appears from boils. When it ripens, he rubs it and it torments him so he flees to the stones until he tears a hole in it, and what is inside it flows out. When it comes out from it, it dries and heals over and the substance returns and collects in it as before." Al-Sīrāfī then explains how the musk hunters locate these scrapings of musk and collect it "and put it inside the pods".

The idea that the blood collects or is forced to collect in the navel of the musk deer as part of its hunting recurs frequently in Islamicate literature. The sources emphasize that it is retained blood; indeed, a folk etymology of Arabic *misk* "musk" ties it to the verb *masaka*, which means to hold or cling.[237] Likewise, al-Jāḥiẓ's informant calls musk *al-dam al-ḥaqīn* "retained blood". I suggest that this is partly to reinforce the idea that musk is not merely spilled blood (and thus *ḥarām*). It also reflects the practice of the Tamang people in Nepal who hunted the musk deer, for they had the idea that spilling the blood of the musk deer was bad luck for the hunt, so they carefully suffocated the animals instead before removing the pods.[238]

Musk must also be aged, either naturally or through human intervention. The notion that musk must be aged, separated from the animal, so that it transforms into musk was found in al-Jāḥiẓ, but al-Jayhānī's account makes that aging a natural process as well. Al-Sīrāfī writes: "The pod might be cut off from the gazelle before the musk has matured inside it. Although when it is cut from its gazelle it has a bad odor for a time until it has dried over long days, as it dries it changes until it becomes musk." He likens musk that has matured properly in the pod to fruit ripened on the tree; that which has ripened later is never quite as good. The Ibn Mandawayh-Ibn Kaysān tradition states that the blood-filled navel must be aged for a year before the blood transforms into musk.[239]

---

237 Al-Sarī al-Raffāʾ 3.139.

238 C. Jest, "Valeurs d'Échange en Himalaya et au Tibet: Le amber et le musc," in *De la voûte céleste au terroir, du jardin au foyer: mosaïque sociographique: textes offerts à Lucien Bernot* (Paris: Éditions de l'École des hautes études en sciences sociales, 1987), 231.

239 Cf. Kāshānī 251.

One of the most striking statements made about the production of musk by al-Tamīmī is that musk was suspended in a privy for forty days, then removed and hung in another place to dry. Far from being a mere marvel, this statement very likely reflects actual practice. The complex scent of musk has urinous elements; the statement that musk was suspended in a privy could be an attempt to explain why.[240] But the urine and the urinous scent of the privy is said to have an enhancing effect on musk. Jan Huyghen Linschoten, writing in the late 16th century, states that when musk was losing its smell, the Chinese would beat it in a mortar with the urine of a child and then seal it in a pot in order to remedy its scent.[241] Robert Boyle, in his discussion of the properties of urine, written in 1671, noted that in China musk was treated in this way:

> [T]he musk being made up, and put into Cods or Bags made of the skin of the same Animal, (in which form I have received Presents of Musk sent me from the *Indies*) they do either before or after hang it in a house of Office, so as it may, without touching the grosser Bodies, receive the fetid Exhalations of that Nasty place; by which Urinous Steams, which tis expos'd to for some dayes, the lesse Active of more immers'd Scent is, as it were, call'd out, and excited or heightened.[242]

Boyle also notes that this procedure was followed by those wishing to restore the quality of musk after the sea voyage. It appears that al-Tamīmī's statement also attests this practice of enhancing the quality of musk.

Hunting the Musk Deer

The methods used for hunting the musk deer as described in Islamicate sources are well paralleled in more recent times. These writers had access to decent information on the production of musk.

The hunter might look for the musk deer in specific locations they prefer. Al-Sīrāfī and al-Mas'ūdī mention the gathering of musk that has been naturally

---

240 Trained perfumers sometimes find it hard to distinguish a certain smell as musky or urinous, see D. M. Stoddart, *The Scented Ape: The biology and culture of human odour* (Cambridge: Cambridge University Press, 1990; repr. with corrections, 1991), 64–5.

241 Jan Huyghen Linschoten, *The Voyage to the East Indes*, ed. A. C. Burnell and P. A Tiele, 2 vols. (London: Hakluyt Society, 1885), 2.95.

242 R. Boyle, "The Usefulness of Natural Philosophy II. Sect. 2," in *The Works of Robert Boyle*. Vol. 6. Ed. M. Hunter and E. B. Davis (London: Pickering and Chatto, 1999), 523–4. Boyle's source was apparently Father Gregorius de Bolivar, whose information was transmitted by Johannes Faber, cf. 523 n. c.

shed by the animal. Al-Jayhānī mentions that musk deer have specific "wallows" where they rub the painfully swollen musk glands off. The idea that there were specific wallows where the musk pods were shed was also recorded by Tao Hongjing, as quoted in Chapter 3. Modern hunters do not hunt for wallows of pods, for the animals do not shed the pod, but for droppings in accustomed places. These favored places are excellent locations to set snares. Hunters may also seek the remains of musk deer killed by predators, who do not eat the musk.[243] Gathering musk from these animals may account in part for the idea that the animals shed the musk pod.[244]

The method of obtaining musk emphasized by al-Sīrāfī, al-Masʿūdī, and al-Jayhānī is to gather musk shed by the animal, either in the pod or as a loose substance. Al-Sīrāfī states that failing this, one could hunt for the animal "with set-up nets and arrows". Al-Jāḥiẓ and his predecessor Cosmas Indicopleustes mentioned the use of snares, which are also mentioned by al-Masʿūdī. As we have seen in the first chapter, these techniques were employed in recent times to hunt for the musk deer as well.

Conclusion

Despite initial doubts, by the 9th century it was mostly clear that musk was derived from an animal like a gazelle, the same idea as our musk deer, rather than from a mouse. In the *Qiṣaṣ al-Anbiyā'* of al-Thaʿlabī (d. 1035), Muḥammad himself, when questioned about musk, is made to relate that it is produced from an animal that resembles a gazelle, which feeds on the fragrant vegetation of India, brought from Eden by Adam, and then God produces musk from that vegetation within its navel.[245] This view is anachronistic for the early 7th century, but it fits the 10th and 11th centuries perfectly. The expansion of geographical knowledge meant that the musk deer could be situated in specific regions of Asia such as Tibet and China rather than the vague India of the early poets, who were more likely to link musk with the port of Dārīn, where it was imported. The concept of the "musk mouse" survived due to the influence of classical poetry and to the real existence of the musk shrew, but the learned

243 Markham 96–7.

244 Markham notes that he once discovered the fleshless skin and skeleton of a musk deer that must have been dead "some months", "but the musk pod was entire".

245 Al-Thaʿlabī, *'Arā'is al-majālis fī qiṣaṣ al-anbiyā'*, published as *Kitāb Qiṣaṣ al-anbiyā'* (Cairo: Al-Maṭbaʿat al-Ummah, 1331/1912), 23 and trans. W. Brinner, *'Arā'is al-Majālis fī Qiṣaṣ al-Anbiyā' or 'Lives of the Prophets'* (Leiden: Brill, 2002), 61.

scholars of the 10th and 11th centuries were quite familiar with the real origin of musk. The belief that musk was derived from blood was a result of the need to explain the true origin of musk in pre-Islamic times. This concept is related to the notion of the heat of musk, for blood is a hot humor. The appearance of grain musk is vaguely like old dried blood. In the codification of Islamic law, this apparent contradiction between unlawful blood and lawful musk had to be dealt with. The explanation that musk was transformed blood, and hence no longer blood, was so compelling that it became a metaphor in Sūfī literature for the alchemical process by which God transforms the life of the believer.

CHAPTER 5

The Merchant World and the Musk Trade

As Chapter 2 has shown, the products brought from Further Asia into the early medieval Middle East were many and varied and included many kinds of spices, drugs, aromatics, and dyes. Chapter 4 has illustrated the level of knowledge the medieval Islamic world possessed of the origins of musk. This chapter considers the facts of the early medieval trade with Further Asia itself, especially as it pertains to musk. This is not a subject about which detailed economic analysis can be written because of the lack of suitable source materials. It can be said, however, that there was flourishing trade with Further Asia as part of the great economic prosperity of the early medieval world. We can divide the economic history of the period covered by this work into two broad phases. The features of the first period, extending from around 700 to 1000, were conditioned at first by the power of the Umayyad and Abbasid caliphates in Iraq, Syria, and Iran. The Abbasid period, in particular, witnessed a great flourishing of Eurasian trade both overland and by sea.[1] The unity of the Abbasid caliphate was, of course, broken from the very beginning as the Umayyads resumed their rule in Iberia in the far west and by the mid-10th century, the Abbasids were mere figureheads. The flourishing of trade greatly accelerated during the 9th and 10th centuries. The burgeoning urban aristocracies craved exotic imports, and the courts of the Abbasids and their numerous successors utilized these same goods. The full effects of this process, as far as the Asian trade is concerned, only manifested themselves with the development in the late 10th century of the powerful Fatimid caliphate in Egypt, rival to the court of the Abbasids in Iraq. Under the Fatimids the maritime trade from India and Southeast Asia that had flowed into the Gulf was increasingly rivaled by the trade that entered the Red Sea.[2] At the same time, the Abbasid caliphate had fragmented into smaller states in the east such as the Samanids and Ghaznavids, nominally

---

1 E.g., C. I. Beckwith, *The Tibetan Empire in Central Asia*, rev. ed. (Princeton: Princeton University Press, 1993), 179–80. Cf. also Beckwith, "Tibet and the Early Medieval *Florissance* in Eurasia: A Preliminary Note on the Economic History of the Tibetan Empire," *CAJ* 21 (1977): 90–2. On the Abbasid sea trade, see D. Whitehouse, "Abbasid maritime Trade: Archaeology and the Age of Expansion," *Rivista degli Studi Orientali* 59 (1987): 339–47.

2 B. Lewis, "The Fatimids and the Route to India," *Revue de la Faculté des sciences économiques de l'Université d'Istanbul* 11 (1949–50): 50–4, and D. Agius, *Classic Ships of Islam from Mesopotamia to the Indian Ocean* (Leiden: Brill, 2008), 105–6.

 | DOI 10.1163/9789004336315_007

dependent on the caliph, but in reality independent. These eastern Islamic powers continued as avid participants in trade with Further Asia.[3]

Another change that marks off the period after 1000 from earlier Islamic times was the decline in the long-haul voyage in the Indian Ocean and the increasing segmentation of trade, with merchant-adventurers minimizing the duration of their voyages and putting in at emporia in India and Sri Lanka rather than sailing all the way to East Asia.[4] This hardly caused a decline in the volume of commerce; Indian and Southeast Asian sailors in particular stepped in, and trade between southern China, Southeast Asia, and India flourished in the 11th and 12th centuries.[5] Our work closes just before the rise of the Mongols, and their unification of the Central Eurasian heartland marks the apogee of the history of Central Eurasian commerce.[6]

The surviving corpus of business documents from the early medieval Islamic world is limited in scope but quite detailed in certain ways. It is mostly

---

3 A. King, "Eastern Islamic Rulers and the Trade with Eastern and Inner Asia in the 10th and 11th Centuries," *BAI* 25 (2011 [2015]): 175–185.

4 K. N. Chaudhuri, *Trade and Civilisation in the Indian Ocean: An Economic History from the Rise of Islam to 1750* (Cambridge: Cambridge University Press, 1985), 37–8; see especially his maps on 40–1.

5 On the South and Southeast Asians, see O. W. Wolters, *Early Indonesian Commerce: A Study of the Origins of Śrīvijaya* (Ithaca: Cornell University Press, 1967), K. R. Hall, *Trade and Statecraft in the Age of the Cōlas* (New Delhi: Abhinav Publications, 1980), and K. R. Hall, *A History of Early Southeast Asia: Maritime Trade and Societal Development, 100–1500* (Lanham, MD: Rowman and Littlefield, 2011), 131, on the 11th century shifts in trade patterns in Southeast Asia owing to the Chola expeditions. For trade in the ports of southern China, see J. Kuwabara, "On P'u Shou-kêng 蒲壽庚, a Man of the Western Regions, who was the Superintendent of the Trading Ships' Office in Ch'üan-chou 泉州 towards the End of the Sung dynasty, together with a General Sketch of Trade of the Arabs in China during the T'ang and Sung Eras," *MRDTB* 2 (1928): 1–79 and 7 (1935): 1–104, H. Clark, *Community, Trade, and Networks: Southern Fujian Province from the Third to the Thirteenth Century* (Cambridge: Cambridge University Press, 1991), and A. Schottenhammer, "The Maritime Trade of Quanzhou (Zaitun) from the Ninth through the Thirteenth Century," in H. P. Ray, ed. *Archaeology of Seafaring: The Indian Ocean in the Ancient Period* (Delhi: Pragati Publications, 1999), 271–90. For a Song handbook of the Sea Trade, see Zhao Rugua's *Zhufan zhi*, ed. and trans. F. Hirth and W. Rockhill, *Chau Ju-kua: His work on the Chinese and Arab Trade in the Twelfth and Thirteenth Centuries, entitled Chu-fan chih*, 2 vols. (St. Petersburg, 1911).

6 See in particular the works of T. Allsen on the Mongols and their commercial interests, "Mongolian Princes and their Merchant Partners," *Asia Major* 3rd ser. 2 (1989): 83–126 and *Commodity and Exchange: A Cultural History of Islamic Textiles* (Cambridge: Cambridge University Press, 1997), and see Beckwith, *Empires*, 201–3.

contained within the numerous documents preserved in the Cairo Geniza.[7] The commercial documents among this collection are the writings of Jewish merchants active in the Mediterranean and sometimes in the India trade over the Indian Ocean. In the absence of the material necessary to write a detailed economic history of the musk trade, it is on the architecture of the trade—the routes, entrepots, and the identities of merchants—that we must focus. Musk was, as shown above, part of the trade of Further Asia, and in many cases where we know of merchants but do not know the specific goods in which they dealt, musk was likely often one of them. Finally, this chapter discusses what is known about the musk trade itself on the basis of the documentary evidence that is available.

Musk Producers and the Trade

Very little information is available on the actual producers of musk. As seen in the previous chapter, musk came from basically three areas: Turkic Central Eurasia, Tibet, and China, with some from the Sino-Tibetan frontier with Southeast Asia as well. No information has been found about the production of musk in early medieval Turkic Central Eurasia, beyond the fact that it certainly came from there as mentioned in Arabic and other sources. Musk only becomes visible within the commercial life of Central Eurasia, as evinced first by the Sogdian Ancient Letters and later by Uyghur business documents, but the collection of musk in the wild is not mentioned in these sources.

So far, no indigenous information on the production of musk in early medieval Tibet has been found. For this we are at the mercy of the Arabic sources, which tell of the people of Tibet gathering musk, sometimes by hunting the animals, sometimes by locating excreted musk upon stones or the little minarets erected for the animals to rub upon. Al-Yaʿqūbī says that they gathered it freely (*mubāḥan*) from these little minarets, and he gives the information that it was then taken into Tibet, presumably meaning the capital of Tibet, and one tenth of its value was taken as a tax.[8] This is the oldest and best indication

7 S. D. Goitein, *Letters of Medieval Jewish Traders, Translated from the Arabic with Introductions and Notes* (Princeton: Princeton University Press, 1973) and S. D. Goitein and M. A. Friedman, *India Traders of the Middle Ages: Documents from the Cairo Geniza: "India Book"* (Leiden: Brill, 2008). See also Goitein's discussion of the economic world of the Geniza documents in his *A Mediterranean Society* 6 vols. (Berkeley: University of California Press, 1967–93), especially vol. 1.

8 Quoted in Nuwayrī 12.4 and see Chapter 4.

from an Islamic source of how musk got to market. If we compare Markham's 19th century notice, mentioned in the Chapter 1, that people who lived upon the land where musk deer roamed paid their rent in musk to the king, who owned all the musk deer, it appears that the situation was somewhat freer in early medieval times. If those gathering musk got to retain nine tenths of it, they could make money from it by selling it to merchants. Just a few decades later, al-Masʿūdī notes that it was carried to the kings, but the best musk, the "naturally ripened musk" that was supposedly stuffed into pods, was used by kings and exchanged as gifts among royalty, and "merchants carry it rarely from their country".[9] All of this information indicates that the government took an interest in the production of musk, but no documentary evidence apparently survives from the early medieval Tibetan side to illustrate it. Once the merchants had their musk, they shipped it out of Tibet by the standard trade routes, south into India for export through Sind or other ports, or overland through Eastern Turkestan into Sogdiana, the great clearinghouse for Tibetan musk. Of course, much musk went east into China as well. By the 11th century at least, Tibetan musk was traded especially for Chinese tea; al-Bīrūnī states that nothing else except musk was accepted for tea.[10]

There is considerable data for the structure of Tibetan trade from the last few centuries,[11] but it is debatable how many parallels can be drawn with the early medieval period. The musk trade, because of the peculiar character of the acquisition of musk and the relatively stable market for it beyond the frontiers of Tibet, would provide an interesting study of continuity and change in Tibetan trade if more data were available for the early medieval period. In the states of the Himalayas during the 19th century, musk deer were considered royal property, and the kings kept men to hunt them; the people were also obliged to provide musk, along with other local products, to the king instead of rent.[12] This is likely what occurred in the early medieval period, but it is

---

9 Al-Masʿūdī, *Murūj al-dhahab wa maʿādin al-jawhar*, ed. B. de Meynard and P. de Courteille, revised by C. Pellat, 5 vols. (Beirut: Manshūrāt al-Jāmiʿah al-Lubnāniyyah, 1966–74), 1.189 (§394).

10 Al-Bīrūnī, *Kitāb al-ṣaydanah*, ed. and trans. H. M. Saʿīd (Karachi: Hamdard National Foundation, 1973), 128, and *Kitāb al-ṣaydanah*, ed. ʿAbbās Zaryāb (Tehran: Markaz-i Nashr-i Dānishgāhī, 1991), 165.

11 See W. van Sprengen, *Tibetan Border Worlds: A Geohistorical Analysis of Trade and Traders* (London: Kegan Paul, 2000), with references to earlier literature.

12 F. Markham, *Shooting in the Himalayas: A Journal of Sporting Adventures and Travel in Chinese Tartary, Ladac, Thibet, Cashmere, &c.* (London: Richard Bentley, 1854), 94. See also M. P. Joshi and C. W. Brown, "Some Dynamics of Indo-Tibetan Trade through Uttarākhaṇḍa (Kumaon-Garhwal), India," JESHO 30 (1987): 307.

not attested in the few accounts of musk that still exist. The early medieval royal beneficiaries of this trade probably became the Buddhist establishment in later centuries; the dominant Dgelugspa order certainly benefited from the musk trade.[13]

The American W. W. Rockhill (1854–1914), who travelled in Mongolia and Tibet, notes that locals would approach with musk to sell in areas with populations of musk deer.[14] He would purchase musk for trading during his travels.[15] This sort of casual trade by a traveler or even a merchant engaged in some other commerce was a significant factor in the transportation of musk, which had the advantages of being easily carried and very valuable. Rockhill also mentions that Chinese traders from Canton and Sichuan travelled to Lusar, east of the Koko Nor, to buy up musk.[16] In his time at Lusar, musk sold for four times its weight in silver. In the 19th century in eastern Tibet, then, musk was collected by the locals and traded with the Chinese.[17]

There are no accounts of musk production in China in early medieval Arabic literature comparable to those for Tibet. However, we do have an early source in Tao Hongjing (452–536), whose account of musk was given in Chapter 3. He mentions that musk was obtained in two ways: by hunting, and, preferably, by locating the places where musk deer scraped off their musk. Beyond this, he gives no details on how musk got into the hands of merchants. We can assume that, similar to the situation in Tibet, much musk was simply acquired by those living in the area where the deer lived and traded to merchants when the opportunity arose. Much musk probably originated in the non-Chinese lands to the north, west, and south, and was acquired by trade from the local hunters.[18] The Ming bureau for the exchange of tea for horses also accepted musk from the western peoples.[19]

---

13 Van Sprengen 135.

14 W. W. Rockhill, *Diary of a Journey through Mongolia and Tibet in 1891 and 1892* (Washington: Smithsonian Institution, 1894), 268, 283.

15 Rockhill 335, 348.

16 Rockhill 71.

17 See also E.-R. Huc and J. Gabet, *Travels in Tartary, Thibet and China 1844–1846*, trans. W. Hazlitt, 2 vols. (London: Routledge, 1928; repr. [2 vols in 1] New York: Dover, 1987), 2.372.

18 This occurred during the 19th century, with the chief suppliers of musk being the Lolo of Yunnan, see "Fragrans", "How Musk is Made," *The China Review* 9:4 (1881): 253–4.

19 L. M. J. Schram, "The Mongours of the Kansu-Tibetan Frontier Part III," *Transactions of the American Philosophical Society* 51:3 (1961): 32.

Tribute and Royal Gift-Giving

In this chapter, we will be largely concerned with the merchants who supplied musk to the markets of the Islamic world. Straightforward commercial exchange was, however, only one means by which musk traveled westward. An additional vector for the arrival of musk in the Islamic world must be mentioned: tribute and gift-giving. Al-Masʿūdī, as noted above, says of the best variety of musk, the musk which naturally ripened upon the animal: "That is what their kings use and give as gifts between them; the merchants carry it rarely from their country."[20] He implies that it was difficult to find the very best musk on the marketplace because of the monopolization of it by kings. This variety of musk is presumably the "treasury" musk (*khazāʾinī*) he mentions elsewhere.[21] Literature about royalty often associates musk with kingship; the symbolic meanings of this relationship with musk will be discussed in the last chapter. Here we consider the practical manifestation: musk was gifted and re-gifted among kings, princes, caliphs, and anyone with status in the Islamic world and beyond.[22]

Musk was a substance fit for offering to royalty.[23] The Sasanian kings supposedly demanded musk in tribute from their subjects who specialized in that substance.[24] Muḥammad himself sent a gift of musk to the Negus of Ethiopia.[25] Most of the records of the presentation of musk to kings are of gifts from rulers in the eastern lands of the Caliphate, for musk was preeminently associated with the east. The historian al-Ṭabarī records al-Maʾmūn's gifts (*al-taʿẓīm wa-l-hadāyā*) from the riches (*ṭuraf*) of Khurāsān to his brother the caliph al-Amīn before their falling out: household goods, vessels, musk, beasts of burden and weapons (*al-matāʿ wa-l-ānīyah wa-l-misk wa-l-dawābb wa-l-silāḥ*).[26] Among such a varied list of generic goods, musk stands out for its particularity.

There are also some records of ʿAmr b. al-Layth al-Ṣaffār's tribute to the court of al-Muʿtamid (r. 870–892), which consisted of money, fifty *manns* of

20 Masʿūdī 1.189 (§394).

21 Masʿūdī 1.309 (§624).

22 M. A. Newid, *Aromata in der iranischen Kultur unter besonderer Berücksichtigung der persischen Dichtung* (Wiesbaden: Reichert, 2010), 96–8, has examples of the gifting of musk from Persian literature. On the royal exchange of aromatics, see S. Anṣārī, *Tārīkh-i ʿIṭr dar Īrān* (Tehran: Wizārat-i Farhang wa Irshād-i Islāmī, 1381/2002–3), 219–22.

23 E.g., Ibn Miskawayh, *Kitāb Tajārib al-umam*, 3 vols. (Baghdad: al-Muthanna, n.d.), 1.67.

24 Al-Jāḥiẓ (attr.) *Kitāb al-tāj*, ed. Ahmed Zeki Pacha (Cairo: Imprimerie Nationale, 1914), 146–50.

25 Ibn Saʿd, *al-Ṭabaqāt al-kubrā*, 9 vols. (Beirut: Dār Ṣādir, 1960), 8.95.

26 Ṭabarī III.775. Another mention of Khurāsānian rarities is at III.817.

musk, fifty of saffron, two hundred of aloeswood, three hundred embroidered garments, gold and silver vessels, animals, and slaves.[27] Another time 'Amr b. al-Layth sent a gift that included money, horses, garments, perfume and falcons.[28] In Gardīzī's account of this gift (in the *Zayn al-Akhbār*, written around 1049–52) the perfumes are specified as aloeswood, musk, and sandalwood.[29] The *Kitāb al-Hadāyā wa-l-tuḥaf* records a gift of money, three thousand *mithqāls* of musk, a thousand of ambergris, and five *manns* of aloeswood.[30] The Karakhanids, because of their position straddling the Central Asian trade routes into China, had access to goods from different parts of Eurasia. In 1001 CE, the Karakhanid Nāṣir al-Ḥaqq replied to an embassy from Maḥmūd of Ghazna bringing precious goods characteristic of India with "exotic imports of the Turks, including ingots of precious metal from the mines, musk pods (*nawāfij al-misk*), leaders of the male horses, reddish-white female camels, attractive male and female slaves, white falcons, black fur (*awbar*), carved *khutū*-ivory, jasper (*yashb*) stones, and other precious goods of China."[31] The Karakhanid Qādir Khān (r. 1026–32) received from Mas'ūd the Ghaznavid (r. 1030–41) gifts including musk, aloeswood, and ambergris.[32] Mas'ūd also sent a gift to the caliph which included among textiles, jewels, horses, and slaves fifty musk pods and one hundred *shamāmahs* of camphor.[33]

The poet Abū al-'Atāhiyah (748–826) once gave the caliph al-Mahdī (r. 774–85) a pot (*barniyyah*) on Nawrūz or Mihrajān containing a garment perfumed with musk upon which were written with ambergris or *ghāliyah* two lines of poetry.[34] Likewise, the caliph would bestow perfumes along with the

---

27 Ṭabarī III.2018.

28 Ṭabarī III.2188.

29 Gardīzī, *Zayn al-akhbār*, ed. 'Abd al-Ḥayy Ḥabībī, as *Tārīkh-i Gardīzī* (Tehran: Dunyā-yi Kitāb, 1363/1984), 186.

30 *Kitāb al-Hadāyā wa-l-tuḥaf*, ed. M. Ḥamīdullāh as Ibn al-Zubayr, *Kitāb al-Dhakhā'ir wa-l-tuḥaf*, Kuwait: Dā'irat al-Maṭbū'āt wa-al-Nashr, 1959), 43. Trans. Ghāda al-Ḥijjāwī al-Qaddūmī. *Book of Gifts and Rarities* (Cambridge, MA: Distributed for the Center for Middle Eastern Studies of Harvard University by Harvard University Press, 1996).

31 Al-'Utbī, *al-Yamīnī fī sharḥ akhbār al-sulṭān Yamīn al-Dawlah wa Amīn al-Millah Maḥmud al-Ghaznawī*, ed. Iḥsān Dhunūn al-Thāmirī (Beirut: Dar al-ṭalī'ah li-l-ṭibā'ah wa-l-nashr, 2004), 257, W. Barthold, *Turkestan down to the Mongol Invasion*, 3rd ed. (London 1968; repr. Taipei: Southern Materials Center, n.d.), 272, and King, "Eastern Islamic Rulers" 179–80.

32 Bayhaqī, *Tārīkh-i Bayhaqī*, ed. 'Alī Akbār Fayyāḍ (Mashhad: Dānishgāh-i Mashhad, 1971), 281.

33 Bayhaqī 389.

34 Mas'ūdī 3. 174 (§2450) with the variant *ghāliyah*, which makes more sense, mentioned in note 10.

robe of honor to those he favored.[35] There are many stories about the splendor of the wedding of al-Maʾmūn and Būrān. He distributed to many important people hazelnut-sized balls of musk with a slip inside with writing, stating what prize the bearer could claim.[36] Coins, musk pods, and eggs of ambergris were scattered before the other people. At night massive candles of ambergris were burned. The custom of distributing hazelnut-sized balls of musk is attested elsewhere.[37] These hazelnut-sized balls could be fondled in the hands to release their aroma.[38]

The practice of exchanging musk as a royal gift was projected back into the past. When Bilqīs, the Queen of Saba, sent gifts to Solomon she included musk and ambergris.[39] This is an anachronistic description that reflects the value of musk as a diplomatic gift in Islamic times.

In the Persian tale of *Wīs u Rāmīn* of Fakhr al-Dīn Jurjānī (11th century), which has its roots in Parthian times, there is an account of the gifts sent by the king of Marw, Mowbad, to Shahrū, queen of Māh. The gifts consisted of many treasures such as camels, horses, slave girls, and jewels. Among these gifts were "a hundred golden caskets full of gems; pure musk and ambergris scattered on them, in color like a sweetheart's hair."[40] The *Wīs u Rāmīn* story goes back to a Pahlavi version, and ultimately was Parthian in origin. None of those earlier versions are extant, and Jurjānī's work, though based on the Pahlavi version, can only be used as a source for the high regard musk and ambergris had in the poet's time. Nevertheless, it is not surprising that the two aromatics singled out for mention should be musk and ambergris, so often associated with royalty in Arabic literature, which in turn had a powerful influence on the imagery of Persian literature.

We have already seen numerous examples of musk offered as a gift. The tradition of offering musk and aromatics as a perfume is old and venerable in Islam. Muḥammad himself never refused a gift of aromatics,[41] and this practice became the accepted custom, hence the excellence of aromatics in Islamic

---

35 Ibn Miskawayh 1.326, 396, etc.

36 Masʿūdī 3.327 (§2752).

37 E.g., al-Zamakhsharī, *Rabīʿ al-abrār wa-fuṣūṣ al-akhbār*, eds. ʿAbd al-Majīd Diyāb and Ramaḍān ʿAbd al-Tawwāb, 2 vols. (Cairo: al-Hayʾah al-Miṣriyyah al-ʿĀmmah li-l-Kitāb, 1992–2001), 2.216.

38 Ibn Abī al-Ḥadīd, *Sharḥ nahj al-balāghah*, 20 vols. (Cairo: ʿĪsā al-Bābī al-Ḥalabī, 1960–4), 19.343.

39 Al-Zamakhsharī, *al-Kashshāf ʿan ḥaqāʾiq al-tanzīl*, 2 vols. (Cairo, 1948), 2.451.

40 Fakhr al-Dīn Jurjānī, *Vis and Ramin*, trans. G. Morrison (New York: Columbia University Press, 1972), 55.

41 E.g., Ibn Saʿd 1.399.

gift-giving. Epistles and poems exchanged among the literati were likened to gifts of musk and aromatics because of the symbolic closeness of sweet speech and sweet scent.

Routes and Emporiums

The Arabic and Persian geographical writings give a good picture of the routes followed by medieval traders, at least as far as they were used by Muslim merchants and their Jewish counterparts. One of the most impressive accounts of Eurasian trade is preserved by the Abbasid postmaster Ibn Khurradādhbih in the abridgement of his *Kitāb al-Masālik wa-l-mamālik*. It is of the Rādhāniyyah, a Jewish community originating in Mesopotamia that operated throughout Eurasia, and it shows how they transported goods from one end of Eurasia to the other:[42]

> They speak Arabic, Persian, Greek, and the languages of the Franks and of al-Andalus, and of the Slavs. They travel from the west to the east, and from the east to the west by land and sea. They import from the west eunuchs, slave girls, boys, brocade, skins of *khazz*,[43] furs, sable, and swords. They set off from the country of the Franks on the western sea,[44] and travel to al-Faramā;[45] then they carry their merchandise on back (*ʿalā al-ẓahr*)[46] and travel by land to Qulzum,[47] a distance of 25 parasangs. They set off on the Red Sea and go from Qulzum to al-Jār[48]and to Juddah.[49] Then they travel to Sind, Hind and from there to China. They bring from China musk, aloeswood, camphor, cinnamon and other things brought from

---

42 Ibn Khurradādhbih, *Kitāb al-Masālik wa-l-mamālik*, ed. M. J De Goeje (Leiden: Brill, 1889, repr. 1967), 153–4, and M. Gil, "The Rādhānite Merchants and the Land of Rādhān," *JESHO* 17 (1974): 299–328.

43 *Khazz*, which later means silk, originally denoted fine woolen textiles; in this case it must designate the pelts used to prepare *khazz*. See Gil, "The Rādhānite Merchants" 312, who compares the "wool of rabbits" that appears in Gaonic sources. It is often translated as beaver pelts, but there is an Arabic word *khuzaz* that denoted young rabbits or male rabbits; LA 5.404 derives *khazz* from this word.

44 This probably refers to the Frankish Italy, cf. Gil, "The Rādhānite Merchants" 310–11.

45 On the eastern Mediterranean coast of Egypt.

46 This expression commonly denotes overland trade as opposed to sea trade.

47 On the northern tip of the Red sea.

48 The port of Madīna, on the coast of Western Arabia.

49 The port of Makka, Jidda.

> those regions. They return to Qulzum, then they carry them to al-Faramā, and then they embark on the western sea. Sometimes they turn towards Constantinople, where they sell their merchandise to the Byzantines; perhaps they might return to the king of the Franks and sell them there.
>
> If they wish, they carry their merchandise from the land of the Franks upon the western sea and then go towards Antioch. They journey by land for three days to al-Jābiyah. There they embark on the Euphrates and reach Baghdād, whence they embark upon the Tigris to Ubullah. From Ubullah they set off for Oman, Sind, Hind and China, all of which are adjacent one after the other.
>
> [A discussion of the trade of the Rus' is omitted here.]
>
> As for [the Rādhānites'] routes upon the sea: when one leaves from al-Andalus or from the Franks, he crosses to Sūs al-Aqṣā[50] and then goes to Tanjir, whence he sets off for Ifrīqiyah[51] and Egypt, and then on to Ramlah, Damascus, Kufa, Baghdād, and Baṣrah, Ahwāz, Fārs, Kirmān, Sind, Hind, and then to China. Perhaps they take the route of beyond the Byzantines, through the Slavs, then to Khamlīj, capital of the Khazars. There they set off on the sea of Jurjān,[52] and then to Balkh and Transoxiana and then on to the Yurt of the Tughuzghuz[53] and from there to China.

Ibn Khurradādhbih's notice illustrates the interconnectedness of the land and sea routes as well as showing the wide scope of the commercial activities of the Rādhānite Jews. There is no comparable account of Muslim (or Persian) merchants, but they were also active on all of these routes as well, even if it cannot be demonstrated that they had the same degree of organization.

As far as Further Asia is concerned, there are broadly two kinds of routes: overland and maritime. As can be seen from Ibn Khurradādhbih's account of the Rādhāniyyah, they were interconnected and used at the same time; the role of the overland trade was complementary to the maritime trade in long distance trade. The overland routes were also especially focused on the supply of the landlocked Central Eurasian realms and Islamic Khurāsān, since maritime routes could not deliver goods there. Goods similar to those traded by sea will have also travelled overland, especially silk, which was in great demand in Central Eurasia.

---

50 "Furthest Sūs," i.e. Southern Morocco.

51 A name of the city of Qayrawān, in what is now Tunisia.

52 I.e., the Caspian Sea.

53 These words in the mss. are corrupted, but can only be restored this way.

In general, it was cheaper to ship goods by sea than by land, but even in areas where the potential use of sea routes paralleled land routes, the land routes did not fall into disuse.[54] Sailors could avoid customs stations, but they also had to pay for the expensive equipment of seafaring. And at the same time, there was a risk of shipwreck. Musk is an interesting case, since musk traded by the overland route was specifically preferred, as we have seen, because it had not been damaged by the moisture of the maritime route. This perhaps applied to some other goods as well.

Arab settlement in Central Asia gave direct access to the overland trade routes. The most important routes, which are traced by Ibn Khurradādhbih, were those followed by the *barīd*, or post.[55] They formed a network of overland routes throughout the Middle East, Central Asia, and Northern Africa. The roots of the *barīd* lie in the ancient Near East and go back to the famous post of the Achaemenids and earlier.[56]

The overland routes by which musk arrived followed the Muslim post routes up to the frontiers of the *Dār al-Islām*. Detailed itineraries of these routes can be constructed on the basis of Arabic and Persian geographical literature, as well as Chinese, Tibetan, and other Central Eurasian sources. Ibn Khurradādhbih discusses two main overland routes to the east, both departing from Marw.[57] One ran by way of Sogdiana to the land of the Tokuz Oghuz Turks (the Uyghurs) through al-Shāsh (Tashkent) into Farghana. To the north, one could travel by way of Tarāz to Nūshajān (Barskhān), a major city of the Tokuz Oghuz, and from there head east by way of Beshbalık and bend south towards Qocho at Turfan.[58] North from the land of the Tokuz Oghuz one could head towards the land of the Kimek, great producers of furs. Merchants could also travel through Farghana and over the mountains into the Tarim

54 Muhsin Yusuf, "Sea versus Land: Middle Eastern Transport during the Muslim Era," *Der Islam* 73 (1996): 232–58, has compared the literary sources for overland and maritime transportation and concluded that overland trade continued to be very significant compared to maritime trade up to the eleventh century, when maritime trade became more and more practical due to improvements in technology.

55 On the *barīd*, see A. J. Silverstein, *Postal Systems in the Pre-Modern Islamic World* (Cambridge: Cambridge University Press, 2007).

56 Herodotus, *The Persian Wars*, ed. and trans. A. D. Godley, Rev. ed., 4 vols., Loeb Classical Libary (Cambridge, MA: Harvard University Press, 1926), 8.98; Silverstein 9.

57 Ibn Khurradādhbih 25–34.

58 See also V. Minorsky, "Tamīm ibn Baḥr's Journey to the Uyghurs," *BSOAS* 12:2 (1948): 280, 283.

basin at Kashgar. These routes were part of the northern route to China and the east, as well as to the steppelands.[59]

The second route from Marw went south, striking along the Oxus by way of Balkh, and entered western Tibet through Ṭukhāristān, Badakhshān, and Wakhān.[60] This road certainly was used for the musk trade.[61] Also at Balkh or further east at Khulm or Warwālīj one could turn south and descend through Ṭukhāristān and take the Khyber Pass into Sind.[62] Bāmiyān, site of the now destroyed famous giant Buddhas, was on the road from Balkh. Sind, in addition to having access to the maritime routes, was linked with overland routes throughout India.[63] From the upper Indus area through Ladakh,[64] a route stretched across Tibet towards the musk-producing lands in the east, the "Changthang Corridor"; the Changthang region appears to have been formerly of a more fertile character.[65] Much of the musk that reached Sind must have come across the Changthang and then down the Indus and its tributaries. Alternately, musk traveled the Northern Route across India, the *Uttarāpatha*, which connected the Indus River Valley with the Ganges River Valley and lands beyond.[66] Al-Bīrūnī discusses the routes in the early 11th century; Kannauj

---

59 The northern route was longer according to Mas'ūdī on 1.186 (§385), taking four months rather than forty days, but it was under the protection of the Turks, and pack animals could easily make the trip, which was difficult for them on the southern route.

60 See also Mas'ūdī 1.116 (§225). The Tibetan army used this route as an egress to the western regions, see Beckwith, *Tibetan Empire*, 30, 91, etc. The plunder of rich Hu merchants by the local tribes in the Pamirs is mentioned by the 8th c. Korean traveler Hyech'o, see *Echōō-Gotenjiku-koku-den Kenkyū*, ed. S. Kuwayama (Kyoto: Kyōto Daigaku Jinbun Kagaku Kenkyūjo, 1992), 25 l. 207.

61 Ibn Ḥawqal, *Kitāb Ṣūrat al-'arḍ*, ed. M. G. De Goeje and rev. J. H. Kramers (Leiden: Brill, 1938; repr. 1967), 449.

62 Mas'ūdī 1.186 (§386) mentions the caravans from India traveling via Sind into Khurāsān. On these routes, see A. Foucher, *La vieille route de l'Inde de Bactres à Taxila*, 2 vols. (Paris: Les éditions d'art et d'histoire, 1942–7) and A. Wink, *Al-Hind: The Making of the Indo-Islamic World. Vol. 1: Early Medieval India and the Expansion of Islam 7th–11th centuries* (Leiden: Brill, 1990; repr. 2002), 171–5 on the great importance of Sind as a center of trade. On the northwestern routes, see now J. Neelis, *Early Buddhist Transmission and Trade Networks: Mobility and Exchange within and beyond the Northwestern Borderlands of South Asia* (Leiden: Brill, 2011), 244–50.

63 See al-Bīrūnī, *Kitāb al-Bīrūnī fī taḥqīq mā li-l-Hind*, ed. E. Sachau (Haydarābād: Dā'irat al-Ma'ārif, 1958), 155–70.

64 On the routes in the Upper Indus area, see Neelis 257–8.

65 P. Denwood, "The Tibetans in the West, Part I," *Journal of Inner Asian Art and Archaeology* 3 (2009): 10.

66 Neelis 186–204.

formed the node linking a road to the northwest with a route further to the east, ending up in the area of Nepal and Bhūtīshar.[67]

The routes into China, not discussed by Ibn Khurradādhbih but mentioned by later writers such as al-Marwazī, went either by way of the Tokuz Oghuz, as mentioned above, or through the Tarim Basin, or north by way of Kucha and Agni to Turfan, or south via Yarkand, Khotan, and Keriya to Shazhou; the latter route is better attested from early medieval Islamic sources.[68] Shazhou, where the important town of Dunhuang is located, was a major crossroads. From there one could go north to Turfan and join up with the route around the northern Tarim Basin, or head east into China.[69] It is unlikely that many Muslim merchants went further than Sogdiana, with the actual journey to China being undertaken mostly by Sogdians and other Central Eurasians during the early medieval period. But as these peoples assimilated to Islamic civilization, Islam went with them to the east. It has been proposed that the spread of Islam had a negative effect on the commerce of the Tarim Basin routes, because pressure from the zealously Islamic Karakhanid Turks upon the Buddhist world of the eastern Tarim Basin turned the region into a war zone.[70] The Karakhanids, however, were interested in commerce and do not seem to have deliberately targeted it.[71] At the same time, trade between the Middle East and East Asia on the maritime routes greatly expanded, reaching the southern Chinese ports of the Song dynasty. In the early 11th century, envoys sent by the Uyghurs feared to bring valuable gifts for Maḥmūd of Ghazna because there was no safe road, yet their Kitan Liao colleagues brought commodities and sought trade relations.[72] The traditional function of the Steppe Empire was securing and promoting trade that enriched the empire and aggrandized its supporters. When Chinggis Khan of the Mongols rose to power at the beginning of the

---

67 Bīrūnī, *Hind*, 159–60.

68 Marwazī, text *7 and trans. 19, Minorsky's comments 68–74. Classic overviews of these routes and their archaeology are found in M. A. Stein, *Serindia: Detailed Report of Explorations in Central Asia and Westernmost China*, 5 vols. (Oxford: Oxford University Press, 1921) and his *Innermost Asia* 4 vols. (Oxford: Oxford University Press, 1928).

69 Marwazī, ibid.

70 Examples of Turkic war poetry from the campaigns against the Buddhists are preserved piecemeal by Maḥmūd al-Kashgārī; they have been collected in R. Dankoff, "Three Turkic Verse Cycles Relating to Inner Asian Warfare," *Harvard Ukrainian Studies* 3 (1979): 159–61. See also Rong Xinjiang, "The Nature of the Dunhuang Library Cave and the Reason for its Sealing," *Cahiers d'Extrême-Asie* 11 (1999–2000): 247–75, which argues that the manuscripts deposited at Dunhuang were hidden from invading Muslims.

71 King, "Eastern Islamic Rulers".

72 Marwazī text *9 and trans. 21; A. King, "Early Islamic Sources" 255–8.

13th century, one of his first concerns was promoting security on the roads so that merchants would be unimpeded.[73]

The Tarim Basin routes were connected with the south as well as the west and north. One could cross the Karakorum through Hunza and travel between Yarkand and Gilgit, descending into Sind.[74] Sogdian merchants, along with travelers from many different places, used this route in late antiquity.[75] To the east of Yarkand lies Khotan, and from there one could enter western Tibet.[76] Thus Tibetan musk was shipped via Khotan, especially in later centuries, and acquired the name *khutanī* "Khotanese" in Persian literature.

The control of these lands was the subject of great contention among the imperial powers of the early medieval world: Tang China, the Turks and then Uyghurs, the Tibetans, and the Arabs all sought to dominate this territory with peculiar insistence.[77] This can partially be explained by the imperial dream of conquering whatever land is available, but strong motivation surely came from the trade that flowed through Central Eurasia as well as from its own indigenous resources. This trade was essentially confined by geography to only a very few practical routes due to deserts and mountains, and its maintenance was an important source of wealth and prestige for both the local potentates as well as the greater powers. As far as the musk trade was concerned, all routes leading from Inner Asia, China, Tibet, and India were potential sources.

The overland trade routes connected the regimes of Central Eurasia with the periphery of the Eurasian landmass. They are particularly important for

---

73 Juwaynī, *Genghis Khan: The History of the World-Conqueror*, trans. J. A. Boyle. 2 vols. (Manchester: Manchester University Press, 1958); New Edition with introduction by D. Morgan in 1 vol., (Seattle: University of Washington Press, 1997), 78: "He ... posted guards ... upon the highways and issued a *yasa* that whatever merchants arrived in his territory should be granted safe conduct ..." See Allsen, "Merchant Princes" 88.

74 This route became better known thanks to a series of publications from the late 1980's and early 1990's: K. Jettmar, ed., *Antiquities of Northern Pakistan: Reports and Studies* (Mainz, 1989–). See also J. Neelis, "*La Vieille Route* Reconsidered: Alternative Paths for Early Transmission of Buddhism Beyond the Borderlands of South Asia," *BAI* 16 (2002): 143–64.

75 N. Sims-Williams, "The Sogdian Merchants in China and India," in A. Cadonna and L. Lanciotti, eds., *Cina e Iran da Alessandro Magno alla dinastia Tang* (Firenze: Olschki, 1996), 45–67 with more bibliography. See also Ma Yong, "The Chinese Inscription of the "Da Wei" Envoy of the 'Sacred Rock of Hunza'," *Antiquities of Northern Pakistan: Reports and Studies* 1 (1989): 139–57.

76 Gardīzī, *Zayn al-akhbār*, ed. 'Abd al-Ḥayy Ḥabībī, as *Tārīkh-i Gardīzī*. (Tehran: Dunyā-yi Kitāb, 1363/1984), 562.

77 The political and military history of the early medieval empires of the Silk Road lands is discussed in Beckwith, *Tibetan Empire*.

Islamic history inasmuch as Central Asia was a thriving center of Islamic civilization. The Abbasid caliphate drew most of its initial support from Khurāsān, and the court of the caliph al-Ma'mūn was located in Marw for years following the defeat of his brother, al-Amīn, in civil war. But after that time, the Central Eurasian Islamic world grew increasingly independent from Baghdād, and governors of local extraction ruled Iran and the east. Eventually, dynasties such as the Samanids, Ghaznavids, and Karakhanids controlled the eastern Islamic lands, and it is through their territory that the overland trade entered the lands to the west.[78] It is remarkable how many merchants known in overland and maritime trade, even as far away as Egypt and the Mediterranean, bear names indicating their origins in Khurāsān.[79] This fact indicates that the merchants of early medieval Central Eurasia operated on both the overland and maritime routes, something which is known to be the case for the Sogdians in the Indian Ocean.[80]

The maritime routes have the most detailed literature extant, ranging from the early *Akhbār al-Ṣīn wa-l-Hind* of 851 and its supplement by Abū Zayd al-Sīrāfī of the early 10th century, to the extensive writings of the geographers and writers on *mirabilia*, such as the book of Buzurg b. Shahriyār, the *Kitāb 'Ajā'ib al-Hind*. The latter works, though really collections of sailors' tales, include many important facts.[81]

The Indian Ocean routes were dependent on the monsoon winds, which gave an annual structure to the trade.[82] From April to September the winds blow from west to east, and the onset of those winds, which could be dangerously strong at their height, was the favored season to undertake the voyage to the east. Conversely, from October to March the winds supported voyaging in the opposite direction.

Ships could sail either the long voyage to the east, all the way to Southeast Asia and China, or stop at emporiums along the way, especially in India and in

78 For their attitudes towards trade, see King, "Eastern Islamic Rulers".

79 A *Marwazī* is noted below. Examples among the Geniza documents include a *Nīshāpūrī*.

80 F. Grenet, "Les marchands sogdiens dans les mers du Sud à l'époque préislamique," *Cahiers d'Asie Centrale* 1–2 (1996): 65–84. Mas'ūdī 1.166 (§336) notes a merchant of Samarkand who travelled to Iraq and then embarked on the sea voyage to China. Cf. also Narshakhī, *Tārīkh-i Bukhārā*, ed. Mudarris Raḍawī, 2nd ed. (Tehran: Intishārāt-i Tūs, 1363/1984), 26, and trans. R. N. Frye, *The History of Bukhara* (Cambridge, MA: Mediaeval Academy of America, 1954), 18.

81 See D. Agius, *Classic Ships of Islam from Mesopotamia to the Indian Ocean* (Leiden: Brill, 2008).

82 See, concisely, L. Casson, *The Periplus Maris Erythraei* (Princeton: Princeton University Press, 1989), 283–91 and Agius 185–8.

Sri Lanka, which was a major emporium since classical times.[83] The Sasanians, who were the pioneers of Near Eastern maritime trade with India, sent their ships to India and Sri Lanka, and do not appear to have usually sailed much further at first.[84] This situation certainly did not continue, for Persian ships eventually plied the seas east of Sri Lanka. The early 8th century Korean pilgrim Hyech'o notes that the Persians were skilled traders and sailed to Sri Lanka with precious goods. He also remarks that they sailed on to Southeast Asia and to China, going to Guangzhou "to buy various sorts of silk gauze and wadding."[85] The Indian Buddhist monk Vajrabodhi (671–741) sailed to China from Sri Lanka with a fleet of 35 Persian ships.[86] Thus, by early Islamic times, "Persian" (*Bosi* 波斯) ships were familiar in the South China Sea.[87] Indian, Chinese, or Southeast Asian merchants will have made the voyage from Indonesia

---

83 On Sri Lanka, D. P. M. Weerakkody, *Taprobanē: Ancient Sri Lanka as known to the Greeks and Romans* (Turnhout: Brepols, 1997), C. E. Bosworth, "Sarandīb" *EI* 2nd ed. s.v., J. Carswell, "The Port of Mantai, Sri Lanka," in V. Begley, and R. D. De Puma, *Rome and India: The Ancient Sea Trade* (Madison: University of Wisconsin, 1991), 197–203, and especially R. Walburg, *Coins and Tokens from Ancient Ceylon* (Wiesbaden: Reichert, 2008), 325–6.

84 This is based on the evidence of Procopius I.20.12 and Cosmas Indicopleustes XI.16. See the discussion of Sasanian maritime trade in Chapter 3. G. F. Hourani minimizes the importance of direct sailing from China in Sasanian times: "Direct Sailing Between the Persian Gulf and China in Pre-Islamic Times," *JRAS* (1947): 157–60. Archaeological evidence for the Sasanians in Sri Lanka appears to be meager, however; see Walburg, 37.

85 Hyech'o 23 lines 165–6.

86 P. Pelliot, "Deux itinéraires de Chine en Inde," *Bulletin de l'École Française d'Extrême-Orient* 4 (1904): 336.

87 The famous pilgrim Yijing took a Bosi ship to Śrivijaya from China in 671; *A Record of the Buddhist Religion as practiced in India and the Malaya Archipelago (AD 671–695) by I-Tsing*, trans. J. Takakusu (Oxford: Clarendon Press, 1896), xxviii. See also Wang Gungwu, "The Nanhai Trade: A Study of the Early History of Chinese Trade in the South China Sea," *JMBRAS* 21:2 (1958): 99–104 and D. Whitehouse and A. Williamson, "Sasanian Maritime Trade," *Iran* 11 (1973): 46–47. The term *Bosi* is unfortunately ambiguous, and can perhaps refer to an area of Southeast Asia as well as genuine Persians. B. Laufer, *Sino-Iranica: Chinese Contributions to the History of Civilization in Ancient Iran, with Special Reference to the History of Cultivated Plants* (Field Museum of Natural History Anthropological Series 15:3 Chicago: Field Museum, 1919; repr. New York: Kraus, 1967), 468–87, was certain that there was a separate Malayan Bosi, while O. Wolters, *Early Indonesian Commerce* (Ithaca: Cornell University Press, 1967), 129–58, etc., believed that the original Bosi were Persians and later were replaced by Malayans even as the genuine Near Eastern cargoes of frankincense and bdellium were replaced by Southeast Asian benzoin and camphor. Whitehouse and Williamson suspect that the 5th century Bosi were Sasanians, and that even in the 8th century it still referred to Persia. Bosi ships basically seem to have been ships of

bringing the goods of Southeast Asia, and the Indian products could be obtained at the emporiums in India, such as Daybul in Sind, which we have already noted as a trading center for musk.

The long-haul voyages to China were an innovation of Islamic times.[88] The golden age of the long-haul voyage was from the 8th to the 10th centuries, and it is vividly described in the *Akhbār al-Ṣīn wa-l-Hind* and its supplement by al-Sīrāfī. During this period the great Muslim merchant diaspora in East Asia thrived and would continue throughout the medieval period. But the considerable risks involved in sea traffic are compounded the longer the voyage becomes, and so after the 10th century the segmented voyages were again favored.[89] Thus none of the Geniza documents, dating mostly to the 11th and 12th centuries, describe the travels of merchants any further than Sumatra, while most of their activity occurred in India.[90] Zakariyyā' b. Muḥammad al-Qazwīnī (c. 1202–83) notes that in his time merchants only sailed to Jāwah (Java) rather than further to China because it was "inaccessible because of the distance of the trip and difference of religions".[91] There they acquired goods of Southeast Asia such as aloeswood, camphor, cloves, nutmeg, and mace, along with medicinal plants from China and porcelain. Sri Lanka also was again an extremely popular terminus, as it had been in Sasanian times. The segmentation of the sea trade is mirrored on the overland routes, as political complications from the 11th century on made that route more difficult to use.[92]

The long voyage to China was especially risky, and to have accomplished it seven times as Buzurg b. Shahriyār's captain 'Abharah (a native of Kirmān in Iran based in Sīrāf on the Persian Gulf) supposedly did was a major achievement.[93] The dangers were great in all voyages in the Indian Ocean, but the profits were sufficient so that it was worthwhile. Allowing for some exaggeration due to its

---

westerners in general, or ships carrying westerners, western cargo, or substitutes for western cargo.

88 G. F. Hourani, *Arab Seafaring in the Indian Ocean in Ancient and Early Medieval Times*, expanded edition ed. J. Carswell (Princeton: Princeton University Press, 1995), 46–50, 61–3.

89 Chaudhuri 40–1 has maps illustrating these patterns.

90 Goitein and Friedman, *India Book* 6–7, 125.

91 Zakarīyā ibn Muḥammad al-Qazwīnī, *'Ajā'ib al-makhlūqāt* and *Āthār al-bilād*, ed. T. Wüstenfeld in *Zakarija ben Muhammed ben Mahmud el-Cazwini's Kosmographie*, 2 vols. (Göttingen, 1848–9; repr. in 1 vol., Wiesbaden: Martin Sändig, 1967), 2.18.

92 Cf. Kuwabara, "P'u Shou-keng" part 2, 14 n. 10.

93 Buzurg b. Shahriyār, *Kitāb 'Ajā'ib al-Hind: Livre des merveilles de l'Inde*, ed. P. A. Van der Lith and trans. L. M. Devic (Leiden: Brill, 1883–6; repr. Tehran: M. H. Asadi's Historical Series, 1966), 85.

preservation as a sailor's yarn, we may cite the story of an Omani merchant, Isḥāq b. Yahūdā, of obviously Jewish background.[94] He left Oman with no more than about 200 *dīnārs*, and returned in his own ship laden with riches from India. Seeking to avoid customs, he paid the governor of Oman more than a million *dirhams*. He also sold a certain Aḥmad b. Marwān a hundred thousand *mithqāls* of musk, which Aḥmad believed was all that Isḥāq had, given the enormous quantity. He discounted it a *dirham* per *mithqāl*, so the discount alone amounted to a hundred thousand *dirhams*. It was said that he had returned with musk worth a million *dīnārs* and silk and porcelain of comparable of value, along with the same in precious stones, and many rarities of China. Even allowing for the exaggeration one expects in a book of sailors' yarns, the moneys involved in the musk trade must have been significant given the cost of the substance. The same story also includes an anecdote that Isḥāq offered Aḥmad b. Hilāl, the governor of Oman, a black porcelain vessel with a lid glinting with gold. Aḥmad was told that it was a dish of *sikbāj* (a vinegared stew) cooked for him in China. Aḥmad was put off because that would have made it two years old from the time Isḥaq was in China, but when the lid was lifted, the vessel contained a fish of gold with eyes of ruby surrounded with fine musk. Undoubtedly most merchants operated on a much smaller scale, but one ship could carry a vast quantity of wealth, and often represented the combined investment of several merchants. The fact that musk figures prominently in these anecdotes illustrates its key place in the discourse of treasures.

The great entrepots were seats of wealth. The overland trade routes formed a network that converged in certain cities and they became major centers of trade. These cities are what M. Rostovtzeff termed "caravan cities" in his book on the Roman Near East.[95] He had in mind cities such as Palmyra, on the frontier between the Roman and Sasanian empires, which was a major center of trade reachable only by overland caravans. Similar cities can be found throughout the Near East into Central Asia and to the frontiers of China. An ideal city was located at the convergence of routes, with easy access to the goods of different regions. In Central Eurasia they often were situated in habitable areas adjacent to steppelands where nomadic peoples could trade. The great

94 Buzurg 107–111. See also Goitein and Friedman, *India Book* 124–5.

95 M. Rostovtzeff, *Caravan Cities* (Oxford: Oxford University Press, 1932). That trade alone was the *raison d'être* of these antique Near Eastern cities has been challenged by F. Millar, "Caravan Cities: The Roman Near East and Long-Distance Trade," in *Modus operandi: Essays in honour of Geoffrey Rickman* (London: Institute of Classical Studies, 1998), 119–37. Undoubtedly the "caravan city" is an ideal type, and actual cities always had more complex economic identities, especially in antiquity.

emporiums of the overland trade included the cities of Bactria/Ṭukhāristān and Transoxiana/Sughd, as well as cities on the western fringes of China and the oasis towns of the the Tarim Basin, Turfan Depression, and Jungharia.

Sughd or Sogdiana was the principal part of the *Mā warā' al-nahr* (Transoxiana) of the Arabic geographers, meaning the land beyond the Oxus River (Amu Darya), called the Jayhān in Arabic. As we have seen, the Sogdians were the earliest Central Eurasian people on record who traded musk; the Sogdian Ancient Letter II mentions the export of musk, probably from China, to Samarkand. Ibn Māsawayh emphasized that the Sogdians played a key role in the musk trade of the 9th century by bringing musk from Tibet into the Islamic world by the overland routes.[96] Al-Iṣṭakhrī (mid 10th century), like Ibn Māsawayh a century before, stressed the role of Sogdiana in the musk trade: "Among the people [of Transoxiana] is musk which is brought to them from Tibet and the Khirghiz; and from it is what is transported to the rest of the metropolises (*amṣār*)."[97]

The entire region of Sogdiana owes its preeminence at least in part to its straddling the routes from Iran into China, and from India into the steppelands.[98] The cities of Sogdiana—Samarkand, Bukhara, and especially its smaller town Paykand—grew rich on the trans-Eurasian trade. The people of Paykand, called Baykand in Arabic, were "all merchants," and when Qutaybah b. Muslim took the city in 705–6, many of them were absent in China and had to ransom their kinfolk.[99] One man offered the Arabs a thousand pieces of Chinese silk supposedly worth a million *dirhams*.[100] In antiquity and the early medieval period, their prosperity was assured by trade as well as through ample access to mineral wealth and the agricultural productivity of the region, both of which declined in later times.[101] This combination of resources for trade and geographical centrality upon the trade routes is the origin of the flourishing Sogdian trade of late antiquity and the early medieval period, in which the Sogdians created networks stretching from the Byzantine Empire to

96 See ch. 4.

97 Al-Iṣṭakhrī, *Kitāb Masālik al-mamālik*, ed. M. J. De Goeje (Leiden: Brill, 1870; repr. 1967), 288. See also Ibn Ḥawqal 465.

98 É. de la Vaissière, *Sogdian Traders: A History*, trans. J. Ward (Leiden: Brill, 2005), 16–7. The mineral wealth of Transoxiana must also be remembered.

99 Narshakhī 26, Frye 18, de la Vaissière 268–71.

100 Ṭabarī II.1188.

101 I. Blanchard, *Mining, Metallurgy and Minting in the Middle Ages Vol. 1. Asiatic Supremacy. Mawara'an-nahr and the Semiryech'ye 425–1125* (Stuttgart: Steiner, 2001).

China, and north into the Steppes, and south into the Indian Ocean.[102] Routes from the east came into Sogdiana both over the Pamirs from Kashgar and from the north through the Semirechiye. Another route came through Wakhān and followed the Oxus, and then entered Sogdiana from the south, via Balkh, and then across the Oxus and through the famous Iron Gates. The Iron Gates (*bāb al-ḥadīd* in Arabic, *dar-i āhanīn* in Persian, and *temir kapıg* in Turkic) were located on the road between Termez and Samarkand at the Buzghāla-khāna Pass, and were built as early as the beginning of the 7th century, for they were described by the Buddhist traveler Xuanzang.[103] The Sogdians were also active on the route between Yarkand and the Indus Valley through the Karakorum; a large number of Sogdian inscriptions are present at Hunza.[104] The wealth encountered by the Arabs in their conquest was great. The Arabs kept these merchant communities intact in order to benefit from their continued economic activity.[105]

For the Arabs, the oasis city of Marw (Merv), now in Turkmenistan, became the most important focus of their settlement in Central Eurasia.[106] It sat upon the routes connecting Transoxiana with Iran and was adjacent to the steppelands as well. Another route led to Balkh and from there one could set off through the Hindukush into Sind. Marw was home to a community of Sogdian merchants; perhaps seeing the potential for profit in becoming part of the Islamic world, they helped finance the Arab expeditions across the Oxus.[107]

The cities of the Tarim Basin and the Gansu Corridor owed much of the reason for their existence to trade. Situated between mountains and the Taklamakan Desert, they survived on the runoff from mountain snows which formed streams that ultimately evaporated away in the heat of the interior desert. Owing to this desert and its changing dimensions, there are impressive

---

102 In addition, de la Vaissièrre, *Sogdian Traders*, and the literature on the Sogdians in South and Southeast Asia cited above, see É. de la Vaissière and É. Trombert, eds., *Les Sogdiens in Chine* (Paris: École Française d'Extrême-Orient, 2005), for a collection of papers on aspects of the Sogdian diaspora in China.

103 T 2087, 872a. See R. N. Frye, "Dar-i Āhanīn," in *EI* 2nd ed. s.v.

104 K. Jettmar, "Sogdians in the Indus Valley," in *Histoire et cultes de l'Asie Centrale préislamique* (Paris: CNRS, 1991), 251–3 & pls. 103–6; N. Sims-Williams, "The Sogdian Merchants in China and India" 52–7.

105 De la Vaissière 278.

106 An international project has directed extensive excavations in Marw; see G. Herrmann, *Monuments of Merv: Traditional Buildings of the Karakum* (London: Society of Antiquaries, 1999) and the bibliography given there for excavation reports.

107 De la Vaissière 273–5; M. A. Shaban, *The Abbasid Revolution* (Cambridge: Cambridge University Press, 1970), 48.

archaeological remains for some of these cities.[108] Kashgar formed the terminus of the route across the Pamirs from Farghana. From there, the route branched around the Taklamakan, either south towards Khotan or north towards Kucha, as discussed above.

Traders using the northern route via Kucha,[109] or traveling north of the Tianshan from the Semirechye through Jungharia, eventually arrived at the Turfan Depression, location of the Qocho (Gaochang) kingdom in the early medieval period.[110] Al-Masʿūdī mentions that *Kawshān* (*Gaochang* 高昌, Early Mandarin *kaw tɕʰiaŋ*)[111] was a kingdom between the land of Khurāsān and China held by the Tughuzghuz (Tokuz Oghuz), who were the Uyghurs. He even knows that Manichaeism was practiced there;[112] Buddhism was also a significant religion in Qocho. Several sites in the Turfan area are known, including ancient Gaochang itself, and excavations have produced numerous documents, coins, and other items making the history of early medieval Turfan well illustrated.[113] Numerous Sasanian and Arab-Sasanian silver coins come from early medieval Turfan and are evidence of its important position on the trade routes.[114] The prosperity of Turfan was a concern of the early medieval powers

---

108 M. Hoyanagi, "Natural Changes of the Region along the Old Silk Road in the Tarim Basin in Historical Times," *MRDTB* 33 (1975): 85–113. There is a vast bibliography on the archaeology of Xinjiang and Northwestern Gansu. For an overview of the archaeological sites in this area, see M. Yaldiz, *Archäologie und Kunstgeschichte Chinesisch-Zentralasiens (Xinjiang)* (Leiden: Brill, 1987) and V. Hansen, *The Silk Road: A New History* (Oxford: Oxford University Press, 2012), for an overview of the history of the region, its sites, and its commerce. For an account of the European exploration of the archaeology of the Tarim Basin lands, see P. Hopkirk, *Foreign Devils on the Silk Road* (Amherst: University of Massachusetts, 1980).

109 A single site near Kucha produced more than a hundred and thirty caravan passes in Kuchean (Tokharian B) dating to the first half of the 7th century, see G. Pinault, "Épigraphie koutchéenne," in *Mission Paul Pelliot: Documents archéologiques VIII. Sites divers de la région de Koutcha* (Paris, 1986): 61–196, esp. 65–121.

110 See V. Hansen, "The Impact of the Silk Road trade on a local community: the Turfan oasis, 500–800," in de la Vaissière and Trombert 283–10, and Hansen, *Silk Road* 83–111 for an overview.

111 E. Pulleyblank, *Lexicon of Reconstructed Pronunciation in Early Middle Chinese, Late Middle Chinese, and Early Mandarin* (Vancouver: UBC Press, 1991), 104, 49.

112 Masʿūdī 1.155 (§312); see also 1.190 (§396), and V. Minorsky, "Tamīm ibn Baḥr's Journey to the Uyghurs," *BSOAS* 12:2 (1948): 288.

113 See H. Härtel, ed., *Along the Ancient Silk Routes: Central Asian Art from the West Berlin State Museums* (New York: Metropolitan Museum of Art, 1982), for an exhibition catalog with many artifacts from Turfan.

114 J. K. Skaff, "Sasanian and Arab-Sasanian Silver Coins from Turfan: Their Relationship to International Trade and the Local Economy," *Asia Major* 3rd series 11:2 (1998): 67–115.

in the region, such as the Turks and Chinese. During the early 7th century Gaochang flourished under the watchful eye of the Turks, who profited from the trade that passed through it.[115]

The routes converged in Shazhou near the famous site of Dunhuang, home to an extensive complex of Buddhist cave temples from which important artwork and a vast quantity of manuscripts have come.[116] From Dunhuang, traders travelled to Lanzhou via Jiuquan and Guzang and crossed the Yellow River into the Chinese heartland. It is upon this route that the first document of the musk trade, the Sogdian letter of Nanai-vandak, would have travelled to Samarkand. During the Tang period the passage of merchants was along official roads with the use of passport permits issued by the state in a well organized system designed to control commerce.[117]

India was linked and networked by a system of well-established overland trade routes since antiquity; al-Bīrūnī provides a description of these routes for the early medieval period. He traces the routes from the city of Kannauj, capital of the Gurjara-Pratīhāras, called al-Jurz by geographers; al-Bīrūnī describes it as the center of India.[118] They adjoined the frontiers of Sind and the Rāshṭrakūṭas and were enemies of the Muslims. Even so, they strove to maintain the security of merchants in their realm.[119] The route to the east headed to Talwat; heading north from there one entered Nepal and passed into Bhūtīshar,[120] which was the frontier of Tibet.[121] In much later times, when leaving the area of Nepal, merchants could continue through the mountains westward to Kabul, where the route split, one half going to Balkh and into Khurāsān, and the other northward into Central Asia.[122] In the 10th century, goods traveling overland

115 Skaff 88–9.

116 There is a vast literature as the study of Dunhuang and its manuscripts and archaeological remains are an integral part of the study of the Silk Roads; see Hansen, *Silk Road*, 167–97 and for a recent exhibition catalog with important essays see N. Agnew, et al., eds. *Cave Temples of Dunhuang: Buddhist Art on China's Silk Road* (Los Angeles: Getty Conservation Institute, 2016).

117 M. Arakawa, "The Transit Permit System of the Tang Empire and the Passage of Merchants," *MRDTB* 59 (2001): 1–21.

118 Bīrūnī, *Hind* 157. On the Gurjara-Pratīhāras, see Wink 277–302.

119 See the *Akhbār aṣ-Ṣīn wa l-Hind: Relation de la Chine et de l'Inde*, ed. and trans. J. Sauvaget (Paris: Les Belles Lettres, 1948), §26.

120 The land of the Bhauṭṭa-īśvara, lord of the Tibetans.

121 Bīrūnī, *Hind* 160.

122 Attested in J. B. Tavernier, *Travels in India*, trans. W. Ball and Ed. W. Crooke, 2 vols. (Oxford: Oxford University Press, 1925; repr. New Delhi: Munshiram Manoharlal, 1995), 2.202–3.

to northwestern India reached Vayhind (Ohind), the capital of Gandhārā.[123] It was under the control of Kannauj and served as a distribution center even though musk was not apparently produced there. The *Ḥudūd al-ʿālam* says: "The cargoes of India arrive more at this region, including valuable musk, jewels, clothing" (*jihāz-hāye hindustān bīshtar badhīn nāḥiyat uftadh az mushk wa gawhar wa jāma-hāye bā qīmat*). Another place mentioned by the *Ḥudūd al-ʿālam* as a musk producer is *B.lhārī*,[124] perhaps referring to the capital of the Rāshṭrakūṭas: "A large and populous town and a residence of merchants from India, Khurāsān, and Iraq. It produces much musk."[125] Once again, this city was a port through which musk was transshipped rather than produced.

The maritime trade likewise was focused on certain port towns.[126] These cities were often termini for long-haul voyages across the Indian Ocean. The port of Dārīn in eastern Arabia was discussed in Chapter 3; it seems to have dwindled in importance in early Islamic times. The major port of Oman in pre-Islamic and early Islamic times was Dabā; it was later eclipsed by Ṣuḥār.[127] Ibn Ḥabīb notes of Dabā: "The merchants of Sind, Hind, and China, and the people of the East and West come to it."[128]

New centers of trade developed under the caliphate. The conqueror of Ubullah, the port of Baṣrah, ʿUtbah b. Ghazwān, wrote to ʿUmar that it was the port of ocean ships from Oman, Baḥrayn, Fārs, Hind, and China.[129] According to the historian al-Ṭabarī, one of al-Manṣūr's motivations in selecting the site for the *Madīnat al-Salām*, Baghdād, was its position on the Tigris with its access to the maritime trade of the Gulf stretching all the way to India and

---

123 Cf. al-Muqaddasī, *Aḥsan al-taqāsīm fī maʿarifat al-aqālīm*, ed. M. J. De Goeje (Leiden: Brill, 1877; repr. 1967), 477, 479.

124 *Ḥudūd al-ʿālam min al-mashriq ilā al-maghrib*, ed. M. Sutūdah (Tehran: Dānishgāh-i Tihrān, 1983), 10:24 and trans. V. Minorsky, *The Regions of the World, A Persian Geography 372 A.H.–982 A.D*, 2nd ed. (London: Gibb Memorial Series, 1970). Cf. also Ibn Bādīs' *Bahārī* musk.

125 Minorsky's trans. 88 slightly modified.

126 Hourani 69–73; Wink 53–5; H. Park, "Port-City Networking in the Indian Ocean Commercial System as Represented in Geographic and Cartographic Works in China and the Islamic West, c. 750–1500," in K. R. Hall, ed., *The Growth of Non-Western Cities: Primary and Secondary Urban Networking, c. 900–1900* (Lanham: Lexington Books, 2011), 21–53.

127 See Yāqūt, *Muʿjam al-buldān*, ed. 5 vols. (Beirut: Dār Ṣādir, n.d.), 2.435–6.

128 Muḥammad b. Ḥabīb, *Kitāb al-Muḥabbar* (Haydarābād, 1942; repr. Beirut: al-Maktab al-Tijārah li-l-Ṭibāʿah wa-l-nashr, n.d.), 265.

129 Al-Balādhurī, *Kitāb Futūḥ al-buldān*, ed. M. J. De Goeje (Leiden: Brill, 1866; repr. 1968), 341 and al-Dīnawarī, *Kitāb al-Akhbār al-ṭiwāl*, ed. V. F. Girgas (Leiden: Brill, 1888–1912), 123.

China.[130] Large ships did not sail all the way to Baghdād, but rather put in at the northern tip of the Gulf. Asian cargoes were taken up the Euphrates as well as the Tigris.[131] Ibn Khurradādhbih's account of the Rādhāniyyah speaks of the importance of Baṣrah with its port at Ubullah. It served the Islamic state well until the revolt of the Zanj disrupted southern Iraq and probably facilitated the greater development of ports such as Sīrāf. In any case, Sīrāf was also a stop on the route to Baṣrah.[132]

On the eastern side of the Persian Gulf was Sīrāf, which, due to literary evidence as well as archaeological excavation, is perhaps the best known of the trading emporiums.[133] As early as 850 CE, the *Akhbār al-Ṣīn wa-l-Hind* gives a discussion of the structure of the shipping of the China trade: "As for the place which they return to and put ashore at, they mentioned that most of the China ships load at Sīrāf, and that their goods are transported from Baṣrah and Oman and other places to Sīrāf, and then they are packed into the China ships in Sīrāf."[134] This is explained by the depth of the water at Sīrāf; the large China ships needed deep water. The great boom of Sīrāf began in the latter part of the 8th century, when the Abbasid caliphate brought stability thoughout the region.[135] The great merchants controlled much wealth and Sīrāf was known for its splendor.[136] Its commerce was especially founded on the trade in perfumes and aromatics.[137] Ibn al-Balkhī recorded that the value of goods in the period 908–32 was no less than 2,530,000 dīnārs per year, a truly vast sum of money.[138] The prosperity of Sīrāf was broken by an earthquake in the mid-10th

130 Ṭabarī III.272, 275. Al-Manṣūr was troubled by the need to find a site for his new capital that would be easily accessible to both the land and sea routes and the site for Baghdād was at a confluence of the Tigris with major canals linking to the Euphrates and also lands further to the East.

131 Mas'ūdī 1.118–9 (§231) notes that a local informant told the conqueror Khālid b. al-Walīd that ships sailed up to Najaf with cargoes of Sind and Hind.

132 Buzurg 16–17.

133 D. Whitehouse, *Siraf: history, topography and environment* (Oxford: Oxbow Books, 2009) and M. Tampoe, *Maritime Trade between China and the West: An Archaeological Study of the Ceramics from Siraf (Persian Gulf), 8th to 15th centuries A.D.* (Oxford: BAR, 1989).

134 Sauvaget §13 on 7. See also al-Ya'qūbī, *Tārīkh*, ed. M. T. Houtsma, 2 vols. (Leiden: Brill, 1883; repr. 1969), 1.207.

135 Tampoe 101.

136 See S. M. Stern, "Rāmisht of Sīrāf, a Merchant Millionaire of the Twelfth Century," *JRAS* (1967): 10–14.

137 Ibn al-Balkhī, *Fārsnāmah*, ed. G. Le Strange and R. A. Nicholson (London: Gibb Memorial Trust, 1921; repr. 1962), 136.

138 Ibn al-Balkhī 171.

century; thus its heyday lasted about a century. But the burgeoning Fatimid caliphate in Egypt reset the center of gravity as well.[139] An emporium developed to the south at the island of Qays, and Hormuz also became an important center as Ṣuḥār declined.

The *Akhbār al-Ṣīn wa-l-Hind* notes that the China ships were loaded at Sīrāf and explicitly states that some of the cargoes came from Oman. The itinerary given in that source states that the China ships left Sīrāf for Masqaṭ as a last stop for water before sailing to India.[140] One could stop at Ṣuḥār, but it does not seem to have had quite the importance it developed over the next century. Ṣuḥār was on the Arabian coast near the entrance to the Persian Gulf in Oman.[141] This port received ships sailing from India and also participated in the regional transit trade with both the Gulf lands and with the west. Its position made it an ideal location for ships to await the monsoons favorable to their voyage. Literary evidence speaks to its prosperity.[142] Oman received many Asian goods from the sea trade including musk.[143] Ṣuḥār was called the antechamber (*dihlīz*) of China, a clear indication of its important status in the trade with Asia.[144] Sailors from Oman and Sīrāf were particularly noteworthy upon the Indian Ocean.[145]

Aden in Yemen was the major port of southern Arabia.[146] Ships from Aden sailed to India, connecting Aden within the Indian Ocean trade.[147] Aden also served as a node connecting the local sea trade with the overland caravan trade of the Arabian Peninsula. It also received ships coming directly from India. From Aden sailors would voyage to Jiddah, the port of Mecca on the Red Sea.[148]

---

139 Tampoe 112.

140 *Akhbār al-Ṣīn wa-l-Hind* §13.

141 A. Williamson, "Sohar and the Sea Trade of Oman," PSAS 4 (1974): 78–96, Tampoe 105–6, and M. Hoffmann-Ruf and A. Al Salimi, eds., *Oman and Overseas*, Studies on Ibadism and Oman, Vol. 2 (Hildesheim: Olms, 2013), for a collection of essays on the maritime aspects of the history of Oman.

142 J. C. Wilkinson, "Ṣuḥār in the Early Islamic Period," *South Asian Archaeology 1977* (Naples: Istituto universitario orientale, 1979), 887–907 and A. A. Ziaee, "Omani Trade and Cultural Relations with East Asian Countries," in Hoffmann-Ruf and Al Salimi, 219–25.

143 Muqaddasī 97.

144 Muqaddasī 92.

145 Mas'ūdī 1.151 (§305).

146 R. E. Margariti, *Aden & the Indian Ocean Trade: 150 Years in the Life of a Medieval Arabian Port* (Chapel Hill: University of North Carolina, 2007) covers a slightly later period, but illustrates how commerce was the defining feature of Aden and other comparable cities in the region. On the role of perfumery in Aden, see D. Jung, *An Ethnography of Fragrance: The Perfumery Arts of 'Adn/Laḥj* (Leiden: Brill, 2011).

147 Margariti 2 and Goitein and Friedman 145–6.

148 Buzurg 16.

In Fatimid times, Aden thrived as the first stop in the Middle East for ships from India before they proceeded into the Red Sea to Egypt. Ibn Khurradādhbih mentions the trade in Aden, saying that "they have ambergris, aloeswood, musk, and goods of Sind, Hind, China, Zanj and Ethiopia (*ḥabashah*)."[149] Al-Idrīsī, probably dependent on Ibn Khurradādhbih, also mentions the musk trade of Aden in his long list of "Chinese" goods traded there that we mentioned in Chapter 2.[150] Musk, at least, was a genuine product of China, unlike much of the Indian and Southeast Asian *materia medica* he mentions. Aden, like Ṣuḥār in Oman, was known as the antechamber (*dihlīz*) of China,[151] and it continued to be a major port for musk in the time of Tomé Pires.[152] This trade must be a continuation of the trade through South Arabia better known in Classical Antiquity; there is no way to know when musk began to reach Arabia this way, but musk traded through Aden and South Arabia never developed a fame like the musk that arrived through Dārīn, so perhaps musk reached Eastern Arabia before the south, an area with a more tenuous relationship to the Sasanian Empire.

The profitability of this trade is related by the geographer al-Muqaddasī.[153] He was on a ship upon the sea off Yemen, and he met a certain Abū ʿAlī al-Ḥāfiẓ al-Marwazī, who, by his name, had been a merchant of the Central Asian city of Marw. Al-Marwazī expresses his concern that al-Muqaddasī will be drawn to the commerce of the sea because of its great profitability: "I fear that when you enter Aden you will hear that a man goes with a thousand *dirhams* and returns with a thousand *dīnārs*, and another enters with a hundred and returning with five hundred, and another with frankincense (*kundur*) and returning with the same quantity of camphor, and you yourself will wish for increase." Frankincense was of little value in Arabia compared to the luxurious imported camphor. Al-Marwazī concludes with a moralizing anecdote as to why this man dissuaded him: his own business partner had died, and then he realized that there was no point in seeking wealth in this world. It is significant that the author emphasizes the importance of the trade in aromatics through his example. This story indicates the great profitibility of the trade with Asia, but it also shows an ambivalent attitude to its value not often evident in Islamic

149 Ibn Khurradādhbih 61.

150 Al-Idrīsī, *Kitāb nuzhat al-mushtāq fī ikhtirāq al-āfāq*, 9 vols. (Naples: Istituto Universitario Orientale di Napoli, 1970–84), 1.54. See above, p. 58.

151 Muqaddasī 34.

152 Tomé Pires, *The Summa Oriental of Tomé Pires and the Book of Francisco Rodrigues*, trans. A. Cortesão, 2 vols. (London: Hakluyt Society, 1944)), 1.16.

153 Muqaddasī 97–8.

literature. Such feelings hardly slowed the overall commerce with Asia, as there was always a market for its textiles, perfumes, and other rarities.

Daybul was perhaps the most important port of the Indus Delta from Late Antiquity into the medieval period; it has been identified with the site of Banbhore in Pakistan. Ibn Māsawayh mentions that Tibetan musk shipped through India was traded overland up to the great port of Sind, Daybul. We have seen how the musk trade is attested for the 6th century through Sind in Cosmas Indicopleustes.[154] Ibn Māsawayh also knew of a musk from Kashmir, which he regarded as inferior even to the imitation varieties and which are discussed later in this chapter. In Chapter 3, the musk trade through Daybul was mentioned.[155] Daybul is attested, under the name Dēb, as a 3rd century port in Manichaean literature.[156] Mānī apparently visited it in his travels; he certainly sent two of his followers there after returning from India.[157] At this time, then, the maritime trade routes with the Sasanian Gulf were established. Daybul itself passed under the rule of the Sasanians when Bahrām I (421–38) married an Indian princess and received the port and surrounding lands as a dowry.[158] Al-Idrīsī noted that "The ships of the Omanis sail to it with their cargoes and goods, and the ships of China and India return to it with garments, goods of China, and Indian simple aromatics (*al-afāwīh al-ʿiṭriyyah al-hindiyyah*)."[159] The 10th century *Ḥudūd al-ʿālam* describes it as a dwelling (*jāygāh*) of merchants and says "The goods of Hindustān and the sea are sent to it in great quantity."[160] This is precisely the function of the emporium town.

Certain cities in Asia beyond the *Dār al-Islām* were even more striking examples of the meetings of cultures. The ports of India and Sri Lanka are particularly noteworthy in this regard. We have seen how al-Bīrūnī described the overland routes of India as converging in Kannauj; from Kannauj the routes

154 Indeed, Muqaddasī 33 notes that no region of the Islamic world had more musk than Sind.

155 On Daybul, see M. Kervran, "Multiple Ports at the Mouth of the River Indus: Barbarike, Deb, Daybul, Lahori Bandar, Dul Sinde," in *Archaeology of Seafaring: The Indian Ocean in the Ancient Period* (Delhi: Pragati, 1999), 70–153, esp. 80–9, and Wink 181–3.

156 The identification of Dēb with Daybul was made by W. Sundermann, "Zur frühen missionarischen Wirksamkeit Manis," *Acta Orientalia Academiae Scientiarum Hungaricae* 24 (1971): 84–5 n. 31.

157 W. Sundermann, *Mitteliranische manichäische Texte kirchengeschichtlichen Inhalts* (Berlin: Akademie-Verlag, 1981), 57 and idem, "Mani, India, and the Manichaean Religion," *South Asian Studies* 2 (1986): 12–13.

158 E.g., Ṭabarī 1.868.

159 Idrīsī 2.167.

160 *Ḥudūd al-ʿālam* §27.4.

also connected to the sea ports. The road to the west took one to Sind, with its port of Daybul, and to the southeast, to the Konkan coast ports. The major ports of the Konkan coast controlled by the Rāshṭrakūṭas were Kambāya, the Cambay of western writings, and Berūj, the Broach of the Greek geographers.[161] Musk acquired from India by the Arabs must have traveled in this way along with the overland route.

The major port of the south western Indian coast was Kūlam-Malay, the Quilon of the Portuguese, now Kollam in Kerala.[162] The *Akhbār al-Ṣīn wa-l-Hind* states that it is the goal for ships setting out from Masqaṭ (Muscat), and that it is a one-month voyage with favorable winds. Kūlam-Malay was a port for "Chinese ships" (*al-sufun al-ṣīniyyah*) that were charged a fee by an armory (*maslaḥah*).[163] These ships could be either Chinese ships or ships engaged in the China trade; the latter is more likely for the mid-9th century. The *Akhbār al-Ṣīn wa-l-Hind* tells us the port charges for these vessels: a thousand *dirhams* for these Chinese ships, and from ten to twenty *dīnārs* for others (there were about 13 ½ *dirhams* per *dīnār*). The higher cost for ships coming from China must reflect the valuable cargoes they carried. As in the north, Muslim merchants formed permanent settlements in Malabar.[164]

Sri Lanka was often the next important stop; it was here that many short-haul voyages stopped and acquired goods brought from further east. Other voyagers would restock and then sail on to Southeast Asia. Arabic sources praise the wealth of the island, called by them Sarandīb; we have already noted it as a source of precious stones and an alleged source of musk along with other aromatics. Sri Lanka was frequented by merchants from the Near East, India, Southeast Asia, and China; it was a true meeting place of cultures.[165] Most of the significant ports of Sri Lanka appear to have been in the south in pre-Islamic times.[166] A port of northern Sri Lanka is at Mantai, a site that has been excavated to reveal occupation from the 5th century BCE up to the 11th century CE Fragments of different varieties of Chinese ceramics illustrate the trade with East Asia.[167] On the southeastern tip of India, opposite

161 *Akhbār al-Ṣīn wa-l-Hind* §25 and 51 n. 6, and Wink 306.

162 Wink 71.

163 *Akhbār al-Ṣīn wa-l-Hind* §14.

164 Wink 70–8.

165 O. Bopearachchi, "Archaeological Evidence on Shipping Communities of Sri Lanka," in *Ships and the Development of Maritime Technology in the Indian Ocean* (London: Routledge, 2002), 92–127.

166 Walburg 288–9.

167 Cf. Bopearachchi 115, M. Prickett-Fernando, "Durable Goods: The Archaeological Evidence of Sri Lanka's Role in the Indian Ocean Trade," in *Sri Lanka and the Silk Road of*

Sri Lanka, the Chola lands rose to prominence in the 11th century while encouraging Muslim traders.[168]

The Andaman and Nicobar Islands (*Lankabālūs*) were visited in the Sea of Harkand, the sea between Sri Lanka and the Sea of Shalāhīṭ, named for the straits between Sumatra and the Malay Peninsula. Here sailors could restock with coconuts, sugarcane, bananas, and coconut toddy, and also acquire ambergris.[169]

There were two ports of particular importance on the western coast of the Malay Peninsula in Arabic sources: Qāqullah and Kalāh. The localization of them is fraught with difficulty; about all that it is certain is that Qāqullah was north of Kalāh.[170] Kalāh has been identified with Kelang, but this remains uncertain though it was surely in the vicinity.[171] The difficulty of identifying it with a single site is reflected in the statement of the *Akhbār al-Ṣīn wa-l-Hind* that "The kingdom and the coast are both called Kalāh-bār". The name *qlh*, Kalah, is first attested in the west, in a Nestorian source, in the middle of the 7th century.[172] Then it was considered the eastern extremity of India. Two centuries later the *Akhbār al-Ṣīn wa-l-Hind* gives Kalāh-bār, the Kalāh land (cf. Zanzibār= the land of the Zanj).[173]

The island of Tiyūmah is off the southeastern tip of the peninsula. It is now called Pulau Tioman, a descendant of its ancient name. It was the next stop on the route to China past Kalāh.[174] We have already discussed in the Chapter 2 the local products to be found in these Southeast Asian ports: camphor, aloeswood, sandalwood, cloves, and mace, as well as foodstuffs like coconuts, bananas, rice, and sugarcane. These ports were also, as in Portuguese and later times, sources of other goods brought to the region for trade. Among the

---

the Sea (Colombo: Sri Lanka National Commission for UNESCO, 1990), 61–84. This volume includes many interesting papers on the role of Sri Lanka in Indian Ocean commerce. See also on the significance of Mantai the more skeptical comments of Walburg, 30–9.

168 Wink 78–80 and K. R. Hall, *Trade and Statecraft in the Age of the Cōḷas* (Delhi: Abhinav, 1980).

169 *Akhbār al-Ṣīn wa-l-Hind* §14.

170 An overview of the problem is in G. R. Tibbetts, *A Study of the Arabic Texts containing material on South-East Asia* (London: Royal Asiatic Society, 1979), 118–36.

171 See S. Q. Fatimi, "In Quest of Kalāh," *Journal of Southeast Asian History* 1:2 (1960): 62–101, B. E. Colless, "Persian Merchants and Missionaries in Medieval Malaya," *JMBRAS* 42:2 (1969): 10–47, and the critical comments in Tibbetts 128.

172 Colless 20–1.

173 *Bār* is Persian, and is an indication of the important role played by the Persians in the exploration of the Indian Ocean.

174 Tibbetts 136–7.

goods found by the European explorers was musk, presumably from China and its Southeast Asian borderlands, and there is no reason to suspect that musk was not traded there before, in the time with which our work is concerned. The great age of Chinese maritime trade in these waters began around the year 1000 under the Song, at about the time the long-haul voyages from the Middle East were in decline.

From Tiyūmah, the next destination was Ṣanf (Champa), home of the Ṣanfī aloeswood we discussed in Chapter 2. Ibn Khurradādhbih mentions that the next stop was Lūqīn, identified with Long Biên, now part of Hanoi.[175] Ibn Khurradādhbih described it as the first port of China, and the goods of China were available there.

In southern China there was a thriving community of foreign merchants. The sailors of the early medieval period who traded in the Indian Ocean had a major trade emporium in southern China at *Khānfū*,[176] attested as early as the so-called *Akhbār al-Ṣīn wa-l-Hind* of 851, which is identified with Guangzhou 廣州.[177] Here communities of Muslim merchants resided as a trade diaspora: a network of resident merchants living abroad to facilitate commerce. Al-Masʿūdī notes that Khānfū was visited by merchants from Baṣrah, Sīrāf, Oman, cities of India, Zābaj, Champa, et al.[178] The number of Muslim, Christian, Jewish, and Zoroastrian merchants killed there in the Huang Chao rebellion (878) is given at either 120,000 or 200,000.[179]

175 Ibn Khurradādhbih 69. See Kuwabara part 1, 16 and E. Schafer, *The Vermilion Bird: T'ang Images of the South* (Berkeley: University of California Press, 1967), 32 and 78, where he notes that during the second half of the 8th century traders from the South Seas frequented Long Biên more than Guangzhou.

176 *Akhbār al-Ṣīn wa-l-Hind* §11. The manuscript has *khānqūā*, but this is emended by Sauvaget to *khānfū*, which is elsewhere attested. Ibn al-Faqīh, *Mukhtaṣar kitāb al-buldān*, ed. M. J. De Goeje (Leiden: Brill, 1885; repr. 1967), 68, has *khānfū* as the name of sweet water encountered off the coast of China. See Pelliot, "Deux itinéraires" 215; Kuwabara part 1, 10–12;

177 The name Arabic name is explained as a transcription of *Guangfu* 廣府, which is unattested as such, but is said to be an abbreviation of a longer name, the exact identity of which is unclear. Cf. Kuwabara part 1, 10–12.

178 Masʿūdī 1.163–4 (§329).

179 Al-Sīrāfī, *Riḥlah*, ed. ʿAbdallāh al-Ḥabashī (Abu Dhabi: Manshūrāt al-Majmaʿ al-Thaqāfī, 1999), 54, gives the former number, Masʿūdī 1.164 (§329) the latter. The discrepency shows how liable numbers are to corruption in literary sources, since al-Sīrāfī was almost certainly Masʿūdī's source for this information. See also H. Levey, *The Biography of Huang Ch'ao* (Berkeley: University of California, 1955), 110–29, for these Arabic accounts, and Hansen 165.

All of these ports were linked together in a network of trade. Given the wide-ranging, and, after the 8th and 9th centuries, increasingly fragmented nature of Islamic rule, there was plenty of room for different routes. The trade which flowed to the Mediterranean world will have gone usually by the Red Sea. That trade reached its high point under the Fatimids, who were especially concerned with its promotion. But the trade through the Persian Gulf should not be discounted after the decline of the caliphate in Baghdad. The dynasties in the heartland of the old Caliphate and Iran also required their goods, and these came by way of the Persian Gulf ports. Goods brought in at Sīrāf could be sent inland into Iran and Khurāsān, up the coast to Ubullah and Baṣrah, or across the Gulf to Arabia itself. It is likely that most of the trade with Arabia entered via the great ports of Oman and Yemen in the south, though, and also from the ports of the western coast, al-Jār and Jiddah. Voyages to both of these ports from Aden and Oman are attested.[180]

Merchants

Merchants from many different parts of Eurasia and the Indian Ocean lands were involved in the musk trade with the Middle East. First were the merchants who imported musk from Asia; these were often merchants who carried wide varieties of goods. From there, musk passed to further merchants, who were often wholesalers dealing in a wide range of commodities, before ending up in the hands of the perfumers and pharmacists. Little specialization can be discerned at any stage of the trade, nor, due to the variety of sources for musk, could any one group dominate the trade.

The Sogdians, Iranic cousins of the Persians, dominated the overland trade in Late Antiquity through the early years of Islam. The Sogdians are the first attested merchants of musk, since musk is mentioned in the Ancient Letter II from the early 4th century. With the conversion of the Sogdians in Sogdiana to Islam beginning in the 8th century, Sogdian commerce became Islamic commerce. In the early Islamic period, it was the Sogdians who acquired musk near the sources and then transported it into Khurāsān, and Sogdian musk was a highly regarded variety. Travelers continued to use the overland routes through Sogdiana; al-Masʿūdī met many (*ʿiddah*) people who traveled from Sogdiana to China by the overland route.[181] Al-Sīrāfī's mention of a merchant

180 E.g., Ibn Khurradādhbih, cited above, and Buzurg 16 and 147; in the last case, it was a merchant of Baṣrah making the voyage from Oman to Jiddah!

181 Masʿūdī 1.186 (§385).

traveling the overland routes from Samarkand carrying musk has already been noted. Whether this individual was of Sogdian family background, or Arab, or something else, his activities represent a continuation of the ancient Sogdian commerce in musk.[182] This overland commerce may have waned in the 10th century in favor of the maritime route, as de la Vaissière has suggested,[183] but overland commerce did not completely die out.

Starting in the 10th century, the Karakhanids began to push the frontier of Islam into Inner Asia and the cities of the western Tarim basin. As noted earlier, warfare and invasion made the trade routes less safe during this period. The Sogdian merchants east of the frontier of Islam assimilated into Turkic culture in the Semirechye and the largely Buddhist Uyghur culture further east.[184] Bilingualism in Uyghur and Sogdian is attested for the late 11th century.[185] Many merchants in the Uyghur commercial documents bear Sogdian names.[186] As we have noted, musk was one of the commodities that was valued and traded among the Uyghur merchants. The prominent role of the Uyghurs in Inner Asia continues up to the days of the Mongols, when the Uyghurs submitted to Chinggis Khan and became part of the Mongol Empire. By that time, Muslim merchants had been plying the roads to the east and probably used Persian and Turkic as their languages of communication.

Islamic Merchants

The mercantile lifestyle was well known in ancient Arabia, even if pre-Islamic Mecca's relationship with long-distance commerce is problematic.[187] Since

182 De la Vaissière 316–17.

183 De la Vaissière 317–21.

184 De la Vaissière 321–2, 325–6.

185 Al-Kashgārī, *Maḥmūd al-Kāšγarī: Compendium of the Turkic Dialects (Dīwān Luγāt at-Turk)*, trans. R. Dankoff and J. Kelly, 3 vols. (Cambridge: Harvard, 1982–5), 1.83 and de la Vaissière 328. See also the mixed-language documents in N. Sims-Williams and J. Hamilton, *Documents turco-sogdiens du IXe–Xe siècle de Touen-houang* (London, 1990).

186 J. Hamilton, *Manuscrits ouïgours du IXe–Xe siècle de Touen-houang*, 2 vols. (Paris: Peeters, 1986), 1.176–7. For Uyghur commerce, see P. Zieme, "Zum Handel im uigurischen Reich von Qočo," *Altorientalische Forschungen* 4 (1976): 235–50.

187 P. A. Crone, *Meccan Trade and the Rise of Islam* (Princeton: Princeton University Press, 1987), which argues for a Mecca rather isolated from the larger trading world. See the critical review by R. B. Serjeant, "Meccan Trade and the Rise of Islam: Misconceptions and Flawed Polemics," *JAOS* 110 (1990): 472–86. Also, see Crone's article pointing out the potential importance of the leather exports mentioned in sources for the Roman army:

antiquity Arab traders were well known. The first few Islamic centuries were a golden age for merchants. Muḥammad encouraged trade as a useful lifestyle and condemned world-denying asceticism and monasticism.[188] In addition, vast quantities of money were available to the Arab conquerors in the form of the booty and taxation revenue from the conquests that were divided among them. This money enabled them to pursue commerce in a spectacular way, leading to the thriving civilization of early medieval Islam.[189]

Besides the incidental references to commerce in various types of literature, a special body of texts on commerce developed. The *Kitāb al-Tabaṣṣur bi-l-tijārah* sometimes attributed to al-Jāḥiẓ contains a wealth of information on trade goods and the importance of commerce, reflecting a society in which trade was a valued way of building wealth. Similar ideas can be found in Jaʿfar b. ʿAlī al-Dimashqī's (11th-12th c.) *Kitāb al-Ishārah ilā mahāsin al-tijārah*.[190] *Ḥisbah* literature also reveals the thriving commerce and its dark sides of the early medieval period from the point of view of the market inspector (*muḥtasib*) charged with keeping the marketplaces trustworthy.[191]

Pre-Islamic Arab merchants have already been noted carrying musk from the eastern Arabian port city of Dārīn inland. Within Arabia was an extensive system of land routes that linked to the port cities, giving access to goods

---

"Quraysh and the Roman Army: Making sense of the Meccan Leather Trade," *BSOAS* 70 (2007): 63–88, and G. Heck, "'Arabia without Spices': An Alternate Hypothesis," *JAOS* 123 (2003): 547–76, who argues for a significant resource base and commercial activities in pre-Islamic Mecca.

188 A. Lambton, "The Merchant in Medieval Islam," in *A Locust's Leg: Studies in Honour of S. H. Taqizadeh* (London: Lund Humphries, 1962), 121–30, G. W. Heck, *Charlemagne, Muhammad, and the Arab Roots of Capitalism* (Berlin: De Gruyter, 2006). For a critique of the "Pirenne thesis", which essentially argues the opposite, see B. Lyon, *The Origins of the Middle Ages: Pirenne's Challenge to Gibbon* (New York, 1972) and Beckwith, *Tibetan Empire* 172–95.

189 S. D. Goitein, "The Rise of the Middle-Eastern Bourgeoisie in Early Islamic Times," *Journal of World History* 3 (1957): 583–604; repr. in S. D. Goitein, *Studies in Islamic History and Institutions* (Leiden: Brill, 1966), 217–41, M. Morony, "Commerce in Early Islamic Iraq," *asien afrika lateinamerica* 20 (1993): 699–720, Heck, *Muhammmad*, Blanchard, *Mining*, and Wink.

190 Jaʿfar b. ʿAlī al-Dimashqī, *Kitāb al-Ishārah ilā mahāsin al-tijārah* (Cairo: Maṭbaʿat al-Mu'ayyad, 1318/1900) and see H. Ritter, "Ein arabisches Handbuch der Handelswissenschaft," *Der Islam* 7 (1917): 1–91.

191 R. P. Buckley's translation of al-Shayzarī's *The Book of the Islamic Market Inspector* (Oxford: Oxford University Press, 1999) contains an introduction treating *ḥisbah* literature in general. For the text, see *Kitāb Nihāyat al-rutbah fī ṭalab al-ḥisbah*, ed. Al-Sayyid al-Bāz al-ʿArīnī (Beirut: Dār al-Thaqāfah, n.d.).

brought from afar.[192] These were linked to a network of regularly held trade fairs at which goods were exchanged.[193]

Those most active in the maritime trade with Asia came from Oman and other port cities. An early Muslim merchant in China was Abū ʿUbaydah ʿAbdallāh b. Qāsim, an Ibāḍite, who was there in the middle of the 8th century, probably before 758.[194] He was a native of Oman and traded in aloeswood, which was available at the Vietnamese port of Lūqīn, then ruled by China.[195] Nevertheless, the account preserved does not actually specify that he bought aloeswood in China. Another Ibāḍite merchant in the China trade, of the late 8th-early 9th century, was al-Naẓar b. Maymūn.[196] By the late 9th century, a substantial community of Muslim merchants existed in China.

Muslim merchants settled in the coastal regions of India, especially in the territories of the Rashṭrakūṭas, called the kings of Ballaharā in Arabic literature, on the northwestern coast of Gujarat, and in Malabar to the south.[197] They intermingled with the locals and created their own hybrid Indo-Muslim cultures. The Bayāsirah, the Indian Muslim community of the north, was governed by Muslims appointed by the Rāshṭrakūṭas. Muslim merchants also carried their faith to new lands. Muslim merchants were instrumental, during the reign of al-Muʿtaṣim, in the conversion of the king of ʿUsayfān, a kingdom "between Kashmir, Mūltān and Kabul," according to al-Balādhurī.[198]

The Arabs had an interest in Central Eurasian commerce as well; their determination to hold the key cities of Khurāsān and Sogdiana is a reflection of this interest. One of the famous battles in the east, the so-called "Battle of the Baggage", was fought between part of the army of the governor Asad b. ʿAbdallāh al-Qasrī and the Türgish under *Suluk (Sulu) in Khuttal. The prize of this battle consisted of caravan loads of valuable goods said to be coming "from China".[199] As we have seen, these goods could have included textiles, porcelain, and musk.

---

192 Heck, *Muhammad* 58 & 289–93.

193 Yaʿqūbī, *Tārīkh* 1.313–4, Ibn Ḥabīb, *Kitāb al-Muḥabbar*, ed. I. Lichtenstadter (Haydarābād, 1942; repr. Beirut: al-Maktab al-Turāthī li-l-Ṭibāʿah wa-l-Nashr wa-l-Tawzīʿ, n.d.), 265–7, R. G. Hoyland, *Arabia and the Arabs from the Bronze Age to the Coming of Islam* (London: Routledge, 2001), 109–10, and Agius 54.

194 See T. Lewicki, "Les premiers commerçants arabes en Chine," *Rocznik Orientalistyczny* 11 (1935): 178–81.

195 Al-Yaʿqūbī, *Kitāb al-Buldān* (Leiden: Brill, 1892), 368.

196 Lewicki 181–2.

197 See Wink 68–80 for an overview.

198 Balādhurī, *Futūḥ* 446.

199 See Beckwith, *Tibetan Empire* 116–17 with references; see also Beckwith, *Empires* 133–4 for *Suluk.

The 11th century Islamic Karakhanid Turkic mirror for princes *Kutadgu Bilig* emphasizes the importance of merchants to the king and singles out the overland China trade: "If the China caravan ceased to raise dust on the roads, how could these countless kinds of silks arrive?"[200] Under the Karakhitāy, Chinese goods such as ceramics and textiles continued to come into Central Asia, probably through the efforts of Muslim merchants.[201] Chinggis Khan's supporters included Muslim merchants from Transoxiana.[202]

The Persians are another people whose commerce, like that of the Sogdians, was largely subsumed into Islamic commerce. As noted above, the Sasanians had a thriving maritime trade in the Indian Ocean. They typically seem to have sailed to the ports of India and Sri Lanka and traded there, while ships from China and Southeast Asia sailed west to those ports for trade as well. Persian merchants are mentioned in the Indian Ocean by the sixth-century writers Cosmas Indicopleustes and Procopius; the latter notes that they basically shut out Byzantine ships from the trade.[203] The struggle for control of the silk trade between Byzantium and the Sasanians is one of the major features of their relations during the 6th century.[204] The Sasanians also had interests in the commerce within Arabia.[205] The market fairs of Aden and Ṣanʿāʾ in Yemen were taxed at a 10% rate by the Persians.[206] As noted, these are ports through which musk certainly arrived in Islamic times.

The importance of the Persian Gulf city of Sīrāf has already been noted. A glance at the names of traders of the early medieval Islamic world shows the importance of Persians, especially in the Indian Ocean trade. The author of the *Kitāb ʿAjāʾib al-Hind* referred to often in these pages, Buzurg b. Shahriyār, has a Persian name and patronymic, and he was from the Persian port city of Rāmhurmuz. Many of the merchants mentioned in his book also bear obviously Persian names. Finally, the famous merchant of Sīrāf, Rāmisht, may also

200 Yūsuf Khāṣṣ Ḥājib, *Kutadgu Bilig, vol. 1*, ed. R. Arat, 2nd ed. (Ankara: Türk Dil Kurumu Yayınları, 1979), 445; R. A. Dankoff's trans. is quoted, *Wisdom of Royal Glory (Kutadgu Bilig): A Turko-Islamic Mirror for Princes* (Chicago: University of Chicago Press, 1983), 184.

201 M. Biran, *The Empire of the Qara Khitay in Eurasian History: Between China and the Islamic World* (Cambridge: Cambridge University Press, 2005), 137–8.

202 Allsen, "Mongolian Princes" 86–94.

203 Whitehouse and Williamson 44.

204 In addition to Procopius, see Menander Protector, *The History of Menander the Guardsman*, ed. and trans. R. C. Blockley (Liverpool: Cairns, 1985), 110–17, who deals with the birth of Turkic relations with Byzantium in the context of the silk trade against Sasanian Persia. See also Chapter 3.

205 M. Lecker, "The Levying of Taxes for the Sassanians in Pre-Islamic Medina," *Jerusalem Studies in Arabic and Islam* 27 (2002): 109–26.

206 Yaʿqūbī, *Tārīkh*, 1.314.

be mentioned.[207] Of course, an Arabic name does not necessarily mean that a merchant was an Arab, for converts to Islam typically adopted Arabic Islamic names, and these were passed down in their lineages gradually hiding their non-Arab origins. Many of the remaining Zoroastrians, escaping from Muslim persecution, moved east to Gujarat in India following the familiar trade routes and formed the Parsee community there.

The importance of Persian vocabulary in the maritime vocabulary of the early medieval period (and later) may also be noted.[208] The ship owner or captain, for the term could have both meanings, was the *nāhkhudā*, lit. "ship-lord".[209] The Persian navigational literature of the Abbasid period formed the foundation for the later works on Indian Ocean navigation.[210]

Jews also played a prominent role within the commerce of the medieval Islamicate Middle East. The most famous Jewish merchants of the early medieval period are the Rādhāniyyah for their extensive trade activities. As the long quotation from the geographer Ibn Khurradādhbih cited earlier in this chapter shows, musk is one of the items specified to have been among the goods traded by the Jewish trans-Eurasian Rādhānite trading network.[211] The Rādhāniyyah took their name from a district of Iraq, in the area of Baghdād.[212] This commerce goes back a long time; Jews conducted a lot of foreign trade under the Parthians and Sasanians,[213] and the trading network of the Rādhāniyyah likely had its roots then, in that very area at the heart of the Sasanian empire.[214]

207 Cf. Stern.

208 Hourani 65–6 and Wink 49.

209 Goitein and Friedman, *India Book* 121–56. See also R. Chakravarti, "Nakhudas and Nauvittakas: Ship-Owning Merchants in the West Coast of India," *JESHO* 43 (2000): 34–64 and R. E. Margariti, *Aden & the Indian Ocean Trade* 143–4, etc.

210 G. Ferrand, "L'élément persan dans les textes nautiques arabes," *JA* (1924:1) : 193–257, Wink 49, and Agius 30, 192–6.

211 M. Gil, *Jews in Islamic Countries in the Middle Ages*, trans. D. Strassler (Leiden: Brill, 2004), has accidently omitted "musk" in his translation of the text on p. 618 but he includes it in his list of goods on p. 626.

212 Gil 630–6. Cf. Gil, "The Rādhānite Merchants and the Land of Rādhān," *JESHO* 17 (1974): 299–328.

213 For the roots of this trade in Parthian times, see J. Neusner, "Some Aspects of the Economic and Political Life of the Babylonian Jewry, Ca. 160–220 C.E.," *Proceedings of the American Academy for Jewish Research* 31 (1963): 165–72.

214 Cf. S. D. Goitein, *Jews and Arabs: A Concise History of their Social and Cultural Relations* (New York, 1955; repr. New York: Dover, 2005), 107.

According to Ibn Khurradādhbih, the Rādhāniyyah imported to the west from China goods including musk, along with aloeswood and camphor.[215] Ibn Khurradādhbih specifies that this musk was brought by the Indian Ocean; however, there is no reason to doubt that the Rādhāniyyah who used the overland routes also carried musk. The later parallel version of this text in Ibn al-Faqīh does not mention musk from China, but mentions cinnamon and "all the goods of China."[216] He does mention musk, however, as one of the goods brought from Faramā and sold by the Rādhāniyyah along with aloeswood. Both these sources show that the China trade involved much more than the products of China, but rather included other goods acquired along the routes, such as Southeast Asian aloeswood and camphor. Cinnamon, of course, was not a product of China, but cassia was, so Ibn al-Faqīh's Chinese cinnamon is either Indian cinnamon traded by "China ships" or cassia.

Apart from this meager information preserved by Ibn Khurradādhbih and Ibn al-Faqīh, there is hardly a trace of the Rādhāniyyah in the literary sources that are generally devoid of information on the history of trade and commerce. The famous Judaeo-Persian letter from Dandan Uiliq along the southern route through the Tarim Basin (8th century), which mentions trade in sheep, textiles, and possibly copper, may be connected with the Rādhāniyyah.[217] But the center of gravity within the Jewish mercantile world tipped to the west during the 9th and 10th centuries as the caliphate declined and the Fatimid caliphate in Egypt prospered, encouraging a westward shift of the eastern trade from the Persian Gulf to the Red Sea.

The largest body of commercial documents extant from the medieval Middle East consists of the documents deposited in the Geniza of a synagogue in Fusṭāṭ in Cairo, called the Cairo Geniza by scholars. Jewish tradition decried the destruction of documents lest they contain the name of God, and they were stored up in this chamber, or *geniza*, until it was possible to bury them. As a result, the documents contained in this collection are of an extremely heterogenous nature. Among them are many documents reflecting the daily life of the Jewish community of Fusṭāṭ in the Fatimid period. It is perhaps the most important body of primary source material for the social history of

215 Ibn Khurradādhbih, *Kitāb al-Masālik wa-l-mamālik*, ed. M. J De Goeje (Leiden: Brill, 1889, repr. 1967), 153.

216 Ibn al-Faqīh 270.

217 See the edition by B. Utas, "The Jewish-Persian Fragment from Dandān-Uiliq," *Orientalia Suecana* 17 (1968), 123–36, and for a possible connection with the Rādhāniyyah see Gil 629. It is most recently illustrated in S. Whitfield, *The Silk Road* 221–2, with references. Another letter has turned up since; see Hansen, *Silk Road*, 218–9, with illustration.

the Middle East in medieval times, even though it concerns mainly the Jewish community.[218]

A resident of Alexandria of Central Asian origin, Isaac Nīsābūrī, a silk wholeseller who lived around the year 1100, dealt in musk along with numerous other goods from many places ranging from the Maghreb to Southeast Asia and China.[219] The list compiled of these goods ranges from dyes, aromatics, medicines, and wax to glass, millstones, coral, and, of course, textiles. Goitein considers the scope of his business to be average for merchants of his situation, and this shows how general merchants were involved in the musk trade even in the Middle East. His connections in his hometown in Khurāsān must have been helpful for the musk trade since the most prized musk came via the overland trade.

From the 11th century also comes the correspondence of the Jewish merchant Nahray b. Nissīm, a native of Qayrawān who settled in Fusṭāṭ. Over three hundred documents relating to his activities are extant.[220] One of these, a letter dated January 27, 1048, was addressed to him by his mentor, Abū Isḥāq Barhūn b. Isḥāq b. Barhūn Tāhertī.[221] It was sent to him from Qayrawān to al-Mahdiyya, from where he was about to depart for Egypt. It contains requests and commercial advice including the following:[222]

> Do not buy hastily. Buying in a hurry has no blessing. Except if you see goods that can be carried as light baggage, such as musk or lapis lazuli, which sells well here because only a little of it is on the market, or *'khrāj* pearls,[223] if they are good.

This document illustrates one of the most important aspects of the commerce in musk: very small quantities of it were valuable, so it was convenient for merchants to carry back and forth.

The heyday of Jewish commerce ran through the 11th century, but by the 13th, rising anti-Jewish sentiment diminished it and marginalized the Jews within the Islamic Middle East. Jews emigrated from the Middle East into Europe in increasing numbers, bringing their commercial skills with them.[224]

---

218 Goitein, *A Mediterranean Society*, is the most important study of this material.

219 Goitein, *Mediterranean Society* 1.153; 455 n. 56.

220 See Goitein, *Mediterranean Society* 1.153–4.

221 Goitein, *Letters* 147–53.

222 Trans. Goitein, *Letters* 151.

223 Goitein, *Letters* 151 n. 15, suggests this means pearls used on saddlebags, meaning of low quality.

224 Cf. Wink 92–3.

As has been mentioned above, much of the musk trade was in Chinese hands during the 19th century at least, and probably throughout many of the previous centuries, too. With the exception of internal trade within China, in the early medieval period it appears that the China trade was in the hands of a diverse group of Arabs, Persians, Indians, and Southeast Asians. Chinese people themselves in the Song period took an increasingly active role in maritime trade, and the government became interested in the financial benefits that could be reaped from trade.[225] This contrasted with the earlier practice of Chinese dynasties to downplay commercial interests and focus on the exchange of diplomatic and tributary missions. The great boom of Song and Yuan commerce gradually led to an increasing presence of Chinese in Southeast Asia that became especially significant by the 13th century.[226] Chinese goods during the Song and Yuan age circulated abundantly in the Indian Ocean lands as far as the eastern coast of Africa.[227] The Chinese foreign trade of the Song period was privately organized rather than government sponsored. Chinese ships and sailors did not generally sail all the way to the Middle East; the early Ming voyages are unparalleled.

The merchants of the Indian Ocean trade were members of different ethnic and religious groups. In pre-Islamic times, Christians and Zoroastrians dominated the trade, while in Islamic times Muslims and Jews became much more powerful, largely eclipsing the Christians, while the Zoroastrian merchants held on a little more tenaciously. Ethnically, the Zoroastrians were Persians, while the Christians were largely Syrians, but also Arabs and Persians.[228] Sooner or later, Persians and Arabs both joined the Islamic fold, both almost completely.

Indians also participated in the direct trade with the Middle East and Central Eurasia.[229] Their influence was strongest in the first half of the first millennium;

---

225 P. Wheatley, "Geographical Notes on Some Commodities involved in Sung Maritime Trade," *JMBRAS* 32:2 (1959): 24–5; D. Heng, *Sino-Malay Trade and Diplomacy from the Tenth through the Fourteenth Century* (Athens, OH: Ohio University Press, 2009), 100–1.

226 On the relations between Southeast Asia and Song and Yuan China, see Heng.

227 M. Horton and J. Middleton, *The Swahili* (Oxford: Blackwell, 2000), 81.

228 In addition to the works of Colless cited in Chapter 3 on the role of Christians in the Indian Ocean trade, see G. Gropp, "Christian Maritime Trade of the Sasanian Age in the Persian Gulf," in *Golf-Archäologie: Mesopotamien, Iran, Kuwait, Bahrain, Vereinigte Arabische Emirate und Oman* (Buch am Elbach: Leidorf, 1991), 83–8, which also discusses the role of Pahlavi as a church language as opposed to Syriac.

229 There is a large bibliography on Indian seafaring, strongest for antiquity. See H. P. Ray, *The Winds of Change: Buddhism and the Maritime Links of Early South Asia* (Delhi: Oxford, 1994) on the period before 400 CE. R. Chakravarti has several important papers on the early medieval period, see especially "Merchants of Konkan," *Indian Economic and Social History Review* 23:2 (1986): 207–15, "Coastal Trade and voyages in Konkan: the early

this period witnessed a great expansion of Indian Buddhist society in the Indian Ocean lands. Later, travelers spread Brahmanic Hinduism when Buddhism was on the decline in India. Indian sailors are known at least since Classical Antiquity in the Red Sea lands of the Near East.[230] Indian traders gave a name to the island of Socotra (Sanskrit *Sukhadhara*) south of Arabia.[231] Their importance in direct trade with the Near East seems to have waned under Islam, but Indians still visited the emporium cities such as Sīrāf in the 10th century.[232] Indian sailors were also active on the eastern maritime routes. The 11th and 12th century hegemony of the Cholas in the eastern Indian Ocean lands was a golden age of commerce there,[233] and even when the Cholas declined, the South Indian trade guilds remained active participants in that trade.[234]

Mention must also be made of the Southeast Asian Malay sailors, who became very important in the trade between China, Southeast Asia, and India during the middle of the first millennium CE.[235] The wide-ranging nature of their seafaring enterprise can be seen in their settlements in Madagascar, dating to around this period, but they do not seem to have traded directly with the Islamic Middle East. At the political level, interest in profiting from maritime trade led to the creation of thalassocracies such as Funan and Śrīvijaya during the first millennium CE.[236] Following the Chola interruption of Śrīvijaya,

---

medieval scenario," *Indian Economic and Social History Review* 35:2 (1998): 97–123, and "Seafarings, Ships and Ship Owners: India and the Indian Ocean (AD 700–1500)," in *Ships and the Development of Maritime Technology in the Indian Ocean* (London: Routledge, 2002), 28–61. Cf. Wink 65 who points out the diminishing role of Indian merchants in direct trade with the Near East.

230 R. Salomon, "Epigraphic Remains of Indian Traders in Egypt," *JAOS* 111 (1991): 731–6 and "Addenda to 'Epigraphic Remains of Indian Traders in Egypt," *JAOS* 113 (1993): 593 and Warmington 75–7. See also R. Tomber, *Indo-Roman Trade: From Pots to Pepper* (London: Duckworth, 2008) and S. Sidebotham, *Berenike and the Ancient Maritime Spice Route* (Berkeley: University of California Press, 2011) for more recent surveys of Indo-Roman trade relations.

231 On the presence of Indians and other groups of sailors on Socotra, see Sidebotham 189.

232 Sīrāfī 93.

233 Hall, *Colas*. H. Kulke, "Rivalry and Competition in the Bay of Bengal in the Eleventh Century and its bearing on Indian Ocean Studies," in Om Prakash and D. Lombard, eds. *Commerce and Culture in the Bay of Bengal, 1500–1800* (New Delhi: Manohar, 1999), 17–35.

234 M. Abraham, *Two Medieval Merchant Guilds of South India* (New Delhi: Manohar, 1988). Musk was hardly the most important good carried by the Ayyāvoḷe and Maṇigrāmam guilds studied by Abraham; it is apparently mentioned once in an Ayyāvoḷe inscription in a less than certain reading, see Abraham 172.

235 Wolters.

236 Wolters and Hall, *Early Southeast Asia*.

trade in the Malay area became more diffuse; rather than proceeding under the sponsorship of a strong state, local ports continued to pursue their own trading interests.[237]

Data on the Commerce in Musk

Musk was ideally purchased by the pod, Ar. *nāfijah*. The reasons for this are that musk in the pod was harder to adulterate, and it presumably stayed fresher in its original vesicle.[238] Al-Tamīmī (in al-Nuwayrī) notes that individual pods were placed inside larger, sewn bags for trade.[239]

Loose musk was also traded by the phial or flask (*qarīrah*, pl. *qawārīr*).[240] The synthetic musks described in the *Kitāb Kīmiyā' al-ʿiṭr wa-t-taṣʿīdāt* were put up into phials for sale;[241] this was the usual vessel for selling perfumes of various types. A document dating to around 1010 mentions a lawsuit concerning a flask valued at 22 *dīnārs*, making it very expensive; Goitein has suggested that this flask contained musk: few other substances that could be placed in a flask could be so pricey.[242] A thirteenth- century document reports that the merchant Mufaḍḍal Ibn Abī Saʿd sent a flask containing musk weighing 12 ½ *mithqāls*,[243] about 53 grams of musk.

A letter from around the year 1100 CE from an Alexandrian merchant, Abū al-Ḥasan son of Khulayf, who had been travelling in Morocco and Spain, to his patron, Abū Saʿīd al-ʿAfṣī in Fusṭāṭ, describes his dealings with Kohen al-Fāsī, who sent ambergris with him so that he could purchase with it *khumāsiyyāt* ("fivers") of musk. These "fivers" were apparently packages containing five dirhams' weight of musk.[244] At about 15 ½ grams of musk, this was somewhat less than the yield of an average musk pod. This was a standard unit for musk traded in the Middle East at the time, though it is known only from the Geniza documents.

---

237 Heng 106, 215.

238 See above, 151–2, 169, 178–9, etc.

239 Nuwayrī 12.3–4.

240 See above, 187.

241 Garbers formulas #2, 5, 6, etc.

242 Goitein, *Letters of Medieval Jewish Traders* (Princeton: Princeton University Press, 1973), 78.

243 Goitein, *Letters* 69.

244 Goitein, *Letters* 50 n. 4.

Not much information is available on the price of musk beyond scattered prices illustrating the fact that it was expensive. A price list was sent by Hārūn al-Ghazzāl to the important merchant Joseph b. ʿAwkal in Fusṭāṭ in 1030; it gives prices for aromatics including musk, which was valued at 4 ½ *dīnārs*; Stillman says this price was per flask (*qinnīnah*).[245] In Alexandria in 1097, according to a court deposition, three phials (*qawārīr*) of musk sold for 13 ½ *dīnārs*, making the price per phial 4 ½ *dīnārs*.[246] In the litigation a year later the court ordered payment of two phials of musk in Alexandria valued at no less than 13 *dīnārs*.[247] In addition to the phials mentioned above there is one more piece of data. A price list from Palermo, Sicily from the middle of the 11th century reports the cost of various goods, mostly spices and aromatics. At the time musk was trading for fifty quarter-dinars per ounce (*ūqiyyah*). The writer notes: "All spices are rare here and almost not to be had."[248] The cost of musk can be compared to the cost of camphor, also imported from Asia, which was trading at thirteen quarter *dīnārs* per ounce. Musk was thus far more expensive there.

Perfumers and Pharmacists

The information on perfumers is scattered. The name for a perfumer is *ʿaṭṭār*, derived from the word *ʿiṭr* "perfume", and it was a common title.[249] Islamic thought reckoned merchants who dealt in perfumes and textiles the best merchants of all.[250] The *ʿaṭṭār*'s duties were often much more numerous than the English word perfumer connotes.[251] They were required to prepare all manner of drugs and medicaments as well as perfumes; these categories were not so fixed in medieval culture as in modern times. Their function was essentially as apothecaries.[252] They also carried out the duties of the homeopath

245 N. A. Stillman, "The Eleventh Century Merchant House of Ibn ʿAwkal (A Geniza Study)," *JESHO* 16 (1973): 50–1; text in M. Gil, *Be-malkuth Yishmaʿel bi-tekufat ha-geʾonim*, 4 vols. (Tel Aviv: University of Tel Aviv, 1997), 2.508.

246 Goitein and Friedman, *India Book* 214, Gil 4.72.

247 Goitein and Friedman, *India Book* 217, Gil 4.74.

248 Goitein, *Letters* 118–9.

249 A. Dietrich, "al-ʿAṭṭār," in *EI* 2nd ed. s.v.

250 E.g., M. A. J. Beg, "Tādjir," in *EI* 2nd ed. s.v.

251 See L. Chipman, *The World of Pharmacy and Pharmacists in Mamlūk Cairo* (Leiden: Brill, 2010), for a study of the social role of pharmacists and perfumers, based on the work of the Jewish al-Kūhīn al-ʿAṭṭār and his famous pharmaceutical handbook *Minhāj al-dukkān*.

252 See G. Levi Della Vida, "A Druggist's Account on Papyrus," in *Archaeologia orientalia in memoriam Ernst Herzfeld* (Locust Valley: Augustin, 1952), 150–5, a short fragment of an account sheet listing some of the aromatics an Egyptian *ʿaṭṭār* dealt in.

(*mutaṭabbib*). Perfumers could be male or female; one of the perfumers cited as the source of a formula by al-Tamīmī was Bunān, a woman, who worked for the caliph al-Wāthiq bi-llāh (r. 842–7).[253] Most of the literature on perfumers deals with the types of adulteration they practiced to defraud their customers.

The Egyptian al-Shayzarī (d. 1193) composed a work on *ḥisbah* that includes a chapter on perfumers (*ʿaṭṭārūn*). It gives some indication of what they sold in the context of explaining how they defrauded their customers.[254] Adulteration and falsification of simple aromatics is the main subject, and he covers musk, ambergris, camphor, saffron, and aloeswood. He mentions that perfumers also adulterated ben oil by mixing it with cheaper oils. He also discusses how the compound perfume *ghāliyah* is falsified, and that civet was adulterated in ways similar to *ghāliyah*. Al-Shayzarī's account thus shows that perfumers sold both simple aromatics and ready-made compounds. He deals separately with pharmacists (*ṣayādilah*); this chapter covers the frauds associated with a variety of *materia medica* excluding the perfumes discussed in the chapter on perfumers.

The works of court physicians reveal the practice of perfumery in its unadulterated aspect. Pharmacists and physicians were as a matter of course expected to know the composition of perfumes along with other forms of medicines. In the case of court physicians such as Muḥammad b. Aḥmad al-Tamīmī or Sahlān b. Kaysān in Egypt, or Ibn Mandawayh at the Buyid court, their high standing in society gave them access to expensive ingredients, including musk, and they also had the time to compose treatises, including the works on perfumery that have been mentioned so often in this work. In their formulas for rich aromatic compounds the use of expensive aromatics such as musk and ambergris is lavish, for they form the basis of most of the perfumes.

Perfumes prepared at home are reflected poorly in our sources. Dinah Jung's study of perfumery in modern Yemen shows clearly that home preparation of perfumed compounds is widespread; it is all but certain that this practice has very old roots. These home-made perfumes were made especially by women, whether wives or household maids, a group largely invisible in the early medieval sources. Perfume making was one of the genteel arts in which respectable ladies could participate.[255] Home production of perfumes has sometimes become more of a cottage industry, with perfumes specifically prepared for use beyond the household.[256] The readership of early medieval treatises on aromatics and perfumery was likely not limited to the professional perfumers, but may have included individuals from other walks of life who pursued

253 Nuwayrī 12.62. Bunān was called *ʿaṭṭārah*, so we know she was female.

254 Shayzarī 48–55, and Buckley's trans. 70–5.

255 Jung 96.

256 Jung 142–5.

perfumery as a hobby at home. Cookbooks such as the *Kitāb Kanz al-fawāʾid fī tanwīʿ al-mawāʾid* often included formulas for perfumes as these were part of the repertory of the kitchen.[257]

Adulteration and Imitation of Musk

A well attested aspect of the commerce in musk was its frequent adulteration. This was mostly accomplished by use of materials quite dissimilar to musk, not using the musky odors of other animals, but by using generally non-odoriferous materials to extend and "bulk up" musk.

It was well known to the scholars of the Middle East that other animals had musky smells. Civet was similar to musk but not highly regarded because of its dubious origin; it comes from glands near the anus of the animal. Castoreum was also known, although like musk, it was an import. Musk shrews were certainly known; al-Jāḥiẓ calls them *jirdhān*, "rats."[258] The musky scent exuded by crocodiles is also noted by the geographer al-Dimashqī (1256–1327),[259] the zoographer al-Damīrī, as well as the encyclopaedist al-Nuwayrī, who says of the crocodile: "there is found in its skin which lies near its belly a cyst (*silʿah*) like an egg, and it is a moisture which has a scent like musk; its scent quits it after months."[260] Al-Jāḥiẓ noted that the thick sweat from the brow of the elephant had a scent resembling the scent of musk, but that this phenomenon only occurred in India.[261] Male elephants in rut are said to be in a state of *musth* (the Hindi word is commonly used in English scientific literature); this phenomenon is indeed more pronounced in Indian elephants than African. One of the features of the *musth* is the secretion of a thick aromatic substance from glands on the head; this is what al-Jāḥiẓ refers to. Far and away the most famous musk scent for the Arabs other than the real item was the musky scent exuded by camels, called *faʾrat al-ibl*.[262] Other than civet, none of these

257 *Kitāb Kanz al-fawāʾid fī tanwīʿ al-mawāʾid*, ed. M. Marín and D. Waines (Beirut: Franz Steiner, 1993), 226–51 and Ibn al-ʿAdīm (attr.), *Al-Wuṣlah ilā al-ḥabīb*, ed. S. Maḥjūb and D. al-Khaṭīb, 2 vols. (Aleppo, 1987–8), 2.481–502.

258 Al-Jāḥiẓ, *Kitāb al-Ḥayawān*, ed. ʿAbd al-Salām Muḥammad Hārūn, 8 vols. (Cairo: Muṣṭafā al-Bābī al-Ḥalabī, 1966), 7.211.

259 Al-Dimashqī, *Kitāb Nukhbat al-dahr fī ʿajāʾib al-barr wa-l-baḥr*, ed. A. Mehren (Baghdād: Yuṭlabu min Maktabat al-Muthannā, n.d.), 92.

260 Nuwayrī 10.315.

261 Jāḥiẓ, *Ḥayawān*, 7.210. Cf. Zamakhsharī, *Rabīʿ*, 2.218.

262 Al-Dīnawarī, *Kitāb al-Nabāt: The Book of Plants: Part of the Monograph Section*, ed. Bernhard Lewin (Wiesbaden: Steiner, 1974), 195.

substances were used as perfumes, although the scent of the camel seems to have been appreciated. This dearth of alternatives left the real article, or various substances designed to imitate musk or at least extend it, as what was desired.

Since musk was very expensive in the Islamic world, it is only natural to expect that it would be extensively adulterated and falsified. Even in China, which had its own nearby supply of musk, Tao Hongjing, as we have seen, bemoaned the extensive adulteration of it. How much worse this problem must have been in the Middle East! Numerous recipes for extending musk go back to the *Kitāb Kīmiyā' al-ʿiṭr wa-l-taṣʿīdāt* attributed to Abū Yūsuf b. Isḥāq al-Kindī. The interpretation of these recipes, and whether they were intended to defraud the customer, depends on how they were sold, and this we do not often know beyond the frequent comment at the conclusion of the formulas: "sell it as you wish". With a substance as expensive as musk, even a small quantity of adulterant added to the real substance could mean a great profit on a large quantity of musk.

The formulas given in *Kitāb Kīmiyā' al-ʿiṭr wa-l-taṣʿīdāt* include a great variety of substances used to extend musk. These substances were chosen so that after being processed together they would approximate the color, texture, and scent of musk. The first formula given runs:

> You take five *mithqāls* of Chinese aristolochia (*zarāwand ṣīnī*),[263] two *mithqāls* of good, fragrant *rāmik*, which is called *rāmik* of musk, two *mithqāls* of good shavings of aloeswood, and a *mithqāl* of dragon's blood (*dam al-akhawayn*).[264] These are carefully crushed. Then you drip upon it a drop of pure lead-colored[265] jasmine oil and crush it together gently.

263 The plants of the genus *Aristolochia* are commonly called birthworts in English. The roots of these plants were used in many medical preparations; it was especially valued for its use in childbirth. On aristolochia see Dymock 3.158–66, and for Islamic medicine, Garbers #131, Dietrich DT III.4, and E. Lev and A. Amar, *Practical* Materia Medica *of the Medieval Eastern Mediterranean According to the Cairo Genizah* (Leiden: Brill, 2008), 359–60.

264 From a species of *Dracaena*, such as the famous *D. cinnabari* of Socotra. See Garbers #, Ibn Juljul #37, M.-C. Simeone-Senelle, "Aloe and Dragon's Blood, some Medicinal and Traditional Uses on the Island of Socotra," *New Arabian Studies* 2 (1994): 189 and Lev and Amar 400. It was called Indian cinnabar by the Greeks and Romans, see E. H. Warmington, *The Commerce between the Roman Empire and India*, 2nd ed. (Cambridge: Cambridge University Press, 1974; repr. New Delhi: Munshiram Manoharlal, 1995), 203. It is a reddish resin that is exuded by the plant. Today dragon's blood resin is still used for incense; it has a rich, herbal scent.

265 *Raṣāṣī*; perhaps an error for *ruṣāfī*, see above, 280, on this jasmine oil type.

> Then you put in a piece of new, heavy linen and rub it gently until the grease of the oil comes out upon the cloth. It turns out well when you take two parts of it and a part of musk and mix them carefully. It sells at one reckoning of good, mature [musk], God willing.[266]

This formula includes ingredients typical of the adulterations in this book: Chinese aristolochia, *rāmik*, and dragon's blood. The third of the formulas in the *Kitāb Kīmiyā' al-ʿiṭr* is also based on Chinese aristolochia. In his third formula he adds some other ingredients:

> You take an *ūqiyah* of good Chinese aristolochia, an *ūqiyah* of *sādawarān* washed with hot water and dried, an *ūqiyah* of boxthorn (*ḥuḍaḍ*),[267] two *mithqāl*s of dragon's blood, and a *dāniq* of sarcocolla (*ʿanzarūt*).[268] You carefully crush each of these individually and then sieve them with silk. Then you take the membranes (*akrāsh*) of musk and wash them with water after they have soaked for a day and a night. Then pour off this liquid and knead it with the herbs and put into it a leaven (*khamīr*) of musk[269] and knead it together gently until it takes on the consistency of ground sesame. Then pour it into a threadbare (*khalaqah*) piece of silk and squeeze it. What comes out from the gaps in the cloth is like sesame seeds. Then mix every two *mithqāl*s of this with a *mithqāl* of musk. Then stuff it into the pod (*nāfijah*). Then seal the hole of the pod with thick gum solution. Then put the pod upon the top of a furnace. Remove it when you come in the morning and take this pod and put it in a perfume box and leave it in it for a month. Then remove it and sell it however you wish.[270]

The substance denoted variously *sādawarān, shādawarān, siyāhdārū,* or *siyāhdāruwān* is often mentioned as an additive to musk. The name comes from Persian *siyāh dārū*, and it refers to a black resin derived from tree sap.[271]

---

266 Garbers formula #1.

267 *Lycium* sp. See Garbers #41, W. Schmucker, *Die pflanzliche und mineralische Materia Medica im Firdaus al-Ḥikma des Ṭabarī* (Bonn: Selbstverlag des Orientalischen Seminars, 1969), 167–8, and Dietrich DT I.69.

268 *Astragalus sarcocolla*; Garbers #6, Schmucker 95, and Dietrich DT III.80.

269 Garbers #43.

270 Garbers formula #3.

271 E. Wiedemann, "Über von den Arabern benutzte Drogen," in *Aufsätze zur arabischen Wissenschaftsgeschichte* 2.236; Garbers #100, A. Siggel, *Arabisch-Deustches Wörterbuch der Stoffe* (Berlin: Akademie-Verlag, 1950), 41, and Dietrich DT I.39 n. 5.

Al-Nuwayrī, for instance, says it is a black resin-like substance that could be produced from walnuts.[272] Any substance with a black color seems to have been considered as a possible adulterant for musk. The author of the *Kitāb Kīmiyā' al-ʿiṭr wa-l-taṣʿīdāt* also includes musk membranes in this formula in addition to the musk it was used to extend. With a substance as valuable as musk, one had to attempt to extract every bit of scent from the pod.

The fourth formula includes crushed lentils along with the aristolochia, and it is mixed with kewda (*kādī*) oil[273] and cologne (*naḍūḥ*) to make a paste that was dried and then ground. The fifth formula calls for spikenard:

> You take ten *mithqāl*s of spikenard (*sunbul al-ṭīb*),[274] the same amount of shavings of aloeswood, the same amount of fine cassia (*qirfah*), a half of a *dirham* of camphor, a *mithqāl* of good cloves, half of a *dirham* of saffron, a *dirham* of black *wars*,[275] and *sādawarān* equivalent to all of the mixture of these. All of them are ground separately and sieved with silk. They are kneaded with Jūrī rosewater and formed into fine pills. A cup (*jām*) or phial is taken, and a perfumed cloth is spread inside it and it is kept from dust by the placement of another cup on top of the cup. It dries in the shade. Then it is crushed and one part of musk is added to one part of it. Sell it as you wish.[276]

It is only natural that the animal-scented spikenard so often associated with musk would appear in a formula for adulterating it, even though this formula is probably earlier than the documents we have that link spikenard with musk. Together with spikenard, other fragrant ingredients are used along with a base of aristolochia. The next few formulas include emblic myrobalan along with other ingredients. The ninth formula is based on a paste prepared from apples and *sādawarān*:

---

272 Nuwayrī 11.317.

273 From the blossom of *Pandanus*; Garbers #53 and Ibn Juljul #36.

274 Spikenard is called for in this recipe because of its strong, peculiar animal odor, which perhaps resembles musk.

275 A dye plant found especially in Yemen, produces a yellow-red dye. Black *wars* is probably the same as the Arabic *al-wars al-ḥabashī*. Usually identified with a species of *Flemingia*, especially *Flemingia grahamiana* (= *F. rhodocarpa*), but sometimes said to be *Memecyclon tinctorium*. See Garbers #127, Schmucker 530–1, Ibn Juljul #61, and H. Schönig, *Schminken, Düfte und Räucherwerk der Jemenitinnen: Lexikon der Substanzen, Utensilien und Techniken*. Beirut: Ergon Verlag, 2002), 297–308.

276 Garbers formula #5.

> You take good, sound Syrian apples and skin them. Discard their cores, and you take the flesh only and chop it, and put it into a cloth and squeeze it forcefully until all of its moisture and flavor comes out. Then it dries in a clean vessel in the shade until it has dried gently. Then it is crushed carefully upon the perfumer's stone. Then to every ten *mithqāls* of it two *mithqāls* of prepared *sādawarān* are added. Its preparation is that you put it into a vessel upon the fire and pour more than enough water over it to submerge it. Then you kindle a gentle fire under it until it boils and its dye emerges from it. Then you take it out, dry it, and crush it gently. Then it is mixed with the apples, as we have said, and crushed together. Its color comes to be the color of musk. Then increase it by another *mithqāl* of *sādawarān* and crush them carefully together again and do not block it with your hand. Then sieve it with a hair sieve which is not coarse and moisten it with as little water as will combine it. Then dry it, and when it has dried gently, add to every ten *mithqāls* of this a *mithqāl* of musk. The more you increase the musk in it, the better it will be. Then dry it and put it into a phial. Be careful that there is no moisture in it, for it will have a rotten smell. Then leave it in the phial for seven days. Then add for one part of this one part of musk. Sell it as you wish. It comes out marvelously. If you want it for *ghāliyah*, I mean, as the base, then knead it with good ben oil before you add the musk to it, and then enhance it with musk, ambergris, and *sukk* as you wish; it produces a fine, wonderful *ghāliyah*.[277]

We have already noted the comparison of the scent of musk to apples; perhaps this is a reason they were included in the base of this imitation musk. The tenth formula is based on aloeswood, and the eleventh on fresh acorn shells (*qushūr al-ballūṭ al-raṭb*). Finally, an anecdote added by the compiler of the manuscript says:

> Muḥammad b. Harthamah said to me, I went to see a famous man in Baghdād, one of the successful merchants of musk who produced musk from plants. He wanted to present musk to the people, so he called for a goat and slaughtered it in a basin (*ṭast*). He took musk and put it into a wide beaker (*fī jāmi qawārīra wāsiʿin*). Then he began to put his palm upon that blood, and then upon the musk, and then that blood came out mixed with musk. He did this several times, and when it was all mixed carefully, he spread it upon a leather mat (*naṭʿ*) and left it a little while.

277 Garbers formula #9.

> It consolidated and matured, and then he presented it to the people and sold it, and he had increased it by a good amount.[278]

This account illustrates the adulteration of musk with blood, something the other recipes in the *Kitāb Kīmiyā' al-ʿiṭr wa-l-taṣʿīdāt* do not include. This passage suggests that the maturation of blood, even a goat's blood, into musk could be accomplished by mixing it with real musk. As in the formulas given above, the object was to extend the musk as much as possible to increase profit. This huckster was able to play on the concept of the production of musk from blood to make his audience believe that he had really generated musk rather than simply adulterating it.

Around this time, the author of the *Kitāb al-Tabaṣṣur bi-l-tijārah* wrote: "Musk is adulterated with lead (*ānuk*), castoreum (*jundbādastar*), dragon's blood (*dam al-akhawayn*), and *sīyāh dārū*."[279] Ibn Māsawayh discussed the adulteration of musk in his *Kitāb Jawāhir al-ṭīb al-mufradah*:

> There is much adulteration of musk; only the one who practices adulteration knows its mysteries; its sellers know only its surface, that the pod resembles gazelle dung, and is like lead, iron, and silver (i.e. in weight?).[280] Musk is adulterated with anything which makes it heavy and with whatever can be added to it or contained within it.... It has reached us that some of the people of India say that there are three types (*alwān*) of musk. The best is the original, well known musk, and then there are two imitation musks. The first of these is of a dried mixture made among them of plants which has no musk at all in it. They enjoin its use and purchase in the places of its origin and neighboring countries. Those that know it and enjoin it are the people of Tibet and the countries adjacent to it. The other is musk which they manufacture, and they forbid its use because it does not keep and it changes and spoils and there is an abasement or lowering of its value. Some of the importers and perfumers know this third kind by some of its odor but most of them do not know it and the mistakes about it are many. The musk imported from the interior of Kashmir and its environs is not good and does not come close to that which is forbidden of the manufactured kind. This kind also

278 Garbers formula #12.

279 *Kitāb al-Tabaṣṣur bi-l-tijārah fī waṣf ma yustaẓraf fī al-buldān min al-amtiʿah al-rafīʿah wa-l-aʿlāq al-nafīsah wa-l-jawāhir al-thamīnah*, ed. Ḥasan Ḥusnī ʿAbd al-Wahhāb al-Tūnisī (Cairo: Maktabat al-Khānjī, 1994), 17.

280 An expert would be able to tell if a pod seemed too heavy for its size.

> comes in both manufactured and non-manufactured kinds. It is half the price or so of the good quality musk.... [Musk's] substitute is castoreum (*jundbādastar*) because this is the closest thing to it.[281]

In the time of Ibn Māsawayh there were thus mixtures of herbal material designed to approximate musk. Ibn Māsawayh suggests that the first plant-based imitation musk is acceptable in its lands of origin and nearby, probably because it would not have had the chance to deteriorate through long transportation. The other manufactured musk did not have the longevity to survive even its lands of origin. Ibn Māsawayh, like the author of the *Kitāb al-Tabaṣṣur bi-l-tijārah*, noted the similarity of castoreum to musk. While the *Kitāb al-Tabaṣṣur bi-l-tijārah* claims it was commonly used to adulterate musk, castoreum itself does not appear to have been common enough to make it worthwhile when there were other, more easily available, ingredients.

The Ibn Mandawayh- Sahlān b. Kaysān tradition gives some information on how to detect adulterated musk:

Ibn Mandawayh[282]

Testing of musk: When there is a doubt about musk and testing is desirable to ascertain whether it is sound, one takes a piece of glass and places it in the fire until it is hot. Then one places a little bit of the musk which needs to be tested upon it. If the scent of musk diffuses, it is musk without a doubt. If there is still doubt after that test, one takes a little of it and bites it with the edge of the teeth, then removes it and places it in a piece of cloth, rubs it with the hand, and examines it. If one finds that all of its

Ibn Kaysān[283]

If a doubt arises as to whether musk is pure or adulterated, then one takes a piece of glass and places it in the fire until it is hot. Then one places a little bit of the musk on it. If the scent of musk diffuses from it, then it is pure. One may test it in another way, which is that one bites a bit of it and then places it in a piece of cloth, rubs it, and then examines it. If any bit of the color of musk clings to the cloth and his saliva remains, then it is pure. If it becomes like an unguent (*ṭilā'*) and none of the saliva remains around it then it is adulterated. There are many

281 Ibn Māsawayh 10–11.

282 Ibn Mandawayh 228–9.

283 Ibn Kaysān 190.

color has transferred to the cloth, and none remains in his saliva, then the musk is adulterated. If it is rubbed and the color of the musk gleams a little bit and it remains in his saliva, then it is pure musk. This is when one does not know what the adulterant itself is. forms of adulteration which the people of discernment only recognize through testing.

Ibn Mandawayh continues with a passage not in Ibn Kaysān's text:

> On the frauds of musk. The frauds for musk are not known by the people of these lands; the knowledge of them has departed from most of the connoisseurs. One must only buy musk from a well-known merchant, because the adulterer likes to sell what he has produced with his own hand. When the merchant is well-known, and the musk is found to have a fault, it can be returned to him. There are none more skilled in the adulteration of musk and its making than the people of Balkh. One should not buy any musk at all from them.[284]

Ibn Mandawayh stresses the importance of trusting one's merchant; fly-by-night musk sellers might offer a better price, but the musk was less likely to be genuine. It is interesting that he specifically condemns the merchants of Balkh, which was the major town on the trade route through Ṭukhāristān into western Tibet. This bias might explain the general preference for musk brought by the Sogdians and their successors through Central Asia. No "Balkhī" musk is listed in the sources otherwise.

The most extensive collection of information on the detection of falsified musk is found in the *Kitāb Nihāyat al-rutbah fī ṭalab al-ḥisbah* of al-Shayzarī. He has an entire chapter on the deceits of the perfumers, as noted above. He treats musk first, then ambergris, camphor, saffron, *ghāliyah*, civet, aloeswood, and ben oil. The material on the adulteration of musk is the most extensive. Since musk was the most important of the Arabic perfumes, as well as perhaps the most expensive, it was the most subject to adulteration. Al-Shayzarī laments the huge number of possible falsifications in perfumery and confines his book only to those most commonly practiced. He writes of musk:[285]

---

284 Ibn Mandawayh 229.

285 Shayzarī 48–50; cf. Buckley's trans. 70–1. My translation differs from Buckley's on a number of points.

Among these are those who make musk pods from the rind of the emblic myrobalan and Indian pepperwort,[286] and likewise *shādawarān*,[287] and they knead it with liquid pine resin. For every four *dirhams* of these, they add a *dirham* of musk, and they stuff it into the pod, close its top with resin, and dry it on top of a furnace.

The way to detect its adulteration, and other adulterations of pods, is for [the *muḥtasib*] to open it and kiss it, like one who is searching within, and if there rises to his mouth from that musk a sharpness (*ḥiddah*) like fire, then it is outstanding (*faḥl*) with no adulteration in it. If it is weak, then it is adulterated.

There are those who make pods from emblic myrobalan and *shādawarān* which has had its dye stripped from it with hot water, together with sarcocolla (*anzarūt*), and kneaded with liquid resin and prepared (*yakhdimuhu*),[288] then for every three *dirhams* of this, a *dirham* of Sogdian musk is added. It is all crushed together and stuffed into the pod, and then it is dried upon a furnace. The way of detecting it is what we have mentioned.

There are those who make pods from acorn rinds (*qushūr al-ballūṭ*) prepared with a fire. For every three *dirhams* of this a *dirham* of musk is mixed in, then the pod is stuffed with it. The way of detecting it is what we have mentioned.

There are those who make musk without the pod from aristolochia, *rāmik*, and dragon's blood; they are kneaded together and a *dirham* of it is mixed with a *dirham* of musk.

There are those who make musk from fragrant spikenard, the shavings of aloeswood, cassia (*qirfah*), and cloves; they are mixed with the equivalent of musk.

There are those who make it from cloves, *shādawarān*, and saffron, which are kneaded together with rosewater, and mixed with its equivalent [of musk] and all of that is stuffed into ambergris.

The way to detect all of these types and the others of the types [of adulteration] of musk is to put a bit of it in your mouth, and then spit it upon a white shirt, and then shake it off. If it is shaken off and does not stain, there is no adulteration of blood and other things in it. If it stains

---

286 *Lepidium latifolium*; its Arabic name, *shiṭraj*, is borrowed from Sanskrit *citraka*, cf. Schmucker 274–5 and Dietrich DT II.157.

287 Apparently the same as *sādawarān*, which has been discussed above.

288 This probably means cooked, because slightly later in the text acorn rinds are described as *makhdūm bi-l-nār*.

> and does not shake off, then it is adulterated. Among them are those who add to pure musk a bit of dragon's blood or goat blood, and those who crush musk with gazelle blood, and then stuff it into its intestine, close it with thread, and then dry it in the shade. Then they break some from it and mix it with other things in phials. There are those who adulterate it with burned liver. The way of detecting all of these adulterations is what we have mentioned. There are those who add to musk lead the size of peppercorns and smaller than that, dyed with ink. These can only be distinguished by crushing.

Al-Shayzarī mentions many of the very same procedures for adulterating musk described in the *Kitāb Kīmiyā' al-ʿiṭr wa-l-taṣʿīdāt* and other sources. These must have been among the most common methods of adulterating musk. In addition, his bite-and-spit test is very similar to Ibn Mandawayh-Ibn Kaysān's procedure.

As we have seen in Chapter 4, the belief that musk was derived from blood was widespread in the medieval Middle East; the blackish color of grain musk somewhat resembles dried blood. This similarity was no doubt also responsible for the widespread adulteration of musk with blood. Many other ingredients, however, were called for in formulas to extend musk. These ingredients were chosen not only for their scent, but especially for their capacity to produce a musk-like base that could be blended with musk so as to be indistinguishable. The extensive information given by our sources on the adulteration of musk and its detection illustrates the high value of musk.

Conclusion

The trade in musk was part of a well-established system of international trade operating in early medieval Eurasia. One of the hallmarks of this system is the diversity of its organization. Goods travelled by different routes, and different communities of people were involved in the trade simultaneously. The overland trade complemented the maritime trade rather than competing with it; both were necessary to supply the specific needs of different regions as well. Musk travelled by both routes, of course, and was handled by different types of merchants, from specialists or generalists who acquired it near its point of origin, to the great general merchants who imported musk into the Middle East along with so many other Asian goods, to the perfumers and pharmacists who were responsible for its actual distribution in the society of the Middle East. The musk trade was of considerable value and importance. Adulterating musk was extremely common because the cost of pure musk was so high.

CHAPTER 6

Musk in Daily Life in the Early Medieval Islamic World

Introduction

Musk had numerous practical applications in the medieval Middle East. Musk is appreciated at the most basic level because of its powerful scent and uses in perfumery. Musk was ubiquitous in medieval Arabic perfumery, and thus it came to be a symbol of perfume in general. These perfumes were worn by members of both sexes. Their use by men was actually more important than their use by women on a social level, because men were expected to be perfumed when properly dressed. The use of perfume by women was a more private affair and was usually confined to the home. Perfuming was not confined to the human frame; food and drink were perfumed as well, for decoration and because it was believed that aromatics could prevent decay. In addition, musk and aromatics were added to ink. Besides its extensive use in perfumery, musk had a major importance in medicine. The introduction of musk into the classical medical tradition was largely accomplished in Islamic times. It was frequently used as a substitute for castoreum; this suggests that its use developed partly as a replacement for that substance.

Arabic and Persian Perfumes

The perfumes used in the medieval Middle East can be divided into several classes.[1] The two most important were the unguents and the incenses. Aromatic powders were also used, and there were scented oils and waters as well. The word "perfume" is perhaps not the best word for this diversity of applications, but no better word is available. "Aromatics" has been reserved for the odoriferous components of the compounds. The Arabs themselves had several words that generically denote perfumes. The most common of these for musk perfumes is *al-ṭīb/ṭuyūb*. This term denoted especially those perfumes compounded with the most important of the aromatics, called the *uṣūl al-ṭīb*, or

1 More detailed coverage of the perfumes will be given in my study in progress on *Medieval Islamicate Perfumery*.

 | DOI 10.1163/9789004336315_008

principal aromatics: musk, ambergris, aloeswood, camphor, and saffron. The word *al-ṭīb* is from the root meaning sweetness, goodness, and purity. Likewise, we have *al-ṭayyibāt*, the good things in life, used especially of food. *Ṭīb* comes close to the notion of perfume; compound perfumes can be described with this term. But the word *ʿiṭr* is also used for perfume. Neither denotes a specific physical form of perfume. *Ṭīb* carries with it etymologically the association of aromatics with purity and incorruptibility and is a judgment of the inherent goodness of the substance.[2]

The earliest extant work dealing exclusively with aromatics is the *Kitāb Jawāhir al-ṭīb al-mufradah* of Ibn Māsawayh. As the title indicates, his work is focused exclusively on simple aromatics and does not discuss compound perfumes. It is divided into two sections: *al-uṣūl* (i.e. *uṣūl al-ṭīb*, the principal aromatics) and *al-afāwīh*, which is another term meaning aromatics or spices.

Ibn Māsawayh's list of *al-uṣūl*:[3]
Musk
Ambergris
Aloeswood
Camphor
Saffron

Ibn Māsawayh's list of *al-afāwīh*
Spikenard (*sunbul*)
Cloves (*qaranful*)
Sandalwood (*ṣandal*)
Nutmeg (*jawzbuwwā*) and mace (*basbās*)[4]
Rose (*ward*)[5]
Falanjah[6]
Yew (*zarnab*)[7]

---

2 Cf. S. Stetkevych, *The Mute Immortals Speak* (Ithaca: Cornell University Press, 1993), 174.

3 Ibn Māsawayh 9 and trans. M. Levey, "Ibn Māsawaih and His Treatise on Simple Aromatic Substances," *Journal of the History of Medicine* 16 (1961): 398. I have not always followed Levey's identifications, but discrepancies have been noted.

4 Levey 404 gives the latter as fennel.

5 P. C. Johnstone, "Ward," EI 2nd ed. s.v., Garbers #126 on 372–6, W. Schmucker, *Die pflanzliche und mineralische Materia Medica im Firdaus al-Ḥikma des Ṭabarī* (Bonn: Selbstverlag des Orientalischen Seminars, 1969), 529–30, Dietrich DT 1.68, E. Lev and Z. Amar, *Practical* Materia Medica *of the Medieval Eastern Mediterranean According to the Cairo Genizah* (Leiden: Brill, 2008), 261–6.

6 Not clearly identifiable. See Levey 398 n. 17, Garbers #25 on 195–6, and Schmucker 323–4.

7 E. Wiedemann, "Über Parfüms und Drogen bei den Arabern," in *Aufsätze zur arabischen Wissenschaftsgeschichte*, vol. 2, 415, following Guignes, suggests *Taxus baccata*, the yew, as do Lev and Amar 507–8, but Levey is skeptical of this identification; see also Garbers #133 on 392–3 and Schmucker 214–15 (*Atriplex odorata*).

Cassia or cinnamon bark (*qirfah*)[8]
Harnuwah[9]
Greater cardamom (*qāqullah*)
Cubeb (*kabābah*)
Lesser cardamom (*hālbawwā*)
Ḥabb al-mīs.m[10]
Fagara (*fāghirah*)[11]
Mahlab (*maḥlab*)[12]
Wars
Costus (*qusṭ*)
Onycha (*aẓfār*)
Bunk[13]
Lentisk (*ḍarw*)[14]
Labdanum (*lādhan*)[15]
Storax (*mayʿah*)[16]
Kamala (*qinbīl*)[17]

8 Garbers #92 on 301–5, Schmucker 341–2, Lev and Amar 143–6.

9 So identified in many medieval texts; it was some sort of pungent seed. Levey calls it "Indian pepper". Schmucker 519–20, Lev and Amar 555: "Guinea pepper".

10 Unknown; Levey 398 and n. 24 gives *mīsum* but on 406 has *mīsam*; he suggests the bark of *Acacia nilotica*.

11 *Zanthoxylum* sp. See Maimonides #307. (The spelling *zanthoxylum* rather than Meyerhof's *xanthoxylum* is preferred today.) Levey cites Meyerhof and translates "Agnus-castus". Fagara was widely used as a peppery spice in East Asia before the introduction of pepper from India; one species at least is still used, *Z. simulans*, called Sichuan pepper. See also E. Shafer, *The Golden Peaches of Samarkand* (Berkeley: University of California, 1963), 149–50.

12 *Prunus mahaleb*: Garbers #64 on 259–60, Schmucker 461, H. Schönig, *Schminken, Düfte und Räucherwerk der Jemenitinnen: Lexikon der Substanzen, Utensilien und Techniken* (Beirut: Ergon Verlag, 2002), 176–7, and Lev and Amar 239–40.

13 Al-Bīrūnī, *Kitāb al-Ṣaydanah*, ed. ʿAbbās Zaryāb (Tehran: Markaz-i Nashr-i Dānishgāhi, 1991), 130 says it is the bark of the root of the acacia (*umm al-ghaylān*), i.e. *Acacia arabica*. See also Levey 407 n. 90, Garbers #20, 189–90, and Dietrich DT I.83 n. 2.

14 Levey 408 and Meyerhof 53, 113, and Schönig 68–9.

15 Levey 408 has accidentally given laudanum, which is a tincture of opium; the word is probably derived ultimately from the Arabic *lādhan*, which refers to the resin of *Cistus*.

16 Resin of *Styrax officinalis* (dry storax). The resin of *Liquidambar orientalis* (liquid storax) is *lubnā*. See Garbers #65 260–1, Schmucker 494–5, and Dietrich DT I.28.

17 *Mallotus philippinensis* (= *Rottlera tinctoria*), Garbers #91 on 300–1, Schmucker 361–2, and Schönig 153–4. It was used as a dye, giving a bright orange color; see also G. Watt, *The Commercial Products of India* (London: John Murray, 1908), 755–6.

We find a nearly identical conception in the *Murūj al-dhahab* of al-Masʿūdī, which is either dependent on Ibn Māsawayh or they both follow the same source.[18] The term *al-afāwīh* thus also refers to aromatics, but its scope is broader and it includes what we think of as spices.[19] It has been translated as "secondary aromatics."[20] Another term, which refers especially to basil, *rayḥān*, pl. *rayāḥīn*, also refers in general to scented herbs and flowers; many of these are not covered by Ibn Māsawayh. We may mention the importance in making scented oils and waters of the different types of jasmine (*zanbaq* and *yasmīn*), wallflower (*khīrī*), lily (*sūsan*), and narcissus (*narjis*); Ibn Māsawayh does mention the rose. Both *al-ṭīb* and *al-afāwīh* can also be considered *al-adwiyah*, which means *materia medica*. The *ṭuyūb* are a much narrower range of aromatics than the *afāwīh* or *rayāḥīn*. *ʿIṭr*, more often than *ṭīb*, seems to refer to compounded aromatics and perfumes, although these can certainly be referred to by the term *ṭīb* as well. The *ṭuyūb*, however, were the luxury perfumes *par excellence*, compounded especially of musk, ambergris, aloeswood, camphor, and saffron: the principal aromatics.

Compound perfumes attested in early Islamicate literature may be divided into four categories: incense, unguents, scented powders, and oils and waters.[21] The types of perfumes favored in the early medieval Islamic world were used in both Arabic and Persian culture. The names for the two most luxurious, *nadd* and *ghāliyah*, are both Arabic, and they seem to be genuinely Arabic inventions. At least they are unlike any previous Near Eastern perfumes and appear to have no real analogues in Further Asia. All of the Arabic perfumes were also made in the Islamic period by the Persians, who had perfumes of their own as well. Those were also used by the Arabs, showing that neither culture had its own peculiar perfumes. All of these perfumes are part of the common heritage of early medieval Islamicate perfumery. Nevertheless, we should not be surprised that there were some regional differences in their composition.

18 Al-Masʿūdī, *Murūj al-dhahab wa-maʿādin al-jawhar*, ed. B. de Meynard and P. de Courteille, revised by C. Pellat, 5 vols. (Beirut: Manshūrāt al-Jāmiʿah al-Lubnāniyyah, 1966–74), 1.194–5 (§407). Cf. Wiedemann, "Über Parfüms," 415.

19 A. Dietrich, "Afāwīh," *EI* 2nd ed. suppl. 1, 42–3.

20 Levey 403.

21 There seem to be significant differences between early Islamic practice and modern Arabian perfumery. A. S. Kanafani's *Aesthetics & Ritual in the United Arab Emirates* (Beirut: American University of Beirut, 1983), 41–51, gives some discussion of practice in the United Arab Emirates, where scented oils and incense predominated in use while the unguents and scented powders are absent. In addition, the repertory of perfume available to men is much more limited and does not include musk, a striking departure from medieval practice.

Another important distinction among medieval Islamicate perfumes was between the high quality and low quality. This distinction is made solely on the basis of the ingredients: a high quality perfume consisted almost entirely of valuable ingredients such as the five principal aromatics and others. Low quality "knock-offs" were designed to imitate the high quality luxury perfumes, but they were made with cheap ingredients easily available by the artifice of the perfumer. We have already seen how the perfumers and pharmacists imitated and adulterated musk; these practices were ubiquitous with all of the expensive ingredients as well as the finished perfumes. Sometimes they were surely intended to defraud, but it is likely that the craft survived because of the great appeal of the perfumes and because only the wealthiest could afford them. The most important source on these imitations is the book of pseudo-al-Kindī, the *Kitāb Kīmiyāʾ al-ʿiṭr wa-l-taṣʿīdāt*.[22]

The psychological impact of aromatics was well understood. More than a millennium before the fashion for aromatherapy in Europe and America, al-Kindī wrote his curious epistle on music. This work includes, besides its discussion of the properties of different musical notes and modes and the effects of colors, a short section on the psychological effects of various perfumes:

> Since we have mentioned what reaches the soul (*yataraqqā ilā al-nafs*) of the sense of hearing and of sight, now we will mention what reaches it of the sense of smell. So we say that the scent of jasmine stirs the capacity for pride. Narcissus stirs the capacity of pleasant flirtation and the feminine impulse, and likewise when it is mixed with the scent of the myrtle, lily, oxeye (*bahār*),[23] and anemones (*shaqāʾiq*). When the scent of jasmine and narcissus is mixed, the capacities of pride and pleasure are stirred. When lily is mixed with rose, the capacity of love is stirred together with boastfulness. When the scent of wallflower is mixed with narcissus, the capacity for generosity is stirred together with affection. When the scent of *ghāliyah* is mixed with the scent of aloeswood, the capacity of royalty and might are stirred together with love, yearning, and pleasure. Wherever the scents of rose, narcissus, and wallflower occur, they stir ardor, pleasure and yearning; they are feminine scents.

22 Al-Kindī [attr.] ed. Garbers.

23 See al-Bīrūnī 135–6 and Dietrich DT III.131, on some of the identifications for this plant. It seems to have been some sort of oxeye daisy (*Buphthalmun*) but another possibility is chamomile. Fragrant chamomile seems to fit the context best. However, al-Kindī's *Aqrābādhīn*, ed. and trans. M. Levey (Madison: University of Wisconsin, 1966), 239, regularly uses the ordinary Arabic word *bābūnaj* for chamomile.

> Wherever the scents of aloeswood, myrtle, violet, jasmine, and marjoram occur, they stir happiness, pride, generosity, and nobility; they are masculine. Musk, *ghāliyah*, and effeminate scents (*al-arāyīḥ al-khanithah*) are feminine. When these masculine scents are mixed with the feminine scents and paired, they stir happiness and pleasure due to the pairing that has occurred. If the combination is regal, it stirs the capacity for royalty, and if it is generous, it stirs generosity. The compound will stir the various compounded faculties that the scents effect in accordance with the proportions of the compound.[24]

Certain scents were held to enhance other scents. An opinion quoted in the *Mustaṭraf* of al-Ibshīhī says that the "mothers" (*ummahāt*) of the herbs are strengthened by the "mothers" of the aromatics: "Narcissus is strengthened by rose, rose is strengthened by musk, violet is strengthened by ambergris, basil is strengthened by camphor, and dog rose (*nasrīn*) is strengthened by aloeswood."[25] It is curious that al-Kindī groups musk and *ghāliyah* as feminine scents; probably these descriptions of scents as masculine and feminine refer to the scent itself, to its essential characteristics, rather than to the people who wear it.[26] In any case, this text throws some light onto the beliefs current in the 9th century about the psychological effects of aromatics and their blending that underlies the pharmacological literature.

Incense

Early medieval Islamicate incense was designed to be smoldered on hot coals. There are three major types: *dukhnah, bakhūr*, and *nadd*. *Dukhnah* and *bakhūr* are both generic terms for incense; they came in a variety of different types. *Nadd* was the luxury version, typically composed of aloeswood, musk, and ambergris.[27] All of these incenses were made of finely crushed or chopped

---

24 Al-Kindī, *Risālat al-Kindī fī ajzā' khabariyyah fī al-mūsīqā* (Cairo: al-Lajnah al-Mūsīqiyyah al-ʿUlyā, 1963?), 34–5; cf. the trans. by H. G. Farmer, "Al-Kindī on the 'Ēthos' of Rhythm, Colour, and Perfume," *Transactions of the Glasgow University Oriental Society* 16 (1955–6): 29–38.

25 Al-Ibshihī, *al-Mustaṭraf fī kull fann mustaẓraf*, ed. Muṣṭafā Muḥammad al-Dhahabī (Cairo: Dār al-Ḥadīth, 2000), 361.

26 See the section on "Musk and Men" below.

27 M. A. Newid, *Aromata in der iranischen Kultur unter besonderer Berücksichtigung der persischen Dichtung* (Wiesbaden: Reichert, 2010), 208–10.

ingredients made into a paste with melted ambergris or a cheaper substitute and dried for use.

The earliest formulas for *nadd* come from the 10th century. No *nadd* formula appears in al-Ṭabarī's *Firdaws al-ḥikmah* or in the *Kitāb Kīmiyā' al-ʿiṭr wa-l-taṣʿīdāt*. The only exception to this is a formula transmitted by al-Tamīmī for a *nadd* made for the Abbasid caliph al-Mustaʿīn (r. 862–6). It was made of a hundred and fifty *mithqāls* of ambergris and fifty each of aloeswood and musk, with a little camphor. This mixture was made into a paste and formed into strips (*shawābīr*).[28] It is not dissimilar to the *nadds* of Ibn Mandawayh-Ibn Kaysān or al-Tamīmī, so we may assume that they had their roots at least back into the 9th century. During the 10th century numerous *nadd* formulas appear, and al-Tamīmī and al-Zahrāwī as well as Ibn Mandawayh-Ibn Kaysān recorded them. Two special varieties of *nadd* are the tripartite (*muthallath*) and five-part (*mukhammas*) *nadds*. These were made of three or five equal parts of aromatic ingredients, such as equal parts of musk, ambergris, and aloeswood, or equal parts of musk, ambergris, saffron, aloeswood, and camphor.

Entire pieces of aromatic wood, especially aloeswood, constituted another type of incense. These were often specially prepared. The Ibn Mandawayh-Ibn Kaysān tradition mentions as well the preparation of larger pieces of aloeswood for censing: the result is called freshened aloeswood (*al-ʿūd al-muṭarrā*). These were "freshened" with ambergris and musk to increase their redolence. Sometimes larger pieces of aloeswood were pierced and stuffed with other aromatics, or sometimes they were rolled or brushed with melted ambergris mixed with musk.

Unguents

Ghāliyah (pl. *ghawālī*) was a dark colored unguent used especially for the anointing of men's facial hair.[29] The *ghāliyah* used by the caliphs and praised by literary sources was made from musk and ambergris made into an ointment with ben oil,[30] from a species of *Moringa*, which itself was scented with fine

---

28 Nuwayrī 12.60–1.

29 Discussion in A. King, "Tibetan Musk and Medieval Arab Perfumery," in A. Akasoy, et al. *Islam and Tibet: Interactions along the Musk Routes*, 145–61; see also S. Anṣārī, *Tārīkh-i ʿIṭr dar Īrān* (Tehran: Wizārat-i Farhang wa Irshād-i Islāmī, 1381/2002–3), 68–9, Newid 195–202.

30 Ibn al-Ḥashshā', *Mufīd al-ʿulūm wa-mubīd al-humūm*, ed. G. S. Colin and H. P. J. Renaud (Paris: Imprimerie économique, 1941), 100. See also King, "Tibetan Musk" 155–6, for an anecdote from the historian al-Ṭabarī which epitomizes the ideal ingredients of *ghāliyah*.

aromatics during its preparation.[31] The name *ghāliyah* means "valuable" or "precious", and it supposedly was given that name by one of the caliphs.

Some of the *ghāliyah* formulas, especially in the *Kitāb Kīmiyā' al-ʿiṭr wa-l-taṣʿīdāt,* are for imitations made of cheap ingredients, such as cedar gum, wax, asphalt, pistachios, or sesame seeds, processed and scented with as little of the expensive real musk as possible.[32] The works of Ibn Mandawayh-Ibn Kaysān, al-Tamīmī, and al-Zahrāwī contain mostly formulas for high-quality *ghāliyah* made from expensive, imported ingredients. According to al-Nuwayrī, al-Tamīmī gave other sorts of formulas, but he only bothered to quote the ones for the caliphs and nobility.[33] One of al-Tamīmī's formulas is quoted by him from the polymath al-Yaʿqūbī:

> A type of *ghāliyah* for the caliphs according to Aḥmad b. Abī Yaʿqūb: a hundred *mithqāl*s of rare Tibetan musk are taken and pounded after being cleansed from its membranes and hair. It is sieved after pounding through thick-weaved Chinese silk, and then the pounding and sieving are resumed, and this is repeated until it becomes like dust. Then a Meccan vessel (*tawr makkī*) or a Chinese dish (*zibdiyyah ṣīnī*) is taken, and what is available of good, rare ben in a sufficient quantity is added. Fifty *mithqāl*s of fatty, blue Shiḥrī ambergris is cut up in it, and the dish with the ben and ambergris which are in it is set on a gentle coal fire without smoke or smell because they would spoil it. It is stirred with a gold or silver spoon until the ambergris melts. Then it is removed from the fire, and when it becomes tepid, musk is cast into it, and it is beaten well with the hand until it has become a single part. Then that is put up on a gold or silver vessel, but the head must be tightly covered so one may seal it, or it can be placed in a clean, glass pot (*barniyyah*) with its head plugged with a stopper that is Chinese silk filled with cotton, so that its scent will not arise from it. He said, this is the best *ghāliyah* of all. If the ambergris is made equal to the musk it is not so strong. This *ghāliyah* with the musk and ambergris equal in it was made for Ḥumayd al-Ṭūsī. It amazed al-Ma'mūn greatly, and this was the *ghāliyah* made for Umm Jaʿfar, except

31 F. Aubaile-Sallenave, "*Bân, un parfum et une image de la souplesse. L'histoire d'un arbre dans le monde arabo-musulman*," in R. Gyselen, ed., *Parfums d'Orient*, 9–27 (Bures-sur-Yvette: Groupe pour l'Étude de la Civilisation du Moyen-Orient, 1998).

32 R. Gottheil, "Fragment on Pharmacy from the Cairo Genizah," *JRAS* (1935): 123–44, has two *ghāliyah* formulas (one incomplete) based on emblic myrobalans and date syrup.

33 Zahrāwī 2.47–9 does include several cheaper versions.

> that they admixed to the ben an amount equal to a fourth of it of Ruṣāfī Naysābūrī[34] jasmine (*zanbaq*) oil.
>
> They made this *ghāliyah* for Muḥammad b. Sulaymān, except that they added with the ben and jasmine a bit of pure oil of balm of Gilead (*balasān*). They also made a *ghāliyah* for Umm Jaʿfar which they called ambergris *ghāliyah*; this was because they made it with ten parts ambergris for every three parts musk; the process of making it is like what we have described above.[35]

This is a very typical formula for caliphal *ghāliyah*. Only the wealthy could afford the expensive ingredients and equipment necessary to make it, but it illustrates the ideal for which the cheaper imitations were striving.

Another type of aromatic unguent is the *lakhlakhah* (pl. *lakhālikh*).[36] A *lakhlakhah* was used for rubbing upon the body.[37] Unlike the *ghawālī*, they were made with many ingredients besides musk, ambergris, and ben oil. Since they were designed to be applied to the skin, usually in association with bathing, they were not as concentrated, and the musk was often used in the form of *sukk* of musk. They were made into balls or lumps that could be rubbed to release their scent.[38] The Ibn Mandawayh-Ibn Kaysān tradition gives three formulas: one made from the usual aloeswood, musk, ambergris, and camphor, but also including *sukk* of musk and jasmine oil.

In addition to the *lahklakhahs*, there were three other similar preparations. These were the *shamāmah*, the *masūḥ*, and the *dastanbū*. They all seem to have been sorts of pomanders. The *dastanbū* ("hand-scent") is an originally Persian aromatic preparation; it conventionally contained ambergris, musk, and other aromatics made into a scented ball for fondling.[39] Al-Mutawakkil (r. 847–61) is once described as having a *dastanbū* in hand for a drinking party.[40]

---

34 There is a village of Nishapur called Ruṣāfah, see Yāqūt, *Muʿjam al-buldān*, 5 vols. (Beirut: Dār Ṣādir, n.d.), 3.49.

35 Nuwayrī 12.53–5.

36 Anṣārī 69–70 and Newid 204–6.

37 Ibn al-Ḥashshāʾ 70.

38 Cf. Muḥammad Abū Rayḥān al-Bīrūnī, *Kitāb al-Jamāhir fī maʿrifat al-jawāhir*, ed. F. Krenkow (Haydarābād: Dāʾirat al-Maʿārif, 1355/1936), 235.

39 *Farhangnāmah-i Adab-i Fārsī*, vol. 2 of *Dānishnāmah-i Adab-i Fārsī* (Tehran: Muʾassasah-i Farhangī wa Intishārāt-i Dānishnāmah, 1996–), 261b.

40 Al-Tanūkhī, *Nishwār al-muḥāḍarah wa-akhbār al-mudhākarah*, ed. ʿAbbūd al-Shāljī, 9 vols. ([Beirut: Dār Ṣādir], 1971), 1.301. Another attestation of the *dastanbū* is at 8.253.

The *shamāmah* has an Arabic name, but seems rather more common in Persian literature than Arabic.[41]

In addition to these unguents, there were two further types used especially by women. One was *ʿabīr*, the other *khalūq*.[42] There were several different types of *ʿabīr*, including varieties used by men, for which we have formulas. A surviving formula for *ʿabīr* in Ibn Mandawayh-Ibn Kaysān is really a scented powder, and was certainly, like the rest of his perfumes, intended for the use of men. The classical forms of *ʿabīr* and *khalūq*, as known from poetry and other literary sources, were saffron-laden aromatic dye-pastes, comparable to the henna pastes that are better known today in the west. Perfumed dye-pastes incorporating saffron were still used in 20th century Arabia.[43]

Scented Powders

The usual term for scented powder is *dharīrah* (pl. *dharāʾir*).[44] They were designed for sprinkling in clothing. The classical *dharīrah* was made with palmarosa (*Cymbopogon martini*), known as *qaṣab al-dharīrah* "the reed of the *dharīrah*", but the formulas are invariably for complex mixtures of dried powdered aromatics. One formula, preserved by ʿAli b. Sahl Rabban al-Ṭabarī, purports to date to Sasanian times and included aloeswood, musk, and ambergris.[45] Other formulas include many different ingredients and are sometimes based on floral and botanical scents. The Ibn Mandawayh-Ibn Kaysān *ʿabīr* formula, as noted above, is really a type of *dharīrah*, and it consists of crushed musk and aloeswood, with a little ambergris.

41 Some Arabic references in R. Dozy, *Supplément aux dictionnaires arabes*, 2 vols., 2nd ed. (Paris: Maisonneuve, 1927), 1.784; it seems much more uncommon in Arabic literature than Persian. Al-Azdī, *Ḥikāyat Abī al-Qāsim al-Baghdādī*, ed. A Mez. (Heidelberg: Carl Winter, 1902), 36 mentions *al-shammāmāt al-q.ṣ.riyyāt.*

42 Newid 187–94 and 203–4.

43 Kanafani 48 and Schönig 321–2, but see also D. Jung, *An Ethnography of Fragrance: The Perfumery Arts of ʿAdn/Laḥj* (Leiden: Brill, 2011), 172, who says that pure saffron is not used in Aden in perfumes.

44 Cf. Newid 194–5.

45 ʿAlī b. Sahl Rabban al-Ṭabarī 611.

Oils and Waters

As noted above, ben oil was the most prized oil for the compounding of aromatics. The preparation of ben oil also involved the use of musk and other aromatics; once the oil was extracted from the ben seed, it was scented, for the ben oil itself has no scent.[46] Jasmine (*zanbaq*), sweet basil (*ḥamāḥim*), wallflower (*khīrī*), and apple oils were particularly important in perfumery, along with compound oils.[47] No musk is called for in the preparation of jasmine or wallflower oil according to al-Tamīmī,[48] but basil and apple oil made use of it in their preparation.[49] The compound oils were made from individual oils and were further processed with other aromatics, usually including musk.

Scented or toilet waters (*nadūḥāt*) were often distilled. The most famous water was of course rosewater (*mā' al-ward*), but other aromatics were used as well. Other scented "waters" (usually denoted *mā' al-x*) were prepared by boiling down water infused with scented material into thickened syrups; these are commonly called for in perfume formulas. Many aromatics were distilled with a little musk, while musk was used in the preparation of the thickened apple and grape juices as well as *maysūs*, a thickened floral water.[50]

In form these perfumes are all very different from modern perfumes, which are alcohol-based. While the distillation of alcohol was known in the Middle East during the period in question, none of the books on perfumery make any use of it to prepare perfumes. Medieval Middle Eastern perfumes were either ointments, oils, powders, or incense. The luxury versions of these compound perfumes often include a very high proportion of musk. The formulas produce substances which are overpowering and even distasteful to modern people. Musk and civet as perfume ingredients in modern times in tiny quantities have not died out completely because of their curious power of enhancing perfumes, but synthetic versions are mostly used. (The even rarer ambergris, obtained from the sperm whale, which has similar properties, has likewise continued in use when it can be obtained; compounds called "ambers" designed to mimic the scent of ambergris are indispensable in perfumery.) Another musky aromatic, spikenard, was quite prized in antiquity, but its strong and rather striking animal scent has made it rare in modern perfumery.

Fashions in perfume change, and the early medieval Islamicate perfumes would have been more appealing to Europeans of earlier times. Musk was once

---

46 See the formula in, e.g., Nuwayrī 12.81–3. See Aubaile-Sallenave 15, 18–20.

47 Nuwayrī 12.78.

48 Nuwayrī 12.92–4, 96–9.

49 Nuwayrī 12.95–6, 99–101.

50 Formulas for all of these compounds are found in al-Nuwayrī's work.

much more important in European culture than its present use in perfumery would suggest. In Elizabethan times it was commonly used, especially mixed with a little rose water. The rise of perfumed gloves began at this time as well, and their preparation usually included musk. Shakespeare refers to "letter after letter, gift after gift, smelling so sweetly, all musk" sent to Mistress Ford in *The Merry Wives of Windsor* 2.2. Robert Boyle prepared a mixture of eight parts ambergris, two parts musk, and one of civet. This was used to "innoble" other perfumes such as benzoin, storax, and sweet flowers and to make "Pastills, Ointments for Leather, Pomander, &c."[51] The fact that the Empress Josephine was fond of musk, and how her boudoir retained the scent of it sixty years later, is frequently repeated.[52]

The changes in the perception of odor that occurred during the late 18th through early 20th centuries ushered in the modern conception of perfumery in which "less is more" when it comes to musky scents, so reminiscent of the scent of the human body that modern hygiene seeks to obliterate.[53] Musk, the most animal and intense of the aromatics, fell from favor and has never since been used very much in perfumery for its own sake, although it has maintained its place as an ingredient in perfume due to its powers of enhancement.[54] It has always maintained a certain erotic importance, though. The famous *Peau d'Espagne* of the turn of the 20th century, a complex mixture of scents, owed its erotic reputation to its inclusion of civet and musk.[55] Synthetic musks continue to be used in perfumery and are an essential part of the perfumer's palette, as are some of the musky plants like *Hibiscus abelmoschus* (called ambrette).

Musk and Men

There is a distinction in Medieval Arabic sources between masculine and feminine perfumes. A ḥadīth says: "the good aromatic for men is that which has an apparent scent and a hidden color, and the good aromatic for women is

51 R. Boyle, "Experiments and Observations about the Mechanical Production of Odours," in *The Works of Robert Boyle*. Vol. 8. Ed. M. Hunter and E. B. Davis (London: Pickering and Chatto, 2000), 386–7.

52 A. Corbin, *The Foul and the Fragrant* (Cambridge, MA: Harvard University Press, 1986), 196.

53 This subject is treated at length in the important work of Corbin.

54 Cf. Corbin 67–9.

55 Cf. Havelock Ellis, *Studies in the Psychology of Sex Volume 4. Sexual Selection in Man* (Philadelphia: F. A. Davis, 1905; repr. 1920), 99–100.

that which has an apparent color and a hidden scent."[56] Much information on the attitudes towards perfumes is given by Muḥammad b. Aḥmad al-Washshā' (d. 936) in his book, the *Muwashshā*. This book is a description of the mores of the elegant aristocrats (*ẓurafā'*); it is a major source of information on the social life of early Medieval Baghdād.[57] He writes of the perfumes used by the elegant men: "they do not use any pungent (*dhafar*) perfume which produces a color and leaves a trace" because of a version of this ḥadīth.[58] Strong scented perfumes like musk and ambergris were masculine, while feminine perfumes were often compounded with saffron, such as *khalūq* and *ʿabīr*, which acted as powerful yellow dyes. Musk was considered a masculine perfume, and men of quality are consistently mentioned as wearing, or smelling of it or of *ghāliyah*, which was a perfume predominantly of musk, as we have seen.

Frequently we find someone attractive, noble or rich described as sweet-smelling.[59] It was a mark of good character. Men anointed their beards and moustaches with musk or *ghāliyah* to achieve a good scent, and they scented their clothes with scented powders. The pre-Islamic poet Ṭarafah b. al-ʿAbd (mid-6th century) describes the nobility in this verse:

> *thumma rāḥū ʿabaqu l-miski bihim*
> *yalḥafūna l-arḍa huddāba l-uzur*[60]

> Then they departed, the redolence of musk with them, covering the earth with the fringes of their *izārs*.

Al-Washshā' describes the perfuming of elegant males:

> It is their fashion to be perfumed and scented with musk crushed and dissolved in rosewater. They use aloeswood anointed with ambergris

56 Quoted by al-Zamakhsharī, *Rabīʿ al-abrār wa-fuṣūṣ al-akhbār*, eds. ʿAbd al-Majīd Diyāb and Ramaḍān ʿAbd al-Tawwāb (Cairo: al-Hay'ah al-Miṣriyyah al-ʿĀmmah li-l-Kitāb, 2 vols., 1992–2001), 2.213. See also Nasā'ī, *Sunan al-Nasā'ī*, 8 vols., Cairo: Muṣṭafā al-Bābī al-Ḥalabī, 1964–5), 8.130, 131. Cf. al-Washshā', *Kitāb al-muwashshā*, ed. R. E. Brünnow (Leiden: Brill, 1886), 126.

57 M. M. Ahsan, *Social Life under the Abbasids 170–289 AH, 786–902 AD*. (London: Longman, 1979), 6–7.

58 Washshā' 126.

59 E.g., al-Tanūkhī, *al-Faraj baʿda al-Shiddah*. 2 vols. (Cairo: Dār al-Ṭibāʿah al-Muḥammadiyyah, 1955), 1.50, 107.

60 Ṭarafah, *Dīwān*, ed. M. Seligsohn (Paris, 1901; repr. Baghdād: Maktabat al-Muthannā, 1968?), 59. Cf. Ibn ʿAbd Rabbih, *al-ʿIqd al-farīd*, 31 vols. (Beirut: Maktabat Ṣādir, 1951–4), 30.148.

> permeated with clove water, and Sulṭānī *nadd*, and Baḥrānī ambergris, *ʿabīr*, and powders enhanced (*maftūqah*) with *ʿabīr*. No scents other than those come near them, and they only use camphor in times of real heat because of its coldness, or for a serious ailment, or placed upon the incense-burner mixed with *ʿabīr* of musk and saffron threads. This is the recipe for the most fragrant (*aṭyab*) incense. The *barmakiyyah* and what resembles it is not prohibited to them if it is good *barmakiyyah* or a redolent incense. Rather the elegant forbid its use when it is what is used by the insignificant people (*al-mutaqallilūn*). Likewise they avoid water of *khalūq* because it is one of the perfumes of women, and *ghāliyah* since it is the perfume of the youths (*ṣibyān*) or slave girls (*imāʾ*).[61]

According to al-Washshāʾ, the perfumes used by the elegant men were largely incense and powders. His concept of *ʿabīr* seems different from that usually understood, i.e. that *ʿabīr* is a perfume with much saffron.[62] Such a perfume would be inappropriate for men because it leaves a stain on the skin. So, al-Washshāʾ probably refers to a scented powder without saffron. He also associates some *ghāliyah* with youths and slave girls; it is impossible to say exactly what sorts of *ghāliyah* are referred to here, but, in general, al-Washshāʾ's elegant men apparently preferred less musky perfumes than the caliphs.[63] Musk crushed and dissolved in rosewater has a different effect from the very heavy *ghāliyah* of musk and ambergris in ben oil.

The second Abbasid caliph al-Manṣūr (r. 754–775) was so notorious for his stinginess that he earned the epithet Abū al-Dawāniq, "father of the small change."[64] However, he insisted his family dress well and perfume themselves to show their prosperity and position as the ruling house. He required the use of the silk brocade called *washy* for their clothing. If he saw someone who had not applied enough musk in his beard, he would say: "O so-and-so, I don't see the flash (*wabīṣ*) of *ghāliyah* in your beard, but I see it gleaming in the beard

61 Washshāʾ 125–6. Cf. Wiedemann, "Über Parfüms", 422–3.

62 Ibn Mandawayh-Ibn Kaysān's *ʿabīr* formula probably would have produced the sort of product referred to here.

63 I suspect that the reason here is that the musky perfumes were characteristic of youth and fertility (qualities which the caliphs needed to possess as part of their symbolic role in society) and the elegant men of al-Washshāʾ sought to grow old gracefully, eschewing the practices of the young.

64 It was well enough known that even the Pahlavi *Shahrestānihā-i Ērān* refers to him with this title; H. S. Nyberg, *A Manual of Pahlavi*, 2 vols. (Wiesbaden: Harrassowitz, 1964–74), I.117 line 21.

of so-and-so."[65] In other words, it was expected that members of the nobility would include perfuming in their daily routine; even a miser like al-Manṣūr expected it. Sometimes this perfuming was taken to excess: a certain Ibn ʿAbbās was said to anoint his body so much that when he passed someone in the street they would say, "Did Ibn ʿAbbās pass by or did musk?"[66] It was also said that the *ghāliyah* on his bald head was like syrup (*rubb*).[67]

Perfuming, especially with musk, was an expected part of proper male grooming.[68] Men attending court would perfume themselves before going. The custom was certainly practiced by the Sasanians: the boy and future king Yazdagird was dressed up and perfumed before his presentation to Kisrā.[69] The *Kitāb al-Ṭabīkh* of Ibn Sayyār al-Warrāq has a chapter on the etiquette expected of courtiers and companions. He gives instructions on washing and then explains the perfuming routine: one must be perfumed with incense, musk, *ghāliyah*, and different kinds of aromatic powders (*dharāʾir*) upon the hair and clothing; the perfume for the clothing is to be incense, while musk and camphor are used for the hair, and the aromatic powders are for the body.[70] The incenses used for personal censing by the wealthy and nobles were often made with musk. As al-Tamīmī said, "The foundation of all incense (*bakhūr*) in goodness is ambergris, musk, aloeswood, camphor, and the fire upon which it is burned."[71]

Most of the references to musk and musk-based perfumes in Classical Arabic literature pertain to the wealthy and nobility if not to the caliphal court itself. Al-Faḍl b. al-Rabīʿ made an inventory of the caliphal treasuries on the accession of al-Amīn in 809. They included one hundred thousand *mithqāls* of musk and the same of ambergris, as well as a thousand baskets (*safaṭ*) of

---

65 Ṭabarī III.538–9.

66 Zamakhsharī, *Rabīʿ*, 2.216 and Ibn Abī al-Ḥadīd, *Sharḥ Nahj al-balāghah*, 20 vols. (Cairo: ʿĪsā al-Bābī al-Ḥalabī, 1960–4), 19.343.

67 Zamakhsharī, *Rabīʿ*, 2.216 and Ibn Abī al-Ḥadīd 19.343. From Arabic *rubb* we get the archaic English word "rob," meaning fruit juice boiled down to the consistency of syrup.

68 Al-Shayzarī, *Kitāb Nihāyat al-rutbah fī ṭalab al-ḥisbah*, ed. al-Sayyid al-Bāz al-ʿArīnī (Beirut: Dār al-Thaqāfah, n.d.), 8, who stresses perfuming with musk as *sunnah*. Cf also al-Kulaynī, *al-Furuʿ min al-kāfī*, 8 vols. (Tehran: Dār al-Kutub al-Islāmiyyah, 1983), 6.514, for the Shiʿite perspective.

69 Ṭabarī I.1044.

70 Ibn Sayyār al-Warrāq, *Kitāb al-Ṭabīkh*, ed. K. Öhrnberg and S. Mroueh. Studia Orientalia 60. (Helsinki: Finnish Oriental Society, 1987), 332. Trans. N. Nasrallah, *Annals of the Caliphs' Kitchens: Ibn Sayyār al-Warrāq's Tenth-Century Baghdadi Cookbook* (Leiden: Brill, 2007), 502–3.

71 Nuwayrī 12.65.

aloeswood, one thousand pots (*barniyyah*) of *ghāliyah* and many other kinds of perfume (*ʿiṭr*).[72] The Fatimid treasury held porcelain jars of Fanṣūrī camphor, skull-shaped vessels (*jamājim*)[73] of ambergris from al-Shiḥr and Aden, phials (*qawārīr*) of Tibetan musk, and aloeswood.[74] These particular varieties of aromatics were all the most esteemed kinds.

The caliph al-Wāthiq (r. 842–7) had *ghāliyah* made that was preserved for decades at his palace. The story is worth quoting in full because of its description of the use of *ghāliyah* and its general interest for the history of that substance:[75]

> One day I [Ṣāfī al-Ḥuramī] was attending on al-Muʿtaḍid [r. 892–902], and he wanted to perfume himself, so he said, "Summon a perfumer," meaning a servant in charge of the treasury of perfumes (*khizānat al-ṭīb*). He said to him, "How many vessels of *ghāliyah* do you have?"
>
> He said, "More than thirty Chinese crocks (*ḥibb ṣīnī*) of what was made by several of the caliphs."
>
> "Which is the most fragrant (*aṭyab*)?"
>
> "That which al-Wāthiq made."
>
> "Have it brought here."
>
> A great crock was brought carried by several servants using a frame to support it, and it was opened. The *ghāliyah* had become white with growth and it had congealed with age. It was intensely fragrant (*fī nihāyat al-dhakā'*). It amazed al-Muʿtaḍid, and he grasped with his hand around the neck of the crock, and took from the smear [around its neck] (*laṭākhatihi*) a small bit lest he disturb the head of the crock. He put it in his beard and said, "I cannot allow myself to strike the growth from this crock; take it away!" So it was removed.
>
> Time passed, and al-Muktafī [r. 902–8] sat to drink one day when he was caliph, and I was standing in attendance. He desired *ghāliyah* and summoned the servant and asked about the different *ghawālī*, and he [the

72 *Kitāb al-Hadāyā wa-l-tuḥaf*, ed. M. Hamīdullāh as Ibn al-Zubayr, *Kitāb al-Dhakhā'ir wa-l-tuḥaf* (Kuwait: Dā'irat al-Maṭbū'āt wa-al-Nashr, 1959), 214–15 (§302). Trans. Ghāda al-Ḥijjāwī al-Qaddūmī, *Book of Gifts and Rarities* (Cambridge, MA: Harvard University Press, 1996), 207–8.

73 These may have had an appearance like the skull-shaped pomanders found in medieval Europe; see an illustration in R. Schmitz, "The Pomander," *Pharmacy in History* 31:2 (1989): 87. Alternately, they might have been shaped like the dome of the skull?

74 *Kitāb al-Hadāyā wa-l-tuḥaf* §384 on p. 255; Qaddūmī 234.

75 Tanūkhī, *Nishwār*, 1.289–91.

servant] told him what he had told his father. He asked for the *ghāliyah* of al-Wāthiq and he brought him the same crock. It was opened, and he deemed it good (*istaṭābahu*) and he said, "Take a little from it." Thirty or forty *mithqāls* were taken from it, and he used immediately what he wanted of it, and called for his perfume-box (*'atīdah*) and put the remainder in it to use over the days.

Al-Muqtadir [r. 908–29] became caliph, and he sat with slave girls drinking one day, and I was present. He wanted to perfume himself, so he summoned the servant, and questioned him, and he told him what he had told his father and brother. He said, "Bring all of the *ghawālī*," and all the crocks were brought. He began to take out from each crock a hundred *mithqāls*, or fifty, or more or less, and he smelled them and distributed them to those in his presence until he came to the crock of al-Wāthiq. He deemed it good, and asked for a perfume-box. They brought him a perfume box, and it was the very perfume-box of al-Muktafī. He looked at the partially emptied vessel and at the perfume-box which had in it a measure of *ghāliyah*, for not much had been used of it. He said, "What is the reason for this?" I explained it to him, and he began to be astonished at the stinginess of the two men and to disparage them for it.

He divided the crock among his family and slave girls, and continued to take *raṭls* out of it. I was torn with anger, and I remembered the story of the grapes,[76] and the speech of my master al-Mu'taḍid, until nearly half the crock had gone. I said to him, "O master, this is the most fragrant and ancient *ghāliyah*; there is no replacement for it. Might you leave some of it for yourself and distribute the remainder from the other crocks?" Tears flowed when I remembered the speech of al-Mu'taḍid, and he felt ashamed in front of me and removed the crock. Only a few years of his caliphate passed before that *ghāliyah* was exhausted and he had to have more prepared at a high cost.

This anecdote is appended to a dialogue of al-Mu'taḍid with his servant Sāfī al-Ḥuramī, who narrated it. One day they had seen the young al-Muqtadir distribute some rare, out-of-season grapes among ten slave boys his age. Al-Mu'taḍid bemoaned the fate of his line, foreseeing that al-Muqtadir would waste the accumulated resources of the caliphate. The story of al-Wāthiq's *ghāliyah* is a moralizing anecdote intended to be read as an assessment of the

76 Tanūkhī, *Nishwār* 1.287–90 and below.

caliphs involved.[77] It shows some important things about *ghāliyah*: namely, that it was specially compounded, and that the different blends had different qualities. It is likely that many of the old caliphal blends were kept secret just as perfumers today keep the precise formulas of their perfumes secret. Some of the difference in quality probably can also be attributed to the availability of quality ingredients; this was a real problem given that the most important ingredients of *ghāliyah* were imported to begin with. This anecdote also shows that *ghāliyah* could be kept for long periods of time without any deterioration; indeed, if anything, it illustrates that it improved with aging.

Ghāliyah was an expensive, luxurious substance. Its use may have been expected among the upper classes, but the more ordinary men very likely seldom got to use it, or at least the higher grades of it. While very fine *ghāliyah* was produced with expensive musk, ambergris, and ben oil only, as we have seen, lesser versions were prepared by enhancing with perfume (*fataqa*) a base and foundation (*jasad* and *rukn*) of some other ingredient. A *ghāliyah* formula preserved in the *Kitāb Kīmiyāʾ al-ʿiṭr wa-t-taṣʿīdāt* describes the preparation of a base made from pistachio and wild thyme censed with aloeswood. Then the directions for enhancing it are given:

> It is enhanced according to three types. A type with musk, ambergris, and ben oil with a part of musk and ambergris for every two parts of the foundation. The patricians of a country (*baṭāriqat al-balad*), the treasurer, judge and his associates, and the postmasters are perfumed (*ghullifa*) with it. The second level of it has three parts of base and a part of musk, a part of good *sukk*, and half a part of ambergris, and it is prepared with ben oil. The officers and those similar to them are perfumed with it. The third level has five parts of the base and a part of musk, two parts of *sukk*, two parts of crushed aloeswood, and a fourth of a part of ambergris, and it is prepared with ben oil like the first. The soldiers, merchants, and other men are perfumed with it.[78]

The provision of fine *ghāliyah* to the military was something notable. During his war with his brother, the future caliph al-Maʾmūn (r. 813–33), the caliph al-Amīn (r. 809–13) anointed the beards of the defectors from Ṭāhir's army with *ghāliyah*. This became their nickname: *quwwād al-ghāliyah*, the "generals of

77 On this form see El-Hibri, *Reintepreting Islamic Historiography* (Cambridge: Cambridge University Press, 1999).

78 Garbers formula #47.

ghāliyah".[79] A caliph, of course, had only the best *ghāliyah*, and it was unusual for him to share it.

It was expected that others would appreciate the good scent of men; conversely, it was a sign of meanness to fail to appreciate fine aroma. The poet Di'bil (765–860) in a lampoon wrote:

lakin ataytu wa rīḥu l-miski yaf'amunī
wa-'anbaru l-hindi mashbūbun 'alā l-nāri
fa-ankara l-kalbu rīḥī ḥīna 'abṣaranī
wa-kānu yu'rifu rīḥa l-zifti wa-l-qāri[80]

But I came and the scent of musk permeated me, and Indian ambergris was kindled on the fire.
The dog did not recognize my scent when he saw me; for he only knew my scent to be that of tar and pitch.

A humorous anecdote on the anointing of the facial hair with *ghāliyah* is preserved by al-Jāḥiẓ in his *Kitāb al-Bukhalā'*. A man named Abū Ja'far al-Ṭarasūsī visited some people who anointed his beard and moustache (*shāribihi wa sabalatihi*) with *ghāliyah*. Unfortunately, he developed an itch on his lip; he put his finger in his mouth and scratched his lip from the inside to avoid rubbing the *ghāliyah* off.[81]

A recipe survives in the *Kitāb Kīmiyā' al-'iṭr wa-l-taṣ'īdāt* for a *ghāliyah* prepared in the time of al-Ma'mūn.[82] It was prepared for Khālid b. Yazīd, the governor of Armenia from whom the line of Sharwān Shāhs originated. It was used at a circumcision party for one of his sons. This is the same formula for which the different levels of enhancement for different social groups are mentioned and that was referred to above.

Perfuming was often done in conjunction with a meal or at a drinking party,[83] and incense would also be burned.[84] This tradition had earlier been

---

79 Ṭabarī III.865.

80 Al-Sarī al-Raffā', *Al-muḥibb wa-l-maḥbūb wa-l-mashmūm wa-l-mashrūb*, 4 vols., ed. Miṣbāḥ Ghalāwinjī (Damascus: Majma' al-Lughah al-'Arabiyyah, 1986–7), 3.163 (#301).

81 Al-Jāḥiẓ, *Kitāb al-Bukhalā'*, ed. Ṭāhā Ḥājirī (Cairo: Dār al-Ma'ārif, n.d. [1958]), 58.

82 Garbers formula #47.

83 E.g., Tanūkhī, *Faraj*, 48 (al-Ḥajjāj distributed *ghāliyah* after a meal). Cf. also Mas'ūdī 4.314 (§2726) and Kai Kā'ūs b. Iskandar, *Qābūs-nāmah*, ed. R. Levy (London: Luzac, 1951), 40: rosewater and *'iṭr* should be provided after handwashing before the meal.

84 Tanūkhī, *Nishwār*, 1.182; al-Mutanabbī, *Dīwān al-Mutanabbī* (Beirut: Dār Ṣādir, 1958), 217.

practiced by the ancient Greeks, who burned incense at their symposia.[85] In the same way al-Maʾmūn had the learned scholars who came to his court for philosophical debate treated to a meal and censing and perfuming before the discussions.[86] The provision of rare aromatics was *de rigueur* at lavish drinking parties. To judge by the literature, drinking parties with famous singing-girls were one of the most desirable diversions available to men. There are numerous descriptions of these parties, so we know what was expected for one of them. While the wine and the singing-girls were undoubtedly the major attraction, scent, along with food, was an important part and is mentioned as one of the expenses of the drinking-party.[87] The use of aromatics at a fine meal is also attested from Persia in the *Shāhnāmah*; shopping for a meal included buying wine, saffron, musk, and rosewater in addition to the food itself.[88] The aroma of wine itself was also associated with musk; this will be discussed below in the section on food and drink.

Some of the most extravagant drinking parties were thrown by the caliphs, who had the wealth to make them fantastically luxurious. It is related that once al-Muqtadir supposedly threw a drinking party in a garden of narcissus (*narjis*). Before the party, al-Muqtadir stopped the gardener from fertilizing the plants with manure. The caliph inquired why this was necessary, and when he was told that it protected the plants and helped them, al-Muqtadir decided to replace the manure with musk. This apparently cost a great deal of money.[89]

The Abbasid caliph al-Mutawakkil (r. 847–61) threw a luxurious circumcision party (*iʿdhār*) for his son al-Muʿtazz. Among the decorations for the party were jeweled gold stands (*marfaʿ*) with figurines (*tamāthīl*) of ambergris, *nadd*, camphor and musk paste (*maʿjūn*).[90] These figurines were likely in the form of fruits; figurines of ambergris in the form of citrons and melons are attested.[91] A description of a particularly lavish *majlis* or assembly held by Sulṭān Maḥmūd of Ghazna (r. 998–1030) mentions the aromatics that were present for this event: "Around the *majlis* were thick plates of gold filled with

85 S. Lilja, *The Treatment of Odours in the Poetry of Antiquity*. Commentationes Humanarum Litterarum 49. (Helsinki: Societas Scientiarum Fennica, 1972), 50–1.

86 Masʿūdī 4.314 (§2726).

87 E.g., Tanūkhī, *Nishwār*, 1.178, 189.

88 Firdawsī, *Abu'l Qasem Ferdowsi: The Shahnameh (The Book of Kings)*, ed. Djalal Khaleghi-Motlagh, 8 vols. (New York: Bibliotheca Persica, 1988–2008), 6.465. Again, Arabic influence cannot be discounted, for saffron is called by its Arabic name *zaʿfarān*.

89 Tanūkhī, *Nishwār*, 1.295.

90 *Kitāb al-Hadāyā wa-l-tuḥaf* 113–15 (§139); Qaddūmī 136–8.

91 Tanūkhī, *Nishwār*, 8.253.

pungent (*al-adhfar*) musk, gray ambergris, fragrant camphor, and redolent aloeswood".[92]

Another luxurious party was thrown by Mufliḥ al-Aswad al-Khādim. He invited al-Muqtadir to his garden (*bustān*) in 924. There were many elaborate preparations: streams were turned into sources of drinks, food was hung from trees, and the seating areas were filled with fruits, flowers, and aromatics. The latter included aloeswood, musk, camphor, *nadd*, ambergris, and saffron. Al-Muqtadir sent to Naṣr al-Qushurī, who could not attend, a bowl of crystal filled with *ghāliyah*, a gold tray of *nadd*, skull-shaped vessels (*jamājim*) of ambergris, musk, etc.[93] Gardens were especially appropriate places to enjoy musky perfumes, for musk is a substance especially associated with the future Garden paradise, as we will see. It was quite appropriate to have a crock of musky perfume present in one's garden; the governor of Baṣrah Muḥammad b. Sulaymān had a Chinese washtub or urn (*ijjānah*) full of *ghāliyah* in his garden.[94]

The use of Chinese porcelain and gold and silver vessels is, as we have seen, quite common with luxury aromatics. This is one more aspect of the high value of these perfumes; one simply didn't store *ghāliyah* worth its weight in gold in common pottery. Perfume formulas likewise specify the use of fine vessels for the manufacture of aromatic compounds.

Musk and Women

During the early Islamic age, musk was an essential ingredient of many types of perfumes and perfumed preparations, and women were expected to wear perfume in the privacy of the home.[95] It was expected that a woman would wear a perfume, even one containing musk, before going to her husband.[96] The poet al-Mutanabbī implied that women regularly wore *ghāliyah* except in

---

92 Al-ʿUtbī, *al-Yamīnī fī sharḥ akhbār al-sulṭān Yamīn al-Dawlah wa Amīn al-Millah Maḥmud al-Ghaznawī*, ed. Iḥsān Dhunūn al-Thāmirī (Beirut: Dar al-ṭalīʿah li-l-ṭibāʿah wa-l-nashr, 2004), 332.

93 *Kitāb al-Hadāyā wa-l-tuḥaf* 107–8 (§126); Qaddūmī 131–2 .

94 Tanūkhī, *Nishwār*, 3.59.

95 Cf. S. D. Goitein, *A Mediterranean Society: The Jewish Communities of the Arab World as Portrayed in the Documents of the Cairo Geniza*. 6 vols. (Berkeley: University of California Press, 1967–1993; repr. 1999), 4.225.

96 Ibn Miskawayh, *Kitāb Tajārib al-umam*, 3 vols. (Baghdad: al-Muthanna, n.d.), 1.78.

time of mourning.[97] Combing the hair could reveal a scent of musk—whether literally or figuratively is another question.[98] There is much less information available on the use of musk perfumes by women, and indeed on the use of perfumes in general by women than for men.[99] There are clear indications that more musky perfumes were indeed used; women wore masculine perfumes in addition to their own feminine perfumes and thus the repertory of scents available to them was greater, as is the case in modern times as well.

As mentioned above, the distinction between masculine and feminine perfumes was based on the physical characteristics of the perfume, as well as its scent. Perfumes that were pastes that dyed the skin were feminine perfumes, while the pungent smelling musk that left no yellowish stain was masculine. Women, however, were able to wear masculine perfumes, even if they were subject to social restrictions. Women could be condemned as prostitutes for wearing perfume that could be smelled by others.[100] The incense burner when carried by a woman was a sign of readiness for sex.[101] Thus, perfume use by women was especially confined to the house, though in that context, it was encouraged. Al-Washshā' writes of elegant women:

> It is of their fashion also to be perfumed with that of which men have no part; they use *lakhlakhah*s, sandalwood, *ṣayyāḥ*,[102] cloves, perfumes

97 Mutanabbī, *Dīwān*, 267.

98 Al-Ma'arrī, *Shurūḥ saqṭ al-zand*, ed. Ṭāhā Ḥusayn, Muṣṭafā Saqqā, et al., 5 vols. (Cairo: al-Dār al-Qawmīyah li-l-Ṭibā'ah wa-al-Nashr, 1964), 4.1614.

99 Here modern anthropological studies can illuminate the uses of aromatics by women. Kanafani's study on women in the United Arab Emirates shows how aromatics were incorporated into daily life. The application of perfumed oils, scented dye-pastes, and the use of incense was an essential part of grooming and social life among women, see esp. 41–51. These usages are certainly ancient as confirmed by the references in medieval literature.

100 E.g., al-Tirmidhī, *Sunan*, 5 vols. (Cairo: Maṭba'at al-Madanī, 1964), 4.194, and G. Van Gelder, "Four Perfumes of Arabia: A Translation of al-Suyūṭī's *Al-Maqāma al-Miskiyya*," in R. Gyselen, ed., *Parfums d'Orient* (Bures-sur-Yvette: Groupe pour l'Étude de la Civilisation du Moyen-Orient, 1998), 205.

101 Ṭabarī II.453 and G. R. Hawting, *The History of al-Tabarī Volume XX. The Collapse of Sufyānid Authority and the Coming of the Marwanids* (Albany: State University of New York Press, 1989), 31 n. 150.

102 A perfumed liquid, perhaps made from *khalūq* which could be used also for the hair, cf. E. Lane, *An Arabic-English Lexicon*. 8 vols. paged continuously (London, 1863–93; repr. Beirut: Librairie du Liban, 1968), 1753b.

> of the night-maker (*al-sāhiriyyah*),[103] *adqāl*,[104] pastes (*maʿjūnāt*), saffron, *khalūq*, water of *khalūq*, camphor, water of camphor, tripartite perfume of the treasuries (*al-muthallathah al-khazāʾiniyyah*), royal Barmakiyyah perfume (*al-barmakiyyah al-sulṭāniyyah*), and other kinds of oils such as violet, jasmine, and ben, except that they avoid the use of *turshutām* (?), and men do not use any of it. Women use all the perfumes of the elegant [men], but the elegant [men] do not use any of the feminine perfumes.
>
> It is of their well-known fashion in the wearing of stringed ornaments that they wear short necklaces (*makhāniq*) censed with cloves, and long necklaces (*murūsil*) with camphor and ambergris . . .[105]

This is one of the most detailed glimpses into the perfumery of women, and it includes perfumes that might contain some musk, although musk is not explicitly mentioned. Not much is known about these feminine perfumes compared with the masculine ones, as the extant formularies concentrate on masculine perfumes. The major source for the scent of women is in poetry, where the image of the musk-scented beauty is so common as to be ubiquitous. This is due to both the association of women with perfume and, in later times, to the association of musk with gazelles, to which women were likened.

Pre-Islamic poets refer to the literal use of musk as a perfume by women. For example, al-Nābighah al-Dhubyānī's famous ode on the wife of the Lakhmid king Nuʿmān describes her:

---

103 An explanation of the *sāhiriyyah* is that its maker stayed up late at night making and perfecting the perfume. A *ghāliyah* formula described as a *sāhiriyyah* is given by Nuwayrī 12.55. They were perfumes made or enhanced at night when it was believed they would become more fragrant. Perhaps the sense of smell is enhanced because of the darkness, and this was believed to reflect a better perfume. Many aromatics are sensitive to light and temperature, however, and deteriorate in unfavorable conditions. Ibn Mandawayh 239–40, Ibn Kaysān 199, and Abū al-Qāsim al-Zahrāwī, *Kitāb al-Taṣrīf li-man ʿajiza ʿan al-taʾlīf*, facsimile of Süleymaniye Beşirağa collection, ms. 502, ed. F. Sezgin (Frankfurt am Main: Institute for the History of Arabic-Islamic Science, 1986), 2.47 (followed by Nuwayrī 12.52) stress the importance of proper conditions for making perfumes. Cf. also G. Ferrand's trans. of the version in Nuwayrī: *Relations de voyages et textes géographiques arabes, persans et turks relatifs à l'Extrême-Orient du VIIIe au XVIIIe sieclès*, 2 vols. (Paris: Leroux, 1913–14), 614.

104 Singular *daql*. This word usually describes a kind of date palm, or the mast of a ship. But it apparently also is a type of dye (*khiḍāb*), though Lane 898b discounts that reading. Certainly neither date palm or mast can fit this context!

105 Washshāʾ 186–7.

muḍammakhatun bi-l-miski makhḍūbata l-shawā
bi-durrin wa-yāqūtin lahā mutaqalladah[106]

Anointed with musk, extremities dyed,
she is garlanded with pearls and corundums

In this case, the dyeing of the extremities surely refers to the painting of the limbs with perfumed dye pastes, similar to the use of henna in modern times. A line from Imru' al-Qays' famous *Mu'allaqah* ode also includes musk in a description of the beloved:

wa-yuḍhī fatītu l-miski fawqa firāshihā
na'ūma l-ḍuḥā lam tantaṭiq 'an tafaḍḍuli[107]

In the forenoon crumbs of musk are on her bedding,
she was a great sleeper in the forenoon, who did
not girdle herself, still clad in a nightgown.

This is an interesting line because, rather than comparing the scent of the beloved to musk, it describes its actual use. The poet has been reminiscing on his romantic relationship with this woman, whose name was 'Unayzah. Perhaps this line, with its eroticism of the woman lingering in bed, might describe the morning after a rendezvous with Imru' al-Qays, and in this context the crumbs of perfume are a symbol and a result of their lovemaking.

Women owned perfumes as part of their possessions; those of high status had large quantities of precious musk. Al-Sayyidah Sitt Miṣr, the daughter of the Fatimid al-Ḥākim bi-'Amr-Allāh, died in 1063 leaving much wealth that is described in the so-called *Kitāb al-Hadāyā wa-l-tuḥaf*. It included more than thirty porcelain (*ṣīnī*) jars (*zīr*) filled with powdered (*masḥūq*) musk.[108] The *zīr* was a large jar with a conical base.[109] Even non-royal women were expected to have a perfume mixing vessel called a *madāf* to prepare their scents.[110] The perfume grinding stone called the *madāk* was also part of the equipment

106 Quoted in al-Ma'arrī, *Risālat al-ghufrān*, ed. 'Ā'isha 'Abd al-Raḥman, 5th ed. (Cairo: Dār al-Ma'ārif, 1969), 207.

107 The poem is in al-Tibrīzī, *Sharḥ al-Qasā'id al-'Ashr*, ed. 'Abd al-Sallām al-Ḥūfī (Beirut: Dār al-Kutub al-'Ilmīyah, 1985), 10–73, this line is on 46.

108 *Kitāb al-Hadāyā wa-l-tuḥaf* 240 (§354); Qaddūmī 222–3.

109 Cf. Qaddūmī 441.

110 Goitein 4.225–6.

of the bride.[111] A marriage contract of 1105 found in the Cairo Geniza lists "a small ointment jar, two very small silver ointment jars and one silver spoon for the labdanum, and another silver spoon for the *ghāliyah*."[112] Labdanum was used, then as now, in making many types of perfumes; in modern times it remains one of the most basic scents in the perfumer's repertory for its rich, balsamic scent shot through with an animal muskiness. It is fundamental for the creation of ambers, compounds designed to mimic the scent of ambergris. Another list preserved in the Geniza records the items brought by one Sarwa to the house of her new husband. It lists gold jewelry, clothing, household utensils, and bedding. The list includes "a vase with perfumes and musk" as well as "a box for ointments ..."[113]

Musk (probably compounded with some form of oil) could be used as a cosmetic. One of al-Mutawakkil's slavegirls wrote his name in musk upon her cheek. He said he had never seen a more beautiful example of black upon white than upon that cheek.[114]

Royal women who could afford it had other uses for musk. Qabīḥah, the mother of the caliph al-Mu'tazz (r. 866–9), had crumbs of musk and ambergris used along with sable hair to stuff quilts (*dawāwīj*).[115] Sandals could be made with cloth soaked in musk and ambergris.[116] The use of musk in footware is also attested for men.[117] These exceptional uses for musk were alleged to occur in the court; it is possible that they reflect an attitude of pious condemnation on the part of those reporting them and thus perhaps should not be taken as the literal truth.

In addition to the literal use of musk by women, the figurative associations of women's scent with musk are a key element of the early medieval Islamic canons of beauty. The scent of women was regularly likened to musk. Al-Damīrī, in his discussion of the meanings of musk in dream interpretation, noted that musk could represent "the beloved (*ḥabīb*) or a girl (*jāriyah*)."[118] Musk is mentioned by most of the early poets in their descriptions of women; the

111 Cf. Imru' al-Qays "Mu'allaqah" line 62 in *Dīwān*, ed. Yāsīn al-Ayyūbī (Beirut: al-Maktab al-Islāmī, 1998) and a verse of 'Abdallāh b. Salimah in al-Mufaḍḍal b. Muḥammad al-Ḍabbī. *Al-Mufaḍḍaliyyāt*. 3 vols., ed. and trans. C. Lyall (Oxford: Clarendon Press, 1918), text 1.193; trans. 2.66.

112 Trans. Goitein 4.226, modified slightly.

113 Goitein 4.315–16.

114 Ibn 'Abd Rabbih 31.37–8.

115 *Kitāb al-Hadāyā wa-l-tuḥaf* 236–8 (§346); Qaddūmī 219–20.

116 Tanūkhī, *Nishwār* 1.293.

117 Zamakhsharī, *Rabī'* 2.216; Ibn Abī al-Ḥadīd 19.343.

118 Al-Damīrī, *Ḥayāt al-ḥayawān*, 2 vols. (Būlāq, 1284/1867; repr. Frankfurt am Main: Institute for the History of Arabic-Islamic Science, 2001), 2.129.

potential references are countless. Musk is used to describe the general scent of women; in the following cases the image of musk can often refer to either the presumed actual use of musk perfume by women, or it could be a description of their own odor:

Imru' al-Qays (6th century):

> *idha qāmatā taḍawwa'a l-misku minhumā*
> *nasīma l-ṣabā jā'at birayyā l-qaranfuli*[119]
>
> When they rose, the fragrance of musk emanated from them as the eastern breeze comes with the fragrance of cloves.

Jamīl (d. 701):

> *ka'anna khuzāmā ʿālijin fī thīyābihā*
> *buʿayda l-karā 'aw fa'ra miskin tudhabbaḥu*[120]
>
> As if there was lavender of the desert sands in her clothing after sleep, or a slit open musk pod.

Al-Akhṭal (c. 640–710):

> *yajrī dhakiyyu l-miski fī ardānihā*
> *wa-taṣīdu baʿda taqattulin wa-dalāli*[121]
>
> Strong musk flows in her sleeves, and she hunts after walking seductively and flirting.

ʿUmar b. Abī Rabīʿah (644–712 or 721):

> *wa-taḍawwaʿa l-misku l-dhakiyyu wa-ʿanbarun*
> *min jaybihā qad shābahu kāfūru*[122]
>
> Strong musk and ambergris mixed with camphor diffuse from her breast.

---

119 Tibrīzī 34. Cf. al-Dīnawarī, *Kitāb al-Nabāt: The Book of Plants: Part of the Monograph Section*, ed. Bernhard Lewin (Wiesbaden: Steiner, 1974), 215.

120 Al-Sarī al-Raffā' 3.160 (#295).

121 Al-Akhṭal, *Dīwān*, ed. A. Ṣāliḥānī, 2nd ed. (Beirut: Dār al-Mashriq, 1969), 158.

122 ʿUmar b. Abī Rabīʿah, *Dīwān* (Cairo: al-Hay'ah al-Miṣriyyah al-ʿĀmmah li-l-Kitāb, 1978), 79.

Bashshār b. al-Burd (c. 714–c. 784):

ʿinnamā ʿaẓmu sulaymā ḥibbatī	*qasabu l-sukkari lā ʿaẓmu l-jamal*
wa-idhā adnayta minhā baṣalan	*ghalaba l-misku ʿalā rīḥi l-baṣal*[123]

The bones of my dear little Salmā are canes of sugar, not bones of camel. When you bring an onion near to her, the musk prevails over the smell of onion.

Bashshār sets up a contrast between the crude and the refined: delicate sugarcanes versus heavy camel bones, and onion versus musk. Sugarcane was known to the Greeks and Romans, but they were not familiar with the actual plant, which grew in India. The beginning of sugarcane culture in the Near East probably began under the Sasanians no earlier than the 7th century.[124] During the early centuries of Islam its cultivation expanded, and perhaps it was still regarded as something exotic and special in the time of Bashshār.

A poem by Abū al-Hindī is introduced by the line:

wa-fārati miskin min ʿidhārin shamamtuhā
yafūḥu ʿalaynā miskuhā wa-ʿabīruhā[125]

I smelled a musk pod from a cheek; her musk and *ʿabīr* wafted over us.

He then goes on to describe his erotic pursuit of the woman described. In the first line the poet plays with the reader by using the feminine enclitic pronoun *hā* to describe the musk; this pronoun refers directly to the *fārat miskin*, "musk pod", but it stands for the beloved as well.[126] The use of *ʿabīr*, a feminine perfume, serves to heighten the erotic mood. In the eighth line, having joined the beloved in bed, the poet says:

123 Al-Iṣbahānī, *Kitāb al-Aghānī*, 24 vols. (Cairo: Dār al-Kutub, 1927–74), 3.156.

124 A. M. Watson, *Agricultural Innovation in the Early Islamic World* (Cambridge: Cambridge University Press, 1983), 26.

125 Ibn al-Muʿtazz, *Ṭabaqāt al-shuʿarāʾ*, ed. ʿAbd al-Sattār Aḥmad Farrāj, 2nd ed. (Cairo: Dār al-Maʿārif, 1968), 140–1.

126 P. F. Kennedy, *The Wine Song in Classical Arabic Poetry: Abū Nuwās and the Literary Tradition* (Oxford: Clarendon Press, 1997), 28–9.

uqabbiluhā fawqa al-firāshi ka-annahā
ṣalāyatu ʿaṭṭārin yafūḥu zarīruhā

I kissed her upon the bedding; it was as if she was a perfumer's stone with purslaine diffusing from it.

Purslaine (*zarīr*) was used as a yellow dye; it reflects the image of the saffron-colored *ʿabīr* in the opening line. The comparison of the beloved to a perfumer's grinding stone makes the act of making love a process of preparing and releasing perfume, continuing the image of the musk pod, which, in the first line, remained to be opened.

Musk is especially used to describe the scent of the mouth of the beloved and the scent of her saliva. The mouth, saliva, and teeth form one of the most common erotic images in Arabic poetry. The saliva is often likened to wine, which was itself perfumed. These comparisons often emphasize the coolness of her saliva, something especially desirable in the Near East, and its freshness, often after sleep when the mouth usually smells foul. These images enhance the beauty of the woman by stressing her superior qualities and purity. Some examples from poetry:

ʿAntarah (6th century CE):

wa-ka-anna faʾrata tājirin bi-qasīmatin
sabaqat ʿawāriḍuhā ilayka min al-fami[127]

As if a merchant's musk in a perfume box
comes to you from her mouth before her teeth.

There is a variant for this line:

wa-kaʾanna rayyā faʾratin hindiyyatin
sabaqat ʿawāriḍuhā ilayka min al-fami[128]

As if the aroma of the Indian "mouse" comes to you from her mouth before her teeth.

127 ʿAntarah b. Shaddād al-ʿAbsī, *Sharḥ dīwān ʿAntara*, ed. Ibrāhīm al-Zayn (Beirut: Dār al-Najjāḥ, 1964), 205.

128 Dīnawarī, *Monograph* 192.

Ṭarafah:

wa-idhā taḍḥaku tubdī ḥababan
ka-ruḍābi l-miski bi-l-māʾi l-khaṣir[129]

When she laughs she shows off drops of dew as if it is saliva of musk with bitterly cold water.

Ḥassān b. Thābit (d. before 661):

tabalat fuʾādaka fī l-manāmi kharīdatun
tashfī l-ḍajīʿa bi-bāridin bassāmi
ka-l-miski takhliṭuhu bi-māʾi saḥābatin
aw ʿātiqin ka-dami l-dhabīḥi mudāmi[130]

A maid wearies your heart in sleep; she satisfies the bedfellow with the cooling of her smile
like musk mixed with cloud water or ancient wine like the blood of the sacrifice.

ʿUmar b. Abī Rabīʿah:

fa-bittu usqā ʿatīqa l-khamri khālaṭahu
shahdu mashārin wa miskun khāliṣun dhafiru
wa-ʿanbar l-hindi wa-l-kāfūri khālatahu
qaranfulun fawqa raqrāqin lahu ushuru
fa-bittu althamuhā ṭawran wa-yumtiʿunī
idhā tamāyala ʿanhu l-bardu wa-l-khaṣaru[131]

I spent the night being given to drink ancient wine mixed with honey from the hive and pure, pungent musk
and Indian ambergris, and camphor mixed with cloves on teeth with glistening drops.
I spent the night kissing her repeatedly, and the chill and cold gave me pleasure when she swayed from it [my kisses].

---

129 Ṭarafah 51.

130 Ḥassān b. Thābit, *Dīwān*, ed. Sayyid Ḥanafī Ḥasanayn and Ḥasan Kāmil al-Ṣayrafī (Cairo: al-Hayʾah al-Miṣriyyah al-ʿĀmmah li-l-Kitāb, 1974), 107.

131 ʿUmar b. Abī Rabīʿah, *Dīwān* 115**.

and also:

> *Man yusqa baʿda l-manām rīqatahā*
> *yusqa bi-miskin wa-bāridin khaṣiri*[132]

> After sleep, whoever is given to drink of her saliva is given musk and chilling coolness.

Dhū al-Rummah (d. 735?)

> *wa-tajlū bi-farʿin min arākin kaʾannahu*
> *min al-ʿanbari l-hindiyyi wa-l-miski yuṣbaḥu*[133]

> She polishes with a tooth-twig[134] and it is as if it had been given a morning drink of Indian ambergris and musk.

By twig is meant a tooth-stick or *miswāk* used to clean the teeth and freshen the mouth; the wood of the *arāk* was especially desirable for this purpose.[135] Dhū al-Rummah turns this concept on its head and instead has the saliva of the beloved refresh the twig with its fragrant perfume.

These are only some examples from what is a vast corpus of poetic references to musk. Many more, from often more obscure poets, have been collected in the anthology *Kitāb al-Mashmūm* of the 10th century poet al-Sarī al-Raffāʾ, who devotes a chapter to verses that include musk or one of the numerous terms associated with it. They clearly illustrate that musk was associated with women as well as with men and cannot be considered to have been simply a masculine perfume.

Musk is also compared with erotic fulfillment, or, in this case, lack of fulfillment. Abū Tammām (c. 728–c. 845) says:

---

132 ʿUmar b. Abī Rabīʿah, *Dīwān* 145**.

133 Dhū al-Rummah, *The Dīwān of Ghailān ibn ʿUqbah known as Dhu ʾr-Rummah*, ed. C. H. H. Macartney, (Cambridge: Cambridge University Press, 1919), #10 line 22 pg. 83.

134 Of the species *arāk*, i.e. *Salvadora persica*.

135 G. Bos, "The *miswāk*, an aspect of dental care in Islam," *Medical History* 37 (1993): 68–79, esp. 69 n. 14, on the wood of the *arāk* tree.

bātat ʿalā l-taṣrīdi illā nāʾilan
illā yakun māʾan qarāḥan yumdhaqi
nazran kamā stakrahta ʿāʾira nafḥatin
min faʾrati l-miski allatī lam tuftaqi[136]

She spent the night giving ungenerously to the obtainer nothing except pure water, unmixed
A trifle, just as one loathes a random whiff from the musk pod (*faʾrat al-misk*) which is unopened.

For Bashshār, musk has become so intertwined with his memory of his beloved that the scent of it reminds him of her:

idhā lāḥa l-ṣuwāru dhakartu salmā
wa-adhkuruhā ʾidhā nafaḥa l-ṣiwāru[137]

When the oryx appear I remembered Salmā,
and I remember her when the musk pod (*ṣiwār*) diffuses its scent.

The imagery of musk applied to women is perhaps even more ubiquitous in Islamic Persian poetry as far back as it goes. The hair of women is described as musky, like the hair of men. In the case of men, it was because of its blackness and association with youth, as well as the pleasant scent of good character, but with women the association with musk becomes eroticized. Firdawsī describes the daughter of Kayd, king of India, as having a crown of black musk upon her head, referring to her hair.[138] Niẓāmī compares the beginning of the tale told by the Indian princess in the *Haft Paykar* to a musk deer opening the pod.[139]

In the end, while musk was the best of perfumes, the scent of the beloved was the best of all:

---

136 Al-Sarī al-Raffāʾ 3.158 (#289) and Abū Tammām, *Dīwān* (Cairo: Dār al-Maʿārif, 1964), 2.407–8.

137 Al-Sarī al-Raffāʾ 3.155 (#280).

138 Firdawsī 6.27 l. 342; more examples of musk imagery and hair in Persian literature in Newid 76.

139 Niẓāmī Ganjavī, *Haft Paykar*, ed. H. Ritter and J. Rypka (Prague: Orientální Ústav, 1934), 32.9 on 121.

Al-Mutanabbī:

law khuliṭa l-misku wa-l-ʿabīru bihā
wa-lasti fīhā la-khiltuhā tafilah[140]

Were musk and *ʿabīr* mixed within
but you were not inside, I'd find it putrid.

Musk and Medicine: Pharmaceutical Specifications of Musk

In the medieval Middle East food and medicine, as well as perfume, were integrally related. Everything that went into the body, or came into contact with the body, was assessed from a medical point of view for its effects. The basis for this assessment was the humoral theory derived from Greek medicine and adopted in Islamicate medicine which held that the body contained four humors: blood, phlegm, yellow bile, and black bile. These four humors were conditioned by the qualities of hot and cold and wet and dry, thus blood was hot and wet, phlegm cold and wet, yellow bile hot and dry, and black bile cold and dry. From quadripartite division of the humors emanated the four temperaments: sanguine, phlegmatic, choleric, and melancholic. Medical theory held that the disorders of the body were caused by imbalance in these humors, and that the correction of them could be accomplished (among other means) through using pharmaceutical substances, which had their own affinities to the humors. Thus, in the humoral system as followed in Islamicate medicine, musk was considered hot and dry, while camphor was cold and dry, the antithesis of musk in the spectrum of heat.

One of the earliest physicians to give extensive information on musk was Ibn Māsawayh; his account of musk is largely devoted to its varieties, but at the end he says the following on its properties: "Musk is hot, mild (*laṭīf*), penetrating (*ghawwāṣ*), good for the heart and stops bleeding when it is placed in a wound. Its substitute is castoreum (*jundbādastar*) because this is the closest thing to it."[141]

The similarity of musk and castoreum was commonly established in Arabic sources. Ibn Māsawayh noted it by the early 9th century, and it was also noted as an adulterant of musk by the anonymous *Kitāb al-Tabaṣṣur bi-l-tijārah* of about the same period. Castoreum was known in antiquity for a long period,

140 Mutanabbī, *Dīwān* 248.
141 Ibn Māsawayh 11.

although it was associated especially with the forested lands to the north of the Mediterranean area, such as Pontus.[142] The glands that produced the castoreum were supposed to be the testicles of the animal; the usually accepted stories were that the beaver bit off its own testicles to escape capture.[143] Modern analysis has confirmed that castoreum contains salicin and salicylic acid, which are the active ingredients of aspirin.[144] Some of the ancient medical preparations were intended to be inhaled, but they are far from being perfumes.[145] Castoreum was sometimes used in European perfumery, but musk and civet were generally preferred.

Castoreum is attested in the Hippocratic corpus in gynecological contexts or in other remedies for women; this is paralleled in Herodotus' notice that the Boudinoi of the Pontic Steppe north of the Black Sea used castoreum to cure the womb.[146] It was renowned as an emmenagogue.[147] Pliny noted that castoreum was a soporific, but he also said that fumigation with it could revive victims of coma.[148] This agrees with Dioscorides, who notes that a preparation of castoreum "arouses both those affected by lethargic fever and those who suddenly fall asleep, and it does the same when burned so as to produce smoke."[149] It was used as a remedy for poisons and bites of scorpions, spiders, and reptiles. It was also used for tremors, spasms, and paralysis. Dioscorides remarked on

142 E.g., Persius, *Satires*, ed. and trans. G. C. Ramsay, Loeb Classical Library (Cambridge, MA: Harvard University Press, 1918), 5.135.

143 Pliny the Elder, *Natural History*, ed. and trans. H. Rackham, et al. 10 vols, Loeb Classical Library (Cambridge, MA: Harvard University Press, 1962), XXXII.26 and Dioscorides, *De Materia Medica*, ed. M. Wellmann, 3 vols. (Berlin: Weidmann, 1958), II.24. Cf. R. Thomas, *Herodotus in Context: Ethnography, Science and the Art of Persuasion* (Cambridge: Cambridge University Press, 2000), 287–8.

144 G. Majno, *The Healing Hand: Man and Wound in the Ancient World* (Cambridge, MA: Harvard University Press, 1975), 210.

145 E.g., Dioscorides II.24; Pliny XXXII.28.

146 See Thomas 286–8. Note her translation of Herodotus 4.109.2 on 286: In a large lake the Boudinoi "catch otters, beavers, and other square-faced animals ... The skins they use for clothing, and the testicles are useful to them for the cure of the womb ..." Presumably beavers' testicles are meant; Thomas points out that the castoreum came from the scent glands. K. Karttunen, *India and the Hellenistic World. Studia Orientalia* 83. (Helsinki: Finnish Oriental Society, 1997), 149 n. 163 has stated that musk was used in two remedies in the *Epidemics* (5.67 and 7.64); in both places castoreum is actually called for.

147 Lucretius, *De rerum natura*, ed. and trans. W. H. D. Rouse, Loeb Classical Library (Cambridge, MA: Harvard University Press, 1937), VI.794.

148 Pliny XXXII.28.

149 Dioscorides II.24; trans. L. Y. Beck, *De Materia Medica*. Altertumswissenschaftliche Texte und Studien Band 38. (Hildesheim: Olms, 2005), 100.

its warming property. The Arabs followed these beliefs; Ibn Sīnā, for instance, considered castoreum hot in the third to fourth degree and dry in the second.[150]

All of these uses are closely paralleled by the uses of musk in Arabic medical literature as described above. It is probably a case of musk being substituted for castoreum and then replacing it, so that finally castoreum was relegated to the position of a substitute for musk. Some confirmation of this can be found in the Sanskrit adoption of the Greek term for castoreum for musk, which shows that musk did serve as a substitute for castoreum in the east, where presumably the genuine article was at the time unavailable.

Later accounts of the medicinal properties of musk are broadly similar to Ibn Māsawayh's, but are often much more detailed. Abū Bakr Muḥammad b. Zakariyyā al-Rāzī (c. 865–925) quotes several authorities including Ibn Māsawayh himself on musk.[151] Among them, Ḥakīm b. Ḥunayn regarded it as hot and dry and stated that it was used in medications for strengthening the eye: it brightened the white and dried its moistures. Masīḥ al-Dimashqī[152] recommended it for strengthening the internal organs because of the sweetness of its scent. He said it was beneficial for chronic headache caused by moisture but that it caused headache in those with hot-tempered brains. Musk was strengthening for the cold brain.

The earliest Persian pharmacopoeia, the *Kitāb al-Abniyah ʿan ḥaqāʾiq al-adwiyah*, was written by Abū Manṣūr Muwaffaq al-Harawī (fl. c. 980–90). Its account of musk is in harmony with the Arabic medical writers, and it emphasizes the use of musk in the context of humoral theory:

> Musk is hot and dry in the third degree. It quickly causes headache to one with a warm nature. Used against cold ailments in the head, they will improve. It strengthens the heart and organs (*andām*). If used in the nose with saffron and camphor it drives away headaches caused by the cold or moisture. It prevents facial paresis also and strengthens the brain in the cold-tempered. It strengthens the weak organs for those with cold humors.[153]

---

150 Ibn Sīnā, *al-Qānūn fī al-Ṭibb* (Dār Ṣādir reprint of Būlāq ed., n.d.), 1.281.

151 Al-Rāzī, *Al-Ḥāwī al-kabīr fī al-ṭibb*, 23 vols. in 25 (Haydarābād: Maṭbaʿat Majlis Dāʾirat al-Maʿārif al-ʿUthmāniyyah, 1958–79), 21 part 2, 516.

152 He lived during the time of Hārūn al-Rashīd, see M. Ullmann, *Die Medizin im Islam* (Leiden: Brill, 1970), 112.

153 Al-Harawī, *Al-Abniyah ʿan ḥaqāʾiq al-adwiyah*, ed. Aḥmad Bahmanyār. 2nd ed. (Tehran: Dānishgāh-i Tihran, 1975; repr. Tehran: Dānishgāh-i Tihrān, 1371/1992), 326. Cf. A.-Ch. Achundow, "Die pharmakologischen Grundsätze (Liber fundamentorum pharmaco-

Ibn Sīnā (980–1037) has a short description of musk in his *al-Qānūn fī al-ṭibb*. He first discusses the different kinds of musk and their sources (that portion is quoted in Chapter 4), and then he gives a concise discussion of it from the medical point of view:[154]

> Properties: It is hot and dry in the second degree, and according to some the dryness is preponderant.
> Effects: It is thin and strengthening.
> Adornment (*zīnah*): It disinfects when it is placed in cooked food.
> The Head: When one snuffs musk with saffron and a little camphor and even by itself it benefits cold headaches because of its property of dissolution and strength, and it fortifies the even-tempered brain.
> The Eye: It strengthens the eyes, dries their moisture, and clears fine leukoma.
> Respiration and the Chest: it strengthens the heart, brings pleasure, and benefits heart palpitations and alienation.
> Poisons: It is a theriac for poisons, especially aconite (*bīsh*).

The Spanish physician Ibn Juljul wrote of musk in his treatise on the *materia medica* absent from Dioscorides:

> It is an Indian drug. It is pods filled with musk which fall from the thighs (*afkhādh*) of an animal which is the shape of a gazelle. Outgrowths appear in its thighs which continue to enlarge and fill with the blood of that animal. When they cease, they fall off, and in them is musk. It is said also [alternatively] that it is in the pods which are in the navel[155] of this animal. Musk is strengthening for the heart, thinning for thick black blood, rousing to the spirits, innately hot, and strengthening to the brain. It is among the exalted aromatics (*rafīʿ al-ṭuyūb*) and their lord (*sayyiduhā*). Its temperament is hot and dry.[156]

---

logiae) des Abu Mansur Muwaffak bin Ali Harawi zum ersten Male nach dem Urtext übersetzt und mit Erklärungen versehen," *Historische Studien aus dem Pharmakologischen Institute der Kaiserlichen Universität Dorpat* 3 (1893): #541 on 277. On Harawī, see L. Richter-Bernburg, "Abū Manṣūr Heravī," *EIr* s.v.

154 Ibn Sīnā, Būlāq ed. 1.360 and ed. Idwār al-Qashsh (Beirut: Mu'asasat 'Izz al-Dīn, 1987), 1.593.

155 *Ṣurrah*, cf. Ibn Juljul #33 n. 1.

156 Ibn Juljul #33.

Perhaps the most detailed discussion of musk is provided by the great pharmacologist Ibn al-Bayṭār (d. 1248) in the *Kitāb al-Jāmiʿ li-mufradāt*. He first quotes al-Masʿūdī's description of the musk deer and its trade and then gives a summary of its uses from a number of medical authorities:[157]

> Al-Qahramān (*ʿ.l.h.mān*):[158] It is hot in the second degree and dry in the third.
>
> Ibn Māsah said,[159] "It perfumes the sweat and strengthens the heart. It emboldens those with excessive black bile and repels cowardice which obstructs them. When it is mixed with medicines it improves their effects. It warms the external members and strengthens them when they are weak when it is placed upon them. It strengthens the internal organs when drunk. A group of the people of Ahwāz and Persia mention that there is in it a moisture because of which it helps the libido, and that when a little bit of it is taken dissolved in oil of wallflower (*khīrī*) and it is coated upon the head of the penis it aids copious intercourse and speeds ejaculation.
>
> Al-Rāzī said in his *Kitāb al-Ijmāʿ* that it censes the mouth when it is dissolved in food. He said in *al-Manṣūrī* that it benefits cold ailments in the head, and it is good for syncope and falling strength.[160]
>
> Al-Ṭabarī: It is thin and strengthens the members with the sweetness of its scent. It is beneficial when it is snuffed mixed with a bit of saffron, a half a lentil's weight (*ʿadas*) each. It benefits cold headaches and strengthens the brain.
>
> Ḥakīm b. Ḥunayn:[161] It is used in strengthening drugs for the eye. It clears fine leukoma and thoroughly dries its moistures.
>
> Isḥāq b. ʿImrān: it is good for the elderly and the moist-tempered especially in cold times and places. It causes headache in the hot-tempered youths, especially in hot places and times. On the whole it benefits all of the cold ailments in the head, and opens obstructions and benefits the winds which appear in the eye and the rest of the body. It restrains

157 Ibn al-Bayṭār, *al-Jāmiʿ li-mufradāt al-adwiyah wa-l-aghdhiyah*, 4 vols. (Būlāq, n.d.), 4.156–7.

158 For this figure, see O. Kahl, *The Sanskrit, Syriac and Persian Sources in the* Comprehensive Book *of Rhazes* (Leiden: Brill, 2015), 52–56.

159 See also Rāzī 516.

160 See also Rāzī 516.

161 See also Rāzī 516.

the belly, and dispels yellowness of the face. It drives out the effects of poison and it is good for heart palpitations. It sharpens the mind and stops premonitions (*taḥdīth li-nafs*).

Ibn Sīnā: It is the most excellent theriac for dryness, the two aconites, and the horns of the spikenard (*sunbul*). It makes one happy and benefits alienation. Its heat can be moderated with camphor and its dryness with moist oils like violet oil and rose oil.

Experiences: When it is used in medications for the four senses it sharpens all of them and strengthens the natural heat. When it is mixed with purgative drugs they are more effective in cleansing. It benefits the flow of blood from the body when added to a purgative drug. When hemiplegiacs and those who have suffered stroke due to the cold use it as snuff, it arouses them and benefits them, and cleanses their brains together with the medicine with which it is taken. When it is dissolved in warm oils and daubed upon the surface of the vertebrae it benefits numbness and hemiplegia with prolonged use. When it is dissolved in ben oil and daubed upon the head it halts catarrh.

Ibn Riḍwān: It benefits the pain of external hemorrhoids when it is daubed upon them. Other than that it benefits thick winds produced in the intestines when it is drunk.

Ibn Rushd: Its substitute is castoreum for pains of the nerves, and it substitutes for it in all uses except in perfume.

As can be seen from these extracts, the hot and dry nature of musk made it valuable for a variety of medical applications in which there was a need to check the influence of cold and moist humors. Pharmacologists frequently point out the danger of musk to those who are already of a hot temperament. Camphor was employed as a way of combating the hotness of musk when necessary. Ibn al-Bayṭār quotes Ibn Sīnā that the dryness of musk could be checked by the moist oils of violet and rose. Physicians believed that real danger could be had from exposure to musk. The Ibn Mandawayh-Ibn Kaysān tradition warns that musk can be quite dangerous: "Every perfumer or person in his presence who opens this musk gets a nosebleed from the potency of its scent." (Ibn Kaysān). Many other sources note that good quality musk induces a nosebleed.[162]

162 The capacity of musk to induce a nosebleed is also mentioned by al-Kāshānī, *'Arāyis al-jawāhir wa-nafāyis al-aṭāyib*, ed. Īraj Afshār. Tehran: Intishārāt-i Anjuman-i Athār-i Millī, 1966), 250. Compare J. B. Tavernier, *Travels in India*, trans. W. Ball and Ed. W. Crooke, 2 vols. (Oxford: Oxford University Press, 1925; repr. New Delhi: Munshiram Manoharlal,

In fact, musk was sometimes called for in formulas for poisons.[163] An interesting anecdote on the dangers of musk preserved by al-Tanūkhī discusses a guest who surreptitiously took a large quantity of *ghāliyah* and hid it under his turban. When he stood up to leave, he found that he had become blind. His host brought a physician for him who removed his turban and discovered the *ghāliyah*. His treatment is then described: "he poured cold water upon it until no trace of it remained. Then he daubed him with sandalwood, rosewater, and camphor, and placed him in a breeze for a time. His vision returned to a state of health."[164] The physician's actions are consistent with treating the heat generated by the *ghāliyah*: cold water removed the physical substance, cooling aromatics, including camphor, helped restore his balance, and the breeze helped cool his body.

The belief that musk was a powerfully warming substance is widespread in Islamicate culture beyond the confines of strictly medical knowledge. For example, al-Tanūkhī tells the story of a young elegant man who beat the cold weather by coating his feet and filling his navel with musk-rich *ghāliyah*. He also put it on his facial hair and head; this enabled him to go about in wintry weather wearing a mere two shirts, turban, hood, and shoes instead of heavy quilted garments, much to the amazement of those he met.[165]

Pharmaceutical Applications of Musk

So far we have considered the somewhat abstract statements on the medicinal properties of musk. In this section, we will consider the practical applications of musk by looking at formulas for actual compounded prescriptions. In order to get an idea of the uses of musk in medicine and its incorporation into

---

1995), 2.114: "If this bladder should be held to any one's nose, blood would immediately issue from it in consequence of the pungency of the odour…" Duarte Barbosa notes that true musk makes one sneeze violently and blood flow from the nostrils: *The Book of Duarte Barbosa*, trans. M. Longworth Dames (London: Hakluyt Society, 1918–21), 2.161, and Ludovico di Varthema (1508) states that musk was tested by its effect on the nose: if it induced a nosebleed, it was genuine: *The Travels of Ludovico di Varthema*, trans. John Winter Jones and ed. George Percy Badger (London: Hakluyt Society, 1863), 102.

163 E.g., M. Levey, *Medieval Arabic Toxicology: The Book of Poisons of Ibn Waḥshīya and its Relation to Early Indian and Greek Texts*. Transactions of the American Philosophical Society New Series 56:7 (Philadelphia: American Philosophical Society, 1966), 42, 56, etc.

164 Tanūkhī, *Nishwār*, 3.59.

165 Tanūkhī, *Nishwār*, 3.58.

the system of Islamicate medicine, three early works have been examined. The first is the *Aqrābādhīn* of Sābūr b. Sahl (d. 869), who was a Christian physician at the famous hospital of Gondēshāpūr,[166] which had its roots in the Sasanian period. This work is a fusion of medical knowledge from Greek and Semitic cultures. Sābūr's predecessor as court physician to al-Mutawakkil was Ibn Māsawayh, the author of the important work on aromatics we have mentioned so often. The second work is the *Aqrābādhīn* of the famous al-Kindī, who has been discussed above. Finally, an anonymous Syriac work, the so-called *Syriac Book of Medicines*, has been examined as it is one of the only surviving medical works from the medieval Syriac tradition, which produced so many of the great physicians.[167]

Sābūr's *Aqrābādhīn* contains 408 complete recipes; of these twenty-one contain musk, meaning that about 5% of the formulas included it. Prescription #25 is for the *dawāʾ al-misk* or "musk remedy" itself, which included musk along with a number of other ingredients; it is a compound called for elsewhere.[168] These formulas were used for a great variety of ailments; the list includes several dozen:

166 I have followed the biography provided by Kahl in his edition of the text: Sābūr b. Sahl. *Dispensatorium Parvum (al-Aqrābādhīn al-Ṣaghīr)*, ed. O. Kahl (Leiden: Brill, 1994), 33–4. See also Kahl's translation, *Sābūr ibn Sahl: The Small Dispensatory: Translated from the Arabic Together with a Study and Glossaries* (Leiden: Brill, 2003) and also his edition of *Sābūr ibn Sahl's Dispensatory in the Recension of the ʿAḍudī Hospital* (Leiden: Brill, 2009).

167 *The Syriac Book of Medicines*, ed. and trans. by E. A. Wallis Budge, *Syrian Anatomy, Pathology and Therapeutics or "The Book of Medicines"*, 2 vols. (London Oxford University Press, 1913). On this work, the history of its study, and importance, see P. Gignoux, "Le traité syriaque anonyme sur les médications," *Symposium Syriacum VII* (Rome, 1998), 725–33 and his other papers "On the Syriac Pharmacopoeia," *The Harp* 11–12 (1998–9): 193–201, "Les relations interlinguistiques de quelques termes de la pharmacopée antique," in D. Durkin-Meisterernst, et al., eds., *Literarische Stoffe und ihre Gestaltung in mitteliranischer Zeit* (Wiesbaden: Harrassowitz, 2009), 91–8, "Les relations interlinguistiques de quelques termes de la pharmacopée antique II, " in W. Sundermann, et al., eds., *Exegisti Monumenta: Festschrift in Honour of Nicholas Sims-Williams* (Wiesbaden : Harrassowitz, 2009), 117–26, and "La pharmacopée syriaque exploitée d'un point de vue linguistique," *Le Muséon* 124 (2011): 11–26.

168 E.g., Kahl, *Sanskrit, Syriac and Persian Sources*, 180, 184, and 218.

Diseases & Symptoms

Colic (*qawlanj*)	Prescription #8,64,66
Quartan fever	342
Vomiting	8,66,242[169]
Phlegm	8,66,254
Heart palpitations (*khafaqān*)	25,26,57,244[170]
Sickness due to the black bile (*amrāḍ al-mirra al-sawdāʾ*)	25
Tumors in the throat	26
Moisture of the stomach and its coldness and weakness	26, 38
Pain in the belly	57, 342
Joint pain	57
Gout	57
Laxity of stomach	240
Hemorrhoids (*bawāsīr*)	240
Abdominal disorder	242, 316
Coldness of the Stomach	244, 316, 342
Indigestion (*sūʾ al-haḍm*)	244, 342
Weakness of the stomach	254, 316, 339
Weakness of the heart	316
Liver pain	58

Psychological & Neurological

Listlessness (*balādah*)	8, 66
Strengthening the brain	8, 66
Epilepsy (*ṣarʿ*)	57, 64
Apoplexy (*saktah*)	57
Hemiplegia (*fālij*)	57, 64
Facial paresis (*laqwah*)	57, 64
Spasm (*tashannuj*)	57, 64
Forgetfulness	57
Tremors (*irtiʿāsh*)	57
Anxiety (*fazaʿ*)	57
Paranoia (*ḥadīth al-nafs*)	57

169 Sābūr b. Sahl #8,66,242. Cf. ʿAlī b. Rabban al-Ṭabarī, *Firdaws al-Ḥikmah fī al-Ṭibb*, ed. M. Z. Siddiqi (Berlin: Sonne, 1928), 309, who recommends it for use against vomiting.

170 Cf. Ṭabarī, *Firdaws*, 454 for a musk remedy used to treat palpitations.

Vertigo (*duwār*)	57
Migraine (*shaqīqah*)	57
Headache (*ṣudāʿ*)	57
Syncope (*ghashy*)	316
Gynecology	
Hemorrhage in women (*nazf al-nisāʾ*)	8
Period pains (*al-riyāḥ allatī taʿriḍu la-hunna fī al-arḥām*)	8, 66
Premature delivery (*isqāṭ*)	8, 48, 57
Strengthening the womb	8, 66
Cauterization of female wounds from all ailments (*ladhʿ awjāʿ al-nisāʾ min jamīʿ al-amrāḍ*)	48
Protection of the fetus in the womb	57, 64
Ophthamology	
Strengthening the brightness of the eye (*muqawwī li-l-ʿayn jalālahā*)	367
Extinguishing the heat of the eye	369
Miscellaneous	
Rectifying the body	8, 66
Opening obstructions (*yaftaḥu al-sadad*)	58
Dispelling heavy vapors (*yuḥallilu al-riyāḥ al-ghalīẓah*)	58
Children's ailments	64
Sweetening the breath	139, 140
Rottenness of temperament (*fasād al-mizāj*)	240
Bad color (*samāḥat al-lawn*)	240
Aphrodisiac	240
Beneficial for the aged	244, 342

The second early work is the *Aqrābādhīn* of al-Kindī, which includes 226 recipes for medications for many different conditions.[171] Musk is used in the four following prescriptions (the translations are Levey's):

---

171 M. Levey, *The Medical Formulary or Aqrābādhīn of al-Kindī* (Madison: University of Wisconsin, 1966).

#16. Another sternutator for an enlarged head [in infants—A. K.].
#56. Ointment for baldness.
#93. For it [a pustule] also.
#147. A clyster that is introduced into the urethra as an effective remedy for the dripping of urine, its strong odor, and calculi.

Formula #102, a recipe for a "Jewish tooth medicine," calls for "false yellow musk" in Levey's translation. In the facsimile of the manuscript accompanying his translation one can see that the Arabic is *rāmik aṣfar*, "yellow *rāmik*." *Rāmik* is called for in another recipe, #106, also a tooth medicine. *Sukk* also appears in formula #102 and in several more formulas; it is more commonly used than musk, which only appears in the four prescriptions above. Since *sukk* recipes are extant in the *Kitāb Kīmiyā' al-ʿiṭr wa-l-taṣʿīdāt* of the school of al-Kindī, it is likely that his *sukk* would have included musk since those formulas do. Al-Kindī called for musk in one of his prescriptions to increase sexual vigor in his *Kitāb al-Bāh*.[172] It is evident that al-Kindī, despite his familiarity with musk, did not use it as extensively in compounding medicines as Sābūr b. Sahl.

A medical work that might explain much about the transition from ancient to Islamicate medicine through the medium of Syriac is the so-called *Syriac Book of Medicines*. Unfortunately, it is a problematic source. The text is incomplete at the beginning and lacks the first two chapters and thus lacks a title or author. It can be dated only in a general way; the consensus favors the seventh or eighth century. The work is also extremely heterogeneous in character. It compiles material from a variety of sources, including Greek works with direct parallels in the Galenic corpus. On the other hand, it includes a chapter on *sammāne arʿānāyā*. Budge translated this term as "medicines of the country" and understood it as signifying native medicines as opposed to the Hellenized prescriptions found in the rest of the book.

Musk appears only once in the section of "native medicines." It is used in a formula for night-blindness in the form of oil of musk (*mešḥā d-mušāk*).[173] There is evidently an old tradition of using musk in eye medications; we have seen it so used by the pharmacologists. Muḥammad recommended scented *kuhl* for the eye; this was explained as meaning that it included musk.[174]

The remainder of the *Syriac Book of Medicines* includes musk in a variety of compound remedies, some paralleled closely in the Arabic-Islamicate

172 In G. Celentano, *Due scritti medici di al-Kindī* (Naples: Istituto orientale di Napoli, 1979), 23–4.

173 Budge 1.558 l.22.

174 Ibn Qayyim al-Jawziyyah, *al-Ṭibb al-Nabawī* (Cairo: Dār al-Turāth, 1978), 331.

tradition. It is used in a remedy that also includes castoreum for a chronic headache.[175] Another headache remedy includes camphor (*qafūr*) and ambergris (*ambar*) along with musk and other substances.[176] It is called for in a Persian remedy named *afsarīšn* that was for the eyes; it consisted mainly of kohl.[177] A medicine for throat diseases contained musk;[178] so too did a medicine to clean the teeth and freshen the breath.[179] Four more medicines for bad breath, one of which was said to have been used by a queen, included musk along with other aromatics.[180] A medicine made of musk that included numerous other ingredients was for heart palpitations, shortness of breath, and diseases arising from black bile.[181] Its heading, like the heading for the throat disease medicine, specifically signals its inclusion of musk. The "antidote of Caesar," which treated a variety of ailments including heart palpitations, bad winds, fevers, indigestion, coughing up of blood, shortness of breath, hiccups, liver and spleen pain, stomach complaint, delayed menses, headache, poison, and insect and reptile bites, also included a little musk.[182] Another musk antidote was used for fear, falling sickness at the beginning of each month, and heart palpitations.[183] There follows in the text two formulas for "pearl antidotes," made with ground pearls and used for heart diseases. The second, also said to be good for diseases arising from black bile, included musk.[184] The first included castoreum only, but the second added musk to it along with many precious substances such as coral, gold, and silver. A wine made from lilies (*šušane*) and including some musk was used for numerous internal problems especially in the circulatory system and digestion.[185] Musk was used in a formula for oil of nard (*nard*),[186] and it was used in a medicine of rhubarb (*rāwan*) for liver and stomach ailments and obstructions.[187] The *Syriac Book of Medicines* includes a formula for a "Persian Pīlōnyā" (a type of medicine credited to the ancient physician Philo) which corresponds to that of Sābūr b. Sahl. Like the Arabic version,

175 Budge 1.55; trans. 2.57 .

176 Budge 1.56; trans. 2.58–9.

177 Budge 1.92; trans. 2.102 and Gignoux "On the Syriac…" 196.

178 Budge 1.161; trans. 2.175.

179 Budge 1.174; trans. 2.191.

180 Budge 1.174; trans. 2.191–2.

181 Budge 1.262–3; trans. 2.297.

182 Budge 1.263; trans. 2.297–8. See also J. M. Riddle, *Contraception and Abortion from the Ancient World to the Renaissance* (Cambridge, MA: Harvard University Press, 1992), 105.

183 Budge 1.264–5; trans. 2.300–1.

184 Budge 1.265; trans. 2.301–2.

185 Budge 1.301–2; trans. 2.339–40 .

186 Budge 1.326; trans. 2.375 .

187 Budge 1.359; trans. 2.411–2 .

it includes musk.[188] Budge's "fillet of musk" (*esplēnyā d-mosīkon*), really a kind of compress (from Greek σπληνίον), is something other than musk, since the word he takes to be musk is spelled strangely with *semkath* instead of *shīn* and no musk is called for in the formula.[189]

As can be seen by the great range of uses to which musk was put, it approaches being a universal remedy.[190] Patterns that emerge show the importance of musk as a stimulant and cure for ailments originating in the cold and wet humors. In the Islamic Middle East, phlebotomists were supposed to keep musk pods and pills on hand in case of fainting.[191] Musk was used in a large number of applications relating to the head and nervous system. Musk also clearly was thought to have an affinity for the heart and could strengthen many organs when they grew weak.[192] It was also noted for its gynecological importance. The use of musk in eye remedies is especially prominent as well.

Another early attested use of musk in the Near Eastern tradition is for snakebite and poisoning in general.[193] This application appears in Ibn Qayyim al-Jawziyyah's (d. 1350) *al-Ṭibb al-Nabawī*. Jābir b. Ḥayyān, the great 9th century alchemist, also used musk in a remedy for viper bite.[194] This usage is paralleled by the numerous traditions on the efficacy of musk for snakebite from China, Tibet, and Nepal as well as modern medical discoveries, as discussed in Chapter 3. There was a belief in the classical Mediterranean lands that the deer was an antagonist of the snake.[195] Musk was used in the Arabic version of the *shīlthā* (from Syriac *she'lthā*, "request [for a cure]") that was commonly given for poisons.[196] Part of the procedure for scorpion poisoning included

---

188 Budge 1.425–6; trans. 2.501–2; Sābūr b. Sahl #8, 44–5.

189 Budge 1.151–2, trans. 2.163–4.

190 See A. Akasoy and R. Yoeli-Tlalim, "Along the Musk Routes: Exchanges between Tibet and the Islamic World," *Asian Medicine* 3 (2007): 230 and 234–5 on the connections between musk and theriac.

191 'Abd al-Raḥmān b. Naṣr al-Shayzarī, *Kitāb Nihāyat al-Rutbah fī Ṭalab al-Ḥisbah*, ed. al-Sayyid al-Bāz al-'Arīnī (Beirut: Dār al-Thaqāfah, n.d.), 91.

192 Akasoy and Yoeli-Tlalim 230.

193 Cf. Akasoy and Yoeli-Tlalim 231.

194 Jābir b. Ḥayyān, *Kitāb al-Sumūm*, facsimile of manuscript with translation by A. Siggel, *Das Buch der Gifte des Gābir ibn Ḥayyān*, Wiesbaden: Steiner, 1958), 168b and trans. p. 175. Cf. also Van Gelder/Suyūṭī 208.

195 Cf. Lilja 160–1.

196 Arabic formulas: Sābūr b. Sahl #57 and al-Ṭabarī, *Firdaws* 463. The *shīlthā* was extensively used against poisons, examples in Levey, *Medieval Arabic Toxicology* 34, 68, 81. The formula in the *Syriac Book of Medicines* does not call for musk, but for castoreum: Budge 1.263–4 and trans., 2.298–9. Cf. also J. Schleifer, "Zu Sobhys General Glossary der Dahīra," *Orientalistische Literaturzeitung* 39 (1936): 670.

aromatics: "Have him smell musk and ambergris or both, continuously. His body is guarded with perfumed aloe."[197]

Musk, as an aromatic substance and ingredient of perfume, was also believed to act as a disinfectant. The belief in the disinfecting qualities of perfume and incense is ancient and ties in with the concept of disease being borne by pestilential air; sweet scents were the natural enemy of the fetid.[198] An interesting anecdote preserved by al-Suyūṭī tells of a merchant who came with musk from Khurāsān to Damascus during the plague of 716–7. He entered the house of Ayyūb b. Sulaymān b. ʿAbd al-Malik, who would later die from the plague, and his musk was seized by the inhabitants of the house for its powers.[199] Musk was used in medieval Europe as well along with ambergris in the pomanders worn against the plague.[200] ʿAlī b. Rabban al-Ṭabarī gave a formula for a musk remedy that was used against recent and long standing leprosy due to black bile.[201]

Many of the usages of these early texts are attested in later works. Modern research has demonstrated that musk acts as an antihistamine in laboratory animals, and that it can raise blood pressure in some also.[202] Thus, many of these uses are reasonable and likely developed through practical experience. The warming, stimulating qualities of musk and its antiphlegmatic properties are stressed by the Chinese, Indian, and Tibetan medical traditions as well, as discussed in Chapter 3.[203]

Our sources close to the school of Gondeshapur and Syriac culture, the formulary of Sābūr b. Sahl and the *Syriac Book of Medicines*, make more extensive use of musk in medicine than the Arab al-Kindī, even though he was born in Iraq and had an extensive knowledge of aromatics, as his work on the subject shows. A tentative conclusion is that musk, while certainly known to the pre-Islamic Arabs and probably used in their medicine, in such functions as an eye medication or a remedy for snakebite, was extensively integrated into

---

197 Levey, *Toxicology* 72; Arabic text unseen.

198 For the ancient and early Christian aspects, see B. Caseau, "Les usages médicaux de l'encens et des parfums. Un aspect de la médecine populaire antique et de sa christianisation," in *Air, miasmas et contagion: les épidémies dans l'Antiquité et au Moyen Age* (Prez-sur-Marne: D. Guéniot, 2001), 75–85.

199 M. Dols, *The Black Death in the Middle East*, 2nd ed (Princeton: Princeton University Press, 1979), 140–1. I have not seen the Arabic text.

200 J. M. Riddle, "Pomum ambrae: Amber and Ambergris in Plague Remedies," *Sudhoffs Archiv* 48 (1964): 113. See also R. Schmitz, "The Pomander," *Pharmacy in History* 31:2 (1989): 86–90.

201 Ṭabarī, *Firdaws* 455.

202 S. D. Seth et al., *Pharmacodynamics of Musk* (New Delhi: Central Council for Research in Indian Medicine and Homoeopathy [sic], 1975), 71–3.

203 And see Akasoy and Yoeli-Tlalim 231–3.

the emerging system of Islamicate medicine in Sasanian Mesopotamia by the Syriac physicians. While much of the *Syriac Book of Medicines* consists of translations of Galen, musk is called for in formulas that have not been traced to Greek prototypes. Indeed, in one case it is used in a formula that claims a Persian origin. We would do well to suspect an important role of the Persians in the complex process of the formation of the medicinal uses of musk as well.[204]

Musk in Food and Drink

Aromatics were added to food for several reasons.[205] They enhanced the presentation of the food and contributed to its overall appeal. In this sense adding perfumes was a part of the decoration of food. Aromatics were also believed to aid in the preservation of food (and there is some biological basis for this belief),[206] so vessels were often censed before being used.[207] Today musk is being studied as a flavor additive in food again.[208] Musk also contributed to the overall medicinal impact of a beverage or dish.

Here we use the *Kitāb al-Ṭabīkh* of Ibn Sayyār al-Warrāq as a source for musk in cuisine. This work is the earliest extant Arabic cookbook; it dates to the late 10th century.[209] It is very much a product of the upper classes and includes many recipes associated with the caliphs. Its level of luxury and use of rare

204 On the impact of Persian in the *Syriac Book of Medicines*, see the works of Gignoux cited above.

205 See M. Marín, "The Perfumed Kitchen: Arab Cookbooks from the Near East," in R. Gyselen, ed., *Parfums d'Orient* (Bures-sur-Yvette: Groupe pour l'Étude de la Civilisation du Moyen Orient, 1998), 159–66 and Marín, "Beyond Taste: the complements of colour and smell in the medieval Arab culinary tradition," in R. Tapper, ed., *Culinary Cultures of the Middle East* (London:1994), 205–14.

206 Musk is singled out as a substance that can disinfect food by Ibn Sīnā, Būlāq ed. 1.360, critical ed., 1.593. For a modern scientific study of the antibiotic role of aromatic spices, see J. Billing and P. W. Sherman, "Antimicrobial Functions of Spices: Why Some Like it Hot," *Quarterly Review of Biology* 73 (1998): 3–49.

207 Marín, "Perfumed Kitchen" 161. Nevertheless, the notion that the heavy use of spices in pre-modern cooking was an attempt to mask "tainted meat" is untrue. See T. Peterson, "The Arab Influence on western European cooking," *Journal of Medieval History* 6 (1980): 317–40, B. Laurioux, "Spices in the medieval diet: a new approach," *Food and Foodways* 1 (1985): 43–76, G. Riley, "Tainted Meat," in *Spicing up the Palate: Studies of Flavourings-Ancient and Modern. Proceedings of the Oxford Symposium on Food and Cookery 1992* (Totnes: Prospect, 1993), 1–6, and Billing and Sherman 28.

208 K. McLaughlin, "The Next Big Flavor," *Wall Street Journal*, April 29–30, 2006, P5–6.

209 Ibn Sayyār al-Warrāq, *Kitāb al-Ṭabīkh*, ed. K. Öhrnberg and S. Mroueh. Studia Orientalia 60 (Helsinki: Finnish Oriental Society, 1987), viii. See N. Nasrallah's translation, *Annals of the*

and expensive ingredients surely did not reflect their use in the diet of the ordinary people. More so than later cookbooks,[210] the *Kitāb al-Ṭabīkh* constantly emphasizes the qualities of food and ingredients in terms of humoral theory. The book is also much more literary than some of the better known later works; Ibn Sayyār al-Warrāq extensively quotes poetry describing the dishes while later cookbooks such as the *Kitāb al-Ṭabīkh* of Muḥammad b. al-Ḥasan al-Kātib al-Baghdādī or anonymous *Kanz al-fawāʾid* are content to give recipes only.

The oldest traditions of musk and food concern beverages. Aromatics have been added to wine since classical times to aid in its preservation; in premodern times the risk of wine turning to vinegar was great because it was not treated with sulphur or other preservatives as is modern wine.[211] Vessels themselves were censed with frankincense in the 20th century United Arab Emirates to provide a pleasant smell for the water stored in them and to disinfect them.[212] As mentioned above, the scent of wine is often compared with musk.[213] Al-Muraqqish the Younger mentioned "red wine like musk in scent."[214] Al-Aʿshā described some wine as:

> *kumaytin ʿalayhā ḥumratun fawqa kumtatin*
> *yakādu yufarrī l-miska minhā ḥamātuhā*[215]

> Black wine with red upon it; above it more black; its heat almost makes the scent of musk burst forth from it.

Here the musk is considered an integral part of the wine. It can be seen as either an aspect of its flavor or a literal ingredient. Al-Aʿshā also compares the effect of wine with the action of musk on a cold:

Caliphs' Kitchens: Ibn Sayyār al-Warrāq's Tenth-Century Baghdadi Cookbook (Leiden: Brill, 2007).

210 Such as the *Kanz al-fawāʾid fī tanwīʿ al-mawāʾid*, ed. M. Marín and D. Waines (Beirut: Franz Steiner, 1993) or al-Baghdādī, *Kitāb al-ṭabīkh*, ed. D. Chelebi (Mosul: Umm al-Rabīʿayn, 1934) and trans. A. J. Arberry, "A Baghdad Cookery Book" *Islamic Culture* 13 (1939); repr. in *Medieval Arab Cookery* (Totnes: Prospect Books, 2001, repr. 2006), 19–89.

211 G. Majno, *The Healing Hand: Man and Wound in the Ancient World* (Cambridge, MA: Harvard University Press, 1975), 221–4.

212 Kanafani 40.

213 Cf. also F. Harb, "Wine Poetry (*khamriyyāt*)," in *ʿAbbasid Belles-Lettres* (Cambridge: Cambridge University Press, 1990), 220, 224.

214 *Mufaḍḍaliyyāt* 1.495.

215 Al-Aʿshā, Abū Baṣīr Maymūn b. Qays, *Gedichte von Abû Baṣîr Maimûn ibn Qais al-Aʿšâ*, ed. R. Geyer (London: Gibb Memorial Series, 1928), #1 l. 11.

wa-adkana ʿātiqin jaḥlin sibaḥlin
ṣabaḥtu bi-rāḥihi sharban kirāmā
min allātī ḥumilna ʿalā l-rawāyā
ka-rīḥi l-miski tastallu l-zukāmā[216]

And many a tar-covered jar, ancient, great, and big from which I offered the noble drinkers morning wine
Carried on the back of the water-carrying animals, with a scent like the scent of musk that clears a stuffy nose.

The famous Abbasid poet of wine Abū Nuwās described some very old and fine wine as follows:

tuhdī ilā l-sharbi ṭīban ʿinda nakhatihā
ka-nafḥi miskin fatīqi l-faʾri maftūti[217]

It gives the drinkers a perfume in its breath like a diffusion of musk when the pod is slit open and crumbled.

The poet Ḥārithah b. Badr also compared the scent of wine with musk. A poem of his includes the following lines:

(3) *sa-ashrabuhā ṣahbāʾa ka-l-miski rīḥuhā*
wa-ashrabuhā fī kulli nādin wa-mashhadi

I will drink red wine with a scent like musk, I will drink it at every gathering and assembly.

(8) *muʿattaqatan ṣahbāʾa ka-l-miski rīḥuhā*
idhā hiya fāḥat adhabat ghullata l-ṣadā[218]

Aged red wine with a scent like musk; when it diffuses its scent, it drives away burning thirst.

The most famous instance of the association of musk with wine, of course, is in the Qurʾān, where its single mention of musk describes wine. It is 83:25–26: (25) *yusqawna min raḥīqin makhtūmin* (26) *khitāmuhu miskun* ... "They will be

216 Aʿshā #29 l. 16–7.

217 Kennedy 277 l. 20.

218 Kennedy 162. The translations are mine.

given to drink sealed fine wine (*raḥīq*), its seal of musk." Here the musk is said to be the "seal" (*khitām*) of the pure wine which the inhabitants of paradise drink. The commentator al-Ṭabarī gives a variety of explanations for it, citing three lines of interpretation in his commentary.[219] The first approach was that the musk was an admixture to the beverage. The second approach was that the musk was the scent present at the end of drinking. The third interpretation was that the wine was physically sealed with a seal consisting of musk. Al-Ṭabarī himself favored the second approach. But in the end, it is a multivalent image, because the scent of wine could both be compared to musk and influenced by musk which might be among its ingredients. The seal could also be understood literally, as seals could be made with musk (probably compounded with other ingredients), at least in later times.[220] In any case, the image of the seal of musk reflects the ancient association of aromatics with purity and immortality that we will discuss in the last chapter. The literal meaning of the words shows the importance of aromatics such as musk in beverages, although less explicitly than the single Qurʾānic mention of camphor (76:5) which states that it is the admixture (*mizāj*) of the wine of the Garden.[221] But this is only a facet of the image, which emphasizes the purity and excellence of the Garden by including musk; like the wine of the future paradise, musk confers immortality.[222]

Ibn Sayyār al-Warrāq includes a *dāniq* of musk in a recipe for a *nabīdh* (wine) made from raisins.[223] In a recipe for a wine made from white sugar (*al-sukkar al-ṭabarzad*) he writes: "If it is during the winter, you can add a bit of musk to it, and if it is during the summer, you can add a bit of camphor to it."[224] This clearly shows that the addition of aromatics to alcoholic beverages

219 Abū Jaʿfar Muḥammad al-Ṭabarī, *Jāmiʿ al-bayān ʿan taʾwīl āy al-Qurʾān*, 30 vols. (Cairo: Muṣṭafā al-Bābī al-Ḥalabī, 1954), 30.105–7.

220 Hilāl al-Ṣābiʾ, *Rusūm Dār al-Khilāfah* (Baghdād: Maṭbaʿat al-ʿĀnī, 1964), 178.

221 Even here there was much discussion among the commentators whether camphor would be a suitable admixture to wine, or whether the usage was metaphorical in some way, or the name of the spring from which water mixed with the wine came. See H. Schönig, "Camphor," in *The Encyclopaedia of the Quran*, vol. 1. (Leiden: Brill, 2001), 287–8, and al-Ṭabarī, *Jāmiʿ*, 29.206–7. The cookbooks, however, do mention perfuming food and drink with camphor, e.g., Warrāq 310 and Nasrallah's trans. 471.

222 See S. Stetkevych, "Intoxication and Immortality: Wine and Associated Imagery in al-Maʿarrī's Garden," in *Homoeroticism in Classical Arabic Literature*, ed. J. W. Wright and E. K. Rowson (New York: Columbia University Press, 1997), 210–32.

223 Warrāq 309; Nasrallah 470.

224 Warrāq 310; Nasrallah 471.

could be dictated by humoral theory, where musk and camphor represented the extremes of hot and cold. It also shows that the addition of fragrance could be optional and depend on the circumstances for the consumption of the finished beverage.

Musk is especially associated with wine in Persian mystic poetry.[225] This follows directly from the Qur'ānic tradition, but it also reflects the practice of scenting wine and beverages with musk. Ḥāfiẓ says:

biyār az ān may-i gulrang-i mushkbū jāmī
sharār-i rashk wa ḥasad dar dil-i gulāb andāz[226]

Bring a cup of that rose-colored musk-scented wine; cast sparks of jealousy and envy into the heart of rosewater.

This wine was so excellent that it had a beautiful rose-color and a musky scent, and thus made rosewater jealous. He also says:

chu lālah dar qadaḥam rīz sāqiyā may wa mushk
ki naqsh-i khāl-i negāram namīrawad zi ḍamīr[227]

Saqi, in my cup, like a tulip, pour wine and musk, for the image of the mole of the beauty does not leave my mind.

The poet requests wine in a cup, so that the color of the wine in the cup has the appearance of the cup of the tulip flower; the musk in the bottom is like the black center of the tulip flower, which is like the mole of the beloved. This is a reflection of the practice of adding a bit of musk to wine to scent it.

Musk was also used in recipes for beer (*fuqqāʿ*).[228] A poem by Abū al-Ḥasan al-Kātib on beer quoted by Ibn Sayyār al-Warrāq includes the following line:

---

225 Persian poetry also contains imagery of wine in general scented with musk; citations in Newid 88.

226 Quoted by S. ʿAnṣārī, *Tārīkh-i ʿIṭr dar Īrān* (Tehran: Wizārat-i Farhang wa Irshād-i Islāmī, 1381/2002–3), 247.

227 Quoted by *Farhangnāmah-i Adab-i Fārsī*. Volume 2 of *Dānishnāmah-i Adab-i Fārsī* (Tehran: Muʾassasah-i Farhangī wa Intishārāt-i Dānishnāmah, 1996–), 257a.

228 Warrāq 298, 299 (the latter calls for *sukk* of musk).

lahu rīḥu kāfūrin wa-miskin wa-ʿanbarin
yafūḥu kamā fāḥat muqaddimatu l-ʿiṭri[229]

It has the aroma of camphor, musk, and ambergris
diffusing from it just as the first whiff of perfume diffuses.

This can be understood both as an image of the delightful scent of the beer and as the literal reflection of its supposed ingredients. Musk was also used in a perfumed fruit and honey beverage;[230] a honey beverage perfumed with musk and camphor is also mentioned in *al-ʿIqd al-farīd*.[231] According to Ibn Qayyim al-Jawziyyah, honey mixed with water was a beverage Muḥammad drank; with religious sanction it was inevitable that musk, which was the most important aromatic with the richest associations with religious purity, would be mixed with it.

Musk was used in elegant food. In the *Kitāb al-Ṭabīkh*, Ibn Sayyār al-Warrāq lists the spices and perfumes (*al-abzār wa-l-ʿiṭr*) that can be used to perfume the pot for cooking food in the following order: musk, ambergris, rosewater, saffron, cinnamon (*dārṣīnī*), galangal, spikenard, clove, mastic (*maṣṭakā*),[232] nutmeg (*jawzbuwwā*), greater cardamom (*qāqullah*), mace, and lesser cardamom (*hāl*).[233] Musk is placed first in the list followed by ambergris; this is yet another reflection of their primacy among all aromatics. Later al-Warrāq describes the properties of musk: "It is hot and dry. When it is added to the pot it gives a headache to the hot-tempered quickly. It is beneficial for ailments caused by cold in the head. It preserves the heart, strengthens the stomach, is good for syncope and fallen strength."[234] One practical effect of the use of musk in food was to perfume the mouth.[235] Al-Warrāq cites the same properties we have seen in the medical literature. In medieval times everything that went into the body was assessed for its impact on health, and thus a cookbook had to take the medicinal properties of ingredients into account.

229 Warrāq 301; Nasrallah 459.
230 Warrāq 316; Nasrallah 480.
231 Ibn ʿAbd Rabbih 26.44. Cf. also Warrāq 302; Nasrallah 461.
232 Schmucker 479–80 and Dietrich DT I.34.
233 Warrāq 14; Nasrallah 91.
234 Warrāq 48; Nasrallah 138.
235 Rāzī vol. 21 part 2, 516.

Musk was used in a variety of dishes. A tiny bit of musk was included in an almond *sawīq* (a kind of porridge).[236] A special musk *sikbāj* (a sort of sour stew) prepared for the Abbasid caliph Hārūn al-Rashīd contained saffron and a *qīrāṭ* of musk.[237] Recipes for *lawzinaj*, a kind of almond sweetmeat, included some musk, sometimes along with ambergris or camphor.[238] A fancy dish, such as the beef *sikbāj* cooked for the caliph al-Amīn by the renowned slave girl cook Bidʿah, was served on a platter (*ṭayfūriyyah*) that was censed and then had a little musk sprinkled on it.[239]

Musk is specified in only a few of the many recipes in the *Kitāb al-Ṭabīkh*. It is probable that musk and other aromatics could also be added *ad libitum* to appropriate dishes not specified by the cookbook as part of their finishing touches. The use of musk in food is greatly expanded in later cookbooks such as the *Kanz al-fawāʾid fī tanwīʿ al-mawāʾid*, which also includes numerous recipes for perfumes and other fragrant preparations.[240] Musk continued to be used in elegant food throughout the medieval period and later; Chardin wrote: "The Turks put it in their fine *Sherbets*, and particularly in that which they call *Sultani*, as much as to say Royal. The Persians neither put it in their Meat or Drink, but they use abundance of it in several sorts of their Sweet-meats and Confections…"[241]

Conclusion

Musk was an important part of life in the Early Medieval Middle East. In court life musk, in the form of a perfume, or an incense, or in food or drink, was commonly used. While extensive use of it was undoubtedly confined to the upper classes who could afford it, it is likely that it was available to a wider population in certain circumstances, especially in medicine. Real musk was expensive and could be replaced by adulterated or imitation musk. But the real musk

236 Warrāq 37–8; Nasrallah 126 has "mastic".

237 Warrāq 136; Nasrallah 254–5.

238 Warrāq 265–6; Nasrallah 410–12.

239 Warrāq 133; Nasrallah 250. Cf. Marín, "Perfumed Kitchen" 162.

240 Marín, "Perfumed Kitchen" 163–4. The index to the edition by Marín and Waines, 51–2, lists the dozens of occurrences of musk in this text; while it does include chapters on the making of perfumes, most of these references are to musk in many different kinds of recipes.

241 J. Chardin, *Travels in Persia 1673–1677* (London: Argonaut, 1927; repr. New York: Dover, 1988), 229. Examples of musk-scented food in poetry are cited by Newid 88–9.

compounded in medicines would often likely be used in spite of the high cost. The "Medicine of the Poor" (*Ṭibb al-fuqarāʾ*) of Ibn al-Jazzār includes musk in a folk-remedy to stimulate conception in an infertile woman. It was a suppository consisting of the dried and pulverized penis of a wolf mixed with musk, clove, and saffron.[242] Thus for something as important as conception, even the poor required musk.

242 G. Bos, "Ibn al-Jazzār on Medicine for the Poor and Destitute," *JAOS* 118:3 (1998): 371.

CHAPTER 7

The Symbolic Importance of Musk in Islamic Culture

Just as in India or China, good scent was extremely important in the medieval Middle East. Perfuming had several important practical functions in society, which we have already considered. Aromatics also had profound symbolic and religious associations. Since ancient times, aromatics symbolized purity and holiness. They were products of plants such as woods and resins that did not spoil; their good scent was the opposite of the stench of death and decay. They were believed to protect against corruption, and they were extensively used in religious offerings. Smell is the most intangible of the senses, and this elusiveness caused it to be associated with the spiritual realms. The odor of sanctity, epitomized by the bodies of saints which escaped putrefaction and emitted a sweet smell, demonstrated the holiness of a person. It symbolized the perfection of physical being into an incorruptible, immortal form. Most aromatics of animal origin, like musk, came late into use. They had to overcome the stigma associated with being animal products and were sometimes regarded as less pure, as is the case with musk in Judaism (Chapter 3). Islamic belief also contained many strictures on animal products, but musk transcended them. No other aromatic had the same kind of significance in Islamic belief as musk.

The Primacy of Musk among Aromatics in Islamic Culture

In every listing of the aromatics in Medieval Arabic literature musk is first, and it is discussed first in every book that deals with the perfumes. Ibn Māsawayh's work on aromatics and al-Masʿūdī's, as mentioned above, list musk first. Musk perfumes and ways to adulterate and imitate musk are discussed first in pseudo-al-Kindī's *Kitāb Kīmiyāʾ al-ʿiṭr wa-l-taṣʿīdāt*. The Ibn Mandawayh-Ibn Kaysān tradition also discusses musk first. Later, the encyclopaedists al-Nuwayrī (1279–1332) and al-Qalqashandī (1355–1418), who calls musk "the most exalted," also discuss it first of all the aromatics in their respective sections on perfumes.[1]

1 Al-Qalqashandī, *Ṣubḥ al-aʿshā*, 14 vols. (Cairo: Al-Muʾassasah al-Miṣriyyah al-ʿĀmmah li-l-Taʾlif, 1964), 2.119.

 | DOI 10.1163/9789004336315_009

Muḥammad himself said that musk was the best aromatic, as will be discussed below, and it thus became part of Islamic tradition. Ibn Qayyim al-Jawziyyah (d. 1350) said of musk in his book *Al-Ṭibb al-Nabawī*: "It is the king of the varieties of aromatics, their noblest and sweetest (*aṭyab*). It is that about which proverbs are coined, and other things are compared to, while it is not compared to anything else. It is the sand-dunes (*kuthbān*) of the Garden [of Paradise]."[2]

Al-Jāḥiẓ supposedly wrote a debate, which is now lost, between musk and the similar-smelling civet.[3] There is a *maqāmah* by al-Suyūṭī (d. 1505) called *al-Maqāmah al-miskiyyah* or *maqāmat al-ṭīb*. It has been studied and translated by G. J. H. van Gelder.[4] This text features a contest among four perfumes: musk, ambergris, saffron, and civet. It is a different list of perfumes from the ones favored in earlier times; aloeswood and camphor are gone, and civet has been added. Aloeswood is likely not included because it is used mainly in incense, which does not seem to be a concern of the text. The Ibn Mandawayh-Ibn Kaysān tradition considers camphor low-class[5] and al-Nuwayrī does not discuss camphor among the perfumes (*al-ṭīb*); his account of it is placed instead among the resins (*al-ṣumūgh*).[6] Meanwhile, civet had become more popular. It was known from at least the mid-9th century, but there were doubts about it, perhaps due to its source.[7] Evidently by al-Suyūṭī's time, these perfumes were sometimes considered peers, but his *maqāmah* explodes this idea, placing civet fourth among them. Al-Suyūṭī assigns primacy to musk among these perfumes for both religious and worldly reasons. He cites the mentions of it in the Qur'ān and Ḥadīth and quotes traditions about it. He also mentions its numerous medical properties. Furthermore, he discusses its use in poetry.

---

2 Muḥammad b. Abī Bakr b. Qayyim al-Jawziyyah, *al-Ṭibb al-nabawī* (Cairo: Dār al-Turāth, 1978), 437. Cf. also al-Ghazālī, *Iḥyā' 'ulūm al-dīn*, 5 vols. (Beirut: Dār al-Kutub al-'Ilmiyyah, 2005), 3.283.

3 C. Pellat, "Nouvel essai d'inventaire de l'œvre Gāḥiẓienne," *Arabica* 31 (1984): 148.

4 G. Van Gelder, "Four Perfumes of Arabia: A Translation of al-Suyūṭī's *Al-Maqāma al-Miskiyya*," in R. Gyselen, ed., *Parfums d'Orient*, 203–12. (Bures-sur-Yvette: Groupe pour l'Étude de la Civilisation du Moyen-Orient, 1998), 203–12.

5 Ibn Mandawayh 242; Ibn Kaysān 202.

6 Nuwayrī 11.292–5.

7 Al-Jāḥiẓ, *Kitāb al-Ḥayawān*, ed. 'Abd al-Salām Muḥammad Hārūn, 8 vols. (Cairo: Muṣṭafā al-Bābī al-Ḥalabī, 1966), 5.304–5; Cf. al-Zamakhsharī, *Rabī' al-abrār wa-fuṣūṣ al-akhbār*, eds. 'Abd al-Majīd Diyāb and Ramaḍān 'Abd al-Tawwāb (Cairo: al-Hay'ah al-Miṣriyyah al-'Āmmah li-l-Kitāb, 2 vols., 1992–2001), 2.218.

The Abbasid caliph al-Ma'mūn was sent a gift by one of the Byzantine emperors, and to reciprocate the caliph asked what was the most precious (*aʿazz*) thing for them. These were deemed to be musk and sable (*sammūr*). He sent 200 *raṭls* of musk in addition to other gifts.[8] An expression of the value of musk can also be seen in the following expression: "The Arabs say of a thing to be held tenaciously because of its preciousness (*maḍnūn*), that it is like musk: if you hide it its scent lingers (*ʿabiqa*), and if you set it out it becomes exhausted (*nafiqa*)," meaning, its scent dissipates when exposed.[9] A Persian expression notes that musk may be scattered, but its scent remains evident, a thought intended to give hope to nobility when they had fallen on bad times.[10] Numerous proverbs in Arabic and especially Persian confirm the value of musk. Persian has an expression like the English "cat guarding the canary": "to entrust musk to the wind".[11]

The great value of musk can be seen in the Egyptian zoographer al-Damīrī's (1349–1405) discussion of the meanings of musk in dream interpretation: "It . . . indicates wealth because it is higher in value than gold and other things. It also indicates the goodness of life (*ṭīb ʿaysh*). The best scent (*khayr ṭīb*) accrues to the one who smells it and owns it. It indicates the innocence of the accused."[12] In the same way, Muḥammad uses musk to illustrate the importance of keeping good company by comparing someone who has musk with a blacksmith: "the suitable (*ṣāliḥ*) sitting-companion is like someone who has musk; if nothing else from him reaches you, some of his scent will. The bad sitting-companion is like one who operates bellows; if none of his blackness reaches you, some of his smoke will."[13] There are many different versions of this ḥadīth with slight variations; some do not include musk but rather have generic perfume. It is a distant echo of a verse from the *Pañcatantra*:

---

8 *Kitāb al-Hadāyā wa-l-tuḥaf.* ed. M. Ḥamīdullāh as Ibn al-Zubayr, *Kitāb al-Dhakhā'ir wa-l-tuḥaf* (Kuwait, 1959), sec. 31 p. 28. Trans. Ghāda al-Ḥijjāwī al-Qaddūmī, *Book of Gifts and Rarities* (Cambridge, MA: Harvard, 1996), 77.

9 Al-Sarī al-Raffā', *Al-Muḥibb wa-l-maḥbūb wa-l-mashmūm wa-l-mashrūb*, ed. Miṣbāḥ Ghalāwinjī, 4 vols. (Damascus: Majmaʿ al-Lughah al-ʿArabīyah, 1986–7), 3.160.

10 S. Anṣārī, *Tārīkh-i ʿIṭr dar Īrān* (Tehran: Wizārat-i Farhang wa Irshād-i Islāmī, 1381/2002–3), 229.

11 Anṣārī 229.

12 Al-Damīrī, *Ḥayāt al-Ḥayawān*, 2 vols. (Būlāq, 1284/1867; repr. Frankfurt am Main: Institute for the History of Arabic-Islamic Science, 2001), 2.129.

13 E.g., Muslim b. al-Ḥajjāj, *Saḥīḥ*, 5 vols. (Beirut: Dār al-Kutub al-ʿIlmiyyah, 1998), 4.198 (*Kitāb al-Birr wa-l-Ṣillah* ch. 45).

labhate puruṣas tāṇs tān guṇadoṣān sādhvasādhusaṃparkāt
nānādeśavicārī pavana iva śubhāśubhān gandhān[14]

A man acquires qualities and faults from association with good and bad people,
just as the wind which wanders in various regions acquires pleasant and unpleasant smells.

This was translated in the early Syriac version: "For whoever follows evil brings evil with him, and who follows the righteous brings righteousness with him; in like manner the wind, when it touches a stench, brings stench with it, and when it touches perfume, brings a sweet fragrance."[15] This particular bit has disappeared from the Arabic translation, but obviously the conception was well known in the early medieval Middle East.

Symbolic Meanings of Musk in Medieval Islamic Culture

Musk acquired many symbolic associations because of its value and striking origin. The poet al-Mutanabbī (915–965) sums up the importance of musk:

fa-in tafuqi l-anāma wa-anta minhum
fa-inna l-miska baʿḍu dami l-ghazāli[16]

If you surpass the human race, and yet you are of them,
well surely musk is part of the blood of the gazelle.

Musk is surpassingly great; the difference between the blood of gazelles and musk seems insurmountable, yet that is the origin of musk.

The excellence of the scent of musk made it an important symbol of excellence in literature. In another verse from a panegyric al-Mutanabbī plays on the relationship between the patron and the poet:

14 F. Edgerton, *The Panchatantra Reconstructed*, vol. 1 (New Haven: American Oriental Society, 1924), 176.

15 *Kalila und Dimna. Syrisch und Deutsch*, ed. and trans. F. Schulthess. 2 vols. (Berlin: Reimer, 1911), 1.49.

16 Abū al-Ṭayyib Aḥmad b. al-Ḥusayn al-Mutanabbī, *Dīwān* (Beirut: Dār Ṣādir, 1958), 268; al-Sarī al-Raffāʾ 3.159 (#292) has *nafuqi* instead of *tafuqi*, but the latter seems to make better sense.

wa-dhāka l-nashru ʿirḍuka kāna miskan
wa-hādhā l-shiʿru fihrī wa-l-madākā[17]

That diffusion, your honor, is musk, and this poem my pestle and stone.

Al-Mutanabbī compares the honor of the patron, ʿAḍud al-Dawlah, to musk, and so he implies his own importance to his patron by likening his poem to the perfumer's pestle and grinding stone: he is comparing what he does to what the person does who prepares perfumes and thus releases their fragrance. At the same time, by likening the spread (*nashr*, which can mean both promulgation and the diffusion of a perfume) of ʿAḍud al-Dawlah's fame to musk he has sought the best perfume for his comparison. The job of the poet is thus to create beautiful poetry from the qualities of the patron, even as the perfumer creates a wonderful perfume from aromatics. A disciple of the poet Abū al-ʿAlā al-Maʿarrī (973–1057) eulogized him:

sayyarta dhikraka fī l-bilādi ka-annahu
miskun fa-sāmiʿatan yuḍammikhu aw famā[18]

You sent out your reputation in the lands as if it
were musk; it perfumes a hearer or mouth

Thus al-Maʿarrī's mention or reputation (*dhikr*), his verses, perfumed both those who recited his poetry as well as those who heard it. Good words in general were likened to aromatics, especially musk. In al-Maʿarrī's *Epistles* (*Rasāʾil*), the letters to which the poet replied he likened to various fragrant substances including musk and ambergris,[19] and he frequently likened his own responses to scents. For example, he closes a letter with the wish that it diffuses scent like the diffusion of pungent musk.[20] As noted in Chapter 4, Muʿizz b. Bādīs included a discussion of musk in his work on bookmaking and ink. It is likely that writers who speak of the delightful scent of received correspondence,

17 Al-Mutanabbī, *Dīwān al-Mutanabbī* (Beirut: Dār Ṣādir, 1958), 568.

18 Abū al-Ḥassan ʿAlī b. Hammam, quoted in Ibn Khallikān, *Wafāt al-aʿyān*, ed. I. ʿAbbās, 8 vols., 4th ed. (Beirut: Dār Ṣādir, 2005), 1.115.

19 E.g., Abū al-ʿAlā al-Maʿarrī, *The Letters of Abū ʾl-ʿAlā of Maʿarrat al-Nuʿmān*, ed. and trans. D. S. Margoliouth. (Oxford: Clarendon Press, 1898), text 63 and trans. 70.

20 Maʿarrī, *Letters*, text 49 and trans. 56.

such as al-Maʿarrī, do not intend it exclusively as a metaphor.[21] Scenting inks is attested also in East Asia, as noted in Chapter 3.

The Persian expression *nāfah-yi mushk yāftan*, literally, "to get a musk pod" means "to get a great name".[22] In Persian literature also good speech can be likened to musk. In the *Haft Paykar* of Niẓāmī, the poet describes the Indian princess, beginning her story by comparing it to the opening of a musk pod:

> *āhū-yi turk-i chashm-i hindū zād*
> *nāfah-yi mushk-rā garah bikashādh*[23]

> The Indian-born Turk-eyed deer unloosed the knot of the musk pod.

The poet plays on the word *āhū*, meaning both "deer" and "beautiful woman" but here evoking the *āhū-yi mushk*, the musk deer. Thus the princess is compared to a musk deer, further characterized by being called Turk-eyed, thus evoking the lands which produced musk, and the story itself is the musk which she diffuses. To have a "musky mouth" means that one speaks sweetly.[24]

Persian poetry extols the quality of musk perhaps even more than Arabic. ʿUnṣurī (1006–88) says:

> *agar khalq-i ū-rā kasi waṣf gūyad*
> *birīzad bi kharwār mushk az dahānash*[25]

> If someone describes his good qualities, a whole donkey-load of musk pours from his mouth.

This continues the idea that praise is metaphorically musk but the imagery has become even more extravagant. The thirteenth century Persian poet Saʿdī says:

> *ṣabā gar bi-gdharad bar khāk-e pāyat*
> *ʿajab gar dāmanash mushkīn nabāshad*[26]

---

21 Examples of musk-scented writing in Persian poetry are cited by Newid 90–1.

22 F. Steingass, *A Comprehensive Persian-English Dictionary*, (London, 1892; repr. New Delhi: Munshiram Manoharlal, 1996), 1376a.

23 Nizāmī 32 l.9 (p. 121).

24 *Farhangnāmah-i Adab-i Fārsī*, vol. 2 of *Dānishnāmah-i Adab-i Fārsī* (Tehran: Mu'assasah-i Farhangī wa Intishārāt-i Dānishnāmah, 1996–), 261a.

25 *Farhangnāmah* 257a.

26 *Farhangnāmah* 257a.

> If Zephyr passed over the dust of your feet, it would be a wonder if its skirts did not become musky.

The lowliest part of the subject is thus likened to musk to emphasize his good qualities.

Another illustration of the excellence of the scent of musk for early medieval Muslims is the inverted comparison of its scent to the foul stench of corpses. Abū Tammām uses it twice. He wrote a famous poem about the capture of the city of Amorium from the Byzantines in 838.[27] This poem contains striking imagery describing pleasure taken by the victors in the horrors of war suffered by the enemy. One of these images involves musk:

> (60) *yā rubba ḥawbā'a lammā jtuththa dābiruhum ṭābat*
> *wa-law ḍummikhat bil-miski lam taṭibi*[28]

> How many a soul when it was uprooted became sweet; it would not have been as sweet even if it had been anointed with musk.

In this verse, musk is set up as the sweetest of perfumes, the very essence of good fragrance, but it is not as sweet to the victors as the smell of the enemies killed on the battlefield. Musk is an ultimately purified form of blood, and this intensifies the image. As Stetkevych points out, this verse illustrates the imagery of vengeance: impurity rendered pure like an aromatic (exemplified by the verb *ṭābat*) through purification by vengeance.[29] Abū Tammām also describes the scent smelled by the spectators at the immolation of the Afshīn, who had fallen from favor,[30] by comparing the smell of his roasting body to the aroma of fine musk and ambergris used as incense:

> *wa-stansha'ū minhu qutāran nashruhu*
> *min ʿanbarin dhafirin wa-miskin dārī*[31]

---

27 S. Stetkevych, *The Poetics of Islamic Legitimacy* (Bloomington: Indiana University Press, 2002), 152.

28 Abū Tammām, *Dīwān Abī Tammām bi-Sharḥ al-Khaṭīb al-Tibrīzī* (Cairo: Dār al-Maʿārif, 1964), 70.

29 S. Stetkevych, *Abū Tammām and the Poetics of the Abbasid Age* (Leiden: Brill, 1991), 209.

30 See C. E. Bosworth "Afšīn," in *EIr* s.v.

31 Abū Tammām 2.204; see also Stetkevych, *Abū Tammām* 226 with emphasis on how the appearance of these scents ties in with the concept of ritual purification.

And they smell from it an aroma, its diffusion
of pungent ambergris and musk of Dārīn.

The image can be traced even earlier, to an anecdote about the first Abbasid caliph al-Saffāḥ, "the butcher" (r. 749–754). He did not mind the smell of the bodies of the Umayyads he had overthrown, and instead remarked "By God, this is more pleasing to me than the smell of musk and ambergris".[32] In all of these cases, it is the smell of the vanquished enemy which is sweeter than musk.

The penetrating scent of musk came to be associated with the idea of revealing secrets since musk could not be hidden because of its strong smell. The scent of musk was especially associated with the detection of royalty, as will be seen below. The idea that musk could reveal the unknown is rather common in the Middle East as elsewhere. Al-Damīrī noted in his discussion of musk in the interpretation of dreams that: "A thief who carries musk will be seized because the potent aroma shows its possessor and carrier and reveals his secret."[33] In Chapter 3 an Indian proverb hinging on this notion was mentioned. A similar example, featuring jasmine, comes from the *Pañcatantra*:

guṇā anuktā api te svayaṃ yānti prakāśatām
chādyamānā 'pi sāugandhyam udvamaty eva mālatī[34]

Even your unspoken qualities proclaim your renown by themselves;
Even when covered, jasmine emits a sweet scent.

This work was translated into Pahlavi in Sasanian times. From Pahlavi, a Syriac translation was made, and there is also the famous Arabic version of Ibn al-Muqaffaʿ, the *Kalīlah wa Dimnah*. The Syriac translation of this particular verse of the Sanskrit original is: "Even if you have not confessed it (your virtue), it shows by itself, like musk (*mūshk*); even if it is concealed, it is revealed by its scent."[35] The Arabic translation likewise has musk in this image.[36] The replacement of jasmine with musk probably dates to the Pahlavi translation

32 Abū al-Faraj al-Iṣbahānī, *Kitāb al-Aghānī*, 24 vols. (Cairo: Dār al-Kutub, 1927–74), 4.351.
33 Damīrī 2.127.
34 F. Edgerton, *The Panchatantra Reconstructed*, vol. 1 (New Haven: American Oriental Society, 1924), 197.
35 *Kalila und Dimna. Syrisch und Deutsch*, ed. and trans. F. Schulthess, 2 vols. (Berlin: Reimer, 1911), 55.
36 Ibn al-Muqaffaʿ, *Kalīlah wa Dimnah* (Beirut: Dār Maktabat al-Hayāh, [1966]), 239.

and illustrates the preference for musk in the Middle East. The Andalusian poet Ibn Zaydūn (1003–71) says that the gifts (*nawāfiḥ*—which can also mean perfumes!) of his patron can be hidden only in the way that musk is hidden in bags (*ṣurar*),[37] meaning that they cannot be hidden at all.[38] This motif occurs in Islamic Persian literature;[39] the *Haft Paykar* of Niẓāmī has:

> *garch dar nāfah mushk hast nihān*
> *āshkār ast būy-i ū bi-jihān*[40]

> Even if musk is kept concealed in the pod, its scent is manifest in the world.

The *Ṭuṭī-nāmah* notes that musk and love cannot be hidden.[41] The Arabic poet Abū ʿUbādah al-Buḥturī (821–97) makes the musk scent of women reveal their departure from the campsite:

> *ẓaʿāʾinu aẓʿanna l-karā ʿan jufūninā*
> *wa-ʿawwaḍnanā minhā suhādan wa-admuʿā*
> *wa-ḥāwalna kitmāna l-taraḥḥuli fī l-dujā*
> *fabāḥa bihinna l-misku lammā taḍawwaʿā*[42]

> The howdahs made sleep depart from our eyelids and exchanged it for insomnia and tears
> They desired to conceal their departure in the darkness but the musk, when it diffused, revealed them.

Likewise, Ibn Zaydūn notes that a woman leaving at night could not conceal her departure because of her scent of aloeswood and *nadd*.[43]

---

37 *Ṣurar* is also a playful evocation of the navels (*surar*) of the musk deer.

38 Aḥmad b. ʿAbdallāh b. Zaydūn, *Dīwān*, ed. Karam al-Bustānī (Beirut: Dār Ṣādir, 1964), 150.

39 Bayhaqī, *Tārīkh-i Bayhaqī*, ed. ʿAlī Akbār Fayyāḍ (Mashhad: Dānishgāh-i Mashhad, 1971), 265, refers to excellence becoming known just like the scent of musk.

40 Niẓāmī Ganjavī, *Haft Paykar*, ed. H. Ritter and J. Rypka (Prague: Orientální Ústav, 1934), 37 l. 215 (p. 234).

41 Cf. Anṣārī 230.

42 Al-Sarī al-Raffāʾ 3.158 (#290).

43 Ibn Zaydūn, *Dīwān*, 209.

The dark color of musk and men's perfumes also became a metaphor for the color black or for darkness in Islamic literature.[44] In this sense it is contrasted with camphor, which is white and bright.[45] Ibn Zaydūn says:

aḍḥā l-zamānu nahāruhu kāfūratun
wa-l-laylu miskun min khilālika ʿāṭirū[46]

Time brought to light a camphored day,
and the perfuming night musk from your friendship.

Persian can use *mushk rang* "musk colored" for black, as well as *mushkīn* "musky".[47] Niẓāmī associates musk with the black pavilion for the Indian princess in the *Haft Paykar*.[48] The association of musk with the darkness of night is also found in Persian literature.[49] Firdawsī describes the onset of night:

chu az bākhtar tīrah shud rūy-i mihr bipūshīd dībā-yi mushkīn sipihr[50]

When from the west the face of the sun became dark, the sphere donned musky brocade.

Likewise, he describes the dawn as the sun rending a musk-scented veil; the image of night as a musky veil is common in the *Shāhnamah*.[51]

---

44 For Arabic examples, see al-Maʿarrī, *Shurūḥ saqṭ al-zand*, ed. Ṭāhā Ḥusayn, Muṣṭafā Saqqā, et al., 5 vols. (Cairo: al-Dār al-Qawmiyyah li-l-Ṭibāʿah wa-al-Nashr, 1964), 2.853–4 and 2.777. Cf. also C. van Ruymbeke, *Science and Poetry in Medieval Persia: The Botany of Nizami's Khamsa* (Cambridge: Cambridge University Press, 2007), 50–1; Newid 81.

45 For example, Maʿarrī's depiction of the celestial reincarnation of the black librarian Tawfiq states that she had become "whiter (*anṣaʿ*) than camphor; *Risālat al-ghufrān*, ed. ʿĀʾisha ʿAbd al-Raḥman, 5th ed. (Cairo: Dār al-Maʿārif, 1969), 287. Cf. on camphor as whiteness, Anṣārī 229.

46 Ibn Zaydūn, *Dīwān* 196.

47 *Farhangnāmah* 259b; 257a.

48 Niẓāmī 31.7 and 32.5 on 120.

49 *Farhangnāmah* 260a.

50 *Farhangnāmah* 257a; Firdawsī, *Abu'l Qasem Ferdowsi: The Shahnameh* (*The Book of Kings*), ed. Djalal Khaleghi-Motlagh, 8 vols. (New York: Bibliotheca Persica, 1988–2008), 17 l.229 on 5.505.

51 Firdawsī 6.472; cf. 6.567, etc.

As a nickname, Arabic *misk* connotes darkness of complexion, which was frequently considered undesirable in medieval Islamic culture, and so was often jocularly applied because *misk* also indicated the valuable musk.[52] The most famous application of the name is the 10th century eunuch ruler of Egypt, Kāfūr Abū Misk, "Camphor, father of Musk", erstwhile patron of the poet al-Mutanabbī and later target of his most violent invective.[53]

With its deep associations with the erotic, which we discussed in the preceding chapter, musk also comes to be associated with youth (as well as springtime):

wa-l-misku ashbahu shay'in bi-l-shabābi fa-hab
baʿḍa l-shabābi li-baʿḍi l-maʿshari l-shaybi[54]
(al-Ṣanawbarī)

Musk is the thing which most resembles youth, so grant some of this youth to the assembly of the aged.

The same concept is clearly stated in the *Shāhnāmah*; the young Shāpūr is described as follows:

chu gulbarg rukhsār wa chūn mushk mūy
bih rang-i ṭabarkhūn wa chūn mushk būy[55]

His cheek like rose petals, his hair musk with the tint of the red Tabaristan willow, and his scent musk.

Persian has the expression "to substitute camphor for musk" (*mushk bih kāfūr badal shudan*), meaning "to grow old".[56] The white hair of the aged was likened to camphor.[57] Firdawsī has a striking pair of verses in which he compares the bloom of youth with musk and the white of camphor with old age:

---

52 S. D. Goitein, "Nicknames as Family Names," *JAOS* 90 (1970): 522. See also B. Lewis, "The Crows of the Arabs," *Critical Inquiry* 12:1 (1985): 95 for verses by poets of African descent that deny any ugliness associated with blackness because it is the color of musk.

53 M. Larkin, *al-Mutanabbi: Voice of the 'Abbasid Poetic Ideal* (Oxford: Oneworld, 2008), 63–78.

54 Al-Sarī al-Raffā' 3.162.

55 Firdawsī 6.295, l. 52.

56 *Farhangnāmah* 260b.

57 For an example not from poetry, see Bayhaqī 457.

chu yak mūy gardad bih sar bar sapīd
bibāyad gasastan 'z shādī umīd
chu kāfūr shud mushk maʿyūb gasht
bih kāfūrbar tāj nākhūb gasht[58]

When one hair upon the head turns white, one must turn from happiness and hope
When musk has become camphor it has become imperfect (*maʿyūb*) and for the camphor-haired the crown becomes unseemly.

A king in the Near Eastern tradition was an epitome of fertility;[59] it was through his rule that nature renewed itself in spring, hence the great importance placed on the king's participation in Nawrūz festivities in ancient Iran. Musk, as the preeminent aromatic of youth and fertility, was naturally associated with kingship.

Musk and Kingship

Musk, along with ambergris, was especially associated with kingship. This is no doubt in large part due to its high cost and scarcity. If the aristocracy was expected to make use of musk and other rare and expensive perfumes, this was even more true of royalty. Ibn Kaysān states in the introduction of the *Mukhtaṣar fī al-ṭīb* that he mentions in his book the perfumes that are "indispensable to the treasuries of the kings";[60] he begins his book with a discussion of musk. Al-Masʿūdī noted that the best musk was mostly used by kings and was rarely available to merchants.[61] Verses quoted by al-Zamakhsharī contrast the lowly beginnings of musk perfume with its royal destination:

wa-l-misku baynā tarāhu mumtahanan
bi-fihri ʿaṭṭārihi wa-sāḥiqati

58 Firdawsī 6.511, lines 1269–70.

59 S. Sperl, "Islamic Kingship and Arabic Panegyric Poetry in the early 9th century," *Journal of Arabic Literature* 8 (1977): 29–30 and Newid 80–1. Similar conceptions are found in early medieval Europe, for example, among the Franks.

60 Ibn Kaysān 187.

61 Al-Masʿūdī, *Murūj al-dhahab wa maʿādin al-jawhar*, ed. B. de Meynard and P. de Courteille, revised by C. Pellat. 5 vols. (Beirut: Manshūrāt al-Jāmiʿah al-Lubnaniyyah, 1966–74), 1.189 (§ 393).

ḥattā tarāhu ʿāriḍay malikin
aw mawḍiʿa l-tāji min mafāriqihi[62]

Even though one sees musk humbled by the pestle of its perfumer and the crusher,
eventually you see it on the cheeks of a king, or in the place of the crown in the part of his hair.

Musk was apparently valued by royalty as soon as it became available in the Near East, and its use was projected backwards into the past. Solomon, the epitome of the great sovereign, is repeatedly associated with musk and ambergris. He is said to have been the first to reveal the use of "musk, ambergris, camphor, sandalwood, aloeswood, and different kinds of aromatics".[63] We have already discussed the parallel role of Jamshīd as discover of perfumes in the Iranian tradition in Chapter 3. There is a striking description of the throne of Solomon in the *ʿArāʾis al-majālis fī qiṣaṣ al-anbiyāʾ* of Abū Isḥāq Aḥmad b. Muḥammad b. Ibrāhīm al-Thaʿlabī (d. 1036). Solomon's throne had mechanisms that sprinkled him with musk and ambergris.[64] The throne, which was made of ivory studded with jewels, was surrounded by four date palm trees of gold, with clusters of dates made from ruby and emerald. Two of these trees held gold peacocks, the other two gold eagles (*nasrān*).[65] Two lions of gold sat beside the throne.[66] When Solomon began to mount the throne, the birds spread their wings and the lions opened their paws and struck the ground with their tails. When he reached the top of it, the eagles took musk and ambergris and crumbled (*yufattitu*) them over him, and then a gold dove standing upon the jeweled

62 Zamakhsharī, *Rabīʿ*, 2.219.

63 Ibn Rustah, *Kitāb al-ʿAlāq al-nafīsah*, ed. M. G. De Goeje (Leiden, 1892; repr. Leiden: Brill, 1967), 198.

64 Abū Isḥāq Aḥmad b. Muḥammad b. Ibrāhīm al-Thaʿlabi, *ʿArāʾis al-majālis fī qiṣaṣ al-anbiyāʾ*, published as *Kitāb Qiṣaṣ al-anbiyāʾ* (Cairo: Al-Maṭbaʿah al-Ummah, 1331/1912), 195–7; trans. W. Brinner, *ʿArāʾis al-Majālis fī Qiṣaṣ al-Anbiyāʾ or 'Lives of the Prophets'* (Leiden: Brill, 2002), 512.

65 Perhaps these eagles are a reflection of the Biblical cherubim?

66 This is probably a reflection of the throne of the Sasanian kings, which was supported by griffins- a combination of the eagle and lion, cf. H. P. L'Orange, *Studies on the Iconography of Cosmic Kingship in the Ancient World* (Oslo: H. Aschehoug & Co., 1953; repr. New Rochelle: Caratzas, 1982), 65. The association of eagles with the throne is also found in the *Shāhnāmah*; eagles were the vehicle of the monarch to the heavens, an idea found in Latin literature as well; see L'Orange 69–70. The eagle was a *psychophorus*, a symbol of the elevation to heaven.

posts of the throne took up the Torah and opened it for Solomon. This anointing is thus a perpetual renewal of the heaven-mandated authority of the monarch.

In the story of the People of the Cave (the Seven Sleepers) as recorded by al-Thaʿlabī we find a parallel to Solomon's mechanical throne. The story tells of the court of the Roman emperor Duqyānūs (i.e. Decius, ruled 249–51 CE), who would later cause the pious to go into self-imposed exile in their cave.[67] At his court, three slaves held three items: the first, a gold cup (*jām*) of musk, the second, a silver cup of rosewater, and the last, a bird. Upon command, the bird would dip itself in rosewater, then in the musk, soaking it up, and would then descend upon the crown of Duqyānūs and dribble the rosewater and musk on him. According to the text, this regimen prevented the emperor from contracting headache, pain, fever, drivel, spittle, and mucus; in short, it provided some of the principal medical benefits associated with musk. This process made him unusually healthy, and he became arrogant and was worshipped as a god. There is of course no Roman attestation of Decius using musk in this manner. Musk was not known in the Roman Empire at that time. The story is important not for its historical veracity but for its symbolic meaning. The mixture of musk and rosewater dribbling on the king is a symbolic anointing, continually renewing the king's sacral authority. The medical benefits of this regimen are merely the effects of musk well attested in early medieval medical literature.

In addition to the mythic tales of the use of musk, we have accounts with a much better chance of being historical. Al-Aṣmaʿī (d. 828) records that ʿAbd al-Muṭṭalib visited the last ruler of Yemen, Sayf b. Dhī Yazan, and found him perfumed with ambergris and musk from head to toe.[68] Another source describes him perfumed with ambergris and with the gleam of musk in the part of his hair.[69] In the early 7th century, a visitor to the Ghassānid court described King Jabalah as follows: "When he would sit upon a blanket for drinking, myrtle (*ās*) and jasmine and other sorts of aromatics (*rayāḥīn*) were under it, and he was daubed with ambergris and musk in vessels (*ṣiḥāf*) of silver and gold, and genuine (*ṣaḥīḥ*) musk was brought to him in silver vessels, and moistened[70] aloeswood was kindled for him when it was winter, and in the summer he had ice all around."[71] The Sasanian king Khusraw received musk and ambergris from

67 Thaʿlabī 266–7; Brinner 694.

68 Al-Aṣmaʿī, *Tārīkh al-ʿArab qabl al-Islām*, ed. Muḥammad Ḥasan āl Yāsīn (Baghdād, Maṭbaʿat al-Maʿārif, 1959), 52.

69 Ibn ʿAbd Rabbih, *al-ʿIqd al-farīd*, 31 vols. (Beirut: Maktabat Ṣādir, 1951–4), 4.38.

70 The text has *al-munaddī* and there is *al-hindī* in a variant.

71 Iṣbahānī 17.166–7.

his governor in Yemen, as mentioned in the previous chapter. These customs evidently continued; the throne room of the Ghaznavids included a display of camphor, musk, aloeswood, and ambergris.[72]

Unlike food and drink, the king was not obliged to share perfumes with his courtiers. The *Kitāb al-Tāj* of pseudo-al-Jāḥiẓ says that the retinue and boon companions of a king do not share in the application of perfume or censing because it raises the station of the king above all others. Likewise, they do not apply scents when he is perfumed.[73] It is doubtful if this rigid distinction was always observed, but it is important evidence of the special scent of royalty which is as much symbolic as literal. The caliph al-Mutawakkil gave Ibn Abī Fanan a musk pod, and he replied:

> *la'in kāna hādhā ṭīban wa-hwa ṭayyibun*
> *la-qad ṭayyabathu min yadayka al-anāmilu*[74]

> Even if this is an aromatic, and it is good, it was perfumed by the fingertips of your hands.

This association with kingship continued in Islamic times. At the culmination of the war over the caliphate with his brother al-Ma'mūn, the fleeing Caliph Muḥammad al-Amīn was recognized and captured because of the scent of musk and perfume on him; it bespoke royalty.[75] A poet who went to recite a poem before al-Ma'mūn met him in his army camp; he did not recognize him but noted the smell of ambergris and musk from him; he recited his poem after being offered a reward by al-Ma'mūn, and then he realized the identity of his benefactor in terror as part of the army arrived and saluted him.[76] In short, musk has had a strong association with royalty, both by itself and paired with ambergris. The special anointing oil prepared for the coronations of the British monarchs, most recently for Queen Elizabeth II in 1953, includes musk and ambergris among its ingredients. Anointing oils for the British monarchs

72 Bayhaqī 714.

73 [Pseudo] al-Jāḥiẓ, *Kitāb al-Tāj*, ed. Ahmed Zeki Pacha (Cairo: Imprimerie Nationale, 1914), 46.

74 Quoted by Zamakhsharī, *Rabīʿ*, 2.214. Cf. also Bahā' al-Dīn Muḥammad b. Aḥmad al-Ibshihī, *Al-Mustaṭraf fī kull fann mustaẓraf*, ed. Muṣṭafā Muḥammad al-Dhahabī (Cairo: Dār al-Ḥadīth, 2000), 360, and Ibn Abī al-Ḥadīd 19.342.

75 Masʿūdī 4.294 (§2684).

76 Ṭabarī III.1145–6.

including musk date back to the coronation of Charles I.[77] Curiously, the anointing vessel (the Ampulla) is in the form of an eagle; this particular vessel dates in its original form to the coronation of Henry IV in 1399, but the original vessel was said to have been given to Thomas Becket by the Virgin Mary.[78]

Musk and Islam

Good scent has been associated with religion for a very long time. The immediate antecedents of the concept of the divine good aroma in Islam can be found in the pre-Islamic cultures of the Near East, including Judaism, Christianity and Zoroastrianism. Many of the uses and images associated with scent are part of a common stock drawn upon by the various religions of the world. The central place of musk especially but also the other members of the principal aromatics in religious practice and imagery makes Islam unique. Islam came of age in a time when musk was still relatively novel in the world's perfume-box. Judaism, Christianity, and Zoroastrianism had enjoyed the scents of Arabia-frankincense and myrrh.[79] The pre-Islamic Arabs also used these resins in their incense, but they were inexorably replaced in priority by the principal aromatics, especially musk. By the time of Islam, frankincense and myrrh had no significance within Arabia as prestige perfumes; that role was now filled by musk, ambergris, and aloeswood in particular.[80]

The close connection of the soul with scent is emphasized in ancient Christianity. For the Apostle Paul, knowledge of God is characterized by a scent

---

77 Anonymus, "The Anointing of the Queen: Some Notes on the Coronation Oil," *Pharmaceutical Journal* 170 (Jan.–June 1953), 405.

78 W. Ullmann, "Thomas Becket's Miraculous Oil," *Journal of Theological Studies* n.s. 8 (1957): 129 and J. W. McKenna, "The Coronation Oil of the Yorkist Kings," *The English Historical Review* 82:322 (Jan. 1967): 102–4.

79 For an overview of scent in Rabbinic Judaism, see D. Green, *The Aroma of Righteousness: Scent and Seduction in Rabbinic Life and Literature* (University Park: Pennsylvania State University Press, 2011). For Christianity, S. A. Harvey, *Scenting Salvation: Ancient Christianity and the Olfactory Imagination* (Berkeley: University of California Press, 2006) and on the reintroduction of incense into worship, see B. Caseau, "Incense and Fragrances: from House to Church: A Study of the Introduction of Incense in the Early Byzantine Christian Churches," in M. Grünbart, et al., eds., *Material Culture and Well-Being in Byzantium (400–1453)* (Vienna: Akademie der Wissenschaften, 2007), 75–92. We have discussed the Zoroastrian uses of good scent briefly in Chapter 3.

80 A. King, "The Importance of Imported Aromatics in Arabian Culture: Illustrations from Pre-Islamic and Early Islamic Poetry," *JNES* 67:3 (2008): 175–189.

that is associated with the believers. The believers are a good scent (εὐωδία) to God, which is the odor of life (ὀσμὴ ἐκ ζωῆς), while the unbelievers have the odor of death (ὀσμὴ ἐκ θανάτου).[81] In Syriac, as in Arabic, the fact that the words for spirit (*rūḥ*) and scent (*rīḥ*) are from the same root was one more factor leading to the association of good scent with the holy. We find this word-play in the 8th century mystic John of Dalyatha: "Blessed is he who breathes Holy Spirit (*rūḥ*) and who has mixed also the smell (*rīḥ*) of his own body into the one who has taken delight in his aroma."[82] Likewise, according to Ephrem, at Pentecost, the Apostles experienced the perfumed scent of Paradise as the Holy Spirit descended upon them.[83] Islam inherited Syriac Christianity's fondness for good scent along with much else. Ibn Qayyim al-Jawziyyah plays on the words for spirit and scent like the Syriac writers, explaining that a good aroma is the food of the spirit.[84] In addition to the word play of *rūḥ* and *rīḥ*, the word for perfume *ṭīb*, discussed in Chapter 5, comes from the root ṭ-y-b from which numerous words describing goodness are derived. Perfume (*ṭīb*), as the quintessential goodness (*ṭībah*), has a meaning of purity and immortality in pre-Islamic Arabian culture that is carried into Islam. Indeed, it was even possible for one to swear an oath upon perfume in pre-Islamic times.[85]

In the analysis of musk and religion we must contend with two distinct facets: practical uses of musk with a religious sanction or significance, and the imagery and symbolism of musk. The practical uses of musk with religious sanction are intimately connected with many of the practical uses of musk referred to above inasmuch as authority from the Sunnah was sought for all aspects of life. Ḥadīth texts have already been mentioned as the sanction for many of these uses of musk. The practical uses of musk we will consider in this chapter center on the use of musk in worship and purification.

As mentioned previously, in pre-Islamic Iran men perfumed their beards with musk before worship. Abū ʿAmr is quoted in the *Kitāb al-Nabāt* of al-Dīnawarī: "I entered one of the shrines (*miḥrāb*) of the shrines of the *jāhiliyyah* and musk

81 II. Corinthians 2:14–6 and Harvey 18–19.

82 Quoted by S. Seppälä, *In Speechless Ecstasy: Expression and Interpretation of Mystical Experience in Classical Syriac and Sufi Literature* (Helsinki: Finnish Oriental Society, 2003), 88.

83 Ephraem Syrus, *Hymnen de paradiso und contra julianum*, ed. E. Beck, 2 vols., (Louvain, 1957), XI.14. Cf. Harvey 64–5.

84 Ibn Qayyim al-Jawziyyah 330.

85 See Ibn Hishām, *Kitāb Sīrat Rasūl Allāh*, vol. 1,1, ed. F. Wüstenfeld (Frankfurt-am-Main: Minerva, 1961), 85.

diffused in my face."[86] Incense was extensively used in pagan worship in pre-Islamic times. We have incense-burners from South Arabia that are inscribed with the names of aromatics, and alongside characteristically Arabian products such as *lbny* "frankincense" we find two terms of possibly foreign origin. The first is *rnd*, which has been interpreted as a metathesis of Hebrew *nērd*, meaning spikenard, an import from India.[87] Second we have *qsṭ*, which is Arabic *qusṭ* and Greek κόστος and Latin (and hence English) *costus*.[88] The name is derived from Sanskrit *kushṭha*.[89] The trade in costus can be traced far into antiquity.

It is likely that some of the goods exported from Syria, where Muḥammad traded before the beginning of his prophetic mission, were perfumes. His uncle, Abū Ṭālib, is said to have traded in them. But from the beginning of Islam we find the new aromatics, such as musk, that are characteristic of the medieval period rather than classical antiquity. Islam absorbed these aromatics from its outset and gave them a special role unparalleled anywhere else.

Musk had a potent role in Islam. No aromatic is more associated with the religion of Islam than musk even to this day, when musk is not so appreciated as an aromatic elsewhere. Even the Chinese knew of the use of musk in Muslim funerary practice from the book of Ma Huan 馬歡, the *Yingya shenglan* 瀛涯勝覽 (1433); its author was an interpreter for the expedition of the Muslim Ming envoy Zhenghe 鄭和, who led the famous voyages in the Indian Ocean.[90]

The role of musk in Islam starts with Muḥammad himself: "The Prophet, Peace be upon him and his family, often was perfumed with musk and other varieties of aromatics."[91] Muḥammad said he only held dear two things of the

---

86 Dīnawarī, *Kitāb al-Nabāt: The Book of Plants: Part of the Monograph Section*, ed. Bernhard Lewin (Wiesbaden: Steiner, 1974), 187.

87 Cf. K. Nielsen, *Incense in Ancient Israel* (Leiden: Brill, 1986), 18. Rejected by W. Müller, "Namen von Aromata im antiken Südarabien," in A. Avanzini, ed., *Profumi d'Arabia* (Rome: Bretschneider, 1997), 203–4. Arabic *rand* can denote a variety of aromatics, especially myrtle and aloeswood; see al-Dīnawarī, *The Book of Plants of Abū Ḥanīfa ad-Dīnawarī: Part of the Alphabetical Section* (Alif-Zayn), ed. Bernhard Lewin, *Uppsala Universitets Årsskrift* 1953:10 (Wiesbaden: Harrassowitz, 1953), #422 on 185–6.

88 Nielsen 18; Müller 204.

89 Monier Williams 297c.

90 Ma Huan 馬歡, *Yingya shenglan* 瀛涯勝覽, in *Congshu ji cheng chu bian* 叢書集成初編 3274 (Shanghai: Shang wu yin shu guan, Minguo 26 [1937]), 81. Trans. J. V. Mills, *Ying-yai Sheng-lan: The Overall Survey of the Ocean's Shore* (London: Hakluyt Society, 1970; repr. Bangkok: White Lotus, 1997), 166.

91 Ibn Abī al-Ḥadīd, *Sharḥ Nahj al-balāghah*, 20 vols. (Cairo: ʿĪsā al-Bābī al-Ḥalabī, 1960–4), 19.341.

world: women and perfume: "Women and perfume (*ṭīb*) were made dear to me in this world, and they were made the delight of my eye in prayer."[92] There is a great deal of evidence in the Ḥadīth literature on Muḥammad's use of musk; indeed, no other aromatic is mentioned with anything remotely resembling the frequency of musk. There are many different versions of these ḥadīths, so here will be mentioned some basic types. There is a type of ḥadīth where musk is mentioned, and Muḥammad says it is the most fragrant or best of the aromatics (*aṭyab al-ṭīb*).[93] Another version closely related to this has it as "the best of your aromatics" (*aṭyab ṭībikum*).[94] There is yet another type praising musk that goes, "musk is among the best of your aromatics" (*min khayri ṭībikum al-misk*).[95] All of these examples show clearly the great importance of musk as the best aromatic in the early tradition.

In addition to these general commendations of the quality of musk above other aromatics there are many ḥadīths that indicate that Muḥammad himself did use musk. Many of these are attributed to his wife ʿĀʾishah. ʿĀʾishah said that he perfumed himself with masculine perfumes of musk and ambergris (*bi-dhikārati l-ṭībi l-miski wa-l-ʿanbari*).[96] As we have seen, both of these aromatics are associated with kingship in Arabic literature describing pre-Islamic times, and we may consider this ḥadīth a reflection of that tradition beyond the simple statement that Muḥammad used musk and ambergris. We have a further ḥadīth in which he anointed (*ḍamakha*) his head with musk.[97] Muḥammad is also said to have kept *sukk* with him and perfumed himself.[98]

The next group of ḥadīths expands on this theme and brings musk squarely into the realm of Islamic ritual practice. Muslims were to perform ritual ablutions (*ightasala*) every seven days and if they had perfume, they were to apply it.[99] Muḥammad entered the state of *iḥrām* or ritual purity on *yawm al-naḥr* before circumambulating the Kaʿbah by being perfumed by ʿĀʾishah with

---

92 Aḥmad b. Shuʿayb al-Nasāʾī *Sunan al-Nasāʾī*, 8 vols. (Cairo: Musṭafa al-Bābī al-Halabī, 1964–5), 7.57. Cf. Ibn Qayyim al-Jawziyyah 330; Ibn Saʿd, *al-Ṭabaqāt al-kubrā*, 9 vols. (Beirut: Dār Ṣādir, 1960), 1.398 and Ibn Abī al-Ḥadīd 19.341.

93 Aḥmad b. Muḥammad b. Ḥanbal, *Musnad*, 6 vols. (Beirut, 1969: reprint of Bulaq edition), 3.31, 47, 87. Cf. Ibn Qayyim al-Jawziyyah 437; Ibn Saʿd, 1.399.

94 Tirmidhī, *Sunan*, 5 vols. (Cairo: Maṭbaʿat al-Madanī, 1964), 2.230; Nasāʾī 4,32–3; Abū Dāwud, *Sunan Abī Dāwūd*, 4 vols. (Dār Iḥyāʾ al-Sunnah al Nabawiyyah, n.d.), 3.200.

95 Nasāʾī 4.33.

96 Nasāʾī 8.130. A note explains that *dhikārah* means perfume "suitable for the use of men like musk, ambergris, aloeswood and camphor." Cf. also Ibn Saʿd 1.399.

97 Ibn Ḥanbal 1.234,344.

98 Ibn Saʿd 1.399 and Ibn Qayyim al-Jawziyyah 331.

99 Ibn Qayyim al-Jawziyyah 331. See also Nasāʾī 3.75 .

musk.[100] ʿĀʾishah also said, "It is as if I can [still] see the flash (*wabīṣ*) of musk in the hair part of the Messenger of God, Peace be Upon Him, when he was in a state of *iḥrām*."[101] Other versions similar to this merely refer to perfume (*al-ṭīb*) or oil used in his hair and beard.[102] ʿĀʾishah herself perfumed him,[103] and she is also said to have been perfumed with musk and ambergris in *iḥrām*.[104] Not only was the use of musk generally acceptable for Muslims, it was used by Muḥammad in a state of ritual purity; this has had an enormous sacralizing effect on musk, despite the fact that the use of aromatics during *iḥrām* came to be condemned by the four great schools of Sunni Islamic law.[105] The resumption of perfuming following *iḥrām* was one of the symbols of the reintegration of the individual into worldly society after the experience. It remained permissible, however, to scent the Kaʿbah with musk and other aromatics[106] and, as we have seen, musk is frequently associated with Islamic places of worship.

Perfuming with musk for the mosque evidently did not end with Muḥammad, at least in the Shiʿite tradition, for ʿAlī b. al-Ḥusayn, the 4th Imam, kept a phial of musk in his mosque "and when he entered it for prayer, he took some from it and anointed himself with it."[107] Jaʿfar al-Ṣādiq, the 6th Imam, said "One prayer while perfumed is better than seventy prayers without perfume."[108]

Other traditions also attribute the scent of musk to Muḥammad himself. That Muḥammad had appeared was known by the scent of his perfume.[109] The sweat on Muḥammad's face was like pearls and was more fragrant than pungent musk (*al-misk al-adhfar*).[110] The mother of Anas b. Mālik would collect it in a phial saying it was the most fragrant of perfumes.[111] A tradition reports: Anas said, "I never touched silks (*khazzah* and *ḥarīrah*) softer than the palm of

---

100 Nasāʾī 5.106, Muslim 2.203, and Ibn Qayyim al-Jawziyyah 437.

101 Muslim 2.202 and Ibn Abī al-Ḥadīd 19.343.

102 Muslim 2.202–3.

103 Nasāʾī 5.105 and Muslim 2.203.

104 Ibn Saʿd 8.486.

105 Cf. J. Schacht, *The Origins of Muhammadan Jurisprudence* (Oxford: Clarendon Press, 1950), 155, and R. Burton, *Personal Narrative of a Pilgrimage to al-Madinah and Meccah*, 2 vols. (London: Tylston and Edwards, 1893; repr. New York: Dover, 1964), 2.285.

106 E.g., Burton 300, quoting Burckhardt.

107 Al-Kulaynī, *al-Furuʿ min al-kāfī*, 8 vols. (Tehran: Dār al-Kutub al-Islāmiyyah, 1983), 6.515.

108 Kulaynī 6,510–11.

109 Ibn Saʿd 1.398–9.

110 Abū Ḥāmid Muḥammad b. Muḥammad al-Ghazālī, *Iḥyāʾ ʿulūm al-dīn*, 5 vols. (Beirut: Dār al-Kutub al-ʿIlmiyyah, 2005), 2.513.

111 Muslim 4.43, Zamakhsharī, *Rabīʿ*, 2.214, and al-Ghuzūlī, *Maṭāliʿ al-budūr fī manāzil al-surūr*, 2 vols. bound as one (Cairo: Maktabat al-Thaqāfah al-Dīniyyah, n.d. [2000]), 1.75.

the Messenger of God, Peace be upon him, and I never smelled musk (*miskah*) or *ʿabīrah* sweeter in scent (*aṭyab rāʾiḥatan*) than the scent of the Messenger of God, Peace be upon Him."[112] Versions quoted in the *Musnad* of Ibn Ḥanbal have different wording and have *ʿanbarah* instead of *ʿabīrah*, which was an easy change; the original was probably *ʿanbarah*. Niẓāmī, in the *Haft Paykar*, says that the breath of the Prophet diffuses into the air like musk.[113]

Though piety-minded Muslims may have worried about the luxury associated with expensive perfumes, their use most certainly continued. A Shiʿite ḥadīth tells that the people were questioning the use of expensive perfumes by ʿAlī al-Riḍā, the 8th Imam.[114] He justified his use of musk by analogy with Joseph, who was said to have worn gold brocade and sat upon golden thrones "and that did not decrease his wisdom at all." ʿAlī al-Riḍā had musk prepared with ben oil for his use, and he ordered four thousand dirhams of *ghāliyah*.

In a well-known ḥadīth concerning fasting, it is said that the foul smell of the mouth (*khalūf*)[115] of a fasting person is more pleasing to Allāh than the scent of musk.[116] Again, musk is portrayed as the best of perfumes by its contrast to something unpleasant. Devotion has its own good scent which has the potential to outstrip even musk.[117]

The Kaʿbah itself was perfumed; this practice goes back before Islam.[118] Aromatics were also kept in the Kaʿbah.[119] In Islamic practice it was censed each day, and twice the amount of incense was employed on Fridays.[120] The Kaʿbah is the center of the world for Muslims. Picking up on the ancient imagery

---

112 Muslim 4.43 and Abī ʿAbdallāh al-Bukhārī, *Ṣaḥīḥ Abī ʿAbdallāh al-Bukhārī bi-sharḥ al-Kirmānī*, 25 vols. (Cairo: Al-Maṭbaʿah al-Bahiyyah al-Miṣriyyah, 1933–62), 3.109–10. Cf. Ibn Ḥanbal 3.259, 267, and 280 with different wording and *ʿabīr* replaced by *ʿanbar*, and also Ibn Saʿd 1.413.

113 Niẓāmī, *Haft Paykar*, 2.26 on 7.

114 Kulaynī 6.516–7.

115 E. Lane, *An Arabic-English Lexicon*, 8 vols. (London, 1863–93; repr. Beirut: Librairie du Liban, 1968), 793 used for the effect of souring milk, a foul mouth, etc. He reads *khulūf*.

116 E.g., Muslim 2.171 and Bukhārī 7.530–1, 3.71, 9.434, and 9.473.

117 Compare the story of the young Rabbi Eliezer given by Green 1–2, who was so poor that he had to eat dirt which gave him bad breath. His teacher chided the other students who complained of his breath, telling Rabbi Eliezer that "the fragrance of your learning" would be renowned. Foul odors are also a mark of piety in the ascetic traditions of Christianity, see Harvey 213ff.

118 R. B. Serjeant, *South Arabian Hunt* (London: Luzac, 1976), 62. Ma Huan also noted this, cf. *Yingya shenglan* 88.

119 Ṭabarī II.537 notes that Ibn al-Zubayr removed aromatics (*ṭīb*) from the Kaʿbah when renovating it.

120 Serjeant 62.

of the center of the world as a navel,[121] the Persian poet Khāqānī, who made the pilgrimage in the 12th century, compares it to a musk pod:

> *īn kaʿbah nāf-i ʿālam wa az ṭīb sāḥatash*
> *āfāq-i waṣf nāfah-yi mushk-i tatār kard*[122]

> This Kaʿbah is the navel of the world, and its courtyard with its perfume has made the horizons of [my] descriptions a pod of Tatar musk.

In other words, the Kaʿbah, as a symbolic musk pod, has given its scent to the poet's verses, making them as fragrant as a pod of Tatar musk. By his use of the word *āfāq*, "horizons", Khāqānī reminds the listener of the Kaʿbah being the navel of the world, the symbolic center of it, and his verse, now like a musk pod, is a mirror of this.

The Dome of the Rock, Islam's answer to Solomon's temple,[123] was also censed and the rock itself anointed. A formula for the incense *nadd* that was burned in the Kaʿbah and at the Dome of the Rock in Jerusalem has been preserved by al-Nuwayrī. It was transmitted by al-Tamīmī and is credited to the mother of the Abbasid Caliph al-Muqtadir; she had it burned at these places every Friday. The formula runs:

> One takes a hundred *mithqāls* of Tibetan musk that has been cleansed of membranes (*akrāsh*). It is crushed and sieved. Shiḥrī ambergris is melted for it, and it is removed from the fire. When it is tepid, the musk is cast upon it alone without any aloeswood or anything else. It is beaten well, then it is spread out on a marble slab and cut into strips (*shawābīr*) and then one may cense with it.[124]

This is an unusual *nadd* formula because it includes no aloeswood or anything other than musk and ambergris. Al-Tamīmī wrote of it: "The head of the servants in Jerusalem gave some of this *nadd* to my father, and he melted it with ben oil. No *ghāliyah* is more pleasant than it." It is not surprising that this

---

121 Cf. the Omphalos or navel of the world at Delphi in ancient Greece.

122 *Farhangnāmah* 258a. Newid 82 cites a verse of Jāmī with a similar image, and see also Newid 89.

123 M. Sharon, "The 'Praises of Jerusalem' as a Source for the Early History of Islam," *Bibliotheca Orientalis* 49 1/2 (1992): 61.

124 Nuwayrī 12.64.

formula, prepared for burning at the most sacred of holy places on Fridays, consists of the two aromatics most associated with kingship.

The *Faḍāʾil bayt al-muqaddas* of al-Wāsiṭī, which dates to the early 11th century, contains material on the anointing and censing rituals performed at the Dome of the Rock in its early years.[125] On every Monday and Thursday saffron was crushed and ground and prepared overnight with musk, ambergris, and Jūrī rosewater; the compound was left to steep overnight. On Monday and Thursday mornings the custodians of the shrine took this mixture, which the text calls *khalūq*, and anointed the rock with it as far as they could reach, and then, having washed their feet, climbed upon it and anointed the rest. Then censers of gold and silver were brought, along with *nadd* and Qimārī aloeswood[126] freshened (*muṭarrā*) with musk and ambergris. The curtains hung between the pillars were opened, and they censed the stone, circumambulating it until the incense came between the pillars and the dome, and then the curtains were opened and the incense diffused because of its abundance until it reached the head of the marketplace. At this point, the opening of the Dome of the Rock for worship would be announced. Those who worshipped there were known to others by the scent of the incense upon them.[127] Incense and unguent containing musk and ambergris are here associated with one of the holiest places in Islam and are the sign of its availability for prayer. The scent of the incense itself becomes the sign to others that one has been there.

The use of musk was projected back in time to Solomon's temple itself. When building the Temple, Solomon is said to have sent out groups searching for the items needed in its construction, including many precious things. Musk and ambergris and "other kinds of aromatics" were required.[128] Devils (*shayāṭīn*) were commanded to bring them.[129] Of course these substances are not included in the formulas for incense specified in the Torah; they reflect a later time.

Likewise, al-Ṭabarī notes that the pre-Islamic Christian Ethiopian ruler Abrahah's temple designed to draw people away from Mecca was made to surpass the Kaʿbah; it was made from different colors of marble, adorned with

125 Abū Bakr Muḥammad b. Aḥmad al-Wāsiṭī, *Faḍāʾil bayt al-muqaddas* (Beirut: Dār al-Maʿārif, 2001), 76. Cf. also Sharon on this text.

126 The text has *wa lā ʿūd al-qimārī*, which seems to be an error for *wa-l-ʿūd al-qimārī*.

127 See also Sharon 60 for these rituals.

128 Damīrī 1.401.

129 Thaʿlabī 197; Brinner 516.

gold, silver and jewels. In it aloeswood (*mandal*) was burnt and the walls were smeared with musk so deep and dark that the jewels were hidden.[130]

The Mosque of the Prophet in Medina, according to the late 12th century traveler Ibn Jubayr, had walls divided into three parts. The bottom third was covered in fine marble, while the top third was an aloeswood lattice up to the ceiling. But the middle portion was smeared with musk and perfume (*ṭīb*) to a depth of a half span (*shibr*) that was blackened and cracked, having accumulated over a long time.[131] Ibn Baṭṭūṭah records that the shrine with the tombs of Adam, Noah, and ʿAlī in Najaf had gold and silver dishes of rosewater, musk, and different perfumes: "[T]he visitor dips his hand in this and anoints his face with it for a blessing."[132]

The final use of musk was funereal. The use of aromatics for embalming corpses is, of course, much older than Islam. Many different aromatics were used, but in the Islamic tradition camphor was especially favored along with the choice aromatics including musk. The use of musk in embalming probably goes back to pre-Islamic times. The pre-Islamic poet ʿAbīd is perhaps one of the earliest who suggests that musk was used during the washing of corpses for burial.[133] According to al-Ṭabarī, after the last Sasanian king Yazdagird was killed near Marw, the bishop (*usquf*) of Marw wrapped his body in a *ṭaylasān* (a kind of hood worn especially by the learned)[134] perfumed with musk (*mumassak*).[135]

In the *Shāhnāmah*, Alexander gives directions for the preparation of his body for burial, wrapped in a shroud of Chinese gold brocade scented with ambergris, covered in honey, and the coffin sealed with tar, camphor, musk, and the compound perfume *ʿabīr*.[136] The use of ambergris, camphor, musk, and *ʿabīr* here is, of course, anachronistic, but gives an idea of the sort of aromatics deemed suitable for burial in Islamic times. Before burial, Muslim law provided

---

130 Ṭabarī 1.943.

131 Muḥammad b. Aḥmad b. Jubayr, *Riḥlat Ibn Jubayr* (Beirut: Dār Ṣādir, 1959), 169.

132 Ibn Baṭṭūtah, *Riḥlat Ibn Baṭṭūṭah* (Beirut: Dār Ṣādir, 1960), 177; trans. H. A. R. Gibb, *The Travels of Ibn Baṭṭūṭa A.D. 1325–1354*, vol. 1 (London: Hakluyt Society, 1958), 257 is quoted.

133 C. Lyall, *The Dīwāns of ʿAbīd ibn al-Abraṣ, of Asad, and ʿĀmir ibn aṭ-Ṭufail, of ʿĀmir ibn Ṣaʿṣaʿah* (Cambridge: Gibb Memorial Series, 1913; repr. 1980), 22 n. 2, and text 16–17.

134 R. Dozy, *Dictionnaire détaillé des noms des vêtements chez les Arabes* (Amsterdam : 1843; repr. Beirut: Libraire du Liban, n.d.), 278–80.

135 Ṭabarī 1.2881. Cf. al-Tha'ālibī, *Tārīkh Ghurar al-Siyar*, ed. and trans. H. Zotenberg (Paris: Imprimerie Nationale, 1900; repr. Tehran: Maktabat al-Asadī, 1963), 747–8.

136 Firdawsī 6.121 l. 1788. See Newid 100–3 for musk and aromatics in Iranian funerary practice, with many references to the *Shāhnāmah*.

for the washing of the corpse and its perfuming.[137] A visit to the perfumer was a necessary part of the funeral preparations.[138]

Different aromatics were used in the preparation of the body in Islamic funerary practice. The use of camphor and the ground leaves of the lote tree (*Celtis australis*) for embalming is mentioned in ḥadīths.[139] Camphor was particularly favored, but in ḥadīth collections the saying of Muḥammad that musk is the best perfume is often quoted in a section on musk for the dead, implying that it was indeed used.[140] Muḥammad is said to have been embalmed with some musk, and ʿAlī requested that he be embalmed with the remainder of this musk at his own death.[141] Other sources suggest that musk was often used in embalming during the early Islamic period. Salmān al-Fārsī asked that his house be sprinkled with it after his death, saying "The angels will be with me; they will neither eat nor drink, but they will notice the smell."[142] The head of the rebel Yaḥyā b. ʿUmar (d. 864) was stuffed with aloeswood, musk, and camphor for preservation and was publically displayed.[143] The Sūfī poet Rūmī says:

gar miyān-i mushk tan-rā jā shawad
rūz-i murdan gand-i ū paydā shawad
mushk-rā bar tan mazan bar dil bimāl
mushk chi buwad nām-i pāk dhū l-jalāl
ān munāfiq muskh bar tan mīnihad
rūḥ rā dar qaʿr-i gulkhan mīnihad[144]

If a body is placed amidst musk,
the day it dies its stench will appear,
Do not rub musk on the body; anoint the heart with it!
What is musk? The pure Name of the Glorious One.

---

137 On funerary practice in Islam in general, see L. Halevi, *Muhammad's Grave: Death Rites and the Making of Islamic Society* (New York: Columbia University Press, 2007), though he does not dwell on perfuming and embalming in particular.

138 Muḥammad b. Yaḥyā al-Ṣūlī, *Kitāb al-awrāq*, ed. J. Heyworth Dunne (repr. Beirut: Dār al-Massīrah, 1983), 2.183.

139 Halevi 53–4.

140 E.g., Abū Dāwud 3.200.

141 Al-Suyūṭī in Van Gelder 208.

142 Ibn Saʿd 7.25.

143 Ṭabarī III.1521.

144 Jalālu'ddīn Rūmī, *The Mathnawī*, ed. R. A. Nicholson. E. J. W. Gibb Memorial Series, New Series 4.1. (London: Luzac, 1925), 1.262 lines 266–8. Quoted also by S. Anṣārī, *Tārīkh-i ʿIṭr dar Īrān* (Tehran: Wizārat-i Farhang wa Irshād-i Islāmī, 1381/2002–3), 223.

The hypocrite puts musk upon his body,
but puts his soul in the depth of the garbage pail.

Rūmī suggests that even a perfume as fine as musk is ultimately futile to prevent the inevitable decay of the body; only God can do that. God is the true musk that will make the soul fragrant, and a hypocrite may smell nice in this world, but his soul stinks.

It was symbolic of the character of the deceased when musk and other aromatics failed to check the stench of decay. This is an image employed by historians to make their judgments on the morality of individuals. Three hundred *mithqāl*s of musk and six hundred of camphor were applied to the corpse of the Turk Abū Naṣr Muḥammad b. Bughā, ally of the deposed caliph al-Muhtadī, but it did not stop the stench.[145]

Another motif found in historical literature has people perfuming themselves before battle, a reflection of their intention to fight to the death. For example, before his death in battle, the Shīʿite leader al-Mukhtār sent for his wife, and she sent much perfume (*ṭīb*); he washed and embalmed (*taḥannaṭ*) himself and put the perfume on his head and beard.[146] Before the battle of Karbalā, Ḥusayn b. ʿAlī called for musk that was then steeped in a deep dish or a bowl (*jafnah ʿaẓīmah aw ṣaḥfah*). He then went into his tent and was anointed with lime (*taṭallā bi-l-nūrati*); this was used as a depilatory.[147] Presumably after the depilation he was scented with the musk.

These uses of musk show that it had achieved a role as a purifying substance. It could be worn for prayer; it was applied to the dead. Thus it is not surprising that in ḥadīth literature musk is prescribed for the purification of women after menstruation. Menstrual blood rendered women ritually impure and subject to many restrictions; the rites associated with the end of menstruation symbolize the reincorporation of women into society. There are different versions of the ḥadīth; usually a woman inquires about how to wash herself after her menses, and Muḥammad responds to purify or wash herself with a musky cloth (*firṣah mumassakah*).[148] In one version, after being told to purify herself with musk, she asked, "how do I purify myself with it?" He answered, "Purify yourself with it, *Subḥān Allāh*." ʿĀʾishah explained this as meaning that she should run it over the traces of blood.[149] This parallels the purification of the unclean

145 Ṭabarī III.1819.
146 Ṭabarī II.736.
147 Ṭabarī II.327.
148 E.g., Muslim 1.216–7.
149 Muslim 1.216.

blood of the musk deer by its conversion into musk. Similar to menstrual blood was unavenged blood; the obligation of vengence rested upon men until they cleansed it symbolically with further bloodshed. The blood shed in vengence could also be likened to perfume.[150] Perfuming, especially with musk, thus symbolized a return to normalcy. Since musk was believed to be the blood of the musk gazelle transformed into musk, this image is especially potent.

Musk had a deep association with the sacred. An interesting example of the symbolic importance of musk in a religious context is an anecdote preserved by al-Zamakhsharī: "A man found a piece of paper upon which was the Name of God, and so he picked it up. In it was a *dīnār*, so he bought musk with it and perfumed himself. In his sleep he saw a vision of a speaker saying to him, "As you have made My Name fragrant (*ṭayyabta*), so I will surely make your mention (*dhikr*) fragrant."[151] In Islam, Angels were said to be attracted to perfume while devils were repelled by it.[152] At the end of the world Allāh will send a wind that has a scent like musk and feels like silk. This wind will kill everyone with any faith to spare them from the terror of the hour.[153] Another example of the goodness of musk is that in dreams, according to al-Damīrī, it indicates the innocence of the accused.[154] Similarly, one of the miracles associated with the mystic al-Ḥallāj (857–922) was that he produced musk from within his sleeve repeatedly for the daughter of al-Samarrī, who was one of his devotees. He gave the musk to her to put with her perfume for her marriage.[155] The production of musk was thus construed as a sign of his divine favor. These examples illustrate the role of musk as a substance associated with the purity of the immortal world.

A humorous anecdote illustrating the conception of good scent in heavenly encounters is found in the story of the ascetic Abū ʿAbdallāh al-Mazabilī recounted by al-Tanūkhī.[156] This ascetic, who earned his name al-Mazābilī (the man of the trash-piles) by picking through the garbage of Antioch for a living, had a large following because of his piety. He was also considered a little unintelligent. One of his followers complained about his neighbor, a man named

150 S. Stetkevych, *The Mute Immortals Speak* (Ithaca: Cornell University Press, 1993), 225 and 227, with a poem by Muhalhil b. Rabīʿah in in which blood shed in vengence is compared to *ʿabīr*. Cf. also 174–5.

151 Zamakhsharī, *Rabīʿ*, 2.220 and Ibn Abī al-Ḥadīd 19.347.

152 Ibn Qayyim al-Jawziyyah 331.

153 Muslim 3.272.

154 Damīrī 2.129.

155 Ibn Miskawayh, *Kitāb Tajārib al-umam*, 3 vols. (Baghdad: al-Muthanna, n.d.), 1.78.

156 Al-Tanūkhī, *Nishwār al-Muḥāḍarah wa-akhbār al-mudhākarah*, ed. ʿAbbūd al-Shāljī, 9 vols. ([Beirut: Dār Ṣādir], 1971), 2.349–50.

Mūsā b. al-Zukūrī, and al-Mazābilī cursed him. This led all his followers to attack al-Zukūrī who then fled the town. Al-Zukūrī decided to get even by trickery. He wore new, white clothing and burned *nadd* and musk above the cave of al-Mazābilī late one night, and then he spoke in a loud voice claiming to be Gabriel, ordering that the ascetic stop the people of Antioch from attacking al-Zukūrī. Al-Mazābilī was so overcome by it all that he burst into tears and did as the false Gabriel asked. This story clearly shows the great importance of good scent in supposed divine encounters.

The sacralizing of musk kept it in demand through the centuries. Many of the luxuries associated with the life of the Persianate courts, such as silk brocades and wine, had been prohibited, but aromatics were allowed. This also created an interesting situation in Islamic law: the jurisprudents had to justify the use of musk, which was legally questionable, since it was thought to be blood removed from a dead animal.

Musk and the Garden

The importance of musk in Islam goes far beyond its physical use. Musk, especially among perfumes, was a link to the spiritual world. As an aromatic, it was believed to have come originally from Eden itself as a sign of the original Garden, and its presence in this world was a continuing reminder of God's grace.

Scent played a major role as a sign of sanctity and sacrifice and a promise of future felicity in ancient religions ranging from the polytheistic traditions to Zoroastrianism, Judaism, and Christianity.[157] Good scent also had an important symbolic role in Manichaeism.[158] Scent imagery in early Christianity appears to cluster in two broad categories: an aroma of relation, associated with sacrifice and worship, and an attribute of divinities and a sign of divine influence and favor.[159] In Islam, the latter is far more important, although the accounts of the blood of martyrs smelling like musk mentioned below seem to fit in the former category. Good scent in Islam was not attributed to God directly, but it was of his original, paradisical creation and was an attribute of the primordial state of blessedness that will be attained in the Garden of paradise in

157 See D. Green, Harvey.

158 A. van Tongerloo, "An Odour of Sanctity," in *Apocryphon Severini presented to Soren Giversen* (Aarhus: Aarhus University Press, 1993), 245–56.

159 Following Harvey Chapter 1.

the afterlife. The presence of musk in particular in this world functions as a symbol of the Garden and the possibility of its attainment.

Islam's doctrine of the future life of the pious is, as is well known, centered on the concept of the Garden or *al-Jannah*.[160] This Garden is sometimes called the Gardens of Eden (*jannāt ʿadn*) or Everlasting Gardens.[161] The Qurʾān contains numerous, lengthy accounts of its amenities. It is of a vast size, containing rivers and fountains, with pavilions, precious stones, shady trees and fruits, and companions. As noted in Chapter 6, there is a single reference to musk in the Qurʾān in a description of the heavenly Garden: *yusqawna min raḥīqin makhtūmin khitāmuhu miskun* (83:25–26) "They [inhabitants of the Garden] will be given to drink sealed fine wine (*raḥīq*), its seal of musk." Al-Ghazālī (d. 1111) quotes a different explanation of the phrase *khitāmuhu miskun* than al-Ṭabarī's more practical explanations discussed in the previous chapter: "if a man from the people of the world were to put his hand in it and then take it out, there would not remain a single soul that did not notice the scent of the perfume."[162] This idea, found also in later commentators, emphasizes the completely otherworldly character of the aromas of paradise; they can only be compared with musk, the best aromatic in this world, but they still excel it by far.

Smelling the scent of the Garden became a metaphor for salvation;[163] by contrast, a punishment mentioned in the ḥadīth for the killer of someone under protection was that they would not get to smell the Garden.[164] Angels sent to extract the soul from the dying believer come bearing silk in which are musk and sprigs of aromatic herbs (*rayḥān*); they extract the soul from the body like a hair from dough and place it among the musk and herbs and wrap it over with the silk.[165] According to the *Qiṣaṣ Qurʾān Mujīd*, a popular Persian work, when Muḥammad died, the scent of musk and ambergris became evident: "He saw the Messenger, Peace be upon him, in the house hidden with the veil of death being drawn over his face, and the whole house became filled with light, and the scent of the musk and ambergris of paradise became evident."[166]

---

160 A. Afsaruddin, "Garden," *EQ* s.v., and J. I. Smith and Y. Haddad, *The Islamic Understanding of Death and Resurrection* (Albany: State University of New York Press, 1981), 87–97, etc.

161 E.g., Qurʾān 9:72, etc.

162 Ghazālī 4.719; al-Qurṭubī, *al-Jāmiʿ li-aḥkām al-Qurʾān* (Cairo: Dār al-Kātib al-ʿArabī li-l-Ṭibāʿah wa-l-Nashr, 1967), 19.266; Ibn Kathīr, *Tafsīr al-Qurʾān al-ʿaẓīm* (Beirut: Dār al-Andalus, [1966]), 7.243.

163 E.g., Tanūkhī, *Nishwār*, 1.222.

164 Ibn Ḥanbal 5.52.

165 Ghazālī 4.664., and cf. 4.663 where they come bearing the embalming perfumes (*ḥanūṭ*).

166 Quoted by the *Farhangnāmah-i Adab-i Fārsī* 249–50.

The poet Ibn Zaydūn describes the death of the daughter of al-Muʿtaḍid, ruler of Seville:

> *thumma wallat fa-wajadnā*
> *araja l-miski thanāʾ*[167]

> Then she passed away, and we found the scent of musk, her eulogy.

The presence of the scent of musk following death reassured those present of her salvation; it also represented the divine approval at her life and death. The poet Abū ʿAlā al-Maʿarrī wrote in a poem mourning his mother and describing her heavenly ascent:

> *dhakiyyan yuṣḥabu l-kāfūru minhu*
> *bi-mithli l-miski mafḍūḍa l-khitām*[168]

> Potent camphor accompanies [her soul] with the like of musk when the seal is broken.

Like the Islamic garden, which became a mirror of the heavenly Garden, the use of aromatics associated with the Garden reinforced the link between this world and the next for the believer. We may compare the odor of sanctity associated with holy people and beings in other traditions from throughout Eurasia, especially Christianity.[169] The odor of sanctity is also found in the Manichaean tradition; Manichaean manuscripts in Uyghur from Inner Asia refer to Manichaean deities as musk-scented (*yıparlıγ*).[170]

The association of good fragrance with paradise has a long history. The importance of this concept in Zoroastrian literature was noted in Chapter 3. Likewise, the Manichaeans believed in the notion of the divine as a paradisiacal garden.[171] In early Christian literature the idea of paradise as a garden is

167 Ibn Zaydūn, *Dīwān* 135.

168 Maʿarrī, *Saqṭ*, 4.1421.

169 Harvey 203 and Van Tongerloo 252.

170 A. von Le Coq *Türkische Manichaica aus Chotscho II* (Berlin: Verlag der Akademie der Wissenschaften, 1919), 9 and Le Coq, *Türkische Manichaica aus Chotscho III* (Berlin: Akademie der Wissenschaften, 1922), 25, and Van Tongerloo 247, 253.

171 W. Oerter, "Das Motiv vom Garten. Betrachtungen zur manichäischen Eschatologie," in A. Von Tongerloo and S. Giversen, eds., *Manichaica Selecta: Studies presented to Professor Julien Ries on the occasion of his seventieth birthday* (Louvain: International Association of Manichaean Studies, 1991), 263–72 .

abundantly attested. Perhaps the most famous of the Syriac Church Fathers was Ephrem (c. 306–73 CE). Among his numerous writings are a cycle of hymns on paradise; Ephrem describes paradise as a vast garden filled with water.[172] He frequently refers to the scent (*rīḥ*) of paradise, but Ephrem never states that it is a specific type of perfume, a major difference from the Islamic accounts, as we will see. The springs of paradise are said to "cause delight with their aromatics."[173] In one case the springs are said to be springs of perfume.[174] Ephrem says that the scent of paradise alone is nourishment: "the scent of paradise nourishes instead of bread."[175] He strives to attain the treasury of perfumes and the storehouse of sweet scents (*riḥā*) and suggests that scent is satisfying to hunger: "My hunger delights in the breath of its aromatics".[176]

Thus it is evident that the Islamic conception of paradise as a luxurious garden had immediate antecedents in the culture of the Near East. In Islamic literature the image is greatly expanded and elaborated by numerous references to specific aromatics in place of the more generic good scents and incense of Syriac literature. These specific aromatics are, not surprisingly, the most important ones featured in the Islamic technical literature on perfumery and the most prized in daily life, of which musk was the head. Description of the future Garden for the righteous is a key theme within the Qur'ān; the Qur'ānic text is replete with imagery of the luxurious life within the Garden.[177] Of this material, references to musk and camphor appear once each as noted earlier; aloeswood and ambergris are unmentioned. However, the Islamic tradition greatly elaborates upon the Qur'ānic descriptions of the Garden. There is a large body of material among the ḥadīths that discuss the future Garden, and

172 Ephrem, *Hymnen de paradiso und contra julianum*, ed. E. Beck, 2 vols. (Louvain, 1957). For a perspective on the controversy of these hymns as the source for the Islamic depictions of paradise, see Walid A. Saleh, "The Etymological Falacy and Qur'anic Studies: Muhammad, Paradise, and Late Antiquity," in *The Qur'ān in Context: Historical and Literary Investigations into the Qur'ānic Milieu*, ed. A Neuwirth, et al. (Leiden: Brill, 2010), 649–97, 684, who emphasizes the original aspects of the Islamic descriptions.

173 Ephrem II,8: *nāba'in bahrumaw mbasemin* "flowing springs causing delight (*mbasem*) with their aromatics".

174 Ephrem XI,12: *mabu'a d-riḥāne* "spring of perfume".

175 Ephrem IX,17: *riḥeh d-pardaysā matrse ḥlāf laḥmā*.

176 Ephrem XI,15: *kad bāsmā kapnuty b-sawfā dahrumaw*.

177 N. Rustomji, *The Garden and the Fire: Heaven and Hell in Islamic Culture* (New York: Columbia University Press, 2009), 70–1 specifically on aromatics; see also in general Afsaruddin, Saleh, and S. Wild, "Lost in Philology? The Virgins of Paradise and the Luxenberg Hypothesis," in A. Neuwirth, et al., eds., *The Qur'ān in Context: Historical and Literary Investigations into the Qur'ānic Milieu*, 625–47 (Leiden: Brill, 2010), 625–47.

one of its most striking features is the profusion of aromatics associated with it. These aromatics are mostly identical with the principal aromatics known to early medieval Islamicate science and culture, and therefore much musk, ambergris, camphor and saffron are found.

Aromatics, so long associated with purity and immortality, the opposite of corruption and death, make up the physical substance of the Garden. This musk was no longer produced by gazelles as on earth, but was a spontaneous attribute of the Garden.[178] The ground of the Garden was made of musk.[179] "The Messenger of God, Peace be upon Him, said to Ibn Ṣā'id, 'What is the ground (*tarbah*) of the garden?' He replied, 'Fine-grained white flour (*darmakah*[180] *bayḍā'*) and musk'. Muḥammad said, "You have spoken the truth."[181] Another version has Ibn Ṣā'id ask Muḥammad, who answers fine grained white flour and pure (*khāliṣ*) musk.[182] Other ḥadīths also include the idea that the earth of paradise is like musk.[183] In later works other substances are included, e.g., "The dust of [the Garden's] earth is musk, camphor and ambergris, and its rivers are milk, honey and wine."[184] The pebbles (*raḍrāḍ*) of the garden were also said to be like musk.[185] Another common image is that the sand-dunes of the garden are musk.[186]

The rivers of the Garden were said to spring forth from beneath mountains of musk.[187] The mud of the great river Kawthar was like musk. While Muḥammad was travelling in paradise, he came upon a river that was bordered in *yāqūt*, ruby, which was "hollowed" (*mujayyab* or *mujawwaf*, the transmitter is uncertain of the word). The angel who was with him struck his hand upon it and drew out musk. Muhammad then asked about it and was told that it was

---

178 Cf. Ma'arrī, *Risālat al-ghufrān*, 379, where Ibn al-Qāriḥ enjoys "musk which was not collected from the blood of gazelles" (*misk mā junā min dimā' al-fūri*).

179 Ghazālī 4.713 and Ma'arrī, *Risālat* 222–3.

180 *Darmak* (also spelled *darmaq*) refers to fine white flour: S. Fraenkel, *Die aramäischen Fremdwörter im Arabischen* (Leiden, 1886; repr. Hildesheim: Olms, 1962), 32–3 and note on 288.

181 Muslim 4.360 and Ibn Ḥanbal 1.4, etc.

182 Muslim 4.360. Cf. Ghazālī 4.716.

183 Ibn Ḥanbal 5.143–4. Cf. Ghazālī 4.713.

184 *Kitāb Aḥwāl al-Qiyāmah*, ed. M. Wolff in *Muhammedanische Eschatologie nach der Leipziger und der Dresdner Handschrift zum ersten Male arabisch und deutsch mit Anmerkungen herausgegeben* (Leipzig: Brockhaus, 1872), 109.

185 Ibn Ḥanbal 1.399.

186 Ibn Qayyim al-Jawziyyah 437; Ma'arrī, *Risālat al-ghufrān* 372 mentions dunes of ambergris and musk. Ghazālī 4.713 says that there are dunes of camphor.

187 Ghazālī 4.716.

al-Kawthar, one of the rivers of paradise.[188] Another version of this ḥadīth has the river surrounded by domes (*qibāb*) of pearl (*lu'lu'*) instead of the mysterious hollowed *yāqūt*, and the angel struck his hand into the clay (*ṭīnah*) to draw out the musk.[189] Elsewhere, in a ḥadīth where Muḥammad is asked about Kawthar, he answered that it is a river in paradise with soil of musk, water whiter than milk and sweeter than honey.[190] The water of Kawthar was more fragrant than musk.[191] Kawthar emptied into the Pool of Muḥammad: its pebbles were pearls while the flatland (*baṭḥā'*) surrounding it was musk.[192]

Just like the image of musky wine is not confined to religious literature, the pleasant smell of earth is compared to musk in poetry.[193] Dhū al-Rummah says:

taḥuffu bi-turbi l-rawḍi min kulli jānibin
nasīmun ka-fa'ri l-miski ḥīna tufattiḥu[194]

Surrounding the earth of the meadow from every direction was a breeze like a musk pod when it is opened.

This image is also widespread in Persian poetry; Farrukhī describes springtime:

khāk rā chūn nāf-i āhū mushk zāyad bīqiyās
bīd rā chūn parr-i ṭūṭī barg ruyad bīshumār[195]

The soil, like the navel of a deer, brings forth infinite musk
The willow, like the wing of a parrot, grows numberless leaves

The poet Ibn Zaydūn emphasizes the excellence of the beloved by suggesting she is created from musk rather than the clay from which the rest of humanity

188 Abū Dāwūd 4.237–8 (#4748) and Ghazālī 4.703. Tirmidhī 5.119 quotes a version of this ḥadīth in his commentary on Qur'ān 108 (*al-Kawthar*), which begins *innā 'aṭaynāka l-kawthara*, "We have given you al-Kawthar," which means "abundance." It is considered the name of one of the four rivers of paradise.

189 Tirmidhī 5.119. For a similar story, cf. Bukhārī 9.451.

190 Ibn Ḥanbal 1.5, 3.236.

191 Ibn Ḥanbal 2.112, 132.

192 Ghazālī 4.625.

193 Cf. a verse of Ibn Abī 'Uyaynah quoted by Yāqūt, *Mu'jam al-Buldān*, 5 vols. (Beirut: Dār Ṣādir, n.d.), 4.356 and al-Sarī al-Raffā' 3.157.

194 Dhū al-Rummah, *The Dīwān of Ghailān ibn 'Uqbah known as Dhu 'r-Rummah*, ed. C. H. H. Macartney (Cambridge: Cambridge University Press, 1919), #10 line 24, on pg. 83.

195 Quoted in Niẓāmī 'Arūḍī, *Kitāb-i Chahār maqālah*, ed. Muḥammad Qazwīnī (London: Luzac, 1927), 43.

was made;[196] this shows again the association of musk with the soil. Musk was a pure, perfect sort of earth.

The water in the Garden was also compared with musk: there was a (square) pool (*ḥawḍ*) in paradise which is a month's journey around, equal sided, its water whiter than silver (*warq*) and its fragrance more fragrant than musk. When one drinks from it, one is never thirsty again.[197] This recalls Ephrem's springs of fragrance.

All the attributes of the Garden symbolize its immortal nature. Wine, which is corrupting on earth, becomes a symbol of the immortality of the Garden, like the fine silks that were also prohibited on earth but commonly found in the future paradise.[198] Aromatics had long had these associations, yet they were never prohibited on earth. They were ubiquitous in the Garden, as we have seen, and these aromatic qualities also extended to the people of the garden. Martyrs, who enter directly into paradise, were said to have blood the color of blood, but with a scent like musk.[199] This is not a strange conception since it was believed that musk was formed from the blood of the musk gazelle, as discussed in Chapter 4. The musk scent is an attribute of immortality, showing that the martyrs have transcended their corporeal form. The poet Abū Tammām says in a lament on Muḥammad b. al-Faḍl al-Ḥimyarī that the inhabitants of the Garden will have:

khuluqun ka-l-mudāmi aw ka-ruḍābi -l-mis-
ki aw ka-l-ʿabīri aw ka-l-malābi[200]

A nature like fine wine or like the saliva of musk, or like *ʿabīr* or *malāb*.[201]

---

196 Ibn Zaydūn, *Dīwān* 11; S. Lug, "Towards a Definition of Excellence in Classical Arabic Poetry: An Analysis of Ibn Zaydūn's *Nūniyya*," *JAOS* 101:3 (1981): 339.

197 Muslim 4.27.

198 J. D. McAuliffe, "The Wines of Earth and Paradise: Qur'ānic Proscriptions and Promises," in *Logos Islamikos: Studia Islamica in Honorem Georgii Michaelis Wickens* (Toronto: Pontifical Institute of Medieval Studies, 1984), 159–74, and S. Stetkevych, "Intoxication and Immortality: Wine and Associated Imagery in al-Maʿarrī's Garden," in J. W. Wright and E. K. Rowson, eds., *Homoeroticism in Classical Arabic Literature* (New York: Columbia University Press, 1997), 210–32.

199 Blood of martyrs smells of musk: Ibn Ḥanbal 4.185; Blood of martyrs is the color of blood but smells like musk: Ibn Hishām 2.586; Muslim 3.251; Ibn Ḥanbal 2.231,242. Cf. Ibn Ḥanbal 3.299.

200 Abū Tammām 4.45.

201 A compound perfume which typically included saffron; see WKAS 2:3.1652–3 for references.

The purity of the Garden is further indicated by the absence of the corrupting bodily processes of this world. The inhabitants of the Garden do not defecate or urinate but have sweat (*rashḥ*) of musk.[202] "Their bodily needs will be satisfied by a sweat (*ʿaraq*) that flows from their skins like musk, then their belly will become lean."[203] This concept echoes the notion of spiritual perfection in Syriac literature manifesting itself in a sweet smell from the clothing and the whole body.[204] A Shi'ite tradition states that one of the signs of the Imam is that his excrement (*najw*) has a scent like musk; this illustrates that the special status ascribed to the Imam is a reflection of the future perfection of the Garden.[205] We have already noted that Muḥammad's sweat was said to have a musky scent. The transformation of human excretions into musk shows the power of God to perfect the physical processes of the corporeal world in the future paradise. What indicates mortality in the material world becomes a process signifying immortality in the Garden.

Other literature includes ḥadīths not found in the standard collections. Some interesting material is contained in three closely related eschatological handbooks, the anonymous *Kitāb Aḥwāl al-qiyāmah*,[206] the *Kitāb Daqāʾiq al-akhbār*,[207] attributed to a certain 11th century writer ʿAbd al-Raḥīm b. Aḥmad al-Qāḍī, and the *Kitāb Ḥaqāʾiq wa-l-daqāʾiq* of Abū Layth al-Samarqandī (d. 983).[208] These three texts are so closely similar that they are surely recensions of the same original work, containing material which must have been in circulation by at least the middle of the 10th century, but most probably earlier. Describing the buildings of the Garden of Eden/ʿAdn in the future paradise, it is said that they are of "silver adobe (*libna*) and adobe of gold, and their mortar is musk and the dust is ambergris and saffron."[209] Likewise, al-Ghazālī quotes a ḥadīth transmitted by Abū Hurayrah: "The wall of the Garden has silver adobe and gold adobe, and its ground is saffron and its clay is musk."[210] There is also a "great mountain of pungent musk (*misk adhfar*),[211] and from

202 Muslim 4.314–5. Cf. Ghazālī 4.718 and 723.
203 Ghazālī 4.719.
204 Philoxenus of Mabbug, quoted by Seppälä 88.
205 Kulaynī 1.388.
206 Ed. Wolff.
207 Edition printed in Egypt, 1333/1914.
208 This text is edited and summarized in a series of articles published by John MacDonald in the journal *Islamic Studies*, starting with v. 3 in 1964. The section relevant here is in "Islamic Eschatology VI– Paradise," *Islamic Studies* 5 (1966): 331–83.
209 *Kitāb Aḥwāl al-qiyāmah* 106 .
210 Ghazālī 4.716.
211 Cf. Bukhārī 9.369, where *misk adhfar* is mentioned in the garden.

beneath it flows the river Salsabīl."[212] The rivers of the Garden are also said to flow out from beneath hills or mountains of musk.[213]

Musk is also a component of the bodies of the otherworldly *ḥūr*, the "houris" who await the pious in the paradise garden: "[God] created their bodies from saffron, musk, ambergris and camphor, and their hair from cloves (*qaranfīl*). From their toes to their knees is sweet saffron; from their knees to their breasts is musk, from their breasts to their necks is ambergris, and from their necks to the top of their heads is camphor. If they were to spit in this world, it would become musk."[214] The scent of one houri would fill this earth.[215] Moreover, the bodies of the believers are also transformed: "their bodies become fragrant like musk."[216] When the inhabitants of paradise eat, the food is transformed in their bodies, and it "becomes a scent like the scent of musk which emanates from their bodies,"[217] so there is no urination or defecation. No other substance is mentioned so consistently in the accounts of the future garden as musk, and this surely says something very important about the esteem in which it was held.

The association of good scent with the garden is intertwined with the story of the Garden of Eden. A ḥadīth quoted by al-Thaʿlabī records that when Adam was cast from the Garden into India, he wore a garment of leaves from the Garden. They dried and scattered throughout India, and because of those leaves aloeswood, sandalwood, musk, ambergris and camphor are fragrant.[218] Thus the best aromatics of this world are but traces of the aroma of the Garden. The belief that Adam brought perfumes into this world from the Garden is also noted by al-Masʿūdī, who was skeptical of it.[219] It is undoubtedly this legend that is behind the geographer Ibn Khuradādhbih's statement

---

212 *Kitāb Aḥwāl al-qiyāmah* 108.

213 Ghazālī 4.716.

214 *Kitāb Aḥwāl al-qiyāmah* 111.

215 Ghazālī 4.716.

216 *Kitāb Aḥwāl al-qiyāmah* 113. Cf. Bukhārī 4.292–3, from Abu Hurayrah.

217 *Kitāb Aḥwāl al-qiyāmah* 114.

218 Thaʿlabī 23; Brinner 61. Cf. Kulaynī 6.513 and Ṭābarī 1.124–6 for several traditions on this subject. The 10th century Byzantine *Life of St. Andrew the Fool* contains an account of his journeying through the heavens, and there, near the Holy of Holies, is a garden with trees sending forth fragrance more pleasant than myrrh and musk, L. Rydén, ed. and trans., *The Life of St Andrew the Fool I. Introduction, Testimonies and Nachleben* (Uppsala: Uppsala University, 1995), 52–3. The association of India with aromatics was so strong that the account of the musk deer (probably) by al-Yaʿqūbī that we discussed in Chapter 4 claims that they had to travel to India to graze on the plants which would produce the best musk.

219 Masʿūdī 1.37 (§48) and Maʿarrī, *Risālāt* text 8, trans. 8.

that one can find on the mountain upon which Adam descended in Sri Lanka, which is still called Adam's Peak,[220] aloeswood, pepper, and other aromatic plants as well as the animals that produced musk and civet (*dābbat al-misk wa-dābbat al-zabād*).[221] Another source states that the musk deer acquired its capacity for producing musk by eating some of the leaves from the Garden.[222] The association of aromatics with the Garden of Eden also can be traced in earlier literature.[223] The 1st century CE *Life of Adam and Eve* states that Adam was permitted to bring aromatics from Eden. These aromatics became the spices used in foods and the incenses used in divine worship.[224]

The association of musk with the paradise Garden and with royalty merges in the story of Iram Dhāt al-ʿImād (Iram of the Pillars), the city of the ancient people of ʿĀd. Shaddād, the king of ʿĀd, conquered the world, and then read the story of the Garden in ancient books. He undertook to recreate the celestial Garden on earth and had the city Iram Dhāt al-ʿImād built according to the descriptions he had read. The city was constructed of precious materials: gold, silver, and jewels collected from all over the world. Fruit trees and watercourses were constructed below its palaces. The foundation stones were made of Yemeni onyx set with mortar of mahlab and ben oil. But when the city was completed, and Shaddād was but one day away from it, he and his entourage were destroyed by God.[225]

Supposedly, during the reign of al-Muʿāwiyah (661–80), Iram Dhāt al-ʿImād was rediscovered.[226] A man named ʿAbdallāh b. Qilābah was seeking some of his camels that had gone missing and stumbled upon the city, which was intact. He entered and reached a gate of aloeswood studded with yellow sapphires and rubies; beyond this gate he found the heart of the city with its abundance of riches. The floors of the palaces were covered with pearls and

220 Marco Polo, *The Travels of Marco Polo*, trans. and annotated by H. Yule and H. Cordier, 3rd. ed.., 2 vols. (London, 1929; repr. New York: Dover, 1993), 316 and 328 n. 5; P. Pelliot *Notes on Marco Polo*, ed. L. Hambis. 3 vols. (Paris: Imprimerie Nationale, 1959–73), 1.13.

221 Ibn Khuradādhbih, *Kitāb al-masālik wa-l-mamālik*, ed. M. J. De Goeje (Leiden: Brill, 1889, repr. 1967), 64. Cf. al-Idrīsī, *Kitāb nuzhat al-mushtāq fī ikhtirāq al-āfāq*, 9 vols. (Naples: Istituto Universitario Orientale di Napoli, 1970–84), 1.73. Cf. Muqaddasī 13–4, who mentions that an herb that resembles musk grows at Adam's Peak. Balʿamī's translation of al-Ṭabarī's history states that Adam brought musk, along with other aromatics; see Newid 12, 62.

222 Ṭabarī 11.51.

223 Harvey 50–1.

224 Harvey 51–2.

225 Thaʿlabī 91–5.

226 Masʿūdī 2.405–6 (§1414); Thaʿlabī 91–5.

hazelnut-shaped pieces (*banādiq*) of musk and saffron.[227] ʿAbdallāh b. Qilābah believed that he had discovered the Garden. He took pearls and musk and saffron pieces and left the city. As he sold some of the pearls, he came to the attention of Muʿāwiyah, who sent for him. Muʿāwiyah found the pieces of musk to be lacking in scent, but when he had one crushed, the musk smell filled the air. Then he believed the story of ʿAbdallāh b. Qilābah and sought to learn the identity of the city. He said that no one was given anything like what Solomon had, but he doubted that even Solomon had a city like this. A learned man, one Kaʿb al-Aḥbār, explained to Muʿāwiyah that this site must be the lost city of Iram Dhāt al-ʿImād.

Musk is central to the recognition of the city of Iram Dhāt al-ʿImād in the story; Muʿāwiyah did not believe the camel-herder until he smelled the fragrance of the musk recovered from the city. Its excellence, combined with the description of the city, suggested to him that the city must be even more marvelous than any city Solomon, the archetypal great ruler, possessed.

The historian al-Ṭabarī tells another story about a miraculous discovery from the past.[228]

> News came of the opening of a tell by the river Sillah known as the tell of the Banū Shaqīq that there were seven graves and in them were seven perfect bodies. Upon them were new and soft shrouds whose fringes (*ahdāb*) diffused the scent of musk. One of them was a youth who had luxuriant hair and his forehead, ears, cheeks, nose, lips, chin and the edges of the lids of his eyes were perfectly preserved and upon his lips was moisture as if he had just drunk water and he was as if kohl had been applied to him. There was a wound on his waist but his shroud had been replaced over it. One of our associates told me that he tugged on the hair of some of them and he found it strongly rooted like the strong hair of the living. It was mentioned that from the tell excavated from above these graves was a stone of the color of a whetstone (*misann*) resembling a basin (*ḥawḍ*) which had writing on it which no one recognized.

This story is clearly a reflection of the story of the seven sleepers of the cave, that is, those pious individuals who fled the rule of Duqyānūs as mentioned above. The scent of musk could be due either to the perfuming of the bodies with musk or to their own sanctity. Either way, the mention of this detail is an indication of the holiness of the people buried in those graves.

---

227 Thaʿlabī 92.

228 Ṭabarī III.2116.

Persian Sūfī poetry takes the importance of musk to a greater level.[229] As noted above, in the heavenly Garden the bodily functions of defecation and urination are transformed into the production of musk. This has its origin, perhaps, in the ability of the physical musk deer to produce musk. As early as Farīd al-Dīn ʿAṭṭār (late 12th-early 13th century), the capacity of the musk deer to transform grass into musk became a symbol for the transformation of base substances into higher forms:

> *āhū-yi chīnī chu giyāhi khurad*
> *mushk shawad dar shikamash ān giyāh*[230]

> The Chinese deer, when it eats plants, those plants become musk in its stomach.

ʿAṭṭār writes in the *Ilāhīnāme*:

> The blood of the musk gazelle, after it has eaten special herbs for forty days, is transformed into aromatic musk by the breath of the morning breeze. In this way man who consists of earth can also become soul. If the light of God enters the soul, then the body takes on the color of the soul. If you wish to produce an elixir, let it be this elixir![231]

ʿAṭṭār makes the period during which the gazelle eats the special herbs forty days, a highly symbolic number, for it stands for the period required for purification.[232] For Rūmī, the power of God is such that it can transform the blood of the musk deer into musk.[233] The fact that musk, the noblest scent, was mere blood served as a reminder of the inferior life of this world,[234] but also pointed to God's redemptive capacity. Rūmī uses musk frequently in his works as a symbol. The navel of the musk deer containing musk can represent

229 See, in general, on aromatics and Sūfī mysticism Anṣārī 223–7.

230 Farīd al-Dīn ʿAṭṭār, *Dīwān-i qaṣāʾid tarjīʿāt wa ghazaliyāt*, ed. Saʿīd Nafīsī (Tehran: Sanāʾī, 1339/1960), #655 l. 8963 on 469. See A. Schimmel, *The Triumphal Sun: A Study of the Works of Jalāloddin Rumi* (London: East-West Publications, 1980), 357 and 468 n. 60.

231 Quoted and trans. H. Ritter, *The Ocean of the Soul: Man, the World, and God in the Stories of Farīd al-Dīn ʿAṭṭār*, trans. J. O'Kane (Leiden: Brill, 2003), 272; Newid 63–4.

232 A. Schimmel, *The Mystery of Numbers* (Oxford: Oxford University Press, 1993), 248.

233 Schimmel, *Triumphal Sun* 228.

234 Cf. Ghazālī 3.283.

the soul within the physical body, likened to the blood which is transformed into musk.[235] Rūmī says:

> *ṭabʿ-i nāf-i āhuwast ān qawm-rā*
> *az birūn khūn wa darūnishān mushkhā*
> *tu magū kīn māyah bīrūn khūn buwad*
> *khūn rawad dar nāf mushkī chūn shawad*[236]

> Those people have the nature of the navel of a deer (*āhū*); blood from outside but inside them, musk.
> Do not say that this origin outside is blood; when it enters the navel how does it become musky?

Rūmī also compares the adept's seeking of God to the hunter seeking the musk deer. The track of the animal is merely a trace which serves as guidance until he can follow the scent of the musk pod alone, the scent of which is a more reliable guide than the tracks.[237]

Mystical literature's employment of musk as a symbol plays off the base origins of musk, its transformation into a pure substance, and its ubiquitous presence in the Garden. Just as human beings, created originally from clay or dust, rose through stages including a clot of blood and a lump of flesh to become fully developed human beings,[238] so their ultimate evolution in the Garden parallels the development of musk from blood that gradually transformed into musk. This view of human evolution reaches its climax in the transfiguration of human bodily processes into processes which produce musk, paralleling the musk deer's transformation of its impure blood into pure musk. The inhabitants of the Garden eat and drink, but do not defecate or urinate; the bodily processes of earth are replaced by sweat of musk. "[Muḥammad said] 'Indeed, the people of the Garden eat within it and drink, but they do not spit, do not urinate, do not defecate, and do not blow their noses.' They asked, 'what of the residue of the food?' 'A mere burp, and sweat like a sweat of musk; they are inspired to say 'Glory to God' and 'Praise God' as you are made to take a

235 Schimmel, *Triumphal Sun* 106.

236 Jalālu'ddīn Rūmī, *The Mathnawī. Volume 1*, ed. R. A. Nicholson, E. J. W. Gibb Memorial Series, New Series 4.1. (London: Luzac, 1925), 90 (Book 1 lines 1470–1).

237 *Mathnawī* 255–6 (Book 2 lines 161–4).

238 Qur'ān 23:14.

breath'."[239] Likewise, a sign of the martyr's entry into the Garden is that the blood of his wounds smells like musk in this world, an odor of sanctity.[240] Thus the humble musk deer, turning its impure blood into pure musk, has become a metaphor for the spiritually perfected future of pious humans, and even their most impure processes will yield only pure substances in the end.

In considering musk to be the most important aromatic, the early medieval Muslims followed Muḥammad, a true connoisseur of perfume who favored musk. Musk became a symbol in this world of the divine spirit and a reflection of the pleasure of the world to come. Sometimes the aromas of the Garden are said to surpass musk; this is the highest praise possible in Islamic culture. Musk and perfume, however, were only reflections of the divine majesty. When musk was being weighed out for distribution to the Muslims before the pious caliph ʿUmar b. ʿAbd al-Azīz (r. 717–20), he held his nose lest the smell overpower him, and he said, "Is there anything which benefits except for His smell?"[241]

---

239 Muslim 4.315. The verb translated as "inspired" and "made to take" is *ilhama*, a word associated with gulping and gobbling; a translation that would preserve this original sense and still do justice to the meaning in English eludes me, but the gustatory element is present in both clauses.

240 E.g., Muslim 3.251.

241 Ghuzūlī 2.75.

Conclusion: Worldly and Otherworldly

There can be little doubt of the importance of musk in medieval Middle Eastern civilization. Its wide use in perfumery and medicine would be sufficient without the added dimensions it takes on in the religion of Islam. Musk, or at least an imitation of it, was used ubiquitously in Islamicate perfumes; perhaps no other culture used musk to such an extent in its perfumed preparations. The medical uses of musk are extensive and varied, and musk, like other aromatics, was valued as a disinfectant. Islam made musk the most important of the aromatics. While it already had many positive associations in poetic language, Islam compounded them with added religious dimensions. The new religion adopted a new aromatic and made it symbolic of excellence. The production of musk in the lowly navel of a deer (or so it was thought), rather than detracting from the quality of musk, became a metaphor for the omnipotence of God and the possibility of redemption.

It is striking, given the fame of Arabian aromatics such as frankincense and myrrh, that musk would acquire such an exalted position. To a large extent this must reflect the belief that commodities acquired from far away at great expense must be excellent. The use of musk by the nobility only reinforced this belief, creating a fashion for musk, as well as its perennial association with royalty, who could always afford it. Musk was entirely imported from beyond the boundaries of the Islamic world; very few Muslims, even merchants who visited India and Central Eurasia, would have ever seen a musk deer, unless they did a considerable amount of traveling, for the habitat of the musk deer is wild and remote. This speaks to the importance of flourishing commerce in the early medieval Middle East and Central Eurasia.

The commercial side of the early medieval Islamic world is taken for granted by scholars who are more occupied with the study of political or religious history. The great importance of merchants has been overshadowed by the study of the caliphate itself. Yet these merchants formed an important social class. Their wealth enabled them to pursue many interests besides their trade; the scholars of medieval Islam were usually merchants.[1]

The world one encounters on reading writers like Ibn Khurradādhbih or al-Masʿūdī is characterized by a great mobility of goods and people. While not perhaps quite comparable to the exchanges associated with the modern

1 H.J. Cohen, "The Economic Background and the Secular Occupations of Muslim Jurisprudents and Traditionists in the Classical Period of Islam," *JESHO* 13 (1970): 16–61.

 | DOI 10.1163/9789004336315_010

world, they nevertheless reflect a highly cosmopolitan world. The geographer Ibn al-Faqīh writes:

> If God the Almighty had not distinguished with His benevolence each land of the lands and given each clime of the climes a thing which He kept from the others, commerce would be in vain and the handicrafts would go away. No one would journey abroad and no one would travel and they would abandon the exchange of gifts; buying and selling would go away, and receiving and giving. Therefore, God the Almighty gave each locality, in every opportunity, an item of the good things and kept it from the others so that one must travel to the country with it, and people will enjoy the delights of [another] people, so that there is even distribution and the arrangement is well ordered.[2]

For Ibn al-Faqīh the diversity of the world is merely God's way of keeping society operating. It is a world-view in which trade and commerce are the very basis of life.

In such a society, acquiring an aromatic—like musk—from abroad was nothing unusual, even if it was expensive. The mythos of the origin and difficulty of acquiring it only enhanced its reputation. Nowhere has any indication of a long-lasting shortage of musk been found; it seems to have been always available for the right price. What distinguished musk from the numerous other imported aromatics was the extremely wide use it enjoyed and the great importance it attained. In a world of valuable imported scents, musk stood at the very top. It reached that high position by its own qualities in perfumery and by its medical applications; these two reinforced each other. Its prestige value, as a precious gift of kings and caliphs, only enhanced its popularity. The religious sanction of musk assured it would reach a high position as the best of all the aromatics.

Paradise is the idealized mirror of the real world, where all of the negative aspects of this world are transformed. Within this world, the base and ignoble dominate. One of the signs of the inferiority of this world (*al-dunyā*; the word literally means "inferior") is that even its most noble aspects are practically reprehensible in their origins. Thus al-Ghazālī notes "the most noble of the scents is musk, and it is blood."[3] But musk and other aromatics provided a link to the heavenly paradise through the tales of their arrival in this world from the original Eden. God created the musk deer specifically for its capacity

---

2 Ibn al-Faqīh, *Mukhtaṣar kitāb al-buldān*, ed. M. J. De Goeje (Leiden: Brill, 1885; repr. 1967), 251.

3 Al-Ghazālī, *Iḥyā' 'ulūm al-dīn*, 5 vols. (Beirut: Dār al-Kutub al-'Ilmiyyah, 2005), 3.283.

to produce musk.[4] As a sought-after delight, musk maintained its association with the heavenly paradise, which had been conceived as fragrant for centuries. But unlike other paradisical delights such as silk and wine, musk remained lawful in this world. In fact, its use was encouraged. This cannot be solely due to its medical usefulness, for wine, though an intoxicant, was long believed to have medical value, and it even continued to be used in medicine in Islamic times as well as drunk by those who were not ill, yet it had to endure the scorn of the religion. The reason, especially in light of its questionable origin, must be Muḥammad's love of musk. It is not at all certain that he knew much about the real or supposed origins of musk, even if stories were known to someone like Cosmas Indicopleustes long before him. But Muḥammad's love of musk meant that the jurisprudents and scholars would find a way to justify what they already knew to be good and licit due to prophetic sanction.

As a matter of practicality, real musk was hard to get and expensive. Fraud was commonplace. Civet, on the other hand, could be produced in the Middle East, and it eventually became better known than musk; yet there is no evidence for the use of civet in the Middle East before the 8th century.[5] Doubts about its origin kept it from attaining the exalted status of musk, even if it possessed comparable, and in some ways, superior, properties to musk. The origins of civet were all too clear to those who could observe its extraction from the anal glands of the poor civet "cat". Civet also lacked the approval of the Prophet, and perhaps that kept it from the popularity of musk during the medieval period. But most of all, civet could not develop the cachet that musk could, produced as it was by gazelles on remote mountains in the lands of fragrant spices and treasures.

4 Yaʿqūbī *apud* Nuwayrī 12.4.

5 Jāḥiẓ's telling statement in *Kitāb al-Ḥayawān*, quoted in Chapter 4, is perhaps the earliest.

Bibliography

Primary Literary Sources

ʿAbīd ibn al-Abraṣ. *Dīwān*. Ed. and trans. C. Lyall. *The Dīwāns of ʿAbīd ibn al-Abraṣ, of Asad, and ʿĀmir ibn aṭ-Ṭufail, of ʿĀmir ibn Ṣaʿṣaʿah*. Cambridge: Gibb Memorial Series, 1913; repr. 1980.

Abū Dāwūd, Sulaymān b. Ashʿath al-Sijistānī. *Sunan Abī Dāwūd*. 4 vols. Dār Iḥyāʾ al-Sunnah al-Nabawiyyah, n.d.

Abū Tammām, Ḥabīb b. Aws al-Ṭāʾī. *Dīwān*. 4 vols. Cairo: Dār al-Maʿārif, 1964.

Aetius of Amida. *Libri medicinales*. [1–8 only] Ed. A. Olivieri. 2 vols. Berlin: in aedibus Academiae Litterarum, 1935–50.

Aetius of Amida. *Gynaekologie des Aëtios. Aetii sermo sextidecimus et ultimus*. Ed. S. Zervos. Leipzig: A. Fock, 1901.

Aetius of Amida. Trans. J. V. Ricci. *The Gynaecology and Obstetrics of the VIth century A.D. Translated from the Latin ed. of Cornarius, 1542*. [16 only] Philadelphia: Blakiston, 1950,

Agnimahāpurāṇam. Ed. and trans. M. N. Dutt. 2 vols. Delhi: Parimal Publications, 2001.

Akhbār al-Ṣīn wa-l-Hind. Akhbār aṣ-Ṣīn wa l-Hind: Relation de la Chine et de l'Inde. Ed. and trans. J. Sauvaget. Paris: Les Belles Lettres, 1948. (See also al-Sīrāfī)

Al-Akhṭal, Ghiyāth b. Ghawth. *Dīwān*. Ed. A. Ṣāliḥānī. 2nd edition. Beirut: Dār al-Mashriq, 1969.

Alexander of Tralles. *Alexander von Tralles. Original-Text und Übersetzung nebst einer einleitenden Abhandlung. Ein Beitrag zur Geschichte der Medicin*. Ed. and trans. T. Puschmann. 2 vols. Vienna: Braumüller, 1878–9.

Alexander of Tralles. Trans. F. Brunet. *Oeuvres médicales d'Alexandre de Tralles, le dernier auteur classique des grands médecins grecs de l'antiquité*. 4 vols. Paris: Geuthner, 1933–7.

Amoghapāšamantrahṛdayasūtra. T #1095 (vol. 20: 406–9).

Ananias of Širak. *The Geography of Ananias of Širak (Ašxarhacʿoycʿ): The Long and Short Recensions*. Trans. R. H. Hewsen, Wiesbaden: Reichert, 1992.

Ananias of Širak. G. R. Cardona. "L'India e la Cina secondo l'Ašxarhaċoyċ." In *Armeniaca: Melanges d'études armeniennes*. Venice: Saint Lazare, 1969, 83–97.

ʿAntarah b. Shaddād al-ʿAbsī. *Sharḥ dīwān ʿAntara*. Ed. Ibrāhīm al-Zayn. Beirut: Dār al-Najjāḥ, 1964.

Apicius. *A Critical Edition with an Introduction and an English Translation of the Latin Recipe Text* Apicius. Ed. and trans. C. Grocock and S. Grainger. Totnes: Prospect Books, 2006.

Ardā Wirāz Nāmag: The Iranian 'Divina Commedia'. Ed. and trans. F. Vahman. London: Scandinavian Institute of Asian Studies, 1986.
Al-Aʿshā, Abū Baṣīr Maymūn b. Qays. *Gedichte von Abû Baṣîr Maimûn ibn Qais al-Aʿšâ*. Ed. R. Geyer. London: Gibb Memorial Series, 1928.
Al-Aṣmaʿī, Abū Saʿīd ʿAbd al-Malik b. Qurayb al-Bāhilī. *Tārīkh al-ʿArab qabl al-Islām*. Ed. Muḥammad Ḥasan al-Yāsīn. Baghdad: Maṭbaʿat al-Maʿārif, 1959.
Aṭṭār, Farīd al-Dīn. *Dīwān-i qaṣā'id tarjīʿāt wa ghazaliyāt*. Ed. Saʿīd Nafīsī. Tehran: Sanā'ī, 1339/1960.
Al-Azdī, Muḥammad b. Aḥmad Abū al-Muṭahhar. *Ḥikāyat Abī al-Qāsim al-Baghdādī*. Ed. A Mez. Heidelberg: Carl Winter, 1902.
Al-Bakrī, Abū ʿUbayd al-Bakrī. *al-Masālik wa-l-mamālik*. Ed. A. P. Van Leeuwen and A. Ferre. 2 vols. Tunis: al-Dār al-ʿArabiyyah li-l-Kitāb, 1992.
Al-Balādhurī, Abū al-ʿAbbās Aḥmad b. Yaḥyā b. Jābir. *Kitāb Futūḥ al-buldān*. Ed. M. J. De Goeje. Leiden: Brill, 1866; repr. 1968.
Bāṇa. *Harṣacarita*. Ed. P. V. Kane. Bombay, 1918; repr. Delhi: Banarsidass, 1997.
Bāṇa. Trans. E. B. Cowell and F. W. Thomas. *The Harṣacarita of Bāṇa*. Repr. Delhi: Motilal Banarsidass, 1993.
Bar Bahlūl. *Lexicon syriacum auctore Hassano bar Bahlule voces syriacas graecasque cum glossis syriacis et arabicis complectens*. 3 vols. Ed. R. Duval. Paris: Republicae typographaeo, 1888–1901.
Bayhaqī, Abū Faḍl. *Tārīkh-i Bayhaqī*. Ed. ʿAlī Akbār Fayyāḍ. Mashhad: Dānishgāh-i Mashhad, 1971.
Benjamin of Tudela. *The Itinerary of Benjamin of Tudela*. Ed. and trans. M. N. Adler. London, 1907; repr. New York: Philipp Feldheim, n.d.
Bilhaṇa. *Caurapañcāśikā*. Varanasi: Chawkhamba Sanskrit Series, 1971.
Al-Bīrūnī. *Kitāb al-Ṣaydanah*. Ed. and trans. H. M. Saʿīd. Karachi: Hamdard National Foundation, 1973.
Al-Bīrūnī. *Kitāb al-Ṣaydanah*. Ed. ʿAbbās Zaryāb. Tehran: Markaz-i Nashr-i Dānishgāhī, 1991.
Al-Bīrūnī. *Kitāb al-Jamāhir fī maʿrifat al-jawāhir*. Ed. F. Krenkow. Haydarābād: Dā'irat al-Maʿārif, 1355/1936.
Al-Bīrūnī. *Kitāb al-Bīrūnī fī taḥqīq mā li-l-Hind*. Ed. E. Sachau. Haydarābād: Dā'irat al-Maʿārif, 1958.
Al-Bīrūnī. Z. V. Togan. *Bīrūnī's Picture of the World*. New Delhi: Archaeological Survey of India, 1937. [Collection of excerpts in Arabic from al-Bīrūnī's works.]
Al-Bīrūnī. M. Meyerhof. "Das Vorwort zur Drogenkunde des Bērūnī." *Quellen und Studien zur Geschichte der Naturwissenschaften und der Medizin* 3:3 (1932). 52 pp. + 18 p. Arabic text.
Book of Curiosities: An Eleventh-Century Egyptian Guide to the Universe: The Book of Curiosities. Ed. and trans. Y. Rapoport and E. Savage-Smith. Leiden: Brill, 2014.

Book of the Eparch. In I. Dujčev, ed., *The Book of the Eparch*. London: Variorum, 1970.

Boyle, Robert. "The Usefulness of Natural Philosophy II. Sect. 2." In *The Works of Robert Boyle*. Vol. 6. Ed. M. Hunter and E. B. Davis. London: Pickering and Chatto, 1999.

Boyle, Robert. "Experiments and Observations about the Mechanical Production of Odours." In *The Works of Robert Boyle*. Vol. 8. Ed. M. Hunter and E. B. Davis. London: Pickering and Chatto, 2000.

Al-Bukhārī, Abī ʿAbdallāh. *Ṣaḥīḥ Abī ʿAbdallāh al-Bukhārī bi-sharḥ al-Kirmānī*. 25 vols. Cairo: Al-Maṭbaʿah al-Bahiyyah al-Miṣriyyah, 1933–62.

Bundahišn (Iranian). F. Pakzad, ed. *Bundahišn: Zoroastrische Kosmogonie und Kosmologie*, vol. 1: *Kritische Edition*. Tehran: Center for the Great Islamic Encyclopaedia, 2005.

Bundahišn (Iranian). *The Bondahesh, Being a Facsimile Edition of the Manuscript TD1*, Tehran: Iranian Culture Foundation, n.d.

Bundahišn (Iranian). *Dastnivīs-i TD2/ MS. TD2: Iranian Bundahišn & Rivāyat-i Ēmēt-i Ašavahištān*, 2 vols., Shiraz: Asia Institute of Pahlavi University, 1978.

Bundahišn (Iranian). Transcribed and trans. by B. T. Anklesaria. *Zand-Ākāsīh, Iranian or Greater Bundahišn*. Bombay, 1956.

Buzurg b. Shahriyār. *Kitāb ʿAjāʾib al-Hind: Livre des merveilles de l'Inde*. Ed. P. A. Van der Lith and trans. L. M. Devic. Leiden: Brill, 1883–6; repr. Tehran: M. H. Asadi's Historical Series, 1966.

Carakasaṃhitā. Ed. and trans. R. K. Sharma and V. B. Dash. 7 vols. Varanasi: Chowkhamba Sanskrit Series Office, repr. 2004–5.

Chardin, J. *Travels in Persia 1673–1677*. London: Argonaut, 1927; repr. New York: Dover, 1988.

Chu Hong 芻撰. *Xiang Pu* 香譜. *Congshu jicheng* edition. Shanghai: Shang wu yin shu kuan, 1937.

Cosmas Indicopleustes. *Topographie Chrétienne*. Ed. and trans. W. Wolska-Conus. 3 vols. Paris: Les Éditions du Cerf, 1968–73.

Al-Damīrī, Muḥammad b. Mūsā. *Ḥayāt al-ḥayawān*. 2 vols. Būlāq, 1284/1867; repr. Frankfurt am Main: Institute for the History of Arabic-Islamic Science, 2001.

Dhū al-Rummah, Ghaylān b. ʿUqbah. *The Dīwān of Ghailān ibn ʿUqbah known as Dhu ʾr-Rummah*. Ed. C. H. H. Macartney. Cambridge: Cambridge University Press, 1919.

Al-Dimashqī, Jaʿfar b. ʿAlī. *Kitāb al-Ishārah ilā mahāsin al-tijārah*. Cairo: Maṭbaʿat al-Muʿayyad, 1318/1900–1.

Al-Dimashqī, Shams al-Dīn. *Kitāb Nukhbat al-dahr fī ʿajāʾib al-barr wa-l-baḥr*. Ed. A. Mehren. Baghdād: Yuṭlabu min Maktabat al-Muthannā, n.d.

Al-Dīnawarī, Abū Ḥanīfah. *Kitāb al-Nabāt: The Book of Plants: Part of the Monograph Section*. Ed. Bernhard Lewin. Wiesbaden: Steiner, 1974.

Al-Dīnawarī, Abū Ḥanīfah. *The Book of Plants of Abū Ḥanīfa ad-Dīnawarī: Part of the Alphabetical Section* (Alif-Zayn). Ed. Bernhard Lewin. Wiesbaden: Harrassowitz, 1953. (= *Uppsala Universitets Årsskrift* 1953:10)

Al-Dīnawarī, Abū Ḥanīfah. *Kitāb al-Akhbār al-ṭiwāl*. Ed. V. F. Girgas. Leiden: Brill, 1888–1912.

Dioscorides. *De Materia Medica*. Ed. M. Wellmann. 3 vols. Berlin: Weidmann, 1958.

Dioscorides. *De Materia Medica*. Trans. L. Y. Beck. Altertumswissenschaftliche Texte und Studien Band 38. Hildesheim: Olms, 2005.

Draxt-ī Āsūrīg. Māhyār Nawwābī, ed. *Manẓūmah-i Draxt-i Āsūrīg*. Tehrān: Intishārāt-i Bunyād-i Farhang-i Īrān, 1346/1967.

Du You 杜佑. *Tongdian* 通典. Beijing: Zhonghua shuju, 1988.

Duarte Barbosa. *The Book of Duarte Barbosa*. Trans. M. Longworth Dames. 2 vols. London: Hakluyt Society, 1918–21.

Ennin. *Ennin's Diary: The Record of a Pilgrimage to China in Search of the Law*. Trans. E. O. Reischauer. New York: Ronald Press, 1955.

Ephrem the Syrian. *Hymnen de paradiso und contra julianum*. Ed. E. Beck, 2 vols. Louvain, 1957. *Corpus scriptorum christianorum orientalium* 174–5. (Vol. 1 is text, 2 is German translation.)

Erya 爾雅. Recension of Guo Pu 郭璞. *Bai bu cong shu ji cheng* 百部叢書集成 edition.

Evliya Çelebi. *Evliya Çelebi in Diyarbekir*. Ed. and trans. M. van Bruinessen and M. Boeschoten. Leiden: Brill, 1988.

Firdawsī, Abū al-Qāsim. *Abu'l Qasem Ferdowsi: The Shahnameh* (*The Book of Kings*). Ed. Djalal Khaleghi-Motlagh. 8 vols. New York: Bibliotheca Persica, 1988–2008.

Gardīzī, Abū Saʿīd ʿAbd al-Ḥayy. *Zayn al-Akhbār*. Ed. ʿAbd al-Ḥayy Ḥabībī as *Tārīkh-i Gardīzī*. Tehran: Dunyā-yi Kitāb, 1363/1984.

Gardīzī, Abū Saʿīd ʿAbd al-Ḥayy. V. M. Minorsky, trans. "Gardizī on India." *BSOAS* 12 (1948): 625–40.

Gardīzī, Abū Saʿīd ʿAbd al-Ḥayy. A. P. Martinez, trans. "Gardīzī's Two Chapters on the Turks." *Archivum Eurasiae Medii Aevi* 2 (1982): 109–217.

Ge Hong 葛洪. *Bao Puzi* 抱朴子. Taibei: Zhongguo zixue mingzhu jicheng, 1978.

Al-Ghassānī, al-Muẓaffar Yūsuf b. ʿUmar b. ʿAlī. *Kitāb al-Muʿtamad fī al-adwiyah al-mufradah*. Ed. Muṣṭafā Saqqā. 3rd ed. Beirut: Dār al-Maʿrifah, 1975.

Al-Ghazālī, Abū Ḥāmid Muḥammad b. Muḥammad. *Iḥyāʾ ʿulūm al-dīn*. 5 vols. Beirut: Dār al-Kutub al-ʿIlmiyyah, 2005.

Al-Ghuzūlī (also given as Ghazūlī), ʿAlāʾ al-Dīn ʿAlī b. ʿAbdallāh. *Maṭāliʿ al-budūr fī manāzil al-surūr*. 2 vols. bound as one. Cairo: Maktabah al-Thaqāfah al-Dīniyyah, n.d. [2000].

Gurgānī, Fakhr al-Dīn. *Vis and Ramin*. Trans. G. Morrison. New York: Columbia University Press, 1972.

Al-Harawī, Muwaffaq al-Dīn Abū Manṣūr ʿAlī. *Al-Abniyah ʿan Ḥaqāʾiq al-Adwiyah.* Ed. Aḥmad Bahmanyār. 2nd ed. Tehran: Dānishgāh-i Tihrān, 1975; repr. Tehran: Dānishgāh-i Tihrān, 1371/1992.

Al-Harawī, Muwaffaq al-Dīn Abū Manṣūr ʿAlī. Trans. A.-Ch. Achundow. "Die pharmakologischen Grundsätze (Liber fundamentorum pharmacologiae) des Abu Mansur Muwaffak bin Ali Harawi zum ersten Male nach dem Urtext übersetzt und mit Erklärungen versehen." *Historische Studien aus dem Pharmakologischen Institute der Kaiserlichen Universität Dorpat* 3 (1893): 139–414 and 450–81.

Ḥassān b. Thābit. *Dīwān.* Ed. Sayyid Ḥanafī Ḥasanayn and Ḥasan Kāmil al-Ṣayrafī. Cairo: al-Hay'ah al-Miṣriyyah al-ʿĀmmah li-l-Kitāb, 1974.

Herodotus. *The Persian Wars.* Ed. and trans. A. D. Godley. Rev. ed. 4 vols. Loeb Classical Libary. Cambridge, MA: Harvard University Press, 1926.

Ḥudūd al-ʿālam min al-mashriq ilā al-maghrib. Ed. M. Sutūdah. Tehran: Dānishgāh-i Tihrān, 1983.

Ḥudūd al-ʿālam min al-mashriq ilā al-maghrib. Trans. V. Minorsky. *The Regions of the World, A Persian Geography 372 A.H.–982 A.D.* 2nd ed. London: Gibb Memorial Series, 1970.

Hyech'o. *Echōō-Gotenjiku-koku-den Kenkyū.* Ed. S. Kuwayama. Kyoto: Kyōto Daigaku Jinbun Kagaku Kenkyūjo, 1992.

I-Ching [Yijing]. *A Record of the Buddhist Religion as practiced in India and the Malaya Archipelago* (AD *671–695*). Trans. J. Takakusu. Oxford: Clarendon Press, 1896.

Ibn ʿAbd Rabbih, Abū ʿUmar Aḥmad b. Muḥammad. *Al-ʿIqd al-farīd.* 31 vols. Beirut: Maktabat Ṣādir, 1951–4.

Ibn Abī al-Ḥadīd, ʿAbd al-Ḥamīd b. Hibat Allāh. *Sharḥ Nahj al-balāghah.* 20 vols. Cairo: ʿĪsā al-Bābī al-Ḥalabī, 1960–4.

Ibn al-ʿAdīm (attr.). *Al-Wuṣlah ilā al-ḥabīb.* 2 vols. Ed. S. Maḥjūb and D. al-Khaṭīb. Aleppo, 1987–8.

Ibn Bādīs, Muʿizz. *ʿUmdat al-kuttāb wa ʿuddat dhawī al-albāb.* Ed. Najīb Māyil al-Harawī. Mashhad: Majmaʿ al-Buḥūth al-Islāmiyyah, 1409.

Ibn Bādīs, Muʿizz. *Medieval Arabic Bookmaking and its Relation to Early Chemistry and Pharmacology.* Trans. M. Levey. Transactions of the American Philosophical Society New Series 52:4. Philadelphia: American Philosophical Society, 1962.

Ibn al-Balkhī. *Fārsnāmah.* Ed. G. Le Strange and R. A. Nicholson. London: Gibb Memorial Trust, 1921; repr. 1962.

Ibn Baṭṭūtah, Abū ʿAbdallāh Muḥammad b. ʿAbdallāh. *Riḥlat Ibn Baṭṭūṭah.* Beirut: Dār Ṣādir, 1960.

Ibn Baṭṭūtah, Abū ʿAbdallāh Muḥammad b. *The Travels of Ibn Baṭṭūṭa A.D. 1325–1354.* Vol. 1. Trans. H. A. R. Gibb. London: Hakluyt Society, 1958.

Ibn al-Bayṭār. *al-Jāmiʿ li-mufradāt al-adwiyah wa-l-aghdhiyah.* 4 vols. Būlāq, n.d.

Ibn al-Dawādārī, Abū Bakr b. ʿAbdallāh. *Kanz al-durar wa-jāmiʿ al-ghurar*. Vol. 1. Ed. B. Radtke. Wiesbaden: Franz Steiner, 1982.

Ibn al-Faqīh al-Hamadhānī. *Mukhtaṣar Kitāb al-Buldān*. Ed. M. J. De Goeje. Leiden: Brill, 1885; repr. 1967.

Ibn al-Faqīh al-Hamadhānī. *Kitāb al-Buldān*. Ed. Yusūf al-Hāwī. Beirut: ʿĀlam al-Kutub, 1996.

Ibn Ḥabīb, Muḥammad. *Kitāb al-Muḥabbar*. Ed. I. Lichtenstadter. Haydarābād, 1942; repr. Beirut: al-Maktab al-Turāthī li-l-Ṭibāʿah wa-l-Nashr wa-l-Tawzīʿ, n.d.

Ibn Ḥanbal, Aḥmad ibn Muḥammad. *Musnad*. 6 vols. Beirut, 1969 (reprint of Būlāq edition).

Ibn al-Ḥashshāʾ. *Mufīd al-ʿulūm wa-mubīd al-humūm*. Ed. G. S. Colin and H. P. J. Renaud. Paris: Imprimerie économique, 1941.

Ibn Ḥawqal, Abū al-Qāsim. *Kitāb Ṣūrat al-ʿarḍ*. Ed. M. G. De Goeje and rev. J. H. Kramers. Leiden: Brill, 1938; repr. 1967.

Ibn Isḥāq. *Sīrat Rasūl Allāh. Das Leben Muhammed's*, ed. Ibn Hishām. Text ed. F. Wüstenfeld, v. 1 part 1, Göttingen, 1858, repr. Frankfurt am Main, 1961.

Ibn Jazlah, Abū ʿAlī Yaḥyā b. ʿĪsā. *Minhāj al-bayān fī mā yastaʿmiluhu al-Insān*. Cairo: Dār al-Kutub al-Miṣriyyah, 2010.

Ibn Jubayr, Muḥammad b. Aḥmad. *Riḥlat Ibn Jubayr*. Beirut: Dār Ṣādir, 1959.

Ibn Juljul, Abū Dāwūd Sulaymān b. Ḥassan al-Andalusī. *Die Ergänzung Ibn Ǧulǧul's zur Materia medica des Dioskurides*. Ed. and trans. A. Dietrich, Göttingen: Vandenhoeck & Ruprecht, 1993 (= *Abhandlungen der Akademie der Wissenschaften in Göttingen, philosophisch-historische Klasse, dritte Folge* Nr. 202).

Ibn Kathīr, Ismāʿīl b. ʿUmar. *Tafsīr al-Qurʾān al-ʿaẓīm*. 7 vols. Beirut: Dār al-Andalus, [1966].

Ibn Khallikān, Aḥmad b. Muḥammad Abū al-ʿAbbās Shams al-Dīn al-Barmakī al-Irbilī. *Wafāt al-aʿyān*. Ed. I. ʿAbbās. 8 vols. 4th ed. Beirut: Dār Ṣādir, 2005.

Ibn Khurradādhbih. *Kitāb al-Masālik wa-l-mamālik*. Ed. M. J. De Goeje. Leiden: Brill, 1889, repr. 1967.

Ibn al-Muʿtazz, Abū al-ʿAbbās ʿAbdallāh. *Ṭabaqāt al-shuʿarāʾ*. Ed. ʿAbd al-Sattār Aḥmad Farrāj. 2nd ed. Cairo: Dār al-Maʿārif, 1968.

Ibn Mandawayh, Abū ʿAlī Aḥmad b. ʿAbd al-Raḥmān. *Risālah fī uṣūl al-ṭīb wa-l-murakkabāt al-ʿiṭriyyah*. In M. Dānishpazhūh. "*Du risālah dar shinākht-i ʿiṭr.*" *Farhang-i Īrān-Zamīn* 15 (1347/1967): 224–253.

Ibn Manẓūr. *Lisān al-ʿArab*. 18 vols. Beirut: Dār al-Kutub al-ʿIlmiyyah, 2003.

Ibn Māsawayh, Abū Zakarīyāʾ Yūḥannā. *Kitāb Jawāhir al-ṭīb al-mufradah*. Ed. P. Sbath, "Traité sur les substances simples aromatiques." *Bulletin de l'Institut d'Égypte* 19 (1936–7): 5–27.

Ibn Māsawayh, Abū Zakarīyāʾ Yūḥannā. Trans. M. Levey. "Ibn Māsawaih and His Treatise on Simple Aromatic Substances." *Journal of the History of Medicine* 16 (1961): 394–410.

Ibn Miskawayh, Aḥmad b. Muḥammad. *Kitāb Tajārib al-umam*. 3 vols. Baghdad: al-Muthanna, n.d.
Ibn al-Muqaffaʿ. *Kalīlah wa Dimnah*. Beirut: Dār Maktabat al-Hayāh [1966].
Ibn al-Nadīm, Abū al-Faraj Muḥammad b. Isḥāq. *al-Fihrist*. Ed. G. Flügel. repr. Beirut: Maktabat al-Khayyāṭ, 1966.
Ibn al-Nadīm, Abū al-Faraj Muḥammad b. Isḥāq. *The Fihrist of al-Nadīm: A Tenth-Century Survey of Muslim Culture*. Trans. B. Dodge. 2 vols. New York: Columbia University Press, 1970.
Ibn Qayyim al-Jawziyyah, Muḥammad b. Abī Bakri. *al-Ṭibb al-Nabawī*. Cairo: Dār al-Turāth, 1978.
Ibn Qutaybah, Abū Muḥammad ʿAbdallāh b. Muslim al-Dīnawarī. *Kitāb al-Shiʿr wa-l-shuʿarāʾ*. Ed. M. J. De Goeje. Leiden: Brill, 1902.
Ibn Rustah, Abū ʿAlī Aḥmad b. ʿUmar. *Kitāb al-ʿAlāq al-nafīsah*. Ed. M. G. De Goeje. Leiden, 1892; repr. Leiden: Brill, 1967.
Ibn Saʿd, Abū ʿAbdallāh Muḥammad. *al-Ṭabaqāt al-kubrā*. 9 vols. Beirut: Dār Ṣādir, 1960.
Ibn Sayyār al-Warrāq. *Kitāb al-Ṭabīkh*. Ed. K. Öhrnberg and S. Mroueh. Studia Orientalia 60. Helsinki: Finnish Oriental Society, 1987.
Ibn Sayyār al-Warrāq. Trans. N. Nasrallah. *Annals of the Caliphs' Kitchens: Ibn Sayyār al-Warrāq's Tenth-Century Baghdadi Cookbook*. Leiden: Brill, 2007.
Ibn Sīnā. *al-Qānūn fī al-ṭibb*. Reprint of Būlāq ed. Beirut: Dār Ṣādir, n.d.
Ibn Sīnā. *al-Qānūn fī al-ṭibb*. Ed. Idwār al-Qashsh. 4 vols. Beirut: Muʾassasat ʿIzz al-Dīn, 1987.
Ibn Wāfid. *Kitāb al-Adwiya al-mufrada (Libro de los medicamentos simples)*. Ed. and trans. Luisa Fernanda Aguirre de Cárcer. 2 vols. Madrid: Consejo superior de investigaciones cientificas agencia espanola de cooperación internacional, 1995.
Ibn Zaydūn, Aḥmad b. ʿAbdallāh. *Dīwān*. Ed. Karam al-Bustānī. Beirut: Dār Ṣādir, 1964.
Al-Ibshihī, Bahāʾ al-Dīn Muḥammad b. Aḥmad. *al-Mustaṭraf fī kull fann mustaẓraf*. Ed. Muṣṭafā Muḥammad al-Dhahabī. Cairo: Dār al-Ḥadīth, 2000.
Al-Idrīsī, Muḥammad b. Muḥammad. *Kitāb Nuzhat al-mushtāq fī ikhtirāq al-āfāq*. 9 vols. Naples: Istituto Universitario Orientale di Napoli, 1970–84.
Al-Idrīsī, Muḥammad b. Muḥammad. *Kitāb al-Jāmiʿ li-ṣifāt ashtāt al-nabāt wa-ḍurūb anwāʿ al-mufradāt*. Ed. F. Sezgin, et al. 3 vols. Frankfurt am Main: Institute for the History of Arabic-Islamic Science, 1995.
Imruʾ al-Qays. *Dīwān*. Ed. Yāsīn al-Ayyūbī. Beirut: al-Maktab al-Islāmī, 1998.
Isaac of Antioch. *Homilae S. Isaaci Syri Antiocheni*. Ed. P. Bedjan. Leipzig: Harrassowitz, 1903.
Al-Iṣbahānī, Abū al-Faraj. *Kitāb al-Aghānī*. 24 vols. Cairo: Dār al-Kutub, 1927–74.
Al-Iṣṭakhrī, Abū Isḥāq Ibrāhīm b. Muḥammad al-Fārisī. *Kitāb Masālik al-Mamālik*. Ed. M. J. De Goeje. Leiden: Brill, 1870; repr. 1967.

Jābir ibn Ḥayyan. *Kitāb al-Sumūm*. Ed. and trans. by A. Siggel. *Das Buch der Gifte des Gābir ibn Ḥayyān*. Wiesbaden: Steiner, 1958.

Al-Jāḥiẓ al-Baṣrī, Abū 'Uthmān 'Amr ibn Baḥr. *Kitāb al-Bukhalā'*. Ed. Ṭāhā Ḥājirī. Cairo: Dār al-Ma'ārif, n.d. [1958].

Al-Jāḥiẓ al-Baṣrī, Abū 'Uthmān 'Amr ibn Baḥr. *Kitāb al-Ḥayawān*, ed. 'Abd al-Salām Muḥammad Hārūn, 8 vols., Cairo: Muṣṭafā al-Bābī al-Ḥalabī, 1966.

Al-Jāḥiẓ al-Baṣrī, Abū 'Uthmān 'Amr ibn Baḥr. *Rasā'il al-Jāḥiẓ*. ed. 'Abd al-Salām Muḥammad Hārūn. 2 vols. Cairo: Maktabat al-Khānjī, 1964.

Al-Jāḥiẓ al-Baṣrī, Abū 'Uthmān 'Amr ibn Baḥr. (Attr.) *Kitāb al-Tāj*. Ed. Ahmed Zeki Pacha. Cairo: Imprimerie Nationale, 1914.

Al-Jāḥiẓ al-Baṣrī, Abū 'Uthmān 'Amr ibn Baḥr. (Attr.) *Kitāb al-Tabaṣṣur bi-l-tijārah fī waṣf ma yustaẓraf fī al-buldān min al-amti'ah al-rafī'ah wa-l-a'lāq al-nafīsah wa-l-jawāhir al-thamīnah*. Ed. Ḥasan Ḥusnī 'Abd al-Wahhāb al-Tūnisī. Cairo: Maktabat al-Khānjī, 1994.

Al-Jāḥiẓ al-Baṣrī, Abū 'Uthmān 'Amr ibn Baḥr. Trans. Ch. Pellat, "Ǧāhiẓiana, I. Le Kitāb al-Tabaṣṣur bi-l-Tiǧāra attribué à Ǧāḥiẓ." *Arabica* 1 (1954): 153–65.

Jarīr b. 'Aṭiyyah. *Dīwān Jarīr bi-sharḥ Muhammad b. Ḥabīb*. Ed. Nu'mān Muḥammad Amīn Taha. Vol. 1. Cairo: Dār al-Ma'ārif, 1969.

Al-Jawālīqī, Mawhūb ibn Aḥmad. *al-Mu'arrab min al-kalām al-a'jamī 'alā ḥurūf al-mu'jam*. Ed. Abū al-Ashbāl Aḥmad Muḥammad Shākir. Tehrān, 1966.

Jayadeva. *Gītagovinda*. Ed. and trans. B. S. Miller. New York: Columbia University Press, 1977; repr. Delhi: Motilal Banarsidass, 1984.

Al-Jazzār, Abū Ja'far Aḥmad b. Ibrāhīm. *al-I'timād fī al-adwiyah al-'arabiyyah*. Ed. Idwār al-Qashsh. Beirut: Sharikat al-Maṭbu'at li-l-Tawzi' wa-l-Nashr, 1998.

Jerome. *Adversus Jovinianum*. In J. P. Migne, ed., *Patrologia cursus completus, series latina*, vol. 23, Paris, 1883, 221–352.

Jiu Tang shu 舊唐書. Beijing: Zhonghua Shuju, 1975.

Jurjānī, Isma'īl b. Ḥasan. *Dhakhirah-i Khwārazmshāhī*. Ed. Sa'īd Sīrjānī. Tehran: Bunyād-i Farhang-i Īrān, 1976.

Juwaynī, 'Alā' al-Dīn 'Aṭā' Malik. *Genghis Khan: The History of the World-Conqueror*. Trans. J. A. Boyle. 2 vols. Manchester: Manchester University Press, 1958. New Edition with introduction by D. Morgan in 1 vol., Seattle: University of Washington Press, 1997.

Kālidāsa. *Meghadūta*. Ed. M. R. Kale. Delhi: Motilal Banarsidass, 1991.

Kālidāsa. *Kumārasaṃbhava*. Ed. Suryakanta. New Delhi: Sahitya Akademi, 1962.

Kalilag und Damnag. Alte syrische Übersetzung des indischen Fürstenspiegels. Ed. and trans. G. Bickell, Lepizig: Brockhaus, 1876.

Kalilag und Damnag. Kalila und Dimna. Syrisch und Deutsch. Ed. and trans. F. Schulthess. 2 vols. Berlin: Reimer, 1911.

Kanz al-fawā'id fī tanwī' al-mawā'id. Ed. M. Marín and D. Waines. Beirut: Franz Steiner, 1993.

Al-Kāshānī, Abū al-Qāsim 'Abd Allāh b. 'Alī. *'Arāyis al-jawāhir wa-nafāyis al-aṭāyib*. Ed. Īraj Afshār. Tehran: Intishārāt-i Anjuman-i Athār-i Millī, 1966.

Al-Kāshgharī, Maḥmūd. *Maḥmūd al-Kāšyarī: Compendium of the Turkic Dialects* (*Dīwān Luyāt at-Turk*). Trans. R. Dankoff and J. Kelly. 3 vols. Cambridge: Harvard, 1982–5.

Al-Khafājī, Muḥammad 'Abd al-Mun'im. *Shifā' al-ghalīl fīmā fī kalām al-'Arab min al-dakhīl*. Cairo: al-Maktabah al-Azhariyyah li-l-Turāth, [2003].

Al-Kindī, Abū Yūsuf. *The Medical Formulary or Aqrābādhīn of al-Kindī*. Ed. and trans. M. Levey. Madison: University of Wisconsin, 1966.

Al-Kindī, Abū Yūsuf. *Risālat al-Kindī fī ajzā' khabariyyah fī al-mūsīqā*. Cairo: al-Lajnah al-Mūsīqiyyah al-'Ulyā, 1963?.

Al-Kindī, Abū Yūsuf. (Attr.) *Kitāb Kīmiyā' al-'Iṭr wa-t-Taṣ'īdāt: Buch über die Chemie des Parfüms und die Destillationen. Ein Beitrag zur Geschichte der arabischen Parfümchemie und Drogenkunde aus dem 9. Jahrh. P. C.* Ed. and trans. with commentary by K. Garbers. Abhandlungen für die Kunde des Morgenlandes 30. Leipzig, 1948.

Kitāb Aḥwāl al-qiyāmah. Ed. M. Wolff in *Muhammedanische Eschatologie nach der leipziger und der dresdner Handschrift zum ersten Male arabisch und deutsch mit Anmerkungen herausgegeben*. Leipzig: Brockhaus, 1872.

Kitāb al-Hadāyā wa-l-tuḥaf. Ed. M. Ḥamīdullāh as Ibn al-Zubayr. *Kitāb al-Dhakhā'ir wa-l-tuḥaf*. Kuwait: Dā'irat al-Maṭbū'āt wa-al-Nashr, 1959.

Kitāb al-Hadāyā wa-l-tuḥaf. Trans. Ghāda al-Ḥijjāwī al-Qaddūmī. *Book of Gifts and Rarities*, Cambridge, MA: Distributed for the Center for Middle Eastern Studies of Harvard University by Harvard University Press, 1996.

Ge Hong 葛洪. *Bao Puzi* 抱朴子. Taipei: Zhongguo zixue mingzhu jicheng, 1978.

Al-Kulaynī al-Rāzī, *al-Furu' min al-kāfī*. 8 vols. Tehran: Dār al-Kutub al-Islāmiyyah, 1983.

Kuthayyir. *Sharḥ Dīwān Kuthayyir*. Ed. H. Pérès. 2 vols. Paris: Paul Geuthner, 1928–30.

Liaoshi 遼史. Beijing: Zhonghua shuju, 1974.

Li Shizhen 李時珍. *Bencao gangmu* 本草綱目. *Guoxue jiben congshu* edition. Taipei, 1968.

Linschoten, Jan Huyghen. *The Voyage to the East Indes*. Ed. and trans. A. C. Burnell and P. A Tiele. 2 vols. London: Hakluyt Society, 1885.

Lucretius. *De rerum natura*. Ed. and trans. W. H. D. Rouse. Loeb Classical Library. Cambridge, MA: Harvard University Press, 1937.

Ma Huan 馬歡. *Ying ya sheng lan* 瀛涯勝覽. *Congshu jicheng chu bian* 叢書集成初編 edition. Shanghai: Shang wu yin shu guan, Minguo 26 [1937].

Ma Huan 馬歡. *Ying-yai Sheng-lan: The Overall Survey of the Ocean's Shore*. Trans. J. V. Mills. London: Hakluyt Society, 1970; repr. Bangkok: White Lotus, 1997.

Al-Maʿarrī, Abū al-ʿAlā. *The Letters of Abū 'l-ʿAlā of Maʿarrat al-Nuʿmān*. Ed. and trans. D. S. Margoliouth. Oxford: Clarendon Press, 1898.

Al-Maʿarrī, Abū al-ʿAlā. *Shurūḥ saqṭ al-zand*. Ed. Ṭāhā Ḥusayn, Muṣṭafā Saqqā, et al. 5 vols. Cairo: al-Dār al-Qawmiyyah li-l-Ṭibāʿah wa-l-Nashr, 1964.

Al-Maʿarrī, Abū al-ʿAlā. *Risālat al-ghufrān*. Ed. ʿĀʾisha ʿAbd al-Raḥman. 5th ed. Cairo: Dār al-Maʿārif, 1969.

Madanapāla. *Materia Medica of Ayurveda based on Madanapala's Nighaṇṭu*. Ed. Vaidya Bhagwan Dash and K. K. Gupta. New Delhi: Health Harmony, 1991; repr. 2001.

Maimonides, Moses. *Moses Maimonides' Glossary of Drug Names*. Ed. M. Meyerhof. Trans. F. Rosner. Philadelphia: American Philosophical Society, 1979.

Al-Majlisī, Muḥammad Bāqir b. Muḥammad Taqī. *Biḥār al-anwār*. 110 vols. Beirut: Muʾassasat al-wafāʾ, 1983.

Marco Polo. *The Travels of Marco Polo*. Trans. and annotated by H. Yule and H. Cordier. 3rd ed. 2 vols. London, 1929; repr. New York: Dover, 1993.

Al-Marwazī, Sharaf al-Zamān Ṭāhir. *Ṭabāʾiʿ al-ḥayawān*. Partially ed. and trans. V. Minorsky. *Sharaf al-Zaman Tahir Marvazi on China, the Turks and India*. London: Royal Asiatic Society, 1942.

Al-Masʿūdī, Abū al-Ḥasan ʿAlī b. al-Ḥusayn. *Murūj al-dhahab wa-maʿādin al-jawhar*. Ed. B. de Meynard and P. de Courteille, revised by C. Pellat. 5 vols. Beirut: Manshūrāt al-Jāmiʿah al-Lubnāniyyah, 1966–74.

Menander Protector. *The History of Menander the Guardsman*. Ed. and trans. R. C. Blockley. Liverpool: Cairns, 1985.

De Mendoza, Juan Gonzalez. *The History of the Great and Mighty Kingdom of China*. Trans. R. Parke and ed. G. T. Staunton. 2 vols. London: Hakluyt Society, 1853–4.

Al-Mufaḍḍal b. Muḥammad al-Ḍabbī. *Al-Mufaḍḍalīyāt*. 3 vols. Ed. and trans. C. Lyall. Oxford: Clarendon Press, 1918.

Al-Muqaddasī, Shams al-Dīn Abū ʿAbdallāh Muḥammad b. Aḥmad b. Abī Bakr al-Bannāʾ al-Shāmī. *Aḥsan al-Taqāsīm fī Maʿarifat al-Aqālīm*. Ed. M. J. De Goeje. Leiden: Brill, 1877; repr. 1967.

Murasaki Shikibu, *The Tale of Genji*. Trans. Arthur Waley. New York: Modern Library 1960.

Muslim ibn al-Ḥajjāj. *Saḥīḥ*. 5 vols., Beirut: Dār al-Kutub al-ʿIlmiyyah, 1998.

Al-Mutanabbī, Abū al-Ṭayyib Aḥmad b. al-Ḥusayn. *Dīwān al-Mutanabbī*. Beirut: Dār Ṣādir, 1958.

Narshakhī, Abū Bakr Muḥammad ibn Jaʿfar. *Tārīkh-i Bukhārā*. Ed. Mudarris Raḍawī. 2nd ed. Tehran: Intishārāt-i Tūs, 1363/1984.

Narshakhī, Abū Bakr Muḥammad ibn Jaʿfar. Trans. R. N. Frye. *The History of Bukhara*. Cambridge, MA: Mediaeval Academy of America, 1954.

Al-Nasāʾī, Aḥmad b. Shuʿayb. *Sunan al-Nasāʾī*. 8 vols. Cairo: Muṣṭafā al-Bābī al-Ḥalabī, 1964–5.

Niẓāmī ʿArūḍī. *Kitāb-i Chahār maqālah*. Ed. Muḥammad Qazwīnī. London: Luzac, 1927.

Niẓāmī Ganjavī. *Haft Paykar*. Ed. H. Ritter and J. Rypka. Prague: Orientální Ústav, 1934.

Al-Nuwayrī, Shihāb al-Dīn Aḥmad b. ʿAbd al-Wahhāb. *Nihāyat al-arab fī funūn al-adab*. 33 vols. Cairo: Dār al-Kutub al-Miṣriyyah, 1923–97.

Pañcatantra. Ed. F. Edgerton. *The Panchatantra Reconstructed*. Vol. 1. New Haven: American Oriental Society, 1924.

Paul of Aegina. Ed. I. L. Heiberg. 2 vols. Leipzig: Teubner, 1921–4.

Paul of Aegina. Trans. I. Berendes. *Paulos von Aegina. Des besten Arztes Sieben Bücher*. Leiden: Brill, 1914.

Periplus Maris Erythraei. Ed. and trans. L. Casson. Princeton: Princeton University Press, 1989.

Persius. *Satires*. Ed. and trans. G. C. Ramsay. Loeb Classical Library. Cambridge, MA: Harvard University Press, 1918.

Philostratus. *Imagines*. Ed. and trans. A. Fairbanks. Loeb Classical Library. Cambridge, MA: Harvard University Press, 1931.

Pigafetta, Antonio. *Magellan's voyage: a narrative of the first circumnavigation*. Trans. R. A. Skelton, New Haven: Yale, 1969; repr. New York: Dover, 1994.

Pires, Tomé. *The Summa Oriental of Tomé Pires and the Book of Francisco Rodrigues*. Trans. A. Cortesão. 2 vols. London: Hakluyt Society, 1944.

Pliny the Elder. *Natural History*. Ed. and trans. H. Rackham, et al. 10 vols. Loeb Classical Library. Cambridge, MA: Harvard University Press, 1962.

Procopius. *History of the Wars*. 5 vols. Ed. and trans. H. B. Dewing. Loeb Classical Library. New York: Macmillan, 1914–1928.

Pseudo-Callisthenes. *The History of Alexander the Great, being the Syriac version, edited from five manuscripts, of the Pseudo-Callistnenes*. Ed. and trans. E. A. W. Budge. Cambridge, 1889; repr. Amsterdam: Philo Press, 1976.

Al-Qalqashandī, Shihāb al-Dīn Aḥmad b. ʿAlī. *Ṣubḥ al-Aʿshā*. 14 vols. Cairo: Al-Muʾassasah al-Miṣriyyah al-ʿĀmmah li-l-Taʾlif, 1964.

Quan Tangshi 全唐詩. 8 vols. Taibei: Fu xing shuju, 1967.

Al-Qazwīnī, Ḥamdullāh al-Mustawfī. *The Zoological Section of the Nuzhatu-l-Qulūb*. Ed. and trans. J. Stephenson. London: Royal Asiatic Society, 1928.

Al-Qazwīnī, Zakarīyā ibn Muḥammad. *ʿAjāʾib al-makhlūqāt* and *Āthār al-bilād*. Ed. T. Wüstenfeld in *Zakarija ben Muhammed ben Mahmud el-Cazwini's Kosmographie*. 2 vols. Göttingen, 1848–9; repr. in 1 vol.: Wiesbaden: Martin Sändig, 1967.

Al-Qurṭubī, Muḥammad b. Aḥmad. *al-Jāmiʿ li-aḥkām al-Qurʾān*. 20 vols. Cairo: Dār al-Kātib al-ʿArabī li-l-Ṭibāʿah wa-l-Nashr, 1967.

Al-Rāzī, Abū Bakr Muḥammad ibn Zakariyyā. *Al-Ḥāwī al-kabīr fī al-ṭibb*. 23 vols. in 25. Haydarābād: Maṭbaʿat Majlis Dāʾirat al-Maʿārif al-ʿUthmāniyah, 1958–79.

Rgyud bźi. A Reproduction of a set of prints from the 18th century Zuṅ-Cu Ze Blocks from the Collection of Prof. Raghu Vira. Leh: S. W. Tashigangpa, 1975.

Rivāyat Accompanying the Dādestān ī Dēnīg. Ed. and trans. A. V. Williams. 2 vols. Copenhagen: Munksgaard, 1990.
Rūmī, Jalāl al-Dīn. *The Mathnawī*. Ed. R. A. Nicholson. 3 vols. E. J. W. Gibb Memorial Series, New Series 4.1. London: Luzac, 1925–33.
Sa Skya Paṇḍita. *Subhāṣitaratnanidhi*. Tibetan and Mongolian texts ed. and trans. by J. E. Bosson, *A Treasury of Aphoristic Jewels*. Bloomington: Indiana University, 1969.
Al-Ṣābi', Hilāl b. al-Muḥassin. *Rusūm Dār al-Khilāfah*. Baghdād: Maṭba'at al-'Ānī, 1964.
Sābūr b. Sahl. *Dispensatorium Parvum (al-Aqrābādhīn al-Ṣaghīr)*. Ed. O. Kahl. Leiden: Brill, 1994.
Sābūr b. Sahl. *Sābūr ibn Sahl: The Small Dispensatory*. Trans. O. Kahl. Leiden: Brill, 2003.
Sābūr b. Sahl. *Sābūr ibn Sahl's Dispensatory in the Recension of the 'Aḍudī Hospital*. Leiden: Brill, 2009.
Sahlān b. Kaysān, Abū al-Ḥasan. *Mukhtaṣar fī al-Ṭīb*. Ed. P. Sbath "Abrégé sur les arômes." *Bulletin de l'Institut d'Égypte* 26 (1943–4): 183–213.
Sahlān b. Kaysān, Abū al-Ḥasan. *Sahlān ibn Kaysān et Rashīd al-Dīn Abū Ḥulayqa: Deux traités médicaux*. Ed. P. Sbath and C. D. Avierinos. Cairo: Institut Français d'archéologie orientale, 1953.
Al-Sarī al-Raffā'. *Al-muḥibb wa-l-maḥbūb wa-l-mashmūm wa-l-mashrūb*. Ed. Miṣbāḥ Ghalāwinjī. 4 vols. Damascus: Majma' al-Lughah al-'Arabiyyah, 1986–7.
Al-Shayzarī, 'Abd al-Raḥmān b. Naṣr. *Kitāb Nihāyat al-rutbah fī ṭalab al-ḥisbah*. Ed. al-Sayyid al-Bāz al-'Arīnī. Beirut: Dār al-Thaqāfah, n.d.
Al-Shayzarī, 'Abd al-Raḥmān b. Naṣr. *The Book of the Islamic Market Inspector*. Trans. R. P. Buckley. Oxford: Oxford University Press, 1999.
Shennong bencao. 神農本草. *Congshu jicheng* edition. Shanghai: Shangwu yinshu guan, 1937.
Simeon Seth. *Simeonis Sethi syntagma de alimentorum facultatibus*. Ed. B. Langkavel. Leipzig: Teubner, 1868.
Simeon Seth. Trans. M. Brunet. *Siméon Seth, médecin de l'empereur Michel Doucas, sa vie- son œuvre*. Bordeaux: Delmas, 1939.
Al-Sīrāfī, Abū Zayd. *Riḥlah*. Ed. 'Abdallāh al-Ḥabashī. Abu Dhabi: Manshūrāt al-Majma' al-Thaqāfī, 1999.
Suśrutasaṃhitā. Ed. and trans. P. V. Sharma. 3 vols. Varanasi: Chaukhambha Visvabharati, repr. 2004–5.
Suvarṇaprabhāsottamasūtra. Jin guang ming zui sheng wang jing 金光明最勝王經. Taibei: Fo jiao chu ban she, Minguo 85 [1996].
Suvarṇaprabhāsottamasūtra. Sutra zolotogo bleska: Tekst uigurskoi redaktsii. Ed. V. V. Radlov and S. E. Malov. Saint Petersburg: Tipografiia Imperatorskoi Akademii Nauk, 1913–19.
Al-Suyūṭī, Jalāl al-Dīn Abū al-Faḍl 'Abd al-Raḥmān b. Abī Bakr. *al-Muhadhdhab fīmā waqa'a fī al-Qur'ān min al-mu'arrab*. Beirut: Dār al-Kitāb al-'Arabī, 1995.

Syriac Book of Medicines. Ed. and trans. by E. A. Wallis Budge. *Syrian Anatomy, Pathology and Therapeutics or "The Book of Medicines"*. 2 vols. London Oxford University Press, 1913.

Al-Ṭabarī, Abū Jaʿfar Muḥammad. *Jāmiʿ al-bayān ʿan taʾwīl āy al-Qurʾān*. 30 vols. Cairo: Muṣṭafā al-Bābī al-Ḥalabī, 1954.

Al-Ṭabarī, Abū Jaʿfar Muḥammad. *Taʾrīkh al-rusul wa-l-mulūk*. 15 vols. Ed. M. J. De Goeje, et al. Leiden, 1879–1901; repr. Leiden, 1964–1965.

Al-Ṭabarī, Abū Jaʿfar Muḥammad. Trans. G. R. Hawting, *The History of al-Tabarī Volume XX. The Collapse of Sufyānid Authority and the Coming of the Marwanids*. Albany: State University of New York Press, 1989.

Al-Ṭabarī, ʿAlī b. Rabban. *Firdaws al-ḥikmah fī al-ṭibb*. Ed. M. Z. Siddiqi. Berlin: Sonne, 1928.

Taiping yulan 太平御覽. Beijing: Zhonghua Shuju, 1960.

Tamba Yasuyori. *Ishimpō: The Essentials of Medicine in Ancient China and Japan*. Trans. E. C. H. Hsia, et al. 2 vols. Leiden: Brill, 1986.

Al-Tamīmī, Abī ʿAbdallāh Muḥammad b. Aḥmad b. Saʿid. *Ṭīb al-ʿarūs wa-rayḥān al-nufūs fī ṣināʿat al-uṭūr*. Ed. Luṭf Allāh Qārī and Aḥmad Fuʾād Bāshā. Cairo: Maṭbaʿat Dār al-Kutub wa-l-Wathāʾiq al-Qawmiyyah bi-l-Qāhirah, 2014.

Al-Tanūkhī, al-Muḥassin b. ʿAlī. *al-Faraj baʿda al-shiddah*. 2 vols. Cairo: Dār al-Ṭibāʿah al-Muḥammadiyyah, 1955.

Al-Tanūkhī, al-Muḥassin b. ʿAlī. *Nishwār al-muḥāḍarah wa-akhbār al-mudhākarah*. Ed. ʿAbbūd al-Shāljī, 9 vols. [Beirut: Dār Ṣādir], 1971.

Ṭarafah b. al-ʿAbd al-Bakrī. *Dīwān*. Ed. M. Seligsohn. Paris, 1901; repr. Baghdād: Maktabat al-Muthannā, [1968].

Tavernier, Jean-Baptiste. *Travels in India by Jean-Baptiste Tavernier*. Trans. V. Ball and ed. W. Crooke. 2 vols. London, 1889; repr. New Delhi: Munshiram Manoharlal, 1995.

Al-Thaʿālibī, ʿAbd al-Malik ibn Muḥammad. *Laṭāʾif al-maʿārif*. Cairo: Dār Iḥyāʾ al-Kutub al-ʿArabiyyah ʿĪsā al-Bābī al-Ḥalabī wa-Shurakah, n.d.

Al-Thaʿālibī, ʿAbd al-Malik ibn Muḥammad. *The Book of Curious and Entertaining Information*. Trans. C. E. Bosworth. Edinburgh: University Press, 1968.

Al-Thaʿālibī, ʿAbd al-Malik ibn Muḥammad. *Tārīkh Ghurar al-siyar*. Ed. and trans. H. Zotenberg. Paris: Imprimerie Nationale, 1900; repr. Tehran: Maktabat al-Asadī, 1963.

Al-Thaʿlabī, Abū Isḥāq Aḥmad b. Muḥammad b. Ibrāhīm. *ʿArāʾis al-majālis fī qiṣaṣ al-anbiyāʾ*. Published as *Kitāb Qiṣaṣ al-anbiyāʾ*. Cairo: al-Maṭbaʿah al-Ummah, 1331/1912.

Al-Thaʿlabī, Abū Isḥāq Aḥmad b. Muḥammad b. Ibrāhīm. Trans. W. M. Brinner. *ʿArāʾis al-Majālis fī Qiṣaṣ al-Anbiyāʾ or 'Lives of the Prophets'*. Leiden: Brill, 2002.

Al-Tirmidhī, Muḥammad b. ʿĪsā. *Sunan*. 5 vols. Cairo: Maṭbaʿat al-Madanī, 1964.

Al-Ṭūsī, Naṣīr al-Dīn Muḥammad b. Muḥammad. *Tansūkhnāmah-i Īlkhānī*. Ed. Mudarris Raḍawī. Tehran: Intishārāt-i Bunyād-i Farhang-i Īrān, 1969.

ʿUmar b. Abī Rabīʿah. *Dīwān*. Cairo: al-Hay'ah al-Miṣriyyah al-ʿĀmmah li-l-Kitāb, 1978.

Al-ʿUtbī, Muḥammad b. ʿAbd al-Jabbār. *al-Yamīnī fī sharḥ akhbār al-sulṭān Yamīn al-Dawlah wa Amīn al-Millah Maḥmud al-Ghaznawī*. Ed. Iḥsān Dhunūn al-Thāmirī. Beirut: Dar al-ṭalīʿah li-l-ṭibāʿah wa-l-nashr, 2004.

Vāgbhaṭa. *Aṣṭāṅgahṛdayasaṃhitā*. ed. R. P. Das and R. E. Emmerick. Groningen: Egbert Forsten, 1998.

Vāgbhaṭa. *Aṣṭāṅgahṛdayasaṃhitā. The First Five Chapters of its Tibetan Version*. Ed. and trans. C. Vogel. Wiesbaden, 1965.

Varāhamihira. *Bṛhatsaṃhitā*. Ed. and trans. M. Ramakrishna Bhat. 2 vols. Delhi: Motilal Banasidass, 1981–82.

Varthema, Ludovico di. *The Travels of Ludovico di Varthema*. Trans. John Winter Jones and ed. George Percy Badger. London: Hakluyt Society, 1863.

Vidyākara. *Subhāṣitaratnakoṣa*. Ed. D. D. Kosambi and V. V. Gokhale. Cambridge, MA: Harvard University Press, 1957.

Vidyākara. *An Anthology of Sanskrit Court Poetry*. Trans. D. H. H. Ingalls. Cambridge, MA: Harvard University Press, 1965.

Al-Washshā', Muḥammad b. Aḥmad, *Kitāb al-Muwashshā*. Ed. R. E. Brünnow. Leiden: Brill, 1886.

Al-Wāsiṭī, Muḥammad b. Aḥmad. *Faḍā'il al-Bayt al-Muqaddas*. Ed. Muḥammad Zaynahum Muḥammad ʿAzab. Beirut: Dār al-Maʿārif, 2001.

Al-Waṭwāṭ, Muḥammad b. Ibrāhīm. *Mabāhij al-fikar wa-manāhij al-ʿibar*. Ed. ʿAbd al-Razzaq Aḥmad al-Ḥarbī. Beirut: al-Dār al-ʿArabiyyah li-l-Mawsūʿāt, 2000.

Xin Tang shu 新唐書. Beijing: Zhonghua Shuju, 1975.

Xinxiu bencao 新修本草. *Shinshū honzō. Zankan*. Tokyo: Meiji Shoin, 1983.

[Al-Yaʿqūbī] Aḥmad b. Abī Yaʿqūb b. Waḍiḥ al-Kātib. *Kitāb al-Buldān*. Ed. M. J. De Goeje in BGA VII (1892), 231–373.

[Al-Yaʿqūbī] Aḥmad b. Abī Yaʿqūb b. Waḍiḥ al-Kātib. *Tārīkh*. Ed. M. T. Houtsma. 2 vols. Leiden: Brill, 1883; repr. 1969.

Yāqūt, Shihāb al-Dīn b. ʿAbdallāh al-Ḥamawī. *Muʿjam al-buldān*. 5 vols. Beirut: Dār Ṣādir, n.d.

Yūsuf Khāṣṣ Ḥājib. *Kutadgu Bilig, Vol. 1*. Ed. R. Arat. 2nd ed. Ankara: Türk Dil Kurumu Yayınları, 1979.

Yūsuf Khāṣṣ Ḥājib. *Wisdom of Royal Glory (Kutadgu Bilig): A Turko-Islamic Mirror for Princes*. Trans. R. Dankoff. Chicago: University of Chicago Press, 1983.

Zādspram, *Wizīdagīhā-ī Zādspram*. Ed. and trans. P. Gignoux and A. Tafazzoli. *Anthologie de Zādspram*. Paris: Association pour l'avancement des études iraniennes, 1993.

Al-Zahrāwī, Abū al-Qāsim. *Kitāb al-Taṣrīf li-man ʿajiza ʿan al-taʾlīf.* Facsimile of Süleymaniye Beşirağa collection, ms. 502. Ed. F. Sezgin. Frankfurt am Main: Institute for the History of Arabic-Islamic Science, 1986.

Al-Zamakhsharī, Maḥmūd b. ʿUmar. *al-Kashshāf ʿan ḥaqāʾiq al-tanzīl.* 2 vols., Cairo, 1948.

Al-Zamakhsharī, Maḥmūd b. ʿUmar. *Rabīʿ al-abrār wa-fuṣūṣ al-akhbār*. Eds. ʿAbd al-Majīd Diyāb and Ramaḍān ʿAbd al-Tawwāb. 2 vols. Cairo: al-Hayʾah al-Miṣriyyah al-ʿĀmmah li-l-Kitāb, 1992–2001.

Zhang Bangji 張邦基. *Mo zhuang man lü* 墨莊漫錄. 四庫全書 SPTK San-p'ien v. 34.

Zhao Rugua 趙汝适. *Zhufan zhi* 諸蕃志. Ed. and trans. F. Hirth and W. Rockhill. *Chau Ju-kua: His work on the Chinese and Arab Trade in the Twelfth and Thirteenth Centuries, entitled Chu-fan chih*. 2 vols. St. Petersburg, 1911.

Zhoushu 周書. Beijing: Zhonghua Shuju, 1971.

Secondary

Abraham, M. *Two Medieval Merchant Guilds of South India*. New Delhi: Manohar, 1988.

Afsaruddin, A. "Garden." *EQ* s.v.

Agius, D. *Arabic Literary Works as a Source of Documentation for the Technical Terms of Material Culture.* Berlin: Klaus Schwarz, 1984.

Agius, D. *Classic Ships of Islam from Mesopotamia to the Indian Ocean*. Leiden: Brill, 2008.

Agnew, N. et al., eds. *Cave Temples of Dunhuang: Buddhist Art on China's Silk Road.* Los Angeles: Getty Conservation Institute, 2016.

Ahsan, M. M. *Social Life under the Abbasids 170–289 AH, 786–902 AD*. London: Longman, 1979.

Akasoy, A. "Tibet in Islamic Geography and Cartography: A Survey of Arabic and Persian Sources." In A. Akasoy, C. Burnett, and R. Yoeli-Tlalim, eds., *Islam and Tibet: Interactions along the Musk Routes*. Farnham: Ashgate, 2011, 17–41.

Akasoy, A. and R. Yoeli-Tlalim. "Along the Musk Routes: Exchanges between Tibet and the Islamic World." *Asian Medicine* 3 (2007): 217–40.

Allan, J. W. *Persian Metal Technology 700–1300 AD*. London: Ithaca Press, 1979.

Allsen, T. T. "Mongolian Princes and their Merchant Partners." *Asia Major* 3rd. ser. 2 (1989): 83–126.

Allsen, T. T. *Commodity and Exchange: A Cultural History of Islamic Textiles*. Cambridge: Cambridge University Press, 1997.

Amar, Z., and E. Lev. "Trends in the Use of Perfumes and Incense in the near East after the Muslim Conquests." *JRAS* (2013): 11–30.

Amar, Z., Lev, E., and Y. Serri. "Ibn Rushd on Galen and the New Drugs Spread by the Arabs." *JA* 297 (2009): 83–101.

Amar, Z., Lev, E., and Y. Serri. "On Ibn Juljul and the meaning and importance of the list of medicinal substances not mentioned by Dioscorides." *JRAS* (2014): 529–55.

Anderson, K. M. *Kōdō: the Way of Incense*. MA Thesis, Indiana University, Bloomingon, 1984.

André, J., and Filliozat, J. *L'Inde vue de Rome: Textes latins de l'Antiquité relatifs a 'l'Inde*. Paris: Les Belles Lettres, 1986.

Anonymus. "The Town Warehouses of the East and West India Dock Company." *The Chemist and Druggist* 34 (1889): 452–7.

Anonymus. "The Anointing of the Queen: some notes on the coronation oil." *Pharmaceutical Journal* 170 (Jan.-June 1953), 404–5, 415.

Anṣārī, S. *Tārīkh-i 'Iṭr dar Īrān*. Tehran: Wizārat-i Farhang wa Irshād-i Islāmī, 1381/2002–3.

Arakawa, M. "The Transit Permit System of the Tang Empire and the Passage of Merchants." *MRDTB* 59 (2001): 1–21.

Arat, R. A. "Zur Heilkunde der Uiguren." *Sitzungsberichte der preussischen Akademie der Wissenschaften Phil.-hist. Kl.* (1930): 452–73 and (1932): 401–88.

Arctander, S. *Perfume and Flavor Materials of Natural Origin*. Elisabeth, N.J., privately published, 1960.

Arora, R. B., Seth, S. D. S., and Somani, P. "Effectiveness of Musk (Kasturi), an indigenous drug, against *Echis carinatus* (the Saw-Scaled Viper) Envenomation." *Life Sciences* 9 (1962): 453–7.

Arundhati, P. *Royal Life in Mānasōllāsa*. Delhi: Sundeep Prakashan, 1994.

Asbaghi, A. *Persische Lehnwörter im Arabischen*. Wiesbaden: Harrassowitz, 1988.

Ashtor, E. "Ḳaranful" in *EI²* s.v.

Aubaile-Sallenave, F. "*Bân, un parfum et une image de la souplesse. L'histoire d'un arbre dans le monde arabo-musulman.*" In R. Gyselen, ed., *Parfums d'Orient*. Bures-sur-Yvette: Groupe pour l'Étude de la Civilisation du Moyen-Orient, 1998, 9–27.

Avanzini, A., ed. *Profumi d'Arabia*. Rome: Bretschneider, 1997.

Bahār, M. *Vāzhihnāmah-i Bundahish*. Tehran: Iranian Cultural Foundation, 1345.

Bailey, H. W. "Iranian Studies." *Bulletin of the School of Oriental Studies* 6:4 (1932): 945–55.

Bailey, H. "The Profession of Prince Tcūṃ-Ttehi." In E. Bender, ed., *Indological Studies in Honor of W. Norman Brown*. New Haven: American Oriental Society, 1962, 18–22.

Bailey, H. *Indo-Scythian Studies being Khotanese Texts*, vol. 1. Reprinted in one volume with volumes 2 and 3. Cambridge: Cambridge University Press, 1969.

Bailey, H. *A Dictionary of Khotan Saka*. Cambridge: Cambridge University Press, 1979.

Baldwin, B., and A. Cutler. "Kosmas Indikopleustes." In *Oxford Dictionary of Byzantium*, vol. 2, 1151–2.

Balfour-Paul, J. *Indigo in the Arab World*. Richmond: Curzon Press, 1997.

Bang, W., and von Gabain, A. "Türkische Turfan-Texte." *Sitzungsberichte der preussischen Akademie der Wissenschaften zu Berlin, Phil.-hist. Kl.* (1929): 241–68 and pls. 3–4.

Barthold, W. *Turkestan down to the Mongol Invasion*. 3rd ed. London, 1968; repr. Taipei: Southern Materials Center, n.d.

Barthold, W., A. Bennigsen, and H. Carrère-d'Encausse, "Badakhshān." *EI* 2nd ed. s.v.

Baxter, W. H. *Handbook of Old Chinese Phonology*. Berlin: Mouton de Gruyter, 1992.

Bazin, N. "Fragrant Ritual Offerings in the Art of Tibetan Buddhism." *JRAS* (2013): 179–207.

Bazin, L. "Un nom 'turco-mongol' du 'nombril' et du 'clan'." In M. Erdal and S. Tezcan, eds., *Beläk Bitig. Sprachstudien für Gerhard Doerfer zum 75. Geburtstag*. Wiesbaden: Harrassowitz, 1995, 1–12.

Beckwith, C. I. "Tibet and the Early Medieval *Florissance* in Eurasia: A Preliminary Note on the Economic History of the Tibetan Empire." *CAJ* 21 (1977): 89–104.

Beckwith, C. I. "The Introduction of Greek Medicine into Tibet in the Seventh and Eighth Centuries." *JAOS* 99 (1979): 297–313.

Beckwith, C. I. "The Location and Population of Tibet according to Early Islamic Sources." *Acta Orientalia Academiae Scientiarum Hungaricae* 43 (1989): 163–70.

Beckwith, C. I. "The Impact of the Horse and Silk Trade on the Economies of T'ang China and the Uighur Empire: On the Importance of International Commerce in the Early Middle Ages." *JESHO* 34 (1991): 183–98.

Beckwith, C. I. *The Tibetan Empire in Central Asia: A History of the Struggle for Great Power among Tibetans, Turks, Arabs and Chinese during the Early Middle Ages*. Rev. ed. Princeton: Princeton University Press, 1993.

Beckwith, C. I. "The Central Eurasian Culture Complex in the Tibetan Empire: The Imperial Cult and Early Buddhism." In *1000 Jahre Asiatisch-Europäische Begegnung*, ed. R. Erken. Berlin: Peter Lang, 2011.

Beckwith, C. I. *Empires of the Silk Road*. Princeton: Princeton University Press, 2009.

Bedini, S. A. *The Trail of Time: Time measurement with incense in East Asia*. Cambridge: Cambridge University Press, 1994.

Beg, M. A. J. "Tādjir." *EI* 2nd ed. s.v.

Begley, V. and R. D. De Puma, eds. *Rome and India: The Ancient Sea Trade*. Madison: University of Wisconsin Press, 1991.

Bender, E. "An Early Nineteenth Century Study of the Jains." *JAOS* 96:1 (1976): 114–9.

Benveniste, E. *Textes Sogdiens édités, traduits et commentés*. Paris: Geuthner, 1940.

Beyer, S. *The Cult of Tārā: Magic and Ritual in Tibet*. Berkeley: University of California, 1973; repr. 1978.

Billing, J. and P. W. Sherman. "Antimicrobial Functions of Spices: Why Some Like it Hot." *Quarterly Review of Biology* 73 (1998): 3–49.

Biran, M. *The Empire of the Qara Khitay in Eurasian History: Between China and the Islamic World.* Cambridge: Cambridge University Press, 2005.

Biran, M. "Unearthing the Liao Dynasty's Relations with the Muslim World: Migrations, Diplomacy, Commerce, and Mutual Perceptions." *Journal of Song-Yuan Studies* 43 (2013): 221–51.

Bivar, A. D. H. "Trade between China and the Near East in the Sasanian and Early Muslim Periods." In W. Watson, ed., *Pottery and Metalwork in T'ang China.* London: Percival David Foundation of Chinese Art, 1970, 1–11.

Bivar, A. D. H. "The History of Eastern Iran." In E. Yarshater, ed. *The Cambridge History of Iran Volume 3 Part 1.* Cambridge: Cambridge University Press, 1983, 181–231.

Blanchard, I. *Mining, Metallurgy and Minting in the Middle Ages Vol. 1. Asiatic Supremacy. Mawara'an-nahr and the Semiryech'ye 425–1125.* Stuttgart: Steiner, 2001.

Bloom, J. M. "Revolution by the Ream." *ARAMCO World* 50:3 (May/June 1999): 26–39.

Bloom, J. M. *Paper before Print: The History and Impact of Paper in the Islamic World.* New Yaven: Yale, 2001.

Blower, J. "Conservation Priorities in Burma." *Oryx* 19 (1985): 79–85.

Böhtlingk, O. *Indische Sprüche.* 3 vols.Osnabrück: Otto Zeller, 1966; repr. of St. Ptersburg, 1870–3 edition.

Bopearachchi, O. "Archaeological Evidence on Shipping Communities of Sri Lanka." In *Ships and the Development of Maritime Technology in the Indian Ocean.* London: Routledge, 2002, 92–127.

Borschberg, P. "Der asiatische Moschushandel vom frühen 15. bis zum 17. Jahrhundert." In J. M. dos Santos Alves, et al., eds., *Mirabilia Asiatica: Productos raros no comércio marítimo*, vol. 1. Wiesbaden: Harrassowitz, 2003, 65–83.

Borschberg, P. "Der asiatische Ambra-Handel während der frühen Neuzeit (15. bis 18. Jahrhundert)." In J. M. dos Santos Alves, et al., eds., *Mirabilia Asiatica: Productos raros no comércio marítimo*, vol. 2. Wiesbaden: Harrassowitz, 2005, 167–201.

Bos, G. "The *miswāk*, an aspect of dental care in Islam." *Medical History* 37 (1993): 68–79.

Bos, G. "Ibn al-Jazzār on Medicine for the Poor and Destitute." *JAOS* 118:3 (1998): 365–75.

Bosworth, C. E. "Iran and the Arabs before Islam." In E. Yarshater, ed., *The Cambridge History of Iran Volume 3 Part 1.* Cambridge: Cambridge University Press, 1983, 593–612.

Bosworth, C. E. "Afšīn." *EIr* s.v.

Bosworth, C. E. "Kish." *EI* 2nd ed. s.v.

Bovill, E. W. "Musk and Amber." *Notes and Queries* 198 (1953): 487–9, 508–10; n.s. 1 (1954): 24–5, 69–72, 121–3, 151–4.

Bretschneider, E. *Botanicon Sinicum.* 3 vols. 1881–1895; repr. Nendeln: Kraus, 1967.

Brockelmann, C. *Lexicon Syriacum.* Halle: Max Niemeyer, 1928.

Brucker, E. "AI. Nálada= Nardostachys Jatamansi DC. Ein Beitrag zur indischen Pflanzenkunde." *Asiatische Studien* 29 (1975): 131–6.

Brunner, C. J. "The Fable of the Babylonian Tree." *JNES* 39 (1980): 191–202 and 291–302.

Brust, M. *Die indischen und iranischen Lehnwörter im Griechischen*. Innsbruck: Instituten für Sprachen und Literaturen, 2005.

Burkill, I. H. *A Dictionary of the Economic Products of the Malay Peninsula*. 2 vols. Kuala Lumpur: Ministry of Agriculture, 1966.

Burton, R. *Personal Narrative of a Pilgrimage to al-Madinah and Meccah*. 2 vols. London: Tylston and Edwards, 1893; repr. New York: Dover, 1964.

Cahill, S. "Sex and the Supernatural in Medieval China." *JAOS* 105:2 (1985): 197–220.

Capula, M. *Simon & Schuster's Guide to Reptiles and Amphibians of the World*. New York: Simon & Schuster, 1989.

Carles, W. R. "Recent Journeys in Korea." *Proceedings of the Royal Geographical Society* (1886): 307.

Carswell, J. "The Port of Mantai, Sri Lanka." In V. Begley and R. De Puma, *Rome and India: The Ancient Sea Trade*. Madison: University of Wisconsin, 1991, 197–203.

Carswell, J. "China and the Middle East." *Oriental Art* 45:1 (1999): 2–14.

Caseau, B. "Les usages médicaux de l'encens et des parfums. Un aspect de la médecine populaire antique et de sa christianisation." In *Air, miasmas et contagion: les épidémies dans l'Antiquité et au Moyen Age*. Prez-sur-Marne: D. Guéniot, 2001, 75–85.

Caseau, B. "Incense and Fragrances: from House to Church: A Study of the Introduction of Incense in the Early Byzantine Christian Churches." In M. Grünbart, et al., eds., *Material Culture and Well-Being in Byzantium (400–1453)*. Vienna: Akademie der Wissenschaften, 2007, 75–92.

Casson, L. "Rome's Trade with the East: The Sea Voyage to Africa and India." In L. Casson, *Ancient Trade and Society*. Detroit: Wayne State, 1984, 182–98.

Chakravarti, R. "Nakhudas and Nauvittakas: Ship-Owning Merchants in the West Coast of India." *JESHO* 43 (2000): 34–64.

Chakravarti, R. "Seafarings, Ships and Ship Owners: India and the Indian Ocean (AD 700–1500)." In *Ships and the Development of Maritime Technology in the Indian Ocean*. London: Routledge, 2002, 28–61.

Chaudhuri, K. N. *Trade and Civilisation in the Indian Ocean: An Economic History from the Rise of Islam to 1750*. Cambridge: Cambridge University Press, 1985.

Chavannes, É. *Les documents chinois découverts par Aurel Stein dans les sables du Turkestan Oriental*. Oxford: Oxford University Press, 1913.

Chipman, L. N. *The World of Pharmacy and Pharmacists in Mamlūk Cairo*. Leiden: Brill, 2010.

Chipman, L. N., and E. Lev. "Syrups from the Apothecary's Shop: A Genizah Fragment Containing One of the Earliest Manuscripts of the *Minhāj al-Dukkān*." *Journal of Semitic Studies* 50 (2006): 137–68.

Ciancaglini, C. *Iranian Loanwords in Syriac*. Wiesbaden: Reichert, 2008.

Clark, L. V. *Introduction to the Uyghur Civil Documents of East Turkestan (13th–14th cc.)*. PhD. Dissertation. Indiana University, Bloomington. 1975.

Clauson, G. *Turkish and Mongolian Studies*. London: Royal Asiatic Society, 1962.

Clauson, G. *An Etymological Dictionary of Pre-Thirteenth Century Turkish*. Oxford: Oxford University Press, 1972.

Coedès, G. "La stèle de Ta-prohm." *Bulletin de l'École Français d'Extreme-Orient* 6 (1906): 44–85.

Cohen, H. J. "The Economic Background and the Secular Occupations of Muslim Jurisprudents and Traditionists in the Classical Period of Islam." JESHO 13 (1970): 16–61.

Colless, B. C. "Persian Merchants and Missionaries in Medieval Malaya." JMBRAS 42 (1969): 10–47.

Colless, B. C. "The Traders of the Pearl." *Abr-Nahrain* 9 (1969–70): 17–38; 10 (1970–1): 102–21; 11 (1971): 1–21; 13 (1972–3): 115–35; 14 (1973): 1–16; 15 (1974–5): 6–17; 18 (1978–9): 1–18.

The Compact Oxford English Dictionary. 2nd ed. Oxford: Oxford University Press, 1991.

Corbin, Alain. *The Foul and the Fragrant*. Cambridge, MA: Harvard University Press, 1986.

Cornu, G. *Atlas du monde arabo-islamique à l'époque classique, Ixe–Xe siècles*. Leiden: Brill, 1985.

Cribb, J. "Numismatic Evidence for Kushano-Sasanian Chronology." *Studia Iranica* 19 (1990): 151–93.

Crone, P. *Meccan Trade and the Rise of Islam*. Princeton: Princeton University Press, 1987.

Crone, P. "Quraysh and the Roman Army: Making sense of the Meccan Leather Trade." BSOAS 70 (2007): 63–88.

Dalby, Andrew. *Dangerous Tastes: The Story of Spices*. Berkeley: University of California Press, 2000.

Dalby, Andrew. "Some Byzantine Aromatics." In L. Brubaker and K. Linardou, eds., *Eat, Drink, and be Merry (Luke 12:19)- Food and Wine in Byantium*. Aldershot: Ashgate, 2007, 51–8.

Dankoff, R. "Three Turkic Verse Cycles Relating to Inner Asian Warfare." *Harvard Ukrainian Studies* 3 (1979): 151–65.

Dannenfeldt, K. H. "Ambergris: The Search for its Origin." *Isis* 73:3 (1982): 382–97.

Dannenfeldt, K. H. "Europe Discovers Civet Cats and Civet." *History of Biology* 18:3 (1985), 403–31.

Dao Van Tién. "Sur quelques rares mammifères au nord du Vietnam." *Mitteilungen aus dem Zoologischen Museum in Berlin* 53 (1977): 325–30.

Dauzat, A. *Dictionnaire étymologique de la langue française*. Paris: Larousse, 1938.

De la Vaissière, É. *Sogdian Traders: A History*. Trans. J. Ward. Leiden: Brill, 2005.

De la Vaissière, É., and É. Trombert, eds. *Les Sogdiens in Chine*. Paris: École Française d'Extrême-Orient, 2005.

Decker, M. "Plants and Progress: Rethinking the Islamic Agricultural Revolution." *Journal of World History* 20 (2009): 187–206.

Della Vida, G. Levi. "A Druggist's Account on Papyrus." In *Archaeologia orientalia in memoriam Ernst Herzfeld*. Locust Valley: Augustin, 1952, 150–5.

Denwood, P. "Early Connections between Ladakh/Baltistan and Amdo/Kham." In J. Bray, ed. *Ladakhi Histories: Local and Regional Perspectives*. Leiden: Brill, 2005, 31–9.

Denwood, P. "The Tibetans in the West, Part I." *Journal of Inner Asian Art and Archaeology* 3 (2009): 7–21.

Detienne, M. *The Gardens of Adonis: Spices in Greek Mythology*. Trans. J. Lloyd. Princeton: Princeton University Press, 1994.

Di Cosmo, N. *Ancient China and its Enemies: The Rise of Nomadic Power in East Asian History*. Cambridge: Cambridge University Press, 2002.

Dietrich, A. *Dioscurides Triumphans: Ein anonymer arabischer Kommentar (Ende 12. Jahrh. n. Chr.) zur Materia medica*. 2 vols. Göttingen: Vandenhoeck & Ruprecht, 1988.

Dietrich, A. *Die Dioskurides-Erklärung des Ibn al-Baiṭār. Ein Beitrag zurarabischen Pflanzensynonymik des Mittelalters*. Göttingen: Vandenhoeck & Ruprecht, 1991.

Dietrich, A. "al-ʿAṭṭār." *EI* 2nd ed. s.v.

Dietrich, A. "Misk." *EI* 2nd ed. s.v.

Dietrich, A. "Nīl." *EI* 2nd ed. s.v.

Dietrich, A. "Ṣandal." *EI* 2nd s.v.

Dietrich, A. "Afāwīh." *EI* 2nd ed. supplement 1, 42–3.

Dietrich, A., and C. E. Bosworth. "'Ūd." in *EI* 2nd s.v.

Dihkhudā, ʿAlī Akbar. *Lughat Nāmah*. 33 vols. in 42 parts. Tehrān: Dānishgāh-i-Tihrān, 1946–73.

Doerfer, G. *Türkische und mongolische Elemente im Neupersischen*. 4 vols. Wiesbaden: Steiner, 1963–75.

Dols, M. "Plague in Early Islamic History." *JAOS* 94:3 (1974): 371–83.

Dols, M. *The Black Death in the Middle East*. 2nd ed. Princeton: Princeton University Press, 1979.

Donkin, R. A. "The Insect Dyes of Western and West-Central Asia." *Anthropos* 72 (1977): 847–80.

Donkin, R. A. *Beyond Price: Pearls and Pearl-Fishing: Origins to the Age of Discoveries*. Philadelphia: American Philosophical Society, 1998.

Donkin, R. A. *Dragon's Brain Perfume: An Historical Geography of Camphor*. Leiden: Brill, 1999.

Dozy, R. *Supplément aux dictionnaires arabes*. 2 vols. 2nd ed. Paris: Maisonneuve, 1927.

Dozy, R. *Dictionnaire détaillé des noms des vêtements chez les Arabes*. Amsterdam, 1843; repr. Beirut: Libraire du Liban, n.d.

Drompp, M. "Breaking the Orkhon Tradition: Kirghiz Adherence to the Yenisei Region after A.D. 840." *JAOS* 119:3 (1999): 390–403.

Ducène, J.-C. "Al-Ǧayhānī: fragments (Extraits du *K. al-masālik wa l-mamālik* d'al-Bakrī)." *Der Islam* 75 (1998): 259–82.

Dunlop, D. M. *Arab Civilization to A.D. 1500*. London: Longman, 1971.

Dymock, W. *Pharmacographia Indica: A History of the Principal Drugs of Vegetable Origin met with in British India*. London: Kegan Paul, Tench & Trübner, 1890–3; repr. in one volume, Karachi: Hamdard, 1972.

Eberhard, W. *Kultur und Siedlung der Randvölker Chinas*. Supplement to *T'oung Pao* vol. 36 Leiden: Brill, 1946.

Edgerton, Franklin. *Buddhist Hybrid Sanskrit Grammar and Dictionary*. 2 vols. New Haven: Yale, 1953; repr. Delhi: Motilal Banarsidass, 1993.

Eilers, W. "Iranisches Lehngut im arabischen Lexikon: über einige Berufsnamen und Titel." *Indo-Iranian Journal* 5 (1962): 203–32.

Eilers, W. "Iranisches Lehngut im Arabischen." In *Actas IV congresso de estudos árabes e islâmicos. Coimbra-Lisboa … 1968*. Leiden: Brill, 1971, 582–660.

El-Hibri, T. *Reinterpreting Islamic Historiography: Hārūn al-Rashīd and the Narrative of the ʿAbbāsid Caliphate*. Cambridge: Cambridge University Press, 1999.

Ellis, Havelock. *Studies in the Psychology of Sex Volume 4. Sexual Selection in Man*. Philadelphia: F. A. Davis, 1905; repr. 1920.

Erdal, M. "Uigurica from Dunhuang." *BSOAS* 51 (1988): 251–7.

Farhangnāmah-i Adab-i Fārsī. Volume 2 of *Dānishnāmah-i Adab-i Fārsī* (Tehran: Mu'assasah-i Farhangī wa Intishārāt-i Dānishnāmah, 1996-.

Farmer, H. G. "Al-Kindī on the 'Ēthos' of Rhythm, Colour, and Perfume." *Transactions of the Glasgow University Oriental Society* 16 (1955–6): 29–38.

Fatimi, S. Q. "In Quest of Kalāh." *Journal of Southeast Asian History* 1:2 (1960): 62–101.

Fehér, B. "Mysterious Alloys in Early Muslim Metallurgy: On the *Ṭālīqūn* and the *Haft-Ǧūš*." *The Arabist* 23 (2001): 55–63.

Feldbusch, Michael. *Der Brief Alexanders an Aristoteles über die Wunder Indiens. Synoptische Edition*. Beiträge zur klassischen Philologie 78. Meisenheim am Glan: Anton Hain, 1976.

Ferrand, G. *Relations de voyages et textes géographiques arabes, persans et turks relatifs à l'Extrême-Orient du VIIIe au XVIIIe sieclès*. 2 vols. Paris: Leroux, 1913–14.

Ferrand, G. *Voyage du marchand arabe Sulaymân en Inde et en Chine*. Paris: Bossard, 1922.

Ferrand, G. "L'élément persan dans les textes nautiques arabes." *JA* 1924:1, 193–257.

Ferrand, G. Review of D. Campbell Thomson. *The Assyrian Herbal, Journal Asiatique* 206 (Jan.-June 1925), 171–3.

Fiey, J.-M. "Diocèses syriens orientaux du Golfe Persique." In *Mémorial Mgr. Gabriel Khouri-Sarkis*. Louvain: Impr. orientaliste, 1969, 177–219.

Finlay, R. *The Pilgrim Art: Cultures of Porcelain in World History*. Berkeley: University of California, 2010.

Fiorani Piacentini, V. "Merchants- Merchandise and Military Poewe in the Persian Gulf (Sūriyānj/Shahriyāj- Sīrāf)." *Atti della Accademia Nazionale dei Lincei, Memorie*, Series IX, 3:2 (1992): 110–91.

Fischer, W. "Muʿarrab." *EI* 2nd s.v.

Flerov, K. K. *Fauna of USSR Mammals Volume 1 No. 2: Musk Deer and Deer*. Trans. A. Biron and Z. S. Cole. Washington, 1960. (Russian original is from 1952.)

Foucher, A. *La vieille route de l'Inde de Bactres à Taxila*. 2 vols. Paris: Les éditions d'art et d'histoire, 1942–7.

Fraenkel, S. *Die aramäischen Fremdwörter im Arabischen*. Leiden: Brill, 1886; repr. Hildesheim: Olms, 1962.

"Fragrans." "How Musk is Made." *The China Review* 9:4 (1881): 253–4.

Frisk, H. *Griechisches etymologisches Wörterbuch*. Heidelberg: Winter, 1960–70.

Frye, R. N. "Sasanian-Central Asian Trade Relations." *Bulletin of the Asia Institute* 7 (1993): 73–7.

Frye, R. N. "Dar-i Āhanīn." in *EI* 2nd ed. s.v.

Gatten, A. "A Wisp of Smoke: Scent and Character in the Tale of Genji." *Monumenta Nipponica* 32 (1977): 35–48.

Gershevitch, I. *A Grammar of Manichean Sogdian*. Oxford: Blackwell, 1954.

Gharib, B. *Sogdian Dictionary: Sogdian-Persian-English*. Tehran: Farhangan, 1995.

Ghirshman, R. "Un miroir T'ang de Suse." *Artibus Asiae* 19 (1956): 230–3.

Gignoux, P. "Le traité syriaque anonyme sur les médications." *Symposium Syriacum VII*. Rome, 1998, 725–33.

Gignoux, P. "On the Syriac Pharmacopoeia." *The Harp* 11–12 (1998–9): 193–201.

Gignoux, P. "Les relations interlinguistiques de quelques termes de la pharmacopée antique." In D. Durkin-Meisterernst, et al., eds. *Literarische Stoffe und ihre Gestaltung in mitteliranischer Zeit*. Wiesbaden: Harrassowitz, 2009, 91–8.

Gignoux, P. "Les relations interlinguistiques de quelques termes de la pharmacopée antique II." In W. Sundermann, et al., eds. *Exegisti Monumenta: Festschrift in Honour of Nicholas Sims-Williams*. Wiesbaden: Harrassowitz, 2009, 117–26.

Gignoux, P. "La pharmacopée syriaque exploitée d'un point de vue linguistique." *Le Muséon* 124 (2011) : 11–26.

Gil, M. "The Rādhānite Merchants and the Land of Rādhān." *JESHO* 17 (1974): 299–328.

Gil, M. *Be-malkuth Yishmaʿel bi-teḳufat ha-ge'onim*. 4 vols. Tel Aviv: University of Tel Aviv, 1997.

Gil, M. *Jews in Islamic Countries in the Middle Ages*. Trans. D. Strassler. Leiden: Brill, 2004.

Gode, P. K. "Indian Science of Cosmetics and Perfumery. "In P. K. Gode, *Studies in Indian Cultural History* vol. 1. Hoshiarpur: Vishveshvaranand Vedic Research Institute, 1961, 1–8.

Gode, P. K. "History of Ambergris in India Between about A.D. 700 and 1900." In Gode, *Studies in Indian Cultural History Volume 1*, Hoshiarpur: Vishveshvaranand Vedic Research Institute, 1961, 9–14.

Gode, P. K. "Studies in the History of Indian Cosmetics and Perfumery- A Critical Analysis of a Rare Manuscript of Gandhavāda and its Marathi Commentary (Between C. A.D. 1350 and 1550)." *Studies in Indian Cultural History* vol. 1. Hoshiarpur: Vishveshvaranand Vedic Research Institute, 1961, 43–52.

Gode, P. K. "Perfumes and Cosmetics in the Royal Bath." *Studies in Indian Cultural History* vol. 1. Hoshiarpur: Vishveshvaranand Vedic Research Institute, 1961, 53–6.

Gode, P. K. "Verses pertaining to Gandhayukti in the *Agnipurāṇa* (9th Century A.D.) and their relation to the topics dealt with in Gaṅgādhara's Gandhasāra." *Studies in Indian Cultural History* vol. 1. Hoshiarpur: Vishveshvaranand Vedic Research Institute, 1961, 68–73.

Gode, P. K. "The Gandhayukti Section of the Viṣṇudharmottara and its Relation to other Texts on the Gandhaśāstra." *Studies in Indian Cultural History* vol. 1. Hoshiarpur: Vishveshvaranand Vedic Research Institute, 1961, 74–87.

Goitein, S. D. *Jews and Arabs: A Concise History of their Social and Cultural Relations*. New York, 1955; repr. New York: Dover, 2005.

Goitein, S. D. "The Rise of the Middle-Eastern Bourgeoisie in Early Islamic Times." *Journal of World History* 3 (1957): 583–604; repr. in S. D. Goitein, *Studies in Islamic History and Institutions*. Leiden: Brill, 1966, 217–41.

Goitein, S. D. *A Mediterranean Society: The Jewish Communities of the Arab World as Portrayed in the Documents of the Cairo Geniza*. 6 vols. Berkeley: University of California Press, 1967–1993; repr. 1999.

Goitein, S. D. "Nicknames as Family Names." *JAOS* 90 (1970): 517–24.

Goitein, S. D. *Letters of Medieval Jewish Traders, Translated from the Arabic with Introductions and Notes*. Princeton: Princeton University Press, 1973.

Goitein, S. D., and Friedman, M. A. *India Traders of the Middle Ages: Documents from the Cairo Geniza: "India Book"*. Leiden: Brill, 2008.

Golden, P. B. *An Introduction to the History of the Turkic Peoples: Ethnogenesis and State-Formation in Medieval and Early Modern Eurasia and the Middle East*. Wiesbaden: Harrassowitz, 1992.

Gottheil, R. "Fragment on Pharmacy from the Cairo Genizah." *JRAS* (1935): 123–44.

Grami, B. "Perfumery Plant Materials As Reflected in Early Persian Poetry." *JRAS* (2013): 39–52.

Gray, L. H. "Zoroastrian Elements in Muhammedan Eschatology." *Le Muséon* 21 (1902): 153–84.

Green, D. A. *The Aroma of Righteousness: Scent and Seduction in Rabbinic Life and Literature.* University Park: Pennsylvania State University Press, 2011.

Green, M. J. B. "The Distribution, Status and Conservation of the Himalayan Musk Deer *Moschus chrysogaster.*" *Biological Conservation* 35 (1986): 347–75.

Green, M. J. B. "Some ecological aspects of a Himalayan population of musk deer." In *Biology and management of the Cervidae; a conference held at the Conservation and Research Center, National Zoological Park, Smithsonian Institution.* Washington, 1987, 307–19

Green, M. J. B. "Musk production from musk deer." In R. J. Hudson, et al., eds., *Wildlife Production Systems: Economic utilisation of wild ungulates.* Cambridge: Cambridge University Press, 1989, 401–9.

Green, M., and Taylor, R. "The Musk Connection." *New Scientist* 110: 1514 (June 26, 1986): 56–8.

Grenet, F. "Les marchands sogdiens dans les mers du Sud à l'époque préislamique." *Cahiers d'Asie Centrale* 1–2 (1996): 65–84.

Grenet, F. and Sims-Williams, N. "The Historical Context of the Sogdian Ancient Letters." In *Transition Periods in Iranian History*, Paris, 1987, 101–22.

Grenet, F., Sims-Williams, N., and de la Vaissière, E. "The Sogdian Letter V." *Bulletin of the Asia Institute* 12 (1998): 91–104.

Greppin, J. A. "Gk. κόστος: A Fragrant Plant and its Eastern Origin." *Journal of Indo-European Studies* 27 (1999): 395–408.

Groom, Nigel. *Frankincense and Myrrh: A Study of the Arabian Incense Trade.* London: Longman, 1981.

Groom, Nigel. *The New Perfume Handbook.* 2nd ed. London: Blackie Academic and Professional, 1997.

Gropp, G. "Christian Maritime Trade of the Sasanian Age in the Persian Gulf." In *Golf-Archäologie: Mesopotamien, Iran, Kuwait, Bahrain, Vereinigte Arabische Emirate und Oman.* Buch am Elbach: Leidorf, 1991, 83–8.

Groves, C. P., Wang Yingxiang, and Grubb, P., "Taxonomy of Musk Deer, Genus *Moschus* (Moschidae, Mammalia)." *Acta Theriologica Sinica* 15:3 (1995): 181–97.

Grzimek, B. *Grzimek's Animal Life Encyclopedia.* 13 vols. New York: Van Nostrand Reinhold, 1972–5.

Gunter, A. C. "The Art of Eating and Drinking in Ancient Iran." *Asian Art* 1:2 (1988): 7–52.

Halevi, L. *Muhammad's Grave: Death Rites and the Making of Islamic Society.* New York: Columbia University Press, 2007.

Hall, K. R. *Trade and Statecraft in the Age of the Cōḷas.* Delhi: Abhinav, 1980.

Hall, K. R. *A History of Early Southeast Asia: Maritime Trade and Societal Development, 100–1500*. Lanham: Rowman and Littlefield, 2011.

Hamarneh, S. K. "The First Known Independent Treatise on Cosmetology in Spain." *Bulletin of the History of Medicine* 39 (1965): 309–25.

Hamarneh, S. K., and Sonnedecker, G. *A Pharmaceutical View of Abulcasis al-Zahrāwī in Moorish Spain*. Leiden: Brill, 1963.

Hamilton, J. *Manuscrits ouïgours du IXe–Xe siècle de Touen-houang*. 2 vols. Paris: Peeters, 1986.

Hamilton, J. "On the dating of the Old Turkish Manuscripts from Tunhuang." In *Turfan, Khotan und Dunhuang. Vorträge der Tagung "Annemarie v. Gabain und die Turfanforschung"*. Berlin, 1996, 135–46

Haneda, Akira. "On Chinese Rhubarb." In *The Islamic World: Essays in Honor of Bernard Lewis*. Princeton: Darwin Press, 1989, 27–30.

Hannestad, K. "Les relations de Byzance avec la Transcaucasie et l'Asie Centrale aux 5e et 6e siècles." *Byzantion* 25–7 (1955–7): 421–56.

Hansen, V. "The Impact of the Silk Road trade on a local community: the Turfan oasis, 500–800." In É. de la Vaissière and É.Trombert, eds. *Les Sogdiens in Chine*. Paris: École Française d'Extrême-Orient, 2005, 283–10.

Hansen, V. "The Kitan People, the Liao Dynasty (916–1125) and their World." *Orientations* 42:1 (January-February 2011): 34–42.

Hansen, V. *The Silk Road: A New History*. Oxford: Oxford University Press, 2012.

Hansen, V. "International Gifting and the Kitan World, 907–1125." *Journal of Song-Yuan Studies* 43 (2013): 273–302.

Harig, G. "Von den arabischen Quellen des Simeon Seth." *Medizinhistorisches Journal* 2 (1967): 248–68.

Harmatta, J. "The Struggle for the Possession of South Arabia between Aksūm and the Sāsānians." *IV Congresso internazionale di studi etiopici* (*Roma, 10–15 aprile 1972*) (Rome: Accademia nazionale dei Lincei, 1974), 95–106.

Harper, D. J. *Early Chinese Medical Literature: The Mawangdui Medical Manuscripts*. London: Kegan Paul, 1998.

Harris, R. B. "Conservation Prospects for Musk Deer and other Wildlife in Southern Qinghai, China." *Mountain Research and Development* 11:4 (1991): 353–8.

Härtel, H., ed. *Along the Ancient Silk Routes: Central Asian Art from the West Berlin State Museums*. New York: Metropolitan Museum of Art, 1982.

Harvey, Susan Ashbrook. *Scenting Salvation: Ancient Christianity and the Olfactory Imagination*. Berkeley: University of California Press, 2006.

Hauenschild, I. *Die Tierbezeichnungen bei Mahmud al-Kaschgari. Eine Untersuchung aus sprach- und kulturhistorischer Sicht*. Wiesbaden: Harrassowitz, 2003.

Haupt, P. "The Etymology of Egypt. ṯsm, greyhound." *JAOS* 45 (1925): 318–20.

Heck, G. W. *Charlemagne, Muhammad, and the Arab Roots of Capitalism*. Berlin: de Gruyter, 2006.

Heck, G. W. "'Arabia without Spices': An Alternate Hypothesis." *JAOS* 123 (2003): 547–576.

Heikel, A., et al. *Inscriptions de l'Orkhon recueillies par l'expédition finnoise 1890*. Helsinki: Société Finno-Ougrienne, 1892.

Heller, A. "Archaeological Artefacts from the Tibetan Empire in Central Asia." *Orientations* 34:4 (April, 2003): 55–64.

Heng, D. *Sino-Malay Trade and Diplomacy from the Tenth through the Fourteenth Century*. Athens, OH: Ohio University Press, 2009.

Henning, W. B. "Two Central Asian Words." *Transactions of the Philological Society* (1945): 150–62.

Henning, W. B. "The Sogdian Texts of Paris." *BSOAS* 11:4 (1946): 713–40.

Henning, W. B. "The Date of the Sogdian Ancient Letters." *BSOAS* 12 (1948): 601–15.

Henning, W. B. "A Pahlavi Poem." *BSOAS* 13 (1950): 641–8.

Heptner, V. G., et al., *Mammals of the Soviet Union Volume 1*. Trans. P. M. Rao, Washington: Smithsonian Institution, 1988.

Herrmann, G., ed. *Monuments of Merv: Traditional Buildings of the* Karakum. London: Society of Antiquaries, 1999.

Heyd, F. *Histoire du commerce du Levant au Moyen-Âge*. 2 vols. Leipzig: Harrassowitz, 1885–6; repr. Amsterdam: Hakkert, 1959.

Hightower, J. R. "Some Characteristics of Parallel Prose." In *Studia Serica Bernhard Karlgren Dedicata*. Copenhagen: Munksgaard, 1959, 60–91.

Hightower, J. R. "Yüan Chen and 'The Story of Ying-Ying.'" *HJAS* 33 (1973): 90–123.

Hildburgh, W. L. "Notes on some Tibetan and Bhutia Amulets and Folk-Medicines, and a few Nepalese Amulets." *Journal of the Royal Anthropological Institute of Great Britain and Ireland* 39 (1909): 386–96.

Hinz, W. *Islamische Masse und Gewichte*. Leiden: Brill, 1955.

Hoffmann-Ruf, M., and A. Al Salimi, eds. *Oman and Overseas*. Studies on Ibadism and Oman, Vol. 2. Hildesheim: Olms, 2013.

Homes, V. *On the Scent: Conserving Musk Deer- The Uses of Musk and Europe's Role in its Trade*. Brussels: TRAFFIC Europe, 1999.

Hopkirk, P. *Foreign Devils on the Silk Road*. Amherst: University of Massachusetts, 1980.

Horiguchi, S., and D. Jung. "*Kōdō*: Its Spiritual and Game Elements and Its Interrelations with the Japanese Literary Arts." *JRAS* (2013): 69–84.

Horton, M., and J. Middleton, *The Swahili*. Oxford: Blackwell, 2000.

Hourani, G. F. "Direct Sailing Between the Persian Gulf and China in Pre-Islamic Times." *JRAS* (1947): 157–60.

Hourani, G. F. *Arab Seafaring in the Indian Ocean in Ancient and Early Medieval Times.* Expanded edition ed. J. Carswell. Princeton: Princeton University Press, 1995.

Hoyanagi, M. "Natural Changes of the Region along them Old Silk Road in the Tarim Basin in Historical Times." *MRDTB* 33 (1975): 85–113.

Hoyland, R. G. *Arabia and the Arabs from the Bronze Age to the Coming of Islam.* London: Routledge, 2001.

Hoyland, R. G., and B. Gilmour. *Medieval Islamic Swords and Swordmaking : Kindi's treatise "On swords and their kinds".* Oxford: Gibb Memorial Trust, 2006.

Hübschmann, H. *Armenische Grammatik. Erster Teil: Armenische Etymologie.* Heidelberg, 1897.

Huc, Evariste-Régis, and Joseph Gabet. *Travels in Tartary, Thibet and China, 1844–1846.* Trans. W. Hazlitt and intr. P. Pelliot. 2 vols. London: Routledge, 1928; repr. 2 vols. in 1, New York: Dover, 1987.

Hulsewé, A. F. P. "Quelques considérations sur le commerce de la soie au temps de la dynastie des Han." In *Mélanges de Sinologie offerts à Monsieur Paul Demiéville,* Vol. 2. Paris: De Boccard, 1974, 117–35.

The Imperial Gazetteer of India: The Indian Empire Vol. 1: Description. New Edition. Oxford: Oxford University Press, 1909.

Isaacs, H. D. and C. F. Baker. *Medical and Para-Medical Manuscripts in the Cambridge Genizah Collections.* Cambridge: Cambridge University Press, 1994.

Jackson, R. "Aboriginal Hunting in West Nepal with Reference to Musk Deer *Moschus moschiferus moschiferous* and Snow Leopard *Panthera uncia.*" *Biological Conservation* 16 (1979): 63–72.

Jain, M. S. "Observations on the Birth of a Musk Deer Fawn." *Journal of the Bombay Natural History Society* 77 (1980): 497–8.

Jamwal, P. S. "Collection of Deer Musk in Nepal." *Journal of the Bombay Natural History Society* 69:3 (1972): 647–9.

Jäschke, H. *A Tibetan-English Dictionary.* London, 1881; repr. Delhi: Motilal Banarsidass, 1995.

Jeffrey, A. *The Foreign Vocabulary of the Qur'ān.* Baroda: Oriental Institute, 1938.

Jellinek, Paul. *The Psychological Basis of Perfumery.* Ed. and trans. J. Stephan Jellinek. London: Blackie Academic and Professional, 1997.

Jerdon, T. C. *The Mammals of India: A Natural History of all the Animals known to inhabit Continental India.* London: John Wheldon, 1874.

Jettmar, K. "Sogdians in the Indus Valley." In *Histoire et cultes de l'Asie Centrale préislamique.* Paris: CNRS, 1991, 251–3 & pls. 103–6.

Jest, Corneille. "Valeurs d'Échange en Himalaya et au Tibet: Le amber et le musc." In *De la voûte céleste au terroir, du jardin au foyer: mosaïque sociographique: textes offerts à Lucien Bernot.* Paris: Éditions de l'École des hautes études en sciences sociales, 1987, 227–38.

Johnstone, P. C. "Ward." EI^2 s.v.

Joshi, M. P. and Brown, C. W. "Some Dynamics of Indo-Tibetan Trade through Uttarākhaṇḍa (Kumaon-Garhwal), India." JESHO 30 (1987): 303–17.

Jung, D. *An Ethnography of Fragrance: The Perfumery Arts of 'Adn/Laḥj*. Leiden: Brill, 2011.

Jung, D. "The Cultural Biography of Agarwood- Perfumery in Eastern Asia and the Asian Neighbourhood." JRAS (2013): 103–25.

Kahl, O. *The Sanskrit, Syriac and Persian Sources in the* Comprehensive Book *of Rhazes*. Leiden: Brill, 2015.

Kahle, P. "Islamische Quellen zum chinesischen Porzellan." ZDMG 83 (1934), 1–45.

Kahle, P. "Die Schätze der Fatimiden." ZDMG 89 (1935): 329–62.

Kahle, P. "Chinese Porcelain in the lands of Islam." In *Paul Kahle. Opera Minora*, Leiden, 1956, 326–61 (originally published in *Transactions of the Oriental Ceramic Society 1940–41*, London, 1942 and in the *Journal of the Pakistan Historical Society* I (1953), 1–16).

Kanafani, A. S. *Aesthetics & Ritual in the United Arab Emirates*. Beirut: American University of Beirut, 1983.

Kara, G. Review of J. Hamilton, *Manuscrits ouïgours du I^{xe}-Xe siècle de Touen-houang*. 2 vols. Paris: Peeters, 1986. *Acta Orientalia Academiae Scientiarum Hungaricae* 43 (1989): 125–8.

Kara, G. "An Old Tibetan Fragment on Healing from the Sutra of the Thousand-Eyed and Thousand-Handed Great Compassionate Bodhisattva Avalokiteśvara in the Berlin Turfan Collection." In D. Durkin-Meisterernst, et al., eds., *Turfan Revisited-The First Century of Research into the Arts and Cultures of the Silk Road* (Berlin: Reimer, 2004), 141–6.

Karttunen, K. *India in Early Greek Literature*. Studia Orientalia 65. Helsinki: Finnish Oriental Society, 1989.

Karttunen, K. *India and the Hellenistic World*. Studia Orientalia 83. Helsinki: Finnish Oriental Society, 1997.

Käs, F. *Die Mineralien in der arabischen Pharmakognosie*. 2 vols. Wiesbaden: Harrassowitz, 2010.

Kattel, B., and A. W. Alldredge. "Capturing and Handling of the Himalayan Musk Deer." *Wildlife Society Bulletin* 19:4 (Winter, 1991): 397–9.

Kennedy, H. *The Armies of the Caliphs: Military and Society in the Early Islamic State*. London: Routledge, 2001.

Kennedy, P. F. *The Wine Song in Classical Arabic Poetry: Abū Nuwās and the Literary Tradition*. Oxford: Clarendon Press, 1997.

Kervran, M. "Multiple Ports at the Mouth of the River Indus: Barbarike, Deb, Daybul, Lahori Bandar, Dul Sinde." In *Archaeology of Seafaring: The Indian Ocean in the Ancient Period*. Delhi: Pragati, 1999, 70–153.

Kimura, K. "Ancient Drugs Preserved in the Shōsōin." *Occasional Papers of the Kansai Asiatic Society* 1 (February, 1954).

King, A. "The Importance of Imported Aromatics in Arabian Culture: Illustrations from Pre-Islamic and Early Islamic Poetry." *JNES* 67:3 (2008): 175–189.

King, A. "Tibetan Musk and Medieval Arab Perfumery." In A. Akasoy, C. Burnett, and R. Yoeli-Tlalim, eds., *Islam and Tibet: Interactions along the Musk Routes*. Farnham: Ashgate, 2011, 145–61.

King, A. "Some 10th Century Material on Asian Toponymy from Sahlan b. Kaysan." *Archivum Eurasiae Medii Aevi* 16 (2008–9): 121–6.

King, A. "Early Islamic Sources on the Kitan Liao: The Role of Trade." *Journal of Song-Yuan Studies* 43 (2013): 253–271.

King, A. "Eastern Islamic Rulers and the Trade with Eastern and Inner Asia in the 10th and 11th Centuries." *BAI* 25 (2011 [2015]): 175–185.

King, A. "The new *materia medica* of the Islamicate tradition: the pre-Islamic context." *JAOS* 135, no. 3 (2015): 499–528.

King, A. *Medieval Islamicate Perfumery*. Work in progress.

Knauer, E. R. *The Camel's Load in Life and Death*. Zürich: Acanthus, 1998.

Koehler, L. *The Hebrew and Aramaic Lexicon of the Old Testament*. Rev. ed. by Walter Baumgartner. 5 vols. Leiden: Brill, 1994–2000.

Konow, S. *A Medical Text in Khotanese: CH.ii 003 of the India Office Library*. Oslo: Dybwad, 1941.

Kotwal, F. M. and Boyd, J. W. *A Persian Offering. The Yasna: A Zoroastrian High Liturgy*. Paris: Association pour l'avancement des études iraniennes, 1991.

Kovalev, R. "The Infrastructure of the Northern Part of the "Fur Road" between the Middle Volga and the East during the Middle Ages." *Archivum Eurasiae Medii Aevi* 11 (2000–01): 25–64.

Kluge, F. *Etymologisches Wörterbuch der deutschen Sprache*. 20th edition, ed. W. Mitzka. Berlin: De Gruyter, 1967.

Kračkovskiĭ, I. Iu. "Drevneĭshiĭ arabskiĭ dokument iz Sredneĭ Azii." In *Sogdiiskiĭ sbornik*. Leningrad: Akad. NAUK SSSR, 1934, 52–90.

Krahl, R., et al., eds. *Shipwrecked: Tang Treasures and Monsoon Winds*. Washington: Smithsonian Institution, 2010.

Kulke, H. "Rivalry and Competition in the Bay of Bengal in the Eleventh Century and its bearing on Indian Ocean Studies." In Om Prakash and D. Lombard, eds. *Commerce and Culture in the Bay of Bengal, 1500–1800*. New Delhi: Manohar, 1999, 17–35.

Kumekov, B. E. *Gosudarstvo Kimakov IX–XI vv. po arabskim istochnikam*. Alma-Ata: NAUKA Kazakhskoi SSR, 1972.

Kunz, G. F. *The Magic of Jewels and Charms*. Philadelphia: Lippincott, 1915; repr. New York: Dover, 1997.

Kuwabara Jitsuzō, "On P'u Shou-kêng 蒲壽庚, a Man of the Western Regions, who was the Superintendent of the Trading Ships' Office in Ch'üan-chou 泉州 towards the End of the Sung dynasty, together with a General Sketch of Trade of the Arabs in China during the T'ang and Sung Eras." *MRDTB* 2 (1928): 1–79 and 7 (1935): 1–104.

Lamb, A. "A Visit to Sīrāf, an ancient port on the Persian Gulf." *JMBRAS* 37 (1964): 1–19.

Lambton, A. "The Merchant in Medieval Islam." In *A Locust's Leg: Studies in Honour of S. H. Taqizadeh*. London: Lund Humphries, 1962, 121–30.

Lane, E. *An Arabic-English Lexicon*. 8 vols. London, 1863–93; repr. Beirut: Librairie du Liban, 1968.

Larkin, M. *al-Mutanabbi: Voice of the 'Abbasid Poetic Ideal*. Oxford: Oneworld, 2008.

LaRocca, D. J. *Warriors of the Himalayas: Rediscovering the Arms and Armor of Tibet*. New York: Metropolitan Museum of Art, 2006.

Laufer, B. "Indisches Recept zur Herstellung von Räucherwerk." *Verhandlungen der Berliner Gesellschaft für Anthropologie, Ethnologie und Urgeschichte* (July 18, 1896): 394–8.

Laufer, B. "Arabic and Chinese Trade in Walrus and Narwhal Ivory." *T'oung Pao* 14 (1913): 315–64.

Laufer, B. *Notes on Turquois in the East*. Chicago: Field Museum, 1913.

Laufer, B. "Asbestos and Salamander: An Essay in Chinese and Hellenistic Folk-Lore." *T'oung Pao* 16 (1915): 299–373.

Laufer, B. "Supplementary Notes on Walrus and Narwhal Ivory." *T'oung Pao* 17 (1916): 348–89.

Laufer, B. "Loan-words in Tibetan." *T'oung Pao* 17 (1916): 403–552.

Laufer, B. *Sino-Iranica: Chinese Contributions to the History of Civilization in Ancient Iran, with Special Reference to the History of Cultivated Plants*. Field Museum of Natural History Anthropological Series 15:3 Chicago: Field Museum, 1919; repr. New York: Kraus, 1967.

Laufer, B. "History of Ink in China." In F. Wiborg, *Printing Ink: A History*. New York: Harper & Brothers, 1926, 1–52.

Le Coq, A. von. *Türkische Manichaica aus Chotscho II*. Abhandlungen der königlichen preussischen Akademie der Wissenschaften zu Berlin 1919 Nr. 3. Berlin: Verlag der Akademie der Wissenschaften, 1919.

Le Coq, A. von. *Türkische Manichaica aus Chotscho III*. Abhandlungen der königlichen preussischen Akademie der Wissenschaften zu Berlin 1922 Nr. 2. Berlin: Akademie der Wissenschaften, 1922.

Lecker, M. "The Levying of Taxes for the Sassanians in Pre-Islamic Medina." *Jerusalem Studies in Arabic and Islam* 27 (2002): 109–26.

Leibowitz, J. and S. Marcus. *Moses Maimonides on the Causes of Symptoms*. Berkeley: University of California Press, 1974.

Lev, E., and Amar, Z. *Practical* Materia Medica *of the Medieval Eastern Mediterranean According to the Cairo Genizah*. Leiden: Brill, 2008.

Levey, H. *The Biography of Huang Ch'ao*. Berkeley: University of California, 1955.

Levey, M. *Medieval Arabic Toxicology: The Book of Poisons of Ibn Waḥshīya and its Relation to Early Indian and Greek Texts*. Transactions of the American Philosophical Society New Series 56:7. Philadelphia: American Philosophical Society, 1966.

Lewicki, T. "Les premiers commerçants arabes en Chine." *Rocznik Orientalistyczny* 11 (1935): 173–86.

Lewis, B. "The Crows of the Arabs." *Critical Inquiry* 12:1 (1985): 88–97.

Lindegger, P. *Griechische und römische Quellen zum Peripheren Tibet*. 3 vols. Zürich: Tibetan Monastic Institute, 1979–93.

Lokotsch, K. *Etymologisches Wörterbuch der europäischen (germanischen, romanischen und slavischen) Wörter orientalischen Ursprungs*. Heidelberg: Winter, 1927.

Löw, I. *Die Flora der Juden*. 4 vols. Vienna, 1928; repr. Hildesheim, Olms, 1967.

Shih-yü Yü Li. "Tibetan Folk-law." *JRAS* (1950): 127–48.

Lilja, Saara. *The Treatment of Odours in the Poetry of Antiquity*. Commentationes Humanarum Litterarum 49. Helsinki: Societas Scientiarum Fennica, 1972.

Livshits, V. A., and Lukonin, B. G. "Srednepersidskie i sogdiĭskie nadpisi na serebrianykh sosudakh." *Vestnik drevneĭ istorii* 1964:3, 155–76.

L'Orange, H. P. *Studies on the Iconography of Cosmic Kingship in the Ancient World*. Oslo: H. Aschehoug & Co., 1953; repr. New Rochelle: Caratzas, 1982.

Löw, Immanuel. *Die Flora der Juden*. 4 vols. Wien, 1928–1934; repr. Hildesheim: Olms, 1967.

Lug, S. "Towards a Definition of Excellence in Classical Arabic Poetry: An Analysis of Ibn Zaydūn's *Nūniyya*." *JAOS* 101:3 (1981): 331–45.

Mackenzie, D. N. *A Concise Pahlavi Dictionary*. London: Oxford University Press, 1971.

Magariti, R. E. *Aden & the Indian Ocean Trade: 150 Years in the Life of a Medieval Arabian Port*. Chapel Hill: University of North Carolina Press, 2007.

Mahdi, W. "Linguistic and Philological Data Towards a Chronology of Austronesian Activity in India and Sri Lanka." In *Archaeology and Language*, vol. 4: *Language Change and Cultural Transformation*, ed. R. Blench and M. Spriggs. London: Routledge, 1999, 160–242.

Majno, G. *The Healing Hand: Man and Wound in the Ancient World*. Cambridge, MA: Harvard University Press, 1975.

Malov, S. E. *Pamiatniki Drevnetiurkskoĭ Pis'mennosti Mongolii i Kirgizii*. Moscow: Izdatel'stvo Akademii Nauk SSSR, 1959.

Mango, M. M. "Byzantine Maritime Trade with the East (4th-7th centuries)." *ARAM* 8 (1996): 139–63.

Manniche, L. *Sacred Luxuries: Fragrance, Aromatherapy & Cosmetics in Ancient Egypt*. Ithaca: Cornell University Press, 1999.

Marín, M. "Beyond Taste: the complements of colour and smell in the medieval Arab culinary tradition." In R. Tapper, ed., *Culinary Cultures of the Middle East*. London, 1994, 205–14.

Marín, M. "The Perfumed Kitchen: Arab Cookbooks from the Near East." In R. Gyselen, ed., *Parfums d'Orient*. Bures-sur-Yvette: Groupe pour l'Étude de la Civilisation du Moyen-Orient, 1998, 159–66.

Markham, Frederick. *Shooting in the Himalayas: A Journal of Sporting Adventures and Travel in Chinese Tartary, Ladac, Thibet, Cashmere, &c.* London: Richard Bentley, 1854.

Marshak, B. I. "The So-called Zandanījī Silks: Comparisons with the Art of Sogdia." In *Central Eurasian Textiles and Their Contexts in the Early Middle Ages*. Riggisberg: Abegg Stiftung, 2006, 49–60.

Martin, D. "Greek and Islamic Medicines' Historical Contact with Tibet: A Reassessment in View of Recently Available but Relatively Early Sources on Tibetan Medical Eclecticism." In A. Akasoy, C. Burnett, and R. Yoeli-Tlalim, eds. *Islam and Tibet: Interactions along the Musk Routes*. Farnham: Ashgate, 2011, 117–43.

Martin, J. *Treasure of the land of darkness: the fur trade and its significance for medieval Russia*. Cambridge: Cambridge University Press, 1986.

Maue, D., and Sertkaya, O., "Drogenliste und Dhāraṇī aus dem 'Zauberbad der Sarasvatī' des uigurischen Goldglanzsūtra." *Ural-Altaische Jahrbücher* n.F. 6 (1986), 76–99; 10 (1991), 116–27.

Mayrhofer, M., *Kurzgefaßtes etymologisches Wörterbuch des Altindischen*, 4 vols. Heidelberg: Winter, 1956–1980.

McAuliffe, J. D. "The Wines of Earth and Paradise: Qur'ānic Proscriptions and Promises." In *Logos Islamikos: Studia Islamica in Honorem Georgii Michaelis Wickens*. Toronto: Pontifical Institute of Medieval Studies, 1984, 159–74.

McCabe, A. "Imported *materia medica*, 4th-12th centuries, and Byzantine pharmacology." In M. M. Mango, ed. *Byzantine Trade, 4th-12th Centuries: The Archaeology of Local, Regional, and Intercultural Exchange*. Farnham: Ashgate, 2009, 273–92.

McHugh, J. "The Classification of Smells and the Order of the Senses in Indian Religious Traditions." *Numen* 54 (2007): 374–429.

McHugh, J. "The Incense Trees from the Land of Emeralds: The Exotic Material Culture of *Kāmaśāstra*." *Journal of Indian Philosophy* 39 (2011): 63–100.

McHugh, J. *Sandalwood and Carrion: Smell in Indian Religion and Culture*. Oxford: Oxford University Press, 2012.

McHugh, J. "The Disputed Civets and the Complexion of the God." *JAOS* 132:2 (2012): 245–73.

McHugh, J. "*Blattes de Byzance* in India: Mollusk Opercula and the History of Perfumery." *JRAS* (2013): 53–67.

McKenna, J. W. "The Coronation Oil of the Yorkist Kings." *The English Historical Review* 82:322 (Jan. 1967): 102–4.

Melikian-Chirvani, A. S. "*Parand* and *Parniyān* Identified: The Royal Silks of Iran from Sasanian to Islamic Times." *Bulletin of the Asia Institute* 5 (1991): 175–9.

Melikian-Chirvani, A. S. "Iran to Tibet." In A. Akasoy, C. Burnett, and R. Yoeli-Tlalim, eds., *Islam and Tibet: Interactions along the Musk Routes*. Farnham: Ashgate, 2011, 89–116.

Meulenbeld, C. J. *A History of Indian Medical Literature*. 3 vols. in 5. Groningen: Egbert Forsten, 2002.

Meyerhof, M. "ʿAlī aṭ-Ṭabarī's 'Paradise of Wisdom, One of the Oldest Arabic Compendiums of Medicine." *Isis* 16 (1931): 6–54.

Meyerhof, M. *Moses Maimonides' Glossary of Drug Names*. Trans. F. Rosner. Philadelphia: American Philosophical Society, 1979.

Millar, F. "Caravan Cities: The Roman Near East and Long-Distance Trade." In *Modus operandi: Essays in honour of Geoffrey Rickman*. London: Institute of Classical Studies, 1998, 119–37.

Milwright, M. "The balsam of Maṭariyya: an exploration of a medieval panacea." *BSOAS* 66:2 (2003): 193–209.

Minorsky, V. "Tamīm ibn Baḥr's Journey to the Uyghurs." *BSOAS* 12:2 (1948): 275–305.

Miquel, A. *La géographie humaine du monde musulman jusqu'au milieu du 11e siècle*. 4 vols. Paris: Mouton, 1967–88.

Monier-Williams, M. *A Sanskrit- English Dictionary, etymologically and philologically arranged with special reference to cognate Indo-European languages*. Oxford, 1899; repr. Delhi: Motilal Banarsidass, 1993.

Morgenstierne, G. *Indo-Iranian Frontier Languages*. Oslo, 1929–38.

Moriyasu, T. "Notes on Uighur Documents." *MRDTB* 53 (1995): 67–108.

Morohashi, T. *Dai Kan-Wa Jiten*. 13 vols. Tōkyō: Taishūkan Shoten, 1955–60.

Morony, M. "Commerce in Early Islamic Iraq." *asien afrika lateinamerica* 20 (1993): 699–720.

Morony, M. "The Late Sasanian Economic Impact on the Arabian Peninsula." *Namah-i Iran-i Bastan* 1:2 (2002): 25–37.

Morony, M. "Economic Boundaries? Late Antiquity and Early Islam." *JESHO* 47 (2004): 166–94.

Mukerji, B. *The Indian Pharmaceutical Codex: Volume 1- Indigenous Drugs*. New Delhi: Council of Scientific & Industrial Research, 1953.

Müller, C. *Fragmenta Historicorum Graecorum*. Vol. 4. Paris, 1868.

Müller, F. W. K. *Uigurica*. Abhandlungen der preussischen Akademie der Wissenschaften 1908 Nr. 2. Berlin, 1908.

Müller, F. W. K. "Uigurische Glossen." *Ostasiatische Zeitschrift* 8 (1919–20): 310–24.

Müller, W. W. "Weihrauch." In A. F. v. Pauly and G. Wissowa, *Real-encyclopädie der classischen Altertumswissenschaft* Supplement 15. Munich, 1978. 701–777.

Müller, W. W. "Namen von Aromata im antiken Südarabien." In A. Avanzini, ed., *Profumi d'Arabia*, 193–210. Rome: Bretschneider, 1997.

Nadkarni, K. M. *Indian Materia Medica.* 3rd. ed. by A. K. Nadkarni. Bombay: Popular Book Depot, 1955.

Nawata, H. "An Exported Item from Bāḍiʿ on the Western Red Sea Coast in the Eighth Century: Historical and Ethnographical Studies on Operculum as Incense and Perfume." *Papers of the XIIIth International Conference of Ethiopian Studies*, ed. K. Fukui, et al. Kyoto, 1997, vol. 1, 307–25.

Needham, J., and Lu Gwei-djen. *Science and Civilisation in China Volume 5. Chemistry and Chemical Technology: Part II. Spagyrical Discovery and Invention: Magisteries of Gold and Immortality.* Cambridge: Cambridge University Press, 1974.

Neelis, J. *Early Buddhist Transmission and Trade Networks: Mobility and Exchange within and beyond the Northwestern Borderlands of South Asia.* Leiden: Brill, 2011.

Neusner, J. "Some Aspects of the Economic and Political Life of the Babylonian Jewry, Ca. 160–220 C.E." *Proceedings of the American Academy for Jewish Research* 31 (1963): 165–96.

Newid, M. A. *Aromata in der iranischen Kultur unter besonderer Berücksichtigung der persischen Dichtung.* Wiesbaden: Reichert, 2010.

Nielsen, K. *Incense in Ancient Israel*, Leiden: Brill, 1986.

Nikitin, A. "Note on the Chronology of the Kushano-Sasanian Kingdom." In *Coins, Art, and Chronology.* Vienna: Akademie der Wissenschaften, 1999, 259–63.

Nobel, J. "Das Zauberbad der Göttin Sarasvatī." In *Beiträge zur indischen Philologie und Altertumskunde. Walther Schubring zum 70. Geburtstag dargebracht von der deutschen Indologie.* Hamburg: Cram, De Gruyter & Co., 1951, 123–39.

Noonan, T. "Khwārazmian Coins of the Eighth Century from Eastern Europe: The Post-Sasanian Interlude in the Relations between Central Asia and European Russia." *Archivum Eurasiae Medii Aevi* 6 (1986–8): 243–58.

Norbu, Thubten Jigme, and H. Harrer. *Tibet is My Country: Autobiography of Thubten Jigme Norbu, Brother of the Dalai Lama.* London, 1960; repr. London: Wisdom Publications, 1986.

Nowak, R., ed. *Walker's Mammals of the World.* 2 vols. 5th ed. Baltimore: Johns Hopkins University Press, 1991.

Nyambayar, B., H. Mix, and K. Tsytsulina. "*Moschus moschiferus.*" The IUCN Red List of Threatened Species 2015. http://www.iucnredlist.org/details/13897/0. Accessed March 22, 2016.

Nyberg, H. S. *A Manual of Pahlavi.* 2 vols. Wiesbaden: Harrassowitz, 1964–74.

Oda, J., et al., eds. *Sammlung uigurischer Kontrakte*. 3 vols. Osaka: Osaka University Press, 1993.

Oerter, W. "Das Motiv vom Garten. Betrachtungen zur manichäischen Eschatologie." In A. Von Tongerloo and S. Giversen, eds., *Manichaica Selecta: Studies presented to Professor Julien Ries on the occasion of his seventieth birthday*. Louvain: International Association of Manichaean Studies, 1991, 263–72.

Ostrowski, S., et al. "Musk deer *Moschus cupreus* persist in the eastern forests of Afghanistan." *Oryx* 50:2 (2016): 323–8.

Otavsky, K. "Stoffe von der Seidenstrasse: Eine neue Sammlungsgruppe in der Abegg-Stiftung." In Otavsky, K., ed., *Entlang der Seidenstrasse. Frühmittelalterliche Kunst zwischen Persien und China in der Abegg-Stiftung*, Riggisberg, 1998, 13–42.

Park, H. "Port-City Networking in the Indian Ocean Commercial System as Represented in Geographic and Cartographic Works in China and the Islamic West, c. 750–1500." In K. R. Hall, ed., *The Growth of Non-Western Cities: Primary and Secondary Urban Networking, c. 900–1900*. Lanham: Lexington Books, 2011, 21–53.

Peacock, D., and Williams, D., eds. *Food for the Gods: New Light on the Ancient Incense Trade*. Oxford: Oxbow, 2007.

Payne Smith, R. *Thesaurus Syriacus*. 2 vols. Oxford: Oxford University Press, 1879–1901.

Peacock, D. and D. Williams, eds. *Food for the Gods: New Light on the Ancient Incense Trade*. Oxford: Oxbow Books, 2007.

Pellat, C. "al-Djayhānī." In *EI*[2] supplement s.v.

Pellat, C. "Nouvel essai d'inventaire de l'œvre Gāḥiẓienne." *Arabica* 31 (1984): 117–64.

Pelliot, P. "Deux itinéraires de Chine en Inde." *Bulletin de l'École Française d'Extrême-Orient* 4 (1904) : 131–413.

Pelliot, P. "Des artisans chinois à la capitale abbasside en 751–62." *T'oung Pao* 26 (1929): 110–2.

Pelliot, P. *Notes on Marco Polo*. Ed. L. Hambis. 3 vols. Paris: Imprimerie Nationale, 1959–73.

Parry-Jones, R. "TRAFFIC Examines Musk Deer Farming in China." Http://www.traffic.org/traffic-dispatches/traffic_pub_dispatches16.pdf (accessed 6/24/16).

Petech, L. "Nota su *Mābd* e *Twsmt*." *Rivista degli Studi Orientali* 25 (1950); repr. in his *Selected Papers on Asian History*. Rome, 1988, 45–7.

Piesse, Charles H. *Piesse's Art of Perfumery*. 5th ed. London: Piesse and Lubin, 1891.

Pinault, G. "Épigraphie koutchéenne." In *Mission Paul Pelliot: Documents archéologiques VIII. Sites divers de la région de Koutcha*. Paris, 1986, 61–196.

Polunin, O., and A. Stainton. *Flowers of the Himalaya*. Oxford: Oxford University Press, 2000.

Potts, D. T. *The Arabian Gulf in Antiquity Volume II: From Alexander the Great to the Coming of Islam*. Oxford: Clarendon Press, 1990.

Poucher, W. A. *Poucher's Perfumes, Cosmetics and Soaps. Volume 1. The Raw Materials of Perfumery*. Ninth Edition, ed. A. J. Jouhar. London: Chapman & Hall, 1991.

Power, T. *The Red Sea from Byzantium to the Caliphate: AD 500–1000*. Cairo: American Univ. in Cairo Press, 2012.

Prickett-Fernando, M. "Durable Goods: The Archaeological Evidence of Sri Lanka's Role in the Indian Ocean Trade." In *Sri Lanka and the Silk Road of the Sea*. Colombo: Sri Lanka National Commission for UNESCO, 1990, 61–84.

Ptak, Roderich. "Moschus, Calambac und Quecksilber im Handel zwischen Macau und Japan und im ostasiatischen Seehandel ingesamt (circa 1555–1640)." In *Portugal und Japan im 16. und 17. Jahrhundert*. Frankfurt: Verlag der Interkulturelle Kommunikation, 1998, 72–95.

Pulleyblank, E. J. *Lexicon of Reconstructed Pronunciation in Early Middle Chinese, Late Middle Chinese, and Early Mandarin*. Vancouver: UBC Press, 1991.

Pybus, D. *Kodo: The Way of Incense*. Rutland: Tuttle, 2001.

Rachmati, G. R. "Zur Heilkunde der Uiguren I." *Sitzungsberichte der preussischen Akademie der Wissenschaften zu Berlin, Phil.-hist. Kl.* 24 (1930): 451–73.

Rachmati, G. R. "Zur Heilkunde der Uiguren II." *Sitzungsberichte der preussischen Akademie der Wissenschaften zu Berlin, Phil.-hist. Kl.* (1932): 401–48.

Radloff, W. [=V. Radlov]. *Versuch eines Wörterbuches der Türkdialekte*. 4 vols. St. Petersburg, 1893; repr. The Hague: Mouton, 1960.

Radloff, W. [=V. Radlov]. *Uigurische Sprachdenkmäler*. 1928; repr. Osnabrück: Biblio Verlag, 1972.

Raschke, M. G. "New Studies in Roman Commerce with the East." In *Aufstieg und Niedergang der römischen Welt*. Band II.9.2. Berlin: De Gruyter, 1978, 604–1361.

Rawal, A. J. "Society and Socio-Economic Life in the *Brahmavaivartapurāṇa*." *Purāṇa* 15:1 (1973): 6–92.

Ray, H. P. *The Winds of Change: Buddhism and the Maritime Links of Early South Asia*. Delhi: Oxford, 1994.

Reichelt, H. *Die soghdischen Handschriftenreste des Britischen Museums II. Teil*. Heidelberg: Winter, 1931.

Reinisch, L. *Die Somali-Sprache II. Wörterbuch*. Wien: Hölder, 1902.

Rice, B. Lewis. *Epigraphia Carnatica Vol. VII. Inscriptions in the Shimoga District (Part I)*. Bangalore: Mysore Government Central Press, 1902.

Richards, D. S., ed. *Islam and the Trade of Asia*. Oxford: Bruno Cassirer, 1970.

Richter-Bernburg, L. "Abū Manṣūr Heravī." *EIr* s.v..

Riddle, J. M. "Amber in Ancient Pharmacy. The Transmission of Information about a Single Drug. A Case Study." *Pharmacy in History* 15 (1973): 3–17.

Riddle, J. M. "The Introduction and Use of Eastern Drugs in the Early Middle Ages." *Sudhoffs Archiv für Geschichte der Medizin und der Naturwissenschaften* 49 (1965): 185–98.

Riddle, J. M. "Pomum ambrae: Amber and Ambergris in Plague Remedies." *Sudhoffs Archiv* 48 (1964), 111–22.

Riddle, J. M. *Contraception and Abortion from the Ancient World to the Renaissance.* Cambridge, MA: Harvard University Press, 1992.

Rimmel, Eugene. *The Book of Perfumes.* London: Chapman and Hall, 1865; repr. n.p.: Elibron, 2005.

Ritter, H. "Ein arabisches Handbuch der Handelswissenschaft." *Der Islam* 7 (1917): 1–91.

Ritter, H. *The Ocean of the Soul: Man, the World, and God in the Stories of Farīd al-Dīn ʿAṭṭār.* Trans. J. O'Kane. Leiden: Brill, 2003.

Rockhill, W. W. *Diary of a Journey through Mongolia and Tibet in 1891 and 1892.* Washington: Smithsonian, 1894.

Rockhill, W. W. *Land of the Lamas: Notes of a Journey through China, Mongolia, and Tibet.* London, 1891; repr. New Delhi: Asian Educational Services, 1997.

Roerich, Y. N. *Tibetan-Russian-English Dictionary with Sanskrit Parallels.* 11 vols. Moscow: Nauka, 1983–9.

Rong Xinjiang, "The Nature of the Dunhuang Library Cave and the Reason for its Sealing." *Cahiers d'Extrême-Asie* 11 (1999–2000): 247–75.

Rouk, H. F., and Mengesha, H. *Ethiopian Civet (Civettictis civetta).* Addis Ababa: Imperial Ethiopian College of Agriculture and Mechanical Arts, 1963.

Ruska, J., and M. Plessner, "ʿAnbar." *EI*² s.v.

Rustomji, N. *The Garden and the Fire: Heaven and Hell in Islamic Culture.* New York: Columbia University Press, 2009.

Ruymbeke, C. van. *Science and Poetry in Medieval Persia: The Botany of Nizami's Khamsa.* Cambridge: Cambridge University Press, 2007.

Ryckmans, J., Müller, W. W. and Abdallah, Y. M. *Textes du Yémen antique inscrits sur bois.* Louvain-la-Neuve: Institut Orientaliste, 1994.

Sagart, L. *The Roots of Old Chinese.* Amsterdam: Benjamins, 1999.

Saleh, "The Etymological Falacy and Qurʿanic Studies: Muhammad, Paradise, and Late Antiquity." In A. Neuwirth, et al., eds., *The Qur'ān in Context: Historical and Literary Investigations into the Qur'ānic Milieu.* Leiden: Brill, 2010, 649–97.

Salles, J.-F. "Fines Indiae, Ardh el-Hind: Recherches sur le devenir de la mer Érythrée." In *The Roman and Byzantine Army in the East, Proceedings of a colloqium* [sic] *held at the Jagiellonian University, Krakow, in September 1992.* Krakow: Jagiellonian University, 1994, 165–87.

Salomon, R. "Epigraphic Remains of Indian Traders in Egypt." *JAOS* 111 (1991): 731–6.

Salomon, R. "Addenda to 'Epigraphic Remains of Indian Traders in Egypt." *JAOS* 113 (1993): 593.

Sargent, S. "Huang T'ing-chien's 'Incense of Awareness": Poems of Exchange, Poems of Enlightenment." *JAOS* 121:1 (2001): 60–71.

Scarborough, J. "Early Byzantine Pharmacology." *Dumbarton Oaks Papers* 38 (1984): 213–32.

Schacht, J. *The Origins of Muhammadan Jurisprudence*. Oxford: Clarendon Press, 1950.

Schafer, Edward H. *The Golden Peaches of Samarkand: A Study of T'ang Exotics*. Berkeley: University of California Press, 1963.

Schafer, Edward H. *The Vermilion Bird. T'ang Images of the South*. Berkeley: University of California Press, 1967.

Schafer, Edward H., and Wallacker, B. E. "Local Tribute Products of the T'ang Dynasty." *Journal of Oriental Studies* 4 (1957–8): 213–48.

Schimmel, A. *The Triumphal Sun: A Study of the Works of Jalāloddin Rumi*. London: East-West Publications, 1980.

Schimmel, A. *A Two-Colored Brocade: The Imagery of Persian Poetry*. Chapel Hill: University of North Carolina, 1992; repr. Lahore: Sang-e Meel Publications, 2004.

Schimmel, A. *The Mystery of Numbers*. Oxford: Oxford University Press, 1993.

Schleifer, J. "Zu Sobhys General Glossary der Dahīra." *Orientalistische Literaturzeitung* 39 (1936): 665–74.

Schmidt, H.-P. "Ancient Iranian Animal Classification." *Studien zur Indologie und Iranistik* 5–6 (1980): 209–44.

Schmitz, R. "The Pomander." *Pharmacy in History* 31:2 (1989): 86–90.

Schmucker, W. *Die pflanzliche und mineralische Materia Medica im Firdaus al-Ḥikma des Ṭabarī*. Bonn: Selbstverlag des Orientalischen Seminars, 1969.

Schönig, H. "Camphor." EQ s.v.

Schönig, H. *Schminken, Düfte und Räucherwerk der Jemenitinnen: Lexikon der Substanzen, Utensilien und Techniken*. Beirut: Ergon Verlag, 2002.

Schram, L. M. J. "The Mongours of the Kansu-Tibetan Frontier Part III." *Transactions of the American Philosophical Society* 51:3 (1961): 1–117.

Seppälä, S., *In Speechless Ecstasy: Expression and Interpretation of Mystical Experience in Classical Syriac and Sufi Literature*. Studia Orientalia 98. Helsinki: Finnish Oriental Society, 2003.

Serjeant, R. B. *South Arabian Hunt*. London: Luzac, 1976.

Serjeant, R. B. "Meccan Trade and the Rise of Islam: Misconceptions and Flawed Polemics." *JAOS* 110 (1990): 472–86.

Seth, S. D., et al. *Pharmacodynamics of Musk*. New Delhi: Central Council for Research in Indian Medicine and Homoeopathy [sic], 1975.

Sezgin, F. *Geschichte des arabischen Schrifttums*. Leiden: Brill, 1967-.

Shahid, I. *Byzantium and the Arabs in the Sixth Century Volume 1 Part 1: Political and Military History*. Washington: Dumbarton Oaks, 1995.

Shahid, I. *Byzantium and the Arabs in the Sixth Century Volume 2 Part 2: Economic, Social, and Cultural History*. Washington: Dumbarton Oaks, 2009.

Sharon, M. "The 'Praises of Jerusalem' as a Source for the Early History of Islam." *Bibliotheca Orientalis* 49 1/2 (1992), 56–67.

Shatzmiller, M. "Economic Performance and Economic Growth in the Early Islamic World." *JESHO* 54 (2011): 132–84.

Shboul, A. M. H. *Al-Mas'ūdī & his World: A Muslim Humanist and his Interest in non-Muslims*. London: Ithaca Press, 1979.

Shepherd, D. G., and W. B. Henning, "Zandanījī Identified?" In R. Ettinghausen, ed., *Aus der Welt der islamischen Kunst*. Berlin, 1959, 15–40.

Sidebotham, S. *Roman Economic Policy in the Erythra Thalassa 30 BC–AD 217*. Leiden: Brill, 1986.

Sidebotham, S. *Berenike and the Ancient Maritime Spice Route*. Berkeley: University of California Press, 2011.

Siggel, A. *Arabisch-Deustches Wörterbuch der Stoffe*. Berlin: Akademie-Verlag, 1950.

Silverstein, A. J. *Postal Systems in the Pre-Modern Islamic World*. Cambridge: Cambridge University Press, 2007.

Simeone-Senelle, M.-C. "Aloe and Dragon's Blood, some Medicinal and Traditional Uses on the Island of Socotra." *New Arabian Studies* 2 (1994): 186–98.

Sims-Williams, N. "On the Plural and Dual in Sogdian." *BSOAS* 42 (1979): 337–46.

Sims-Williams, N. "The Sogdian Merchants in China and India." In A Cadonna and L. Lanciotti, eds., *Cina e Iran da Alessandro Magno alla dinastia Tang*. Firenze: Olschki, 1996, 45–67.

Sims-Williams, N. "The Sogdian Ancient Letter II." In *Philologica et linguistica: historia, pluralitas, universitas: Festschrift für Helmut Humbach zum 80. Geburtstag*. Trier: Wissenschaftlicher Verlag, 2001, 267–80.

Sims-Williams, N., and J. Hamilton. *Documents turco-sogdiens du IXᵉ–Xᵉ siècle de Touen-houang*. London, 1990.

Sims-Williams, N., and G Khan. "Zandanījī Misidentified." *BAI* 22 (2008)[2012]: 207–13.

Sivin, N. *Chinese Alchemy: Preliminary Studies*. Cambridge, MA: Harvard University Press, 1968.

Skaff, J. K. "Sasanian and Arab-Sasanian Silver Coins from Turfan: Their Relationship to International Trade and the Local Economy." *Asia Major* 3rd series 11:2 (1998): 67–115.

Smith, J. I., and Y. Haddad. *The Islamic Understanding of Death and Resurrection*. Albany: State University of New York Press, 1981.

So, J. F. "Scented Trails: Amber as Aromatic in Medieval China." *JRAS* (2013): 85–101.

Sokolov, V. E., and N. L. Lebedeva. "Commercial hunting in the Soviet Union." In R. J. Hudson, et al., eds., *Wildlife Production Systems: Economic utilisation of wild ungulates*. Cambridge: Cambridge University Press, 1989, 170–85.

Sokolovskaia, L. and A. Rougeulle. "Stratified Finds of Chinese Porcelains from Pre-Mongol Samarkand (Afrasiab)." *BAI* 6 (1992): 87–98.

Sperl, S. "Islamic Kingship and Arabic Panegyric Poetry in the early 9th century." *Journal of Arabic Literature* 8 (1977): 20–35.

Stang, H. "Arabic Sources on the Amdo and a Note on Gesar of Gliṅ." *Acta Orientalia Academiae Scientiarum Hungaricae* 44 (1990): 159–174.

Starostin, S. *Rekonstruktsia drevnekitaiskoi fonologiceskoi Sistemy*. Moscow: Nauka, 1989.

Stein, M. A. *Ancient Khotan*. 2 vols. Oxford: Oxford University Press, 1907.

Stein, M. A. *Serindia: Detailed Report of Explorations in Central Asia and Westernmost China*. 5 vols. London, 1921; repr. Delhi: Banarsidass, 1980.

Stein, O. "Σῦριγξ und suruṅgā." *Zeitschrift für Indologie und Iranistik* 3 (1925), 280–318 and 345–7.

Steingass, F. *A Comprehensive Persian-English Dictionary*. London, 1892; Repr. New Delhi: Munshiram Manoharlal, 1996.

Stern, S. M. "Rāmisht of Sīrāf, a Merchant Millionaire of the Twelfth Century." *JRAS* (1967): 10–4.

Sternbach, L. "Camphor in India." *Vishveshvaranand Indological Journal* 12 (1974): 425–67.

Stetkevych, S. Abū Tammām and the Poetics of the ʿAbbāsid Age. Leiden: Brill, 1991.

Stetkevych, S. *The Mute Immortals Speak*. Ithaca: Cornell University Press, 1993.

Stetkevych, S. "Intoxication and Immortality: Wine and Associated Imagery in al-Maʿarrī's Garden." In *Homoeroticism in Classical Arabic Literature*. Ed. J. W. Wright and E. K. Rowson. New York: Columbia University Press, 1997, 210–32.

Stetkevych, S. *The Poetics of Islamic Legitimacy*. Bloomington: Indiana University Press, 2002.

Stillman, N. A. "The Eleventh Century Merchant House of Ibn ʿAwkal (A Geniza Study)." *JESHO* 16 (1973): 15–88.

Stoddart, D. M. *The Scented Ape: The biology and culture of human odour*. Cambridge: Cambridge University Press, 1990; repr. with corrections, 1991.

Stoneman, R. *The Greek Alexander Romance*. Harmondsworth: Penguin, 1991.

Strickmann, M. *Chinese Magical Medicine*. Ed. B. Faure. Stanford: Stanford University Press, 2002.

Sundermann, W. "Zur frühen missionarischen Wirksamkeit Manis." *Acta Orientalia Academiae Scientiarum Hungaricae* 24 (1971): 79–125.

Sundermann, W. *Mitteliranische manichäische Texte kirchengeschichtlichen Inhalts*. Berlin: Akademie-Verlag, 1981.

Sundermann, W. "Mani, India and the Manichaean Religion." *South Asian Studies* 2 (1986): 11–9.

Tampoe, M. *Maritime Trade between China and the West: An Archaeological Study of the Ceramics from Siraf (Persian Gulf), 8th to 15th centuries A.D.* Oxford: BAR, 1989.

Tardieu, M."Le Tibet de Samarcande et le pays de Kûsh: mythes et réalités d'Asie centrale chez Benjamin de Tudèle." *Cahiers d'Asie Centrale* 1–2 (1996): 299–310.

Tekin, T. *A Grammar of Orkhon Turkic*. Bloomington: Indiana University, 1968.

Thierry, "Sur les monnaies sassanides trouvées en Chine." In R. Gyselen, ed., *Circulation des monnaies, des merchandises et des biens*. Bures-sur-Yvette: Groupe pour l'étude de la civilisation du Moyen-Orient, 1993, 89–139.

Thomas, Rosalind. *Herodotus in Context: Ethnography, Science and the Art of Persuasion*. Cambridge: Cambridge University Press, 2000.

Thomsen, V. *Inscriptions de l'Orkhon déchiffrés*. Helsinki, 1896.

Tibbetts, G. R. *A Study of the Arabic Texts containing material on South-East Asia*. London: Royal Asiatic Society, 1979.

Timmins, R. J., and J. W. Duckworth. "*Moschus cupreus*." The IUCN Red List of Threatened Species 2015, http://www.iucnredlist.org/details/136750/0. Accessed March 22, 2016.

Timmins, R. J., and J. W. Duckworth. "*Moschus leucogaster*." The IUCN Red List of Threatened Species 2015, http://www.iucnredlist.org/details/13901/0. Accessed March 22, 2016.

Tomber, R. *Indo-Roman Trade: From Pots to Pepper*. London: Duckworth, 2008.

Töttössy, C. "Graeco-Indo-Iranica." *Acta Antiqua Academiae Scientiarum Hungaricae* 25 (1977): 130–5.

Ullmann, M. *Die Medizin im Islam*. Leiden: Brill, 1970.

Ullmann, W. "Thomas Becket's Miraculous Oil." *Journal of Theological Studies* n.s. 8 (1957).

Unschuld, P. *Medicine in China: A History of Pharmaceutics*. Berkeley: University of California Press, 1986.

Unschuld, P. *Introductory Readings in Classical Chinese Medicine*. Dordrecht: Kluwer, 1988.

Unvala, J. M. *The Pahlavi Text "King Husrav and his Boy"*. Paris: Geuthner, [1921].

Utas, B. "The Jewish-Persian Fragment from Dandān-Uiliq." *Orientalia Suecana* 17 (1968): 123–36.

Vahman, F. "A Beautiful Girl." In *Papers in Honour of Professor Mary Boyce*, vol. 2. Leiden: Brill, 1985, 665–73.

Van Gelder, G. "Four Perfumes of Arabia: A Translation of al-Suyūṭī's *Al-Maqāma al-Miskiyya*." In R. Gyselen, ed., *Parfums d'Orient*. Bures-sur-Yvette: Groupe pour l'Étude de la Civilisation du Moyen-Orient, 1998, 203–12.

Van Sprengen, W. *Tibetan Border Worlds: A Geohistorical Analysis of Trade and Traders*. London: Kegan Paul, 2000.

Van Tongerloo, A. "An Odour of Sanctity." In *Apocryphon Severini presented to Soren Giversen*. Aarhus: Aarhus University Press, 1993, 245–56.

Vollers, K. "Beiträge zur Kenntniss der lebenden arabischen Sprache in Aegypten. II. Ueber Lehnwörter, Fremdes und Eigenes." *ZDMG* 50 (1896): 607–57.

Waddell, L. A. "Some Ancient Indian Charms, from the Tibetan." *Journal of the Anthropological Institute of Great Britain and Ireland* 24 (1895): 41–4.

Waines, D., and F. Sangustin, "Zaʿfarān" in *EI*² s.v.

Waku Hakuryū. *Bukkyō Shokubutsu Jiten*. Tōkyō: Kokusho Kankōkai, 1979.

Walburg, R. *Coins and Tokens from Ancient Ceylon*. Wiesbaden: Reichert, 2008.

Walker, J. T. "Expeditions among the Kachin Tribes on the North-East Frontier of Upper Burma." *Proceedings of the Royal Geographical Society* (1892): 161–73.

Wang Gungwu. "The Nanhai Trade: A Study of the Early History of Chinese Trade in the South China Sea." *JMBRAS* 21:2 (1958): 1–135.

Wang, Y., and R. B. Harris. "*Moschus anhuiensis*." The IUCN Red List of Threatened Species 2015, http://www.iucnredlist.org/details/136643/0. Accessed March 22, 2016.

Wang, Y., and R. B. Harris. "*Moschus berezovskii*." The IUCN Red List of Threatened Species 2015, http://www.iucnredlist.org/details/13894/0. Accessed March 22, 2016.

Wang, Y., and R. B. Harris. "*Moschus chrysogaster*." The IUCN Red List of Threatened Species 2008, http://www.iucnredlist.org/details/13895/0. Accessed March 22, 2016.

Wang, Y., and R. B. Harris. "*Moschus fuscus*." The IUCN Red List of Threatened Species 2015, http://www.iucnredlist.org/details/13896/0. Accessed March 22, 2016.

Warder, A. K. *Indian Kāvya Literature Volume VII: The Wheel of Time*. 2 parts paged continuously. Delhi: Motilal Banarsidass, 1989.

Warmington, E. H. *The Commerce between the Roman Empire and India*. 2nd ed. Cambridge: Cambridge University Press, 1974; repr. New Delhi: Munshiram Manoharlal, 1995.

Watson, A. M. *Agricultural Innovation in the Early Islamic World*. Cambridge: Cambridge University Press, 1983.

Watt, George. *The Commercial Products of India*. London: John Murray, 1908.

Wayman, A. "Notes on the Three Myrobalans." *Phi Theta Annual* 5 (1954–5): 63–77.

Weerakkody, D. P. M. *Taprobanē: Ancient Sri Lanka as known to the Greeks and Romans*. Turnhout: Brepols, 1997.

Wheatley, Paul. "Geographical Notes on Some Commodities involved in Sung Maritime Trade." *JMBRAS* 32:2 (1959): 3–137.

Whitehouse, D. "Some Chinese and Islamic Pottery from Siraf." In W. Watson, ed., *Pottery and Metalwork in T'ang China*, 35–40. London, 1970.

Whitehouse, D. "Abbasid maritime Trade: Archaeology and the Age of Expansion." *Rivista degli Studi Orientali* 59 (1987): 339–47.

Whitehouse, D. *Siraf: history, topography and environment*. Oxford: Oxbow Books, 2009.

Whitehouse, D., and A. Williamson. "Sasanian Maritime Trade." *Iran* 11 (1973): 29–49.

Whitfield, S., and U. Sims-Williams. *The Silk Road: Trade, Travel, War and Faith*. Chicago: Serindia, 2004.

Wiedemann, E. "Über arabische Parfüms." *Archiv für Geschichte der Medizin* 8 (1915): 83–8.

Wiedemann, E. "Über von den Arabern benutzte Drogen." In *Aufsätze zur arabischen Wissenschaftsgeschichte* 2.230–74.

Wiedemann, E. "Über Parfüms und Drogen bei den Arabern." In *Aufsätze zur arabischen Wissenschaftsgeschichte* 2.415–30.

Wiedemann, E. *Aufsätze zur arabischen Wissenschaftsgeschichte*. Ed. W. Fischer. 2 vols. Hildesheim: Olms, 1970.

Wild, S. "Lost in Philology? The Virgins of Paradise and the Luxenberg Hypothesis." In A. Neuwirth, et al., eds., *The Qurʾān in Context: Historical and Literary Investigations into the Qurʾānic Milieu*. Leiden: Brill, 2010, 625–47.

Wiet, G. "Les Marchands d'épices sous les sultans mamlouks." *Cahiers d'histoire egyptienne* 7 (1955): 81–147.

Wilkinson, J. C. "Arab-Persian Land Relationships in Late Sasanid Oman." *PSAS* 6 (1972): 40–51.

Wilkinson, J. C. "Ṣuḥār in the Early Islamic Period." *South Asian Archaeology 1977*. Naples: Istituto universitario orientale, 1979, 887–990.

Williamson, A. "Sohar and the Sea Trade of Oman." *PSAS* 4 (1974): 78–96.

Wink, A. *Al-Hind: The Making of the Indo-Islamic World. Vol. 1: Early Medieval India and the Expansion of Islam 7th -11th centuries*. Leiden: Brill, 1990; repr. 2002.

Wolters, O. *Early Indonesian Commerce*. Ithaca: Cornell University Press, 1967.

Wright, J. C. "The Supplement to Ludwig Alsdorf's *Kleine Schriften*: A Review Article." *BSOAS* 62 (1999): 529–42.

Wright, W. *A Short History of Syriac Literature*. London: Black, 1894.

Yaldiz, M. *Archäologie und Kunstgeschichte Chinesisch-Zentralasiens (Xinjiang)*. Leiden: Brill, 1987.

Yamada, K. *A Study on the Introduction of An-hsi-hsiang in China and that of Gum Benzoin in Europe*. 2 parts. Osaka: Kinki University, 1954–5.

Yamada, K. *A Short History of Ambergris by the Arabs and Chinese in the Indian Ocean*. 2 parts, Osaka: Kinki University, 1955–6.

Young, G. K. *Rome's Eastern Trade: International Commerce and Imperial Policy 31 BC–AD 305*. New York: Routledge, 2001.

Yule, H., and Burnell, A. C. *Hobson-Jobson: A Glossary of Colloquial Anglo-Indian Words and Phrases, and of Kindred Terms, Etymological, Historical, Geographical and Discursive*. 1903; repr. New Delhi: Munshiram Manoharlal, 2000.

Yusuf, Muhsin. "Sea versus Land: Middle Eastern Transport during the Muslim Era." *Der Islam* 73 (1996): 232–58.

Zaytsev, I. V. "Tatar Musk." In B. Kellner-Heinkele, et al., eds., *Man and Nature in the Altaic World: Proceedings of the 49th Permanent International Altaistic Conferene, Berlin, July 30-August 4, 2006*. Berlin: Klaus Schwarz, 2012, 479–82.

Zhang Baoliang. "Musk deer: their capture, domestication and care according to Chinese experience and methods." *Unasylva* 35 (1983), 16–24; available online at http://www.fao.org/documents/show_cdr.asp?url_file=//docrep/q1093E/q1093e02.htm (accessed 4/26/05).

Ziaee, A. A. "Omani Trade and Cultural Relations with East Asian Countries." M. Hoffmann-Ruf and A. Al Salimi, eds., *Oman and Overseas*. Studies on Ibadism and Oman, Vol. 2. Hildesheim: Olms, 2013, 219–25.

Zieme, P. "Zum Handel im uigurischen Reich von Qočo." *Altorientalische Forschungen* 4 (1976): 235–50.

Zimmermann, F. W. "Al-Kindī." *Cambridge History of Arabic Literature. Religion, Learning, and Science in the 'Abbasid Period*, ed. M. J. L. Young, et al. Cambridge: Cambridge University Press, 1990, 364–9.

Zwalf, W. *The Heritage of Tibet*. London: British Museum, 1981.

Indices

1 Selected Terms: Arabic 414
2 Selected Terms: Pahlavi and Persian 417
3 Literary Sources 418
4 Personal Names and Titles 422
5 Geographical Names 424
6 Subjects 429

1 Selected Terms: Arabic

ʿabīr 281, 284, 285, 298–9, 303, 348, 358
ʿabīrah 345
ʿabīṭ 172, 213
adhfar 156, 292
adqāl 294
afāwih 273, 275
amlaj 67, 187
ʿanbar 62, 122
ʿanbar al-hind 61
ʿanbarah 345
ānuk 180, 267
anzarūt 270
ʿanzarūt 264
aqrāṣ al-misk 180
arāk 301
ās 338
ʿaṭṭār 260
awbar 231
awtād 164
aẓfār 274
aẓfār al-ṭīb 83

bābūnaj 276
bahār 276
bahārī 187, 197
bahīmah 172
bahmanayn 184, 209
bakhūr 277, 286
balasān 280
balīlaj 67
Ballāhārī 197n166
ballūṭ 187, 266, 270
banādiq misk 226, 362
B.lhārī 197, 241
barīd 229
Barmakiyyah 285, 294
barniyyah/barānī 169, 279, 287
basbās 373
basbāsah 73
bayḍah 31
bīsh 206
budd 189
bunk 274
bustān 292

dābbah 161, 174, 183
dābbat al-misk 361
dābbat al-zabād 361
dam al-akhawayn 263, 267
Dārī 141–2
ḍarw 274
dārṣīnī 52, 53, 195, 322
dastanbū 280
dawāwīj 296
dhafar 284
dhafir/adhfar 156
dharīrah/dharāʾir 180, 281
dharīrah mumassakah 34
Dhūsm.t 172, 200
dībā 53
diqāq 152, 179
diraq 193
dukhnah 277
duwaybbah 160–1

fāghirah 274
falanjah 273
faqāḥā 184

fa'r 150
fa'r al-misk 148–9
fārah 141, 147, 149, 179
fa'rat al-ibl 262
fa'rat al-misk 31, 141, 147–51, 160, 302
fārat misk 298
fataqa 97, 289
fihr 329
firind 52, 53, 54, 58, 195
fūl al-bāqilā' 129
fuqqā' 321

ghāliyah 75, 180, 266, 275, 276, 277, 284–90, 345, 346
  Danger of overuse 309
  Formulas for 278–80
  Used in grooming 285, 286

ḥabb al-mīs.m 274
hāl 66, 322
hālbuwwā 66–7, 274
ḥamāḥim 282
ḥarīr 49, 52, 53, 54
ḥarīrah 344
harnuwah 58, 274
ḥibb ṣīnī 287
ḥuḍaḍ 264
ḥūr 360

ihlīlaj 67
ihlīlajāt 58
ijjānah 292
'iṭr 142, 144, 260, 273, 275, 287

jafnah 350
Jabalī 203
jallālah 212
jām 265, 338
jamājim 287, 292
jannah 129, 353
jannāt 'adn 353
jasad 289
jawzbuwwā 73, 273
jawz hindī 67
jilāl 152, 179
jirdhān 262
julūd 153
jundbādastar 22, 267
juradh 149, 161
Jurz 240

kabābah 274
kādhī 185
kādī 265
kāfūr 122
kandasah 159, 174, 208–9
karish 152
Kawshān 239
kdhms 174, 208
khadank 47
khalaqah 264
khalūf 345
khalūq 281, 284, 285, 294, 347
khamīr 264
khāraṣīnī 57
khāwkhīr 53
khazz 227
khazzah 344
khīrī 275, 282
khishf 148, 160
Khiṭā' 51, 201
Khiṭā'ī 180
khitām 353
khiyār shanbar 67
khizānat al-ṭīb 287
khumāsiyyāt 259
khūlanjān 53, 72
khuṣyah 31
khutū 53, 174–5, 177, 182, 225
kilāb 75n265
kirsh/akrāsh 152–3, 264, 346
kīmkhat 58
kīmkhāw 53, 195
k.n.d.s 208
kundur 244
kundus 208
kuthbān 326

lādhan 274
lakhlakhah/lakhālikh 180, 280, 293
Lankabālūs 247
laṭīmah/laṭā'im 154–5
lawzinaj 323
lubnā 274n16
luqaṭ 153

mā' al-ward 282
madāf 295
madāk 295, 329
madhūn 57
maftūqah 285
maḥlab 274
ma'jūn/ma'jūnāt 291, 294
majlis 291
makhāniq 294
malāb 358
ma'mūnī 47
mandal 348
marāghah 167
marāsil 294
marfa'
masaka 35, 215
maṣṭakā 322
masūḥ 280
may'ah 274
maysūs 282
maytah 211
mazāwid 173
misk 28, 35–6, 122, 140, 141, 147, 156, 335, 353
  al-misk al-adhfar 156
  al-misk al-sughdī 114, 140
  misk al-yad 187
miskah 345
miswāk 301
mizāj 320
mukhammas 278
murr 159, 208–9
mushkrāsh/miskrāsh 187
muṭarrā 347
muthallath 278
muthallathah khazā'iniyyah 294

nabīdh 320
nadd 64, 129, 275, 277–8, 333, 346, 347, 352
naḍūḥ 265, 282
nāfijah/nawāfij 148, 151, 152, 160, 171, 173, 179, 259
nāfijat misk 26, 119
nārjīl 58, 67
narjis 275, 291
nashr 329
nasrīn 277
nat' 266
nawāfiḥ 333
nīl 65
nīlūfar 129
nuṣub 174

qāqullah 66–7, 274, 322
qaranful 73, 273
qarīrah/qawārir 186, 187, 259, 287
qaṣab al-dharīrah 281
qinbīl 274
qinnīnah 260
Q.nbārī 203
q.n.d.s 208
qirfah 265, 270, 274
Q.sārī 203
qusṭ 274
Qutāy 201
quwwād al-ghāliyah 289

raḥīq 320, 353
rāmik 155–6, 263, 270, 313
rāwand 52
rayḥān/rayāḥīn 275, 353
rīḥ 144, 341
rubb 286
rūḥ 341
rukn 289

'.ṣmārī 203
sādawarān 264, 265, 266
safaṭ 286
ṣaḥfah 350
sāhiriyyah 294
sāj 68
Sājū 201
ṣandal 273
sawīq 323
ṣayyāḥ 293
shādawarān 264, 270
shadhā 155
shadw 155–6
shāhisfaram 129
shamāmah 225, 280, 281
shaqā'iq 276
shawābīr 278, 346
shīṭaraj 187, 270n286
shiyāf 152, 178–8, 181, 186
sikbāj 236, 323
ṣīlbanj 53, 195

Ṣīn 51
ṣiwār/ṣuwār 153–4, 302
siyāhdārū 264, 267
siyāhdārūwān 264
siyāh dāwrān 187
sukk 153, 266, 280, 289, 313, 343
Definition of 155–6
sukk al-akrāsh 153
ṣumūgh 326
sunbul 11, 63, 159, 208, 273
sunbul al-hind 63
sunbul al-ṭīb 208
surrah/surar/aṣwirah 75n264, 148, 151, 160
ṣurar 150, 333
surūj 195
sūsan 275

ṭabāshīr 69
ṭālīqūn 57
tamāthīl 291
tamr hindī 67
ṭast 266
Tatārī 138, 180, 190, 201
ṭayfūriyyah 323
ṭayyibāt 273
tawr makkī 279
ṭīb/ṭuyūb 142, 272, 275, 341
ṭībah 341
Tughuzghuz 198
Ṭūmanī 200
Ṭūm.s.t 200
turshutām 294
Tūsmat 200
Ṭūsm.t 200

al-ʿūd al-muṭarrā 278
umm al-ghaylān 274n13
uṣūl al-ṭīb 272

wabīṣ 285, 344
*wadyā(*ʾ or *n)ī* 192, 197
ward 273
wardah 121
wars 265, 274
washy 285

yāqūt 69, 356
yashb 225
yasmīn 275

zabād 22
zanbaq 275, 280
zarāwand ṣīnī 263
zarīr 299
zarnab 273
ẓibāʾ 150
ẓibāʾ al-misk 148
zibdiyyah ṣīnī 279
zīr 295
ẓurafāʾ 284
zurunbād 67

2 Selected Terms: Pahlavi and Persian

ʿabīr 132
āhū 330, 364
āhū-yi mushk 157, 330
ambar 62, 122
ʿanbar 122

bān 122
bāranj 47
Bīsh, bīš 32, 33, 209
Bīsh mushk 33
bōy xwaš 123

chīn 209

gund 208
gundbīdastar 208

kāfūr 122
khadang 47
khāl 157, 207
khaṭṭ 157, 206
khāya 31
khūsh-bōy 123
khutanī 232
kurkum 132

mishk 157
mūš 31
mūšak 31
mushk 28, 151, 157
mushk-i adhfar 157
mushk-i nāb 122, 157

mushk (cont.)
mushk rang 334
mushk rāst 187n129
mushk-i sārā 157
mushkīn 157, 334
mušk 28, 31–5, 126–9

nāfag 32, 151
nāfah 151, 157
nāfah-bāf 157
nay shakar 68

panīdh 68

rīwand ṣīnī 46
roshan gulāb 122

shākh-i khutū 199
siyāh dārū 264
sprahm 129

tukhm 31

ʿūd 122, 132

xaz 127

yāsimīn 129

zaʿfarān 132

3 Literary Sources Referred to in the Text

ʿAbīd 348
Abū al-Hindī 298
Abulcasis, see al-Zahrāwī
Abū Nuwās 319
Abū Tammām 301–2, 331–2, 358
Addahamāṇu, *Saṃneharāsayu* 99
Aetius of Amida 36, 135
Agnimahāpurāṇam 98–9
Akhbār al-Ṣīn wa al-Hind 7, 60, 168–9, 202–3, 233, 235, 242, 243, 246, 247, 248
Al-Akhṭal 297
Alexander Romance 34
Alexander of Tralles 36, 135–6
Ananias of Širak 128, 195
ʿAntarah 141, 147–8, 299
Ardā Wirāz Nāmag 123–4
Al-Aʿshā, Maymūn b. Qays 140–1, 153–4, 318–9
Al-Aṣmaʿī, Abū Saʿīd ʿAbd al-Malik b. Qurayb 338
ʿAṭṭār, Farīd al-Dīn 363
Ayādgār-ī Jāmāspīg 128
Al-Azdī, Abū al-Muṭahhar 182, 188, 191–2, 200, 201

Al-Bakrī, Abū ʿUbayd 168n76
Al-Balādhurī, Aḥmad b. Yaḥyā 252
Bāṇabhaṭṭa, *Harṣacarita* 103, 104–5
Bar Bahlūl 34
Bashshār b. Burd 153, 298, 302
Benjamin of Tudela 194
Bilhaṇa, *Caurapañcāśikā* 101
Al-Bīrūnī, Abū Rayḥān 8, 24, 56, 68, 131, 141–2, 144, 159, 177, 196–7, 199, 200, 201, 204, 210, 222, 240, 245
Account of musk 185–187
Book of Curiosities 163, 164n61, 194
Book of the Eparch 136
Brahmavaivartapurāṇa 101
Al-Buḥturī, Abū ʿUbādah 333
Bundahišn 126, 209
Account of *mušk* animals in 32
Buzurg b. Shahriyār, *Kitāb ʿAjāʾib al-Hind* 58, 233, 235, 253

Carakasaṃhitā 94, 102–3
Cakrapāṇi, *Cikitsāsaṃgraha* 104
Cosmas Indicopleustes 36, 138, 145, 211, 217, 253
Account of musk 134–5

Al-Damīrī, Kamāl al-Dīn Muḥammad b. Mūsā 33, 149–51, 210, 327, 332, 351
Danjing yaojue 89
Daqīqī 206
Al-Dawādārī, Abū Bakr b. ʿAbdallah b. 168n76
Dhū al-Rummah, Abū al-Ḥārith Ghaylān b. ʿUqbah 148, 154, 301, 357
Diʿbil b. ʿAlī al-Khuzāʿī 290
Al-Dīnawarī, Abū Ḥanīfah 148–50, 341
Al-Dimashqī, Jaʿfar b. ʿAlī 251
Dioscorides 43, 46, 71, 304
Draxt-ī Āsūrīg 127–8

Ephrem the Syrian 34, 341, 355
Erya 86
Evliya Çelebi 1

Al-Farazdaq, Hammām b. Ghālib 143
Farrukhī 194
Firdawsī, *Shāhnāmah* 122, 131–2, 291, 302, 334, 335–6, 348

Galenic corpus 313, 317
Gardīzī 198–9, 202, 205, 225
Al-Ghassānī, al-Muẓaffar Yūsuf b. ʿUmar b. ʿAlī 168n76
Al-Ghazālī, Abū Ḥāmid Muḥammad 353, 367
Al-Ghuzūlī, ʿAlāʾ al-Dīn ʿAlī b. ʿAbdallāh al-Bahāʾī al-Dimashqī 9

Ḥāfiẓ Shīrāzī 206, 321
Ḥakīm b. Ḥunayn 305, 307
Hangong xiangfang 91
Al-Harawī, Abū Manṣūr Muwaffaq 46
  account of medicinal properties of musk 305
Ḥārithah b. Badr 319
Ḥassān b. Thābit 300
He Ning 89
Herodotus 30, 194, 304
Hippocratic corpus 304
Ḥudūd al-ʿālam 10, 45–6, 53, 59, 60, 67, 68, 69, 70, 78, 193, 196, 197, 198, 200, 203, 241, 245
Husraw-ī Kawādān ud Rēdaq-ē 129–30
Hyech'o 197n164, 230n60, 234

Ibn ʿAbd Rabbih 213
Ibn Abī al-Ḥadīd, ʿAbd al-Ḥamīd b. Hibat Allāh 9
Ibn Bādīs, Muʿizz 159, 187, 188, 189, 197, 329
Ibn al-Balkhī 242
Ibn Baṭṭūṭah 196, 348
Ibn al-Bayṭār 47, 168, 189
  Account of the medicinal properties of musk 307
Ibn al-Faqīh al-Hamadhānī 10, 57, 72, 74, 78, 79, 149, 193, 198, 255, 367
Ibn Ḥabīb, Muḥammad 241
Ibn Ḥanbal, Aḥmad 345
Ibn Ḥawqal, Abū al-Qāsim 48, 78, 193, 199
Ibn Jazlah 185
Ibn al-Jazzār, Aḥmad b. Ibrāhīm 162n58, 164n61, 324
Ibn Jubayr, Muḥammad b. Aḥmad 348
Ibn Juljul 42–3
  Account of musk 306
Ibn Kaysān, Abū al-Ḥasan Sahlān 7, 71, 152–3, 159, 176–7, 196–7, 199, 200, 201, 202, 336
  See also Ibn Mandawayh-Ibn Kaysān tradition
Ibn Khurradādhbih, Abū al-Qāsim ʿUbayd Allāh 10, 53, 56, 70, 74, 78–9, 146, 193, 195, 198, 227, 244, 248, 254, 255, 360, 366
  Itinerary in Southeast Asia 79
Ibn Mandawayh, Abū ʿAlī Aḥmad b. ʿAbd al-Raḥmān 7, 159, 176–7, 200, 201, 269
Ibn Mandawayh-Ibn Kaysān tradition 7, 56, 77, 152, 197, 210, 214, 215, 278, 279, 280, 281, 325, 326
  Account of detection of adulterated musk 268
  Account of musk 176–84
Ibn Māsah 307
Ibn Māsawayh, Abū Zakarīyāʾ Yūḥannā 5, 63, 79, 152, 158, 164, 176n101, 192, 196, 197, 199, 208, 209, 211, 237, 245, 325
  Account of adulteration of musk 267–8
  Account of musk 159–60, 176, 185, 187–8, 190–1
  Account of the medicinal properties of musk 303
  Lists of aromatics 273–4
Ibn al-Muqaffaʿ, Abū Muḥammad b. ʿAbdallāh 332
Ibn al-Nadīm, Abū al-Faraj Muḥammad b. Isḥāq 5, 6
Ibn Qutaybah, Abū Muḥammad ʿAbdallāh b. Muslim 175
Ibn al-Rūmī, Abū al-Ḥasan ʿAlī b. al-ʿAbbās b. Jurayj 143
Ibn Rustah, Abū ʿAlī Aḥmad b. ʿUmar 78
Ibn Sīnā, Abū ʿAlī al-Huṣayn b. ʿAbdallāh 159, 305, 308
  account of the medicinal properties of musk 306
  Account of musk 184–5, 209
Ibn Wāfid 152–3
Ibn Zaydūn, Abū al-Walīd Aḥmad b. ʿAbdallāh 333–4, 354, 357
Al-Ibshīhī, Bahāʾ al-Dīn Muḥammad b. Aḥmad, *Mustaṭraf* 277
Al-Idrīsī, Muḥammad b. Muḥammad 58, 80, 244, 245

Imru' al-Qays 141, 155, 156, 295, 297
Isaac of Antioch 34
Isḥāq b. ʿImrān 307
Al-Iṣṭakhrī, Abū Isḥāq Ibrāhīm b. Muḥammad 49, 78, 198, 237

Jābir b. Ḥayyān 315
Al-Jāḥiẓ, Abū ʿUthmān ʿAmr b. Baḥr 6, 48, 50, 51, 148–50, 158, 211, 214, 215, 217, 251, 290, 326
  Account of musk in the *Kitāb al-Ḥayawān* 148, 160–1, 212–3
  Account of musk in the (attr.) *Kitāb al-Tabaṣṣur bi-l-tijārah* 161, 189
Jamīl b. Maʿmar 297
Al-Jawharī 149
Al-Jawālīqī, Mawhūb b. Aḥmad 35, 151
Al-Jawziyyah, Ibn Qayyim 315, 322, 326
Al-Jayhānī 159, 163, 201, 205, 214–5, 217
  Account of musk 166–7
Jerome, *Adversus Jovinianum* 31, 36, 132, 134
Jiu Tangshu 130
Jīvakapustaka 104, 114
Jurjānī, Fakhr al-Dīn, *Wīs u Rāmīn* 226
Al-Jurjānī, Ismaʿīl b. Ḥasan 177n105

Kālacakratantra 111–12
Kālidāsa 94–5
Al-Khafājī, Aḥmad b. Muḥammad 35
Kalīla wa Dimnah 34, 332
Khāqānī 209, 346
Al-Kāshānī, Abū al-Qāsim ʿAbd Allāh b. ʿAlī 177n105
Al-Kāshgharī, Maḥmūd 26–7, 118–21, 145, 205
Al-Kindī, Abū Yūsuf Yaʿqūb b. Isḥaq 5–6, 48, 263, 310, 312–3
  On the psychology of aromatics 276
  see also *Kitāb Kīmiyāʾ al-ʿiṭr wa-l-taṣʿīdāt*
Kitāb Aḥwāl al-qiyāmah 359
Kitāb al-Hadāyā wa-l-tuḥaf 225, 295
Kitāb Kanz al-fawāʾid fī tanwīʿ al-mawāʾid 262, 323
Kitāb Kīmiyāʾ al-ʿiṭr wa-l-taṣʿīdāt 6, 263–7, 276, 278, 279, 289, 313, 325
  Ghāliyah enhancement procedure 289
Kitāb al-Tabaṣṣur bi-l-tijārah 45, 50, 52, 55, 56, 59, 69, 251, 303
  List of adulterants of musk 267
Kitāb al-Tāj 131, 339
Kou Zongshi 91
Kutadgu Bilig 120–1, 145, 253
Kuthayyir b. ʿAbd al-Raḥmān al-Mulaḥī 143

Lisān al-ʿArab 153, 155
Li Shizhen, *Bencao gangmu* 23, 26, 87–8, 91

Al-Maʿarrī, Abū al-ʿAlā 329–30, 354
Madanapāla, *Nighaṇṭu* 104
Ma Huan 342
Māhuka, *Haramekhalā* 104
Al-Maḥāmilī, *Kitāb al-Lubāb* 150
Al-Marwazī 163, 166–7
  Description of routes into China 166, 231
Masīḥ al-Dimashqī 305
Al-Masʿūdī, Abū al-Ḥasan ʿAlī b. al-Ḥusayn 7, 48, 81, 159, 161, 188, 189, 193, 208, 213, 215, 216, 217, 222, 224, 239, 248, 249, 307, 325, 336, 360, 366
  Account of musk 167–172
  On the principal aromatics 275
Mingyi bielu 88
Minūchihrī 151, 199
Al-Miskī, Muḥammad b. Aḥmad b. al-ʿAbbās 160, 173–6, 177
Al-Muqaddasī, Shams al-Dīn Abī ʿAbdallāh Muḥammad b. Aḥmad b. Abī Bakr al-Bannāʾ al-Shāmī 78, 244
Al-Muraqqish the Younger 318
Murasaki Shikibu, *Tale of Genji* 92
Muṣʿabī 193
Al-Mutanabbī, Abū al-Ṭayyib Aḥmad b. al-Ḥusayn 213, 292, 303, 328–9

Al-Nābighah al-Dhubyānī 294
Al-Nābighah al-Jaʿdī 142
Al-Narshakhī 49
Niẓāmī, *Haft Paykar* 302, 330, 333, 334, 345
Al-Nuwayrī, Shihāb al-Dīn Aḥmad b. ʿAbd al-Wahhāb 5, 7, 8, 22, 56, 162–3, 165–6, 167, 172–6, 177, 279, 325, 326, 346

Pañcatantra 34, 327–8, 332
Paul of Aegina 136
Periplus Maris Erythraei 40, 133
Philostratus 124

Pires, Tomé 203, 244
Pliny the Elder 61, 121, 123, 304
Procopius 145, 253

Al-Qāḍī, 'Abd al-Raḥīm b. Aḥmad, *Kitāb Daqā'iq al-akhbār* 359
Al-Qahramān 307
Al-Qalqashandī, Shihāb al-Dīn Aḥmad b. 'Alī 8, 154, 162, 166, 176, 189, 325
Al-Qazwīnī, Zakariyyā' b. Muḥammad 80, 168n76, 235
Al-Qazwīnī, Ḥamdullāh al-Mustawfī 33, 209
Qiṣaṣ Qur'ān Majīd 353
Qur'ān 2:173 211
6:145 211
76:5 320
83:25–6 319–20, 353

Ravigupta, *Siddhasāra* 104
Al-Rāzī, Abū Bakr Muḥammad b. Zakariyyā 305, 307
Rgyud bźi 108–9
Rūdakī 207
Rūmī 349–50, 363–4

Sābūr b. Sahl 310–12
Sa'dī 202, 206, 330–1
Al-Samarqandī, Abū Layth, *Kitāb Ḥaqā'iq wa-l-daqā'iq* 359
Al-Ṣanawbarī, Abū Bakr Aḥmad b. Muḥammad al-Ḍabbī al-Anṭākī 335
Al-Sarī al-Raffā', Abū al-Ḥasan *Kitāb al-Mashmūm* 35, 77, 82, 301
Saskya Paṇḍita 110
Shāhnāmah, see Firdawsī
Al-Shayzarī, Jalāl al-Dīn 'Abd al-Raḥmān b. Naṣr
Account of adulteration and imitation of musk 270–1
Shennong bencao 86–7
Shuowen jiezi 86
Simeon Seth 136
Al-Sīrāfī, Abū Zayd 7, 52, 67, 81, 159, 188, 208–9, 213, 215, 216, 217, 233, 249
Account of musk 167–72
Sogdian Ancient Letters 237, 240, 249
Sōmēśvara, *Mānasōllāsa* 97–8, 99, 101
Su Song 91
Suśrutasaṃhitā 94, 103
Suvarṇabhāsottamasūtra 26, 115
Al-Suyūṭī, Jalāl al-Dīn Abū al-Faḍl 'Abd al-Rahmān b. Abī Bakr 35, 316, 326
Syriac Book of Medicines 34, 310, 313–17

Al-Ṭabarī, 'Alī b. Rabban 5, 67, 131, 278, 281, 307, 316
Al-Ṭabarī, Abū Ja'far Muḥammad b. Jarīr 48, 75, 138, 224, 241, 320, 348, 353, 362
Taiping Yulan 89, 130–1
Talmud 132–3, 211
Tamba Yasuyori, *Ishimpō* 93
Al-Tamīmī, Muḥammad b. Aḥmad b. Sa'īd 7, 8, 56, 77, 159, 160, 162, 164, 193, 208, 215, 216, 278, 279, 282, 286, 346
Account of musk 172–6
Al-Tanūkhī, Abū 'Alī al-Muḥassin b. 'Alī 81, 351
Tao Hongjing 217, 223, 263
Account of musk 87–8
Ṭarafah b. al-'Abd 284, 300
Al-Tha'ālibī, Abū Manṣūr 'Abd al-Malik b. Muḥammad 35, 47, 50, 76, 79, 129–30, 158
Al-Tha'labī, Abū Isḥāq Aḥmad b. Muḥammad b. Ibrāhīm 217, 337–8, 360
Al-Ṭūsī, Naṣīr al-Dīn Muḥammad b. Muḥammad 177n105
Ṭuṭī-nāmah 333

Al-'Uklī 175
Al-'Ujayr al-Salūlī 76–7, 155
'Umar b. Abī al-Rabī'ah 297, 300
'Unṣurī 330

Vāgbhaṭa 103–4, 110
Varāhamihira, *Bṛhatsaṃhitā* 95–7, 107
Vidyākara, *Subhāṣitaratnakoṣa* 100

Al-Warrāq, Ibn Sayyār 286, 317–8, 320, 321–3
Al-Washshā', Muḥammad b. Aḥmad 284–5, 293–4
Al-Wāsiṭī 347
Al-Waṭwāṭ, Muḥammad b. Ibrāhīm 163–4
Wizīdagīhā-ī Zādspram 33, 34, 128

Xin Tangshu 130
Xinxiu bencao 88–9
Xuanzang 238

Al-Yaʿqūbī, Aḥmad b. Abī Yaʿqūb b. Jaʿfar 7, 8, 77, 152, 158, 167, 172–4, 193, 196, 198–9, 203, 208, 214
Account of musk 161–6
Quoted by al-Qalqashandī 152n21, 164
Ghāliyah formula 279–80
Yāqūt, Shihāb al-Dīn al-Rūmī al-Ḥamawī 77, 80–1
Yijing 118

Al-Zahrāwī, Abū al-Qāsim 8, 278, 279
Al-Zamakhsharī, Abū al-Qāsim Maḥmūd b. ʿUmar 8, 336–7, 351
Zhang Bangji, *Mozhuang manlu* 91

4 Personal Names and Titles

ʿAbharah 235
ʿAbdallāh b. Qilābah 361–2
ʿAbd al-Muṭṭalib 338
ʿAbd al-Rashīd b. Maḥmūd 205
Abrahah 347
Abū ʿAmr 341
Abū al-ʿAtāhiyah 225
Abū al-Dawāniq, see al-Manṣūr
Abū Dāwūd 52
Abū Dulaf 76
Abū al-Ḥasan b. Khulayf 259
Abū Jaʿfar al-Ṭarasūsī 290
Abū Muslim 52
Abū Naṣr Muḥammad b. Bughā 350
Abū Saʿīd al-ʿAfṣī 259
Abū Ṭālib 144, 342
Abū ʿUbaydah ʿAbdallāh b. Qāsim 252
Adam 63, 217, 348, 360–1
ʿAḍud al-Dawlah 176, 329
Afshīn 331
Aḥmad b. Hilāl, governor of Oman 236
Aḥmad b. Marwān 236
ʿĀʾishah 343, 350
Alexander the Great 68, 121, 193, 348
ʿAlī b. Abī Ṭālib 348, 349
ʿAlī b. Ḥusayn 344
ʿAlī al-Riḍā 345
Al-Amīn 48, 233, 286, 323, 339
ʿAmr b. al-Layth al-Ṣaffār 224–5
ʿAmr b. Luḥayy 175
Anas b. Mālik 344
Anklesaria, B. T. 33
An Napantuo 139
Aprin-čor Tegin 118
Arctander, S. 18
Ardashīr I 137
Asad b. ʿAbdallāh al-Qasrī 252
Ayyūb b. Sulaymān b. ʿAbd al-Malik
ʿAzīz Abū Manṣūr Nazār 177

Bādhām 131
Bahrām I, Sasanian king 138, 245
Bazin, L. 26
Becket, Thomas 340
Benveniste, É. 24–5
Bidʿah 323
Bilqīs 226
Boyle, Robert 18, 216, 283
Budge, E. A. W 313, 315
Bumın/*Tümen 139
Būrān 226
Byzantine emperor 327

Candragupta II 94
Chardin, Jean 107, 323
Charles I of England 340
Chinggis Khan 231, 250, 253
Clauson, G. 27–8
Ču 116
Ču Xayšın 116

Darius III 121
Decius, see Duqyānūs
Demiéville, P. 24
Dharmapāla 58
Duqyānūs 338, 362
De la Vaissière, É. 113–4, 250

Elizabeth II 339
Ephrem 341
Erdal, M. 116

Al-Faḍl b. al-Rabīʿ 286

Gabriel 352
Gignoux, P. 35
Goitein, S. D. 256, 259
Green, Deborah 3
Green, M. J. B. 21

Al-Ḥallāj 351
Al-Ḥākim, Fatimid 295
Hamilton, J. R. 116
Harṣa 104
Hārūn al-Ghazzāl 260
Hārūn al-Rashīd 75, 323
Henning, W. B. 24–5, 127–8
Henry IV of England 340
Huc, Evariste-Régis, and Joseph Gabet 210
Ḥumayd al-Ṭūsī 279
Ḥusayn b. ʿAlī 350

Ibn ʿAbbās 286
Ibn Abī Fanan 339
Ibn Ḥamdūn al-Nadīm 64
Ikhshīd 52
Isaac Nīsābūrī 256
Isḥāq b. Yahūdā 236

Jabalah 338
Jaʿfar al-Ṣādiq 344
Jamshīd 122, 337
Jayavarman VII 107
Jellinek, P. 16, 17
John of Dalyatha 341
Joseph 345
Joseph b. ʿAwkal 260
Josephine, Empress 283
Joyce, T. A. 20
Jung, Dinah 3, 261
Justinian 137

Kaʿb al-Aḥbār 362
Kāfūr Abū Misk 335
Kanafani, Aida Sami 3
Khālid b. Yazīd 290
Khāqān of the Türgesh 49
Khāqān of Tibet 193
Khusraw 129, 131, 137, 193, 338
Al-Khwārazmī 78
Kisrā 131, 286
Kohen al-Fāsī 259
Kōmyō, Empress 93
Kumāragupta I 94

Laufer, Berthold 111
Levey, M. 189, 208
Linschoten, J. H. 216
Lisun Tai Sengun 27

Al-Maʿarrī, Abū al-ʿAlā 329
Al-Mahdī, Abbasid caliph 225
Maḥmūd of Ghazna 225, 231, 291
Magellan, Ferdinand 107
Malov, S. E. 28
Al-Maʾmūn, Abbasid caliph 47, 48, 58, 224, 226, 233, 279, 289, 290, 291, 327, 339
Mānī 245
Al-Manṣūr, Abbasid caliph 68, 241, 285–6
Markham, Frederick 21, 210
Al-Marwazī, Abū ʿAlī al-Ḥāfiẓ 244
Mary 340
Masʿūd, Ghaznavid 225
Mayrhofer, M 30
Al-Mazābilī, Abū ʿAbdallāh 351–2
Mir Yegān 117
Moses of Chorene 195
Al-Muʿammil 47
Al-Muʿāwiyah 361–2
Mufaḍḍal b. Abī Saʿd 259
Mufliḥ al-Aswad al-Khādim 292
Muḥammad 1, 3, 144, 156, 212, 217, 224, 226, 251, 313, 326, 342–5, 349, 350, 353, 356, 357, 359, 364, 368
Muḥammad b. al-Faḍl al-Ḥimyarī 358
Muḥammad b. Harthamah 266
Muḥammad b. Sulaymān 280, 292
Muḥammad b. Tughluq 196
Al-Muhtadī 350
Al-Mukhtār 350
Al-Muktafī 287–8
Al-Muqtadī 118
Al-Muqtadir 288, 291, 292, 346
Al-Mustaʿīn 278
Al-Muʿtaḍid, Abbasid 287–8
Al-Muʿtaḍid, ruler of Sevile, his daughter 354
Al-Muʿtamid 224
Al-Mutanabbī 335
Al-Muʿtaṣim 5, 252
Al-Mutawakkil 64–5, 280, 291, 296, 339
Al-Muʿtazz 291, 296

Nahray b. Nissīm 256
Nanai-vandak 25, 113, 240
Nāṣir al-Ḥaqq, Karakhanid ruler 225
Naṣr al-Qushurī 292
al-Naẓar b. Maymūn 252
Negus of Ethiopia 224
Newid, Mehr Ali 3

Nicias 61
Noah 348
Norbu, Thubten Jigme 103
Nuʿmān, Lakhmid king 294

Oγšaγu 116

Paul 340
Pelliot, P. 115–6
Pēsakk 113
Pigafetta, Antonio 107
Ptolemy 78

Qabīḥah 296
Qādir Khān, Karakhanid ruler 225
Qutaybah b. Muslim 237
Qutluγ Arslan El-Tiräk 116

Rāmisht of Sīrāf 253
Rav Ḥisda 132
Rockhill, W. W. 223
Rostovtzeff, Michael 236

Al-Saffāḥ, Abbasid caliph 332
Ṣāfī al-Ḥuramī 287–8
Salmā 302
Salmān al-Fārsī 349
Al-Samarrī 351
Sarwa 296
Sauvaget, J. 000
Sayf b. Dhī Yazan 338
Shaddād 361
Shakespeare, *The Merry Wives of Windsor* 283
Shāpūr 335
Sitt Miṣr, al-Sayyidah 295
Solomon 226, 337–8, 362
Stein, M. A. 20
Stetkevych, S. 331
Stoddard, D. M. 17
Sulu, *Suluk 252
Sun Pingyi 87
Sun Xinyan 87

Taherti, Abū Isḥāq Barhūn b. Isḥaq b. Barhūn 256
Ṭāhir b. al-Ḥusayn 289
Takhsīch-vandak 113, 114
Tcūṃ-Ttehi 114
Tekin, T. 28
Takut 113
Thomsen, V. 27
Tubbaʿ of Yemen 195

ʿUmar b. al-Khaṭṭāb 144, 241
Umm Jaʿfar 279, 280
ʿUnayzah 295
ʿUtbah b. Ghazwān 241

Vajrabodhi 234
Van Gelder, G. J. H. 326

Wan-razmak 113
Al-Wāthiq 47, 287–8
Wirāz 123

Yaḥyā b. ʿUmar 349
Yartaš 116
Yazdagird 286, 348

Zhenghe 342
Al-Zukūrī, Mūsā 352

5 Geographical Names

ʿĀd 361
Adam's Peak 361
Aden 58, 73, 141, 154, 176, 243, 244, 249, 253
Afghanistan 13, 44, 94, 128
Africa 31, 44, 68
Agni 231
Ahwāz 228, 307
Alexandria 256
Altai mountains 198, 205, 206
Amdo 103, 200
Amorium 331
Amu Darya 237
Andalus 227, 228
Andaman Islands 247
Anhui 13
Annam 93
Antioch 228, 351, 352
Arabia 1, 40, 62, 65, 68, 83, 125, 149, 150
Arabian Peninsula 40, 141, 243
Arabian Sea 81
Arakan 203
Armenia 290

Asia 31
ʿ.ṣmār 203
Assam 64, 69
Ava 204

Bāb al-Mandab 137
Babylonia 132
Bactria 138, 237
Badakhshān 48, 230
Baghdād 47, 68, 176, 228, 241, 254, 266, 284
Bago 204
Bahār 197
Baḥrayn 141, 144, 174, 241
Baikal, Lake 201
Balkh 228, 230, 238, 240, 269
Bālūs 75
Bāmiyān 230
Banbhore 245
Banda archipelago 73
Barskhān 229
Baṣrah 6, 75, 228, 242, 248, 249
Bawan 130
Bayāsirah 252
Baykand 237
Berūj, see Broach
Bhutan 13
Bhūtīshar 231, 240
Black Sea 304
B.lhārī 197, 241
Borneo 55
Brazil 74
Broach 246
Bukhara 49, 231
Burma 86, 93, 203
  See also Myanmar
Bushire 137
Buzghāla-khāna Pass 238

Cambay 70, 143, 246
Cambodia 64, 70, 72, 73, 79, 107
Canton 223
Capelanguam 203
Caspian Sea 228n52
Caucasus 22
Central Asia 42, 79, 112–14, 233, 240
Central Eurasia 43
Ceylon 134
Champa 73, 79, 81, 248
Changthang 230
China 1, 12–13, 20, 21, 22, 50–8, 63, 72, 94, 129, 137, 145, 165, 166, 167, 169, 195–6, 200–1, 203, 217, 227, 228, 240, 241, 315
  Goods of 51–8, 244
  History of musk in 85–92
Chu River 205
Constantinople 228
Ctesiphon 131

Dabā 241
Damascus 228, 316
Dandan Uiliq 255
Dārīn 141–3, 174, 217, 241, 244, 251, 332
Daybul 138, 159, 165, 176, 204, 235, 245, 246
Daylam 187
Dēb 245
Diyarbekir 1
Dunhuang 20, 25, 108, 113, 115, 117, 231, 240
  Cave temples 231n70, 240

Egypt 9, 41, 65, 194, 228, 233, 243, 256, 335
Ethiopia 40, 347
  Goods of 244
Euphrates river 242
Eurasia 22
Europe 41

Fanṣūr 73
Al-Faramā 227, 228, 255
Farghana 229, 239
Fārs 197, 228, 241
Funan 258
Further Asia 2, 41–2
Fusṭāṭ 73, 255, 256, 259, 260

Gandhārā 241
Ganesh Himal 19
Ganges River Valley 230
Gansu 12, 108, 139, 166, 201, 238
Gaochang 239, 240
  See also Qocho, Turfan
Gilgit 232
Guangzhou 234, 248
Gujarat 252, 254
Guzang 240

Hanoi 248
Harkand, sea of 247
Heilongjiang 12

Himachal Pradesh 21
Himalayas 12, 15, 20, 21, 22, 59, 63, 95, 108, 138, 196, 204
Hind 58, 59, 62, 163, 164, 196, 204, 227, 241, 248
Goods of 244
See also India
Hindustān 245
Hindu Kush 204 [Hindukush?] 238
Hormuz 165, 243
Hunza 232, 238

India 13, 21, 50, 53, 125, 134, 136, 145, 150, 165, 168, 176, 189, 192, 196–7, 203, 204, 217, 245, 342, 360
Goods of 59–70
History of musk in 93–107
Indian Ocean 40, 43, 61–2, 82, 83, 134, 137, 144, 145, 146, 154, 233–6, 238, 342
Indochina 73
Indonesia 59, 70, 71, 234
Indus River Delta 245
Indus River Valley 59, 196, 230, 238
Inner Asia 43
Iram Dhāt al-ʿImād 361
Iran 22, 34, 48, 64, 168, 249
Iranian Plateau 42
Iraq 48, 62, 142, 149, 170, 188, 241, 242, 254, 316
Iron Gates 238
Ifrīqiyah 228
Iṣfahān 130
Issık Köl 46

Jāba 75
Al-Jābiyah 228
Japan 43, 58
Al-Jār 227, 249
Java 73, 235
Jāwah 235
Jayhān river 237
Jerusalem 346
Jiddah/Juddah 227, 243, 249
Jinxi 88
Jiuquan 139, 240
Juddah see Jiddah
Jungharia 237, 239
Jurjān, sea of 228
Al-Jurz 240

Kabul 240, 252
Kalah 72, 247
Kalāh 247
Kalāh-bār 247
Kambāya, see Cambay
Kanchula Kharak 21
Kannauj 230, 240, 241, 245
Karakorum mountains 140, 232, 238
Karnataka 105
Kashgar 46, 206, 230, 238, 239
Kashmir 13, 111, 159, 194, 197, 208, 252, 268
Kathmandu 19, 109
Kazakhstan 13, 205
Kelang 247
Kerala 246
Keriya 231
Kham 85
Khamlīj 228
Khazars 228
Khānfū 165, 248
Khmer, see Cambodia
Khotan 1, 115, 116, 231, 232, 239
Khulm 230
Khurāsān 44, 48, 79, 114, 136, 146, 149, 172, 173, 176, 192, 204, 206, 233, 239, 240, 241, 249, 252, 256, 316
Khuttal 252
Khwārazm 45, 47
Khyber Pass 230
Kirmān 228, 235
Kish 52
Kögmän mountains 205
Koko Nor 200, 201n179, 223
Kollam 246
Konkan 246
Korea/Korean Peninsula 12, 13, 43, 86, 201
Kucha 231, 239
Kufa 228
Kufri 21
Kūlam-Malay, see Kollam
Kyrgyzstan 205

Ladakh 230
Lahore 64
Lahuwār 64
Lanzhou 166n70, 240
Laos 70, 203
Lārawī, sea of 81
Leipzig 21
Long Biên 248
Longyou 166n70
Luoyang 113

Lūqīn 248, 252
Lusar 223

Madagascar 258
Madīnat al-Salām 241
Maghreb 256
Māh 226
Al-Mahdiyyah 256
Mā'iṭ 79
Makrān 59, 62, 138
Malabar 98, 246, 252
Malacca, Strait of 74
Malaya 259
Malay Peninsula 71, 247
Malaysia 70
Mānabag-lū mountains 205
Manchuria 198, 201
Mandal 64, 79, 81
Manṣūrah 70
Mantai 246
Marw 44, 226, 230, 233, 238, 244, 348
Maskā 130
Masqaṭ 243, 246
Mawangdui 86
Mā warā' al-nahr 44, 237
Mdosmad 172, 200
Mecca 68, 144, 243, 347
Medina 348
Mediterranean Sea 233, 249
Meroli 21
Merv, see Marw
Mesopotamia 41, 42
Minusinsk 199
Mocha 203
Moluccas 73, 107
Mongolia 12, 13, 198, 199, 201
Morocco 228n50
Mosuo 203
Mug, Mt. 50, 56
Mūjah 202–3
Mūltān 165, 196, 252
Mūsa 203
Myanmar 13, 203–4

Najaf 348
Nanshan 201
Nanzhao 203n195
Nepal 13, 19, 20, 85, 90, 109, 196–7, 231, 240, 315
Nicobar Islands 247
Nine Oghuz 198
Nishapur 280n34
Nūshajān 229

Ohind, see Vayhind
Oman 64, 67, 144, 170, 176, 188, 228, 236, 241, 243, 248, 249, 252
Oxus River 237, 238

Pakistan 13, 59, 62, 197, 245
Palermo 260
Palestine 65
Palmyra 236
Pamirs 138, 230n60, 238
Paykand 237
Pegu 204
Persepolis 124
Persia 130, 145, 170, 307
see also Fārs
Persian Gulf 137, 142, 144, 167, 235, 241, 242, 243, 245, 253
Pontus 304
Pulau Tioman 79, 247

Qāmarūn 64, 69
Qāqullah 247
Qays 243
Qayrawān 187, 228n51, 256
Qimār 64, 72, 73, 79, 193
Qinghai 20n39, 200
Q.nbār 165, 203
Q.sār 165, 203
Qocho 229, 239
Quilon, see Kollam
Qulzum 227
Qumm 130

Rakhine, Myanmar 203
Rāmhurmuz 253
Rāmī 72, 74
Ramlah 228
Rayy 56
Red Sea 40, 83, 134, 137, 227, 249, 258
Riau Archipelago 79
Ruṣāfah 280n34
Russia 22, 45

Saba 226
Ṣaghāniyān 47
Sakhalin 12

Samarkand 50, 52, 56, 113, 114, 130, 139, 194, 237, 238, 240, 250
Ṣan'ā' 253
Ṣanf, see Champa
Sarandīb 246
Semirechiye 238, 239
Shahrisabz 52
Shalāhīṭ 72, 73, 74, 247
Sharwān 290
Shazhou 117, 166, 201, 231, 240
Shiḥr 62
Shīrāz 130
Siberia 12, 15, 25, 43, 86, 198, 199
Sichuan 201, 223
Sicily 48
Sillah 362
Sind 44, 58, 59, 62, 63, 70, 138, 140, 149, 165, 192, 196, 204, 227, 228, 230, 232, 235, 238, 241, 246
 Goods of 244
Sīrāf 55, 144, 167, 176, 235, 242, 243, 248, 253
Socotra 258
Sogdiana/Sughd 48, 138, 140, 174, 205, 231, 237, 238, 249, 252
Southeast Asia 44, 58, 59, 64, 70–75, 107, 234
Soviet Union 19
Spain 8
Sri Lanka 53, 63, 67, 220, 234, 235, 245, 246, 253, 361
Śrīvijaya 73, 258
Sughd, see Sogdiana
Ṣuḥār 144, 241, 243
Suijun 88
Sumatra 55, 73, 247
Susa 55
Sūs al-Aqṣā 228
Suzhou 166n70
Swat 197
Syria 342

Tabaristan 130, 335
Tai 88
Ṭā'if 194
Tajikistan 44, 50
Taklamakan Desert 238, 239
Talwat 240
Tanjir 228
Tannu-Ola 199
Tarāz 229
Tarim basin 108, 229, 231, 232, 237, 238, 250, 255
Tarut 141, 144
Tashkent (al-Shāsh) 229
Termez 238
Thailand 70, 203
Tianshan 205, 239
Tibet 44, 48, 63, 107, 114, 136, 145, 149, 159, 160, 162, 163, 164, 165, 166, 167, 169, 170–1, 172, 173, 174, 176, 192–4, 196, 200, 203, 204, 205, 208, 209, 217, 240, 269, 315
 History of musk in 108–12
Tigris river 228, 241, 242
Tiyūmah 79, 247
Transoxiana 44, 50, 114, 199, 205, 228, 237
Ṭukhāristān 127–8, 230, 237, 269
Turfan 113, 229, 231, 237, 239
Turkmenistan 44, 238

Ubullah 144, 165, 228, 242, 249
Udyāna 197
'Usayfān 252
Uttar Pradesh 21, 63
Uzbekistan 44

Vayhind 241
Vietnam 13, 70, 86, 93, 203, 252

Wakhān 193, 230, 238
Wāqwāq 58
Warwālīj 230
Wāshjird 47

Yarkand 231, 232, 238

Xinjiang 12

Yellow River 240
Yemen 35, 40, 131, 132, 137, 142, 144, 193, 243, 249, 338
Yenisei River 199
Yiyang 88
Yizhou 88
Yongzhou 88
Yunnan 203

Zābaj 73, 74, 81, 149, 248
Zanzibār 247

6 Subjects

Abbasids 5, 6, 52, 74, 224, 242
Abelmosk, see *Hibiscus abelmoschus*
Acacia 274n13
Acacia arabica 274n13
Acacia nilotica 274n10
Achaemenids 124
  Postal system 229
Aconite 33, 184, 209–10, 306, 308
Acorns 187, 266, 270
Adab as a source on aromatics and commodities 10, 158
Adulteration of aromatics and perfumes 261, 276
  Of musk 88, 169–70, 173, 263–71
Agallocha, see aloeswood
Agarwood, see aloeswood
Agnus-castus 274n11
Alchemists 5
Alcohol, in modern perfumery 282
Almond 323
Aloeswood 53, 58, 63–5, 71, 76, 82, 91, 92, 99, 100, 102, 105, 122, 125, 133, 155, 195, 225, 235, 244, 247, 252, 255, 265, 276, 277, 284, 287, 292, 326, 333, 337, 339, 340, 346, 347, 355, 361
  As a principal aromatic 273, 275
  As incense 277–8, 286, 338
  Censing aromatic compounds with 289
  Ingredient of scented powders 281
  Ingredient of unguents 280, 289
  Freshened aloeswood incense 278, 347
  In poetry 202
Aloeswood, varieties of
  Indian 76, 79, 130, 196
  Mandalī 76–7, 81, 348
  Qimār (Khmer) 73, 76, 79–82, 193, 347
  Samandrūn 76
  Ṣanf (Champa) 73, 76, 79–81
Alpinia galanga, see galangal
Amber 61
Ambergris 24, 43, 61, 64, 72, 81, 82, 112, 122, 125, 128, 129, 131, 133, 210, 225, 244, 247, 270, 277, 282, 284, 286, 291, 292, 297, 322, 326, 329, 331, 340, 347, 353, 355, 356
  As a gift 226
  As a principal aromatic 273, 275
  Association with royalty 337–9
  Candles of 226
  Eggs of 226
  Ingredient of *ghāliyah* 278–80, 289
  Ingredient of incense 277–8, 286
  Ingredient of scented powders 281
  In poetry 142–3
  In the Garden 359, 360
  Medicinal use 314, 316
  Terminology of 62
  Used by Muḥammad 343, 344
  Use in funeral practice 348
  Use in India influenced by Islamicate culture 107
Ambergris, varieties of
  From Aden 75, 287
  Baḥrānī 285
  Gray 292
  Indian 61–3, 290, 300, 301
  Shiḥrī 62, 75–6, 81, 130, 279, 287, 346
  From Yemen 193
  From Zābaj 81
Ambers 282, 296
Ambrette, see *Hibiscus abelmoschus*
Amomum 125, 134
Amomum subulatum 66
Anemone 276
Angels 349, 351, 353, 356
Angelica 11
Antelope 102
Ants, guarding musk 194
Apothecaries 260
Apostles 341
Apple 207
  Juice of 282
  Lebanese 160, 164, 187, 191
  Oil of 282
  Scent of musk compared to 160, 164
  Syrian 130, 160, 191–2
  Used to adulterate musk 266
Aquilaria agallocha, see aloeswood
Arabs 140, 154
Arabic
  Business document 132
  Terminology for musk in 35–6, 133, 147–56
Arabic literature on aromatics 4–10
Aristolochia, Chinese 263, 264, 265, 270
Armenian, term for musk in 29
Armor, Tibetan 49

Army, Tibetan 230n60
Aromatics
Descended from the leaves of the Garden of Eden 63, 360–1
Indian 245, 360 and n218
In pre-Islamic Arabia 341–2
Islamicate primary sources on 3–10
Major ones imported 82–3
Scholarship on 2–3
Terminology of in Arabic 272–5
Asbestos 57–8
Aspalathus 125
Asphalt 279
Atriplex odorata 273n7
Axillae, human 16, 156

Ballaharā 252
See also Rāshṭrakūṭas
Balm of Gilead 280
Balsam 83
Bamboo 58–9, 69, 74
Spears used to hunt musk deer 20
Bananas 43, 68, 75, 247
Banū Shaqīq 362
Barley 160, 210
Barmakiyyah, Barmakid style perfume 285, 294
Basil 277
Sweet basil 129, 130, 282
Battle of the Baggage 252
Battle of Aṭlakh (Talas) 52, 56
Bdellium, see gum guggulu
Bean, fragrant 129
Beard hair, see hair, facial
Beaver 22, 30, 93, 304
Beer 321–2
Belitung Shipwreck 55
Ben oil 9, 75, 122–3, 125, 278, 279, 282, 289, 294, 308, 345, 346, 361
Benzoin 95, 283
Beryl 69
Betel 98, 101
Bezoar 45, 183
Bilgä Kaghan 27–8
Birch 47
Birds 337, 338
Black, color of musk 155–6, 334
"Black drug" 187
Blackness, musk symbolic of 334
Blacksmith 327
Blood
Adulteration of musk with 169, 266–7, 270
Association with musk 135, 160, 162, 167, 170–1, 173, 182–3, 186, 211–6, 267, 351, 358, 363–4
In Islamic law 211–3
Bosi 234
Boudinoi 304
Box, for perfumes or cosmetics 143, 148, 152, 288, 296, 299
Boxthorn 264
Brazil wood 65, 74
Brocade 52–3, 227, 334
Gold 58, 195, 345, 348
Silk 52, 58, 195, 285, 352
Buddhism 71, 90–1, 108, 114–5, 117, 145, 200n178, 250, 258
Buddhism, and musk trade 223
Buddhist images 161, 189
Byzantine Empire 36, 51, 228, 331
And commerce 39, 134, 137, 139, 253
Sources on musk in 134–6

Calamus, Syrian 125
Caliphs, and *ghāliyah* 278–80
And perfumes in general 285
Camel
As a gift 226
Bones 298
Musky scent of 262
Camphor 56, 58, 64, 71, 72–3, 74, 82, 93, 95, 97, 99, 100, 101, 102, 105, 107, 108, 111, 114, 115, 120, 122, 125, 127, 128, 131, 133, 175, 235, 244, 247, 255, 265, 277, 285, 291, 292, 294, 297, 300, 306, 314, 320, 322, 323, 326, 337, 339, 355, 356, 360
Antithesis of musk 303, 320–1, 334, 335
As a gift 225
As a principal aromatic 275
Attested by Aetius of Amida 135
Counteracts musk 178, 309
Of Fanṣūr 73, 75, 287
Ingredient of incense 91, 278, 286
Ingredient of unguents 280–1
Symbol of whiteness 334–5
Term for in Sogdian 24, 25
Unknown in antiquity 43
Use in funeral practice 348–50, 354
Value of 73
Water of 294

Capsicum 66
Caravan 116, 120, 230n62, 252, 253
"Caravan cities" 236
Cardamom 66–7, 96, 105, 125
Cardamom, greater 58, 66, 274, 322
Cardamom, lesser 66, 274, 322
Carrion 211–2
Cassia 53–4, 65, 95, 96, 114, 125, 255, 265, 270, 273
Cassia fistula, see Cassia, purging
Cassia, purging 67
Castor fiber, see beaver
Castoreum 22–3, 93, 133, 174, 208, 210, 262, 268
  Adulterant of musk 267, 303
  Medicinal properties 303–5
  Substitute for musk 303, 308
  Terminology of 22, 30
  Greek name for the origin of the Sanskrit term for musk 30
Cave, People of the 338, 362
Cedar 210
  Gum 279
Celtis australis, see lote tree
Ceramics, Chinese 55, 246, 253
Chāḷukyas 97
Cham 93
Chamomile 276n23
Champak 99, 100
Chigil 205
China ships 242, 243
Chinese 4
  Documents in 50
  Sources on musk in 86–91
  Skill in handicrafts 51–2
  Terminology of musk in 23, 86
Chionites 138
Cholas 247, 258
Christianity, aromatics and scent in 124, 340–1, 354
Christians 141
Christians, Nestorian 137
Cinnamon 52, 53–4, 65, 195, 255, 273, 322
  Syrian 125
Cinnamomum cassia 53
Cinnamomum camphora 72
Cinnamomum zeylanicum 53
Circumcision parties, aromatics used at 290, 291
Cistus 83, 274n15
CITES 22
Citron 68
  Figurines of ambergris shaped like 291
  Of Ṭabaristān 130
Citrus 68
Civet 4, 18, 22, 107, 112, 149, 188, 210, 212, 262, 282, 283, 326, 361, 368
  See also *Viveridae*, *Viverra zibetha*, and *Viverricula indica*
Cloth, of Yemen/Yamanī 131, 142
Clothing and garments 241, 296
  Perfuming of 281, 284
Cloves 9, 43, 58, 71, 73–4, 82, 91, 105, 121, 125, 133, 235, 247, 273, 293, 297, 300, 322, 324, 360
  Adulterant of musk 187, 265, 270
  Clove water 285
Coconut 43, 58–9, 67–8, 75, 247
Coconut toddy 247
Cods, musk 15n14, 216
Coins 113, 139
  As a gift 226
  Arab-Sasanian 239
  Sasanian 239
Cologne 265
Commerce, and Islam 37, 250–1, 366–7
Commiphora sp., see myrrh
Commodities, of Further Asia 37–84
  In geographical literature 38
Conch 60
Cookbooks 317–8
Copper 255
Coral 256
Coral tree 97
Corpses, scent of likened to musk 331–2
Corundum 69, 295
  See also ruby, sapphire
Cosmetics, utilizing musk 89, 100, 102, 130–1, 296
Costus 63, 82, 91, 95, 125, 138, 274
  Etymology of 342
  Term for in Sogdian 25
Cotton 279
  Garments 49, 70
Cow, impure (*jallālah*) 212
Court, those attending perfumed 286
Crocodiles, musky scent of 262
Cubeb 58, 72, 274
Cuckoo 60
Cucumber 43
Cucurma zedoaria, see zedoary

Cumin 111
Cymbopogon martini, see palmarosa
Cyperus 125
Cyperus rotundus 95
Cypress 87, 210
Cyprus 125

Daoism 89–90
Darkness, musk symbolic of 334
Date palm 127
Deer
  Chew on aloeswood 65
  Produces musk 357, 363
  See also musk deer
Defecation 150
Defecation, absent in the Garden 359–60, 363–4
Delphinium moschatum 24
Deodar 95
Depilation 350
Desman 12
Desmana moschata, see Desman
Devils 347
Dgelugspa 223
Dhow 55
Diamond 69
Disinfectant, musk as 306, 316
Distillation of scented waters 282
Dog 135
  Musk animal described as a 75, 135n265, 161
Dome of the Rock 346–7
Dragon's blood 263, 264, 267, 270
Dreams, interpretation of musk in 327, 332
Drinks, musk used to scent 317–22
Drinking party, use of aromatics at 280, 290–2
Dryobalanops aromatica, see camphor
Dyes 65, 74, 256
  Perfumes as 281, 283–4, 293, 295

Eagle 337, 340
Eagle wood, see aloeswood
Ebony 58–9, 69
Eden, Garden of 217, 360–1
Effeminate scents 277
Eggplant 43
Elephant 59, 60, 81
  Musky scent of 262
  Tusks 58–9, 72, 163
Elettaria cardamomum, see cardamom
Embalming 348
Emporia 234, 236–49
Ermines 205
Ethiopians 137
Ethiopic, term for musk in 29
Eunuchs 227
Excrement of the Imam smells like musk 359
Europe and Eurocentrism 41

Fabric, Zandanījī 49
Fagara 274
Fairs, market or trade 141, 154, 253
Falcons, as a gift 225
Fatimids 56, 219, 243, 244, 249, 255, 287, 295
Feathers 60
Feces 18
Felt 52–3
Feminine scents 276, 277, 283–4
Fennec 79
Fennel 273n4
Figurines, made of aromatics 64, 291
Fish, gold figurine of 236
Flour 356
Fly-whisks 166
Food
  Of the Garden becomes a scent like musk 359–60
  Musk used to scent 317, 322–3
Fox 135
Frankincense 1, 83, 107, 112, 123, 124, 244, 318, 340, 342
Franks 227, 228
French, term for musk in 29
Fruit of Central Asia 47
Furs 45–6, 128, 205, 227, 229
  Black 231

Galangal 53, 58, 72, 195, 322
Gallnut 155
Garden, the (*al-Jannah*) 320, 352–65
  Scent of 353
Gardens, parties in 291–2

Garments 49, 58, 225, 245
Embroidered 225
Imported from India 70
Garments, see also clothing
Gazelle, association with musk 35, 148, 150, 160–1, 163–4, 167, 169–70, 172, 183, 213, 217, 294, 328, 356
Blood of used to adulterate musk 271
Geniza 9, 73, 82, 146, 221, 235, 255–6, 296
Geographical literature as a source on aromatics and commodities 7, 10, 77–82
Gerfalcon 60
German, term for musk in 29
Ghassānids 338
Ghaznavids 219, 233, 339
Gift exchange, musk and 172, 224–7
Ginseng 92
Glassware 139, 170
Goat 127
Blood of used to adulterate musk 266, 271
Gold v, 48, 53, 69, 72, 129, 337, 348, 361
Censer 347
Cup 338
In the Garden 359
Plate 291
Vessel 225, 338
Goshawk 60
Grain musk 16
Grapes 288
Grape, juice of 282
Greek 4
Term for musk in 29, 36, 135, 187
Greeks, burned incense at symposia 291
Greek and Roman trade with India 40–1
Greek medicine 22, 93, 109, 133, 303
Ground of the Garden made of musk 356–7
Gum guggulu (bdellium) 95, 105, 111, 112
Gupta dynasty 93–4, 133
Gurjara-Pratīhāras 240
Gypsophila struthium 208
Gynecology, use of castoreum in 304

Ḥadīth 35, 283, 327, 341, 345, 349
Hair, association with musk 143, 293, 302, 344
Facial 130–1, 157, 206, 278, 284–5, 290, 341, 350
Han dynasty 39, 47, 54, 85–6
Hebrew 133, 342
Terminology for musk in 132–3, 194, 209
Hebrew script 9
Hellebore 208
Hellenistic influence in India 30
Henna 5, 281, 295
Hephthalites 138–9
Hibiscus abelmoschus 11, 99, 283
Hindi, term for musk in 29
Hinduism 71, 258
Ḥisbah literature 251, 269
Holly 210
Honey 60, 91, 93, 97, 112, 125, 183, 300, 322, 348, 356, 357
Horn 60
Horses 52, 54, 223
As a gift 225, 226
Houris 360
Howdah 333
Hu 139, 230n60
Huang Chao rebellion 248
Huadu Monastery 91
Humors 303, 318
Hyrax 33n127
Hyraceum 33n127

Ibaḍites 252
Incense 155, 275, 342, 346–7
Burned at meals and parties 290–1
Burners 124, 342
foundations of 286
Tibetan 111–2
Types of 277
Use in East Asia 28, 90–2
Use in grooming 285, 286
Use in South Asia 95–9
Use in Zoroastrianism 123
Indian medicine, musk in 102–4
Indian pepperwort 187, 270
Indian temples 161
Indigo 65
Indol 18
Ink
Musk used in making 92, 187, 329
Used to dye peppercorns to adulterate musk 271

Insects
Supposedly eaten by musk deer 87
Warded off by musk 90
Inscriptions, Sanskrit 105
Inscriptions, Old Turkic 26–8, 201, 205
Sogdian 238
Tibetan 108
Iron, Chinese 57, 58
Islam
Aromatics in 341–65
Odor of sanctity in 353–4
Islam, spread of 250
Islamicate, usage of term 4–5
Italian, term for musk in 29
Ivory 59, 60, 206, 337
Ivory, walrus and narwhal (*khutū*) 175

Jainism 102
Japan, use of musk in 92–3
Japanese, terminology of musk in 23
Jasmine 9, 18, 43, 154, 275, 276, 277, 332, 338
Oil 263, 280, 282, 294
Oil, Ruṣāfī Naysābūrī 280
Jasper 225
Jerboa 33n122
Jewelry, gold 296
Jewels 69, 115, 129, 195, 241, 337, 348, 361
Jewels, as gift 225, 226
Jin dynasty 51
Judaeo-Persian language 255
Judaism, aromatics and 132–3, 211, 325, 340 and n79
Juniper 111
Jurchen 51

Kaʿbah 343, 344, 345–6
Kamala 274
Karakhanids 118, 120, 205, 225, 231, 233, 250
Karakhitāy 51, 253
Kastūrī, Sanskrit term for musk 29–30
Kawthar, river in the Garden 356–7
Kewda 185, 265
Khwārezmian, term for musk in 24, 185
Khotanese Saka
Sources on musk in 114–15
Term for musk in 24, 114
Kimek 45, 202, 229
Kings, association with musk 172, 336–40
Kirghiz 46, 198–9, 202, 204, 237

Kitan 51, 201, 204, 231
Kohl 174, 314, 362
'khrāj pearls 256
Kumbum Monastery 200n178
Kuphi 134
Kushan Empire 138
Kushano-Sasanians 138

Labdanum 83, 125, 274, 296
Laburnum, Indian 67
Lac 60
Lacquerware, Chinese 57
Lakhmids 294
Lapis lazuli 44, 48, 256
Late Antiquity, introduction of new *materia medica* in 67, 71
Latin 4
Terminology of musk in 29
Laudanum 274n15
Lavanga 114
Lavender 297
Lead, adulterant of musk 180, 267, 270
Leather, scenting of 283
Lemon 43, 68
Lentisk 274
Leopard 59, 60
Lepidium latifolium, see Indian pepperwort
Liao dynasty 51, 201, 231
Lily 275, 276
Lime 350
Lime (fruit) 68
Lion 337
Liquidambar orientalis 274n16
Literati, use of incense by 91–2
Liver, used to adulterate musk 271
Long-haul voyages 38, 220, 233
Lote tree 349
Lotus 125

Mace 58, 71, 73, 82, 235, 247, 273, 322
Mahlab 274, 361
Malabathrum 125
Mallotus philippinensis 274n17
Man people 88, 90
Mango juice 97
Manichaeism 239, 245
Odor of sanctity in 354
Marjoram 125, 277
Marble 346, 347

Marten 45, 127–8
Tukhari 127
Martyrs, blood of smells like musk 352, 358, 365
Marum 125
Materia medica, imported 43
Masculine scents 277, 283–4, 343
Mastic 322
Meals, perfuming at 290
Melon
Figurines of ambergris shaped like 291
Of Central Asia 47
Men, use of musk by 89–90, 283–92
Menstrual blood, musk purifies 350–1
Merchants 37, 165, 172, 174, 249–59
Arab pre-Islamic 250–1
Chinese 241, 257
Christian 248, 257
Indian 142, 164, 173, 234, 241, 257–8
Jewish 146, 194, 227–8, 236, 242, 248, 254, 255–6, 257, 259
Muslim 37, 165, 233, 235, 242, 246, 248–9, 250–3
Muslim, in China 165, 248, 252
Muslim, in India 240, 241, 252
Persian 234, 253
Sogdian 231, 232, 237, 238, 249
Southeast Asian 234, 257
Uyghur 250
Zoroastrian 248, 257
Mercury 48, 57
Metal, Chinese 57
Metalwork 139
Iraqi 64
Mihrajān 225
Milk 114, 356, 357
Millet 117
Ming dynasty 223, 257
Mirrors, Chinese 57
Mongol Empire 4, 39, 116, 202, 220, 250
Mongolian, terminology for musk in 27
Mongols 231, 250
Mole, like musk 157, 207
Mongoose 33
Monsoon winds 40, 233, 243
Moringa 278
Moschus, taxonomy of 12–13
Mosque, Ipariye 1
Mosque of the Prophet 348
Mother of pearl 58–9
Mouse 31, 147–51, 160
Moustache, see hair, facial
Mouth, scent of likened to musk 299
Mulberry 54
Musk
Absent from Graeco-Roman medicine 43, 133
Absolute of 17–8
Adulteration of 88, 267–71
As a gift 224
As a principal aromatic 273, 275
As a sign of the sacred 351
Association with blood see Blood, Association with musk
Association with rodents 31, 134–5, 299
Counted by skins 31, 134
Danger of exposure 308
Earliest use of 85–7
Etymology of term 29–31
Hazelnut-sized balls (*banādiq misk*) 226, 362
History of musk in the west 132–46
Imitations of 173, 266, 267–8
Induces nosebleed 308
Licitness in Islam 211–13, 326
Medicinal properties 87–9, 109, 174, 303, 305–9, 322
Overland trade of 204
Pharmaceutical applications of 309–17, 338
Primacy of in comparison with other aromatics 1, 325–6
Production of in Tibet 221–3
Proverbs concerning 105–6, 110, 119–20, 206, 326, 327, 330
Purity of 150–1, 331, 350–1
Substitute for castoreum 30
Symbolism of 296, 327–36
Terminology of 23–36
Tincture of 17
Use of in China 86–93
Use of in European perfumery 17–8, 145, 283
Use in Islamic funeral practice 348–50
Use of in India 93–107
Value of 19, 113, 223, 236, 259–60, 327
Musk, varieties of 187–207
Adhfar 292
Apple-scented 191

Musk, varieties of (cont.)
  Bahārī 187
  Buddī 161, 189
  Chinese 81, 129, 159, 165–6, 169, 176, 178, 182, 184–6, 188, 195–6, 204, 363
  Of Dārīn 141–3, 174, 332
  h.t.r.s.rī 186, 196
  Indian 159, 165, 176, 185, 187, 196–7, 204
  ʿIrāqī 187, 188
  Jabalī 165–6, 196, 203
  Kashmiri 181–2, 186, 196–7
  Khotan/Khotanī 1, 206, 232
  Kirghiz 165, 180, 182, 184, 185–6, 198–9, 206
  Kitan 180, 182, 185–6
  Of Mdosmad 186
  Nepalese 179, 182, 185–6, 196–7
  Night-black 191
  Oceanic 181–2, 186, 188, 191
  Qawārīrī (musk in phials) 186–7, 191
  From *Q.nbār* 165–6, 203
  Q.ṣārī 165–6, 203
  ʿ.ṣmārī 165–6, 203
  Sogdian 114, 146, 159, 165–6, 176, 188, 192, 269, 270
  Tatar 138, 177, 180, 182, 185–6, 201–2, 206, 346
  Tibetan 75–6, 79, 81, 150–1, 161, 165–6, 169, 178–80, 182, 184, 185, 187, 192–4, 279, 287, 346
  Tughuzghuz 165–6, 198
  Tonquin 196
  "Treasury" (*khazāʾinī*) 193, 224
  Turkish 150, 185, 196, 204
  Udiyākhī 186, 196–7
Musk, warming property of 97, 99–100, 309
Musk deer
  Characterized as a dog 75, 161
  Characterized as a goat 127 and n130
  Characterized as a rabbit 161
  Description of 14–5
  Diet of 15, 159, 175, 208–11
  Hair of 85, 109
  "Horn" 174, 175n95, 182
  Hunting of 19–21, 135, 160, 170, 183, 186, 216–7
  In captivity 21, 103, 105
  Meat of 85
  Range of 12–4
  As royal property 222
  Teeth of 163, 170, 184
  Tongue 175n95
  See also *Moschus*
Musk duck 11
Musk mallow, see *Hibiscus abelmoschus*
Musk melon 11
Musk pod 15–7, 31, 87–8, 105, 143, 150, 151–5, 169, 187–8, 215, 259, 297, 298, 302, 319, 330, 357
  Musk given in 225
Musk rose 11
Musk shrew, Asian 11, 31–3, 149–50, 218, 262
Musk tortoise or turtle 11
Muskone or muscone 16
Muskrat 11
Musky rat-kangaroo 12
Musth 262
Muʿtazilite perfumer 138, 161, 212
Myrobalan 58, 67
  Beleric 67
  Chebulic 67
  Emblic 67, 187, 265, 270
Myrrh 1, 83, 124, 125, 184, 209, 340
Myrtle 276, 277, 338

Nadd 277–8, 291, 292
  five-part 278
  Sulṭānī 285
  tripartite 278
Narcissus 275, 276, 277, 291
  Of Maskā 130
Nard 135, 314
Nardostachys jatamansi or *grandiflora*, see spikenard
Navel 346
  Of the musk deer 15, 32–3, 35, 87, 135, 151, 160, 161, 162, 164, 167, 170–1, 175, 213–5, 217, 363–4
  Perfume, meaning musk 94–5
  Term for in Old Turkic 26
Nawrūz 225, 336
Necklaces, censed with camphor and ambergris 294
  Censed with cloves 294
Nosebleed, caused by musk 308
Nutmeg 43, 58, 71, 73–4, 82, 97, 235, 273, 322
Nuts, Indian 59

Oath, sworn upon perfume 341
Ocean, transport by damages musk 188, 205
Odor, foul, as a sign of piety 345 and n117
Odor, sweet, as a sign of sanctity 128, 353–4, 359, 362
Oenanthe 125, 134
Ondatra zibethicus, see muskrat
Oil
  Anointing 339–40
  Scented 95, 112, 275, 282
Onion 298
Onycha 83, 91, 95, 97, 99, 210, 274
Onyx, Yemeni 361
Opium 274n15
Opobalsam 125
Orange 43, 68
Orris 63
Oryx 153
Oxeye 276

Pahlavi 62, 332
  Terminology of musk in 28, 31–4
Pāla dynasty 58
Palmarosa 281
Panacea 125
Paper 50, 52, 56
Parthian Empire 41, 42, 122, 136, 226, 254
Parthian royal unguent 125
Paradise, as a garden 352–5
Paradise, scent of 353, 355
Parachi word for musk 138
Parrot 60, 357
Parsees 254
Passports and passes 239n109, 240
Patra 95, 96
Peacock 52, 59, 337
Pear 47
Pearl 60, 69, 105, 115, 151, 295, 357, 361–2
Peau d'Espagne 283
Pentecost 341
Pepper 58, 66, 72, 114, 142, 361
  Guinea 274n9
  Indian 274n9
  Long 58, 66
  Long pepper root 66
  Sichuan 274n11
Peppercorns, adulterant of musk 271
Perfume
  As a gift 225
  Compound 275
  For different ages and social classes 285, 285n63, 289
  Gendered aspects of 283–5, 293
  Of the night-maker 294
  Prepared at home 261–2
  Quality of determined by ingredients 276
  Tripartite 130, 294
Perfumers 142, 154, 260–2
Perfumery
  Greek and Roman 125, 134, 210
  Medieval Islamicate 275
  Medieval compared with modern 2, 275n21, 282
Persian, terminology of musk in 28
Persians, perfume among 121–32, 275, 323
Pestle 329, 337
Pharmacists 260–1
Pheromones 17
Phoenix 57
Physicians, interest in aromatics 5
Pig 163, 211
Pine 210
  Resin 95, 270
Piper cubeba, see cubeb
Piper longum, see long pepper
Piper nigrum, see pepper
Pistachio 279, 289
Pitch 290
Poetry, Arabic 35, 140–1
  As a source on aromatics 10
  Associates women and aromatics 294–303
Poetry, Persian mystical 321
Poison
  Musk as an antidote for 32–3, 306, 308
  Musk as a poison 309
  Used to kill musk deer 20
Pomander 280, 283, 287n73
Poplar 47
Porcelain 52, 55–6, 58, 235, 236
  Used to store aromatics 287, 292, 295
Post (*barīd*) 229
Pot, glass 279
Pottery, Indian 137
Powders, scented 95, 275, 281
  Use in grooming 285, 286

Principal aromatics 273
Privy 173
Prunus mahaleb 274n12
Purslaine 299

Qayās 205
Qiang 88
Qing dynasty 87
Qipchaqs 22
Quilts, stuffed with musk and ambergris 296
Qur'ān
  Aromatics in the 355
  Loanwords in 35, 140
  Musk in 35, 319–20
Quraysh 142, 144

Rādhāniyyah 227, 228, 242, 254, 255
Raisin 320
Rāshṭrakūṭas 197n166, 240, 241, 246, 252
Rat 149, 161
Rhinoceros 59, 72, 174
Rhubarb 43, 45, 46, 52, 314
Rice 67, 75, 91, 247
Rivers of the Garden 356–7
Rockrose 83
Rodents 32, 149, 150
  See mouse
Roman Empire 236
Rose 1, 9, 121, 154, 202, 273, 276, 277
  Oil 308
  Persian 130
Rose, dog 277
Rosewater 9, 64, 122–3, 129, 178, 270, 282, 284, 285, 290n83, 309, 321, 322, 338, 347, 348
  Jūrī 265
Rottlera tinctoria, see *Mallotus philippinensis*
Royalty, betrayed by musk 332, 339
Ruby 44, 48, 59, 69, 204, 236, 337, 356, 361
Rush, Syrian 125
Russian, term for musk in 29

Sable 33n128, 79, 92, 227, 296, 327
  Central Eurasian 45
  Chinese 52–3, 195
Saddles, Chinese 52–3, 56–7, 58, 195
Safflower 129n237
Saffron 47, 82, 99, 101, 102, 104, 105, 108, 110, 111, 112, 115, 117, 125, 225, 270, 281, 284, 291, 292, 306, 307, 322, 324, 347, 356, 359, 360, 362
  As a principal aromatic 273, 275
  Chinese musk resembles 178
  Ingredient of incense 278
  In the Garden 359, 360
  Of Qumm and Bawan 130
  Term for in Sogdian 25
Śailendras 73
Sailing, in the Indian Ocean 55
Salamander 57
Sal ammoniac 48, 118
  Term for in Sogdian 25
Śāla 99
Salicin 304
Salicylic acid 304
Saliva, scent of likened to musk 299–301, 360
Salsabīl, river of the Garden 360
Salt 155
Samanids 166, 219, 233
Sandals 70, 143, 296
Sandalwood 43, 65, 73, 74, 95, 99, 100, 101, 102, 105, 107, 108, 111, 114, 115, 119, 123, 125, 225, 247, 273, 293, 309, 337
  In Old Turkic 27–8
  Term for in Sogdian 24, 25
  White 59, 65, 91
Sanskrit 4, 26
  Literature on perfumes 97
  Terminology of musk in 24, 26, 29–31, 93–4, 133, 135, 185, 305
Santalum album, see sandalwood
Sappanwood, see brazilwood
Sapphire 69, 361
Saponaria officinalis 208
Sarcocolla 264, 270
Sasanian empire 32, 34, 42, 51, 62, 68, 122, 145, 234, 235, 236, 244, 245, 253, 254, 332
  Arabia and the 131, 140, 253
  Aromatics under the 125–32, 281, 286
  Commerce 39, 41, 136–41, 253
Saussurea costus, see costus
Scent
  as a mark of character 284, 302, 350
  as a sign of sanctity 341, 352, 353
Scorpion, sting of treated with musk 315
Seal, of musk 320
Secrets, musk reveals 332–3
Seleucids 42
Serichatum 125
Serpent 32

Sesame 114, 279
Sheep 103, 255
Sherbets 323
Shields, Tibetan 49, 79, 193
Shiʿites, attitude toward musk 212n231
Shipbuilding 68
Ships 142
  Byzantine 253
  China 245, 246
  Of India 245
  Omani 245
  Persian 137, 145, 234
Shōsō-in 93
Silk 49, 120, 137, 183, 236, 256, 344, 351, 353, 358, 368
  Chinese 39, 41, 52–5, 195, 237, 279
  Indian 70
Silk Road 42, 54
Silkworm 183
Silver 48, 57, 72, 348, 361
  Censer 347
  Cup 338
  In the Garden 359
  Vessel 225, 338
Skin, association with musk 31
Slave girls 226, 227
Slaves 195, 225, 227
Slavs 227, 228
Snake 33, 87, 90
Snake, bite of treated with musk 90, 109, 315
Snares, hunting 19–21
Soapwort 208
Sogdian, Term for musk in 24–5
Sogdian Ancient Letters 25, 112–4, 221
  Documents in 50
Sogdian trade networks 114, 139–40, 146, 250
Sogdians 39, 44, 112–4, 126, 139–40, 205, 231, 232, 249, 250
Soil, association with musk 148, 356–8
Solomon's temple 346
Solomon's throne 337–8
Somali 122
Song dynasty 51, 116, 231, 248, 257
Southeast Asia, goods characteristic of 235
Spanish 9
  Term for musk in 29
Spices 41, 62, 275, 322
Spikenard 11, 63, 125, 273, 282, 308, 322, 342
  Adulterant of musk, 187, 265
  Musk deer grazes on 159, 169, 182–3, 208–9
Spinach 43
Spoon, gold or silver 279
  Silver 296
Squirrel 45, 205
Stater 116
Steppes 238
Steppe empires and trade 39–40, 230n59, 231–2, 240
Stone, perfumer's grinding (*ṣalāyah* or *madāk*) 299, 329
Stones, precious 236, 246
  See also jewels
Storax 91, 274, 283
Styrax 125
Styrax officinalis 274n16
Sūfis 213, 349–50, 363–4
Sugar 68, 95, 112, 118, 320
Sugarcane 68, 75, 247, 298
  History of 298
Sulphur 318
Sumbul, see spikenard
Suncus murinus, see musk shrew
Sweat 156, 344, 359, 359, 364
Sword hilts 174
Swords 70, 227
Syriac 4
  Terminology of musk in 29, 31, 34–5, 141, 187
  Translation of the *Pañcatantra* 328, 332
  Words for spirit (*rūḥ*) and scent (*rīḥ*) 341

Tabasheer 69
Tamang 19, 85, 215
Tamarind 67
Tang dynasty 40, 51, 89–91, 232, 240
Tar 290, 348
Tatars 201–2, 204
Taxation, of musk 105, 165, 221–2
Taxation, of merchants and trade 229, 236, 246, 253
Tea 90
  Musk traded for 222, 223
Teak 68
Testicle 29–31
Textiles 255
  As a gift 225
  Chinese 253
Throne 337–8 and 337n66, 345

Tibetan 4
  Sources on musk in 108–12
  Terminology of musk in 23–4
Tibetan empire 108
Tibetan medicine, musk in 108–10
Tiger 59, 60
Tin 58, 72
Tokuz Oghuz 198, 204, 229, 231, 239
Tooth-twig 301
Torah 338, 347
Thyme, wild 289
Tibetan empire 230n60, 232
Trade
  Fatimid 219
  Maritime 55, 159, 169
  Overland 159
  Periodization 250
  Profitability of 236, 244
  Sogdian 237; see also Sogdian trade networks
Trade diaspora 248
Trade routes, with Further Asia 228
Trade routes, maritime 233–35
Trade routes, overland 229–33
Transport, by sea versus land 181, 229
Tribute, musk as 222, 224
Trojan War 121
Tughuzghuz 165, 198, 204, 239 see also Tokuz Oghuz
Tukhs 205
Tulip 321
Türgish 205, 252
Turk Kaghanate, First 39, 139, 240
Turk Kaghanate, Second 199, 201
Turkic, Old 4
  Terminology of musk in 25–8, 115, 119, 185
Turks 51, 54, 59, 196, 229, 323
Turks, goods characteristic of 48, 79, 225
Turquoise 48
Tutty 69

Umayyads 44, 332
Unguents 275
United Arab Emirates 275n21, 293n99, 318
Urine 15, 16, 216
Urination 150, 359, 360, 363–4
Utensils, household 296
Uttarāpatha 230

Uyghur 26
  Business documents 115–7, 221, 250
  Medical literature 117–8
Uyghurs 229, 231, 239, 250
Uyghur Steppe Empire 39–40, 54, 198, 199, 232

Valerian 63, 111
Valeriana jatamansi 63
Vase, for perfumes 296
Veil, night as a musky 334
Velvet garments 58, 70
Vessel 48
  Chinese 52–3, 195, 279
  Gold or silver 56, 64, 279, 348
  Meccan 279
  Porcelain 56
  Skull-shaped 287, 292
Vetiver 95, 96
Vinegar 212
Violet 63, 129, 277
  Of Iṣfahān 130
  Oil 294, 308
Viveridae 22
Viverra zibetha 22
Viverricula indica 22
Voyages, long haul 38, 50, 248
Voyages, segmented 50–1, 235

Walnut 265
Wallflower 275, 276, 282, 307
Wars 265, 274
Water of the Garden 358
Waterlily 129
  Of Shīrāz 130
Waters, scented 95, 275, 282
Wax 99, 256, 279
Weapons, of Central Asia 48
Weapons, Tibetan 49
Weasel 33n128
Western Wei dynasty 139
Whale, sperm 61, 282
Whetstone 69
White lead 180
Willow 335, 357
Wine 18, 117, 125, 142–3, 212–3, 291, 300, 356, 358, 368
  Associated with musk 291, 318–21, 353
  Khusrawānī 130

Wolf, penis of 324
Women
Use of aromatics by 292–303
Use of *ghāliyah* by 292
Use of musk by 89, 100–1, 292–303

Xiongnu Empire 39, 54

Yak 166
Yew 273
Yi 88
Youth, musk symbolic of 335–6
Yuan dynasty 257

Zanj 68, 242, 247
Goods of 244
Revolt of the 242
Zanthoxylum 274n11
Zedoary 67
Zephyr 121, 331
Zinc 57
Zoroastrianism, aromatics and 123–4, 126, 128, 130–1, 340,
Odor of sanctity in 352
Zoroastrians 254